AF678145

Development and Modification of New or Recycled Materials and Technological Processes toward Sustainable Development

Development and Modification of New or Recycled Materials and Technological Processes toward Sustainable Development

Editors

Pavol Liptai
Jaroslav Briančin
Maroš Halama

MDPI

Basel • Beijing • Wuhan • Barcelona • Belgrade • Novi Sad • Cluj • Manchester

Editors
Pavol Liptai
Institute of Recycling Technologies
Technical University of Košice
Košice
Slovakia

Jaroslav Briančin
Institute of Geotechnics
Slovak Academy of Sciences
Košice
Slovakia

Maroš Halama
Institute of Materials and Quality Engineering
Technical University of Košice
Košice
Slovakia

Editorial Office
MDPI AG
Grosspeteranlage 5
4052 Basel, Switzerland

This is a reprint of articles from the Special Issue published online in the open access journal *Materials* (ISSN 1996-1944) (available at: www.mdpi.com/journal/materials/special_issues/RK6MGU298A).

For citation purposes, cite each article independently as indicated on the article page online and as indicated below:

Lastname, A.A.; Lastname, B.B. Article Title. *Journal Name* **Year**, *Volume Number*, Page Range.

ISBN 978-3-7258-1858-7 (Hbk)
ISBN 978-3-7258-1857-0 (PDF)
doi.org/10.3390/books978-3-7258-1857-0

© 2024 by the authors. Articles in this book are Open Access and distributed under the Creative Commons Attribution (CC BY) license. The book as a whole is distributed by MDPI under the terms and conditions of the Creative Commons Attribution-NonCommercial-NoDerivs (CC BY-NC-ND) license. (https://creativecommons.org/licenses/by-nc-nd/4.0/).

Contents

materials

Editorial

Development and Modification of New or Recycled Materials and Technological Processes towards Sustainable Development

Pavol Liptai

Institute of Recycling Technologies, Faculty of Materials, Metallurgy and Recycling, Technical University of Košice, Letná 9, 042 00 Kosice, Slovakia; pavol.liptai@tuke.sk

Quality and environmental sustainability are fundamental to the healthy and proper functioning of society on Earth. The current human population should want to leave the environment in a better condition or in the best possible condition for future generations. Therefore, each person must contribute to this goal, regardless of origin, race, religion, or social and occupational status. Then, this social consensus can be achieved. Sustainability should be the overriding and common goal of human civilization, whatever the context and how we approach it.

Many scientists and researchers are trying to develop various recycling technologies to recover high-quality raw materials, while others are working on ways to reduce or prevent waste or developing technologies to eliminate polluted water, soil, or air. Many are looking for ways to optimize technological processes to minimize their environmental impact or to ensure that the outputs of these processes are products that will be efficiently recycled at the end of their life cycle. However, it should be noted that these efforts will be meaningful only if everyone contributes and does not rely on others to solve the problems. Often, it is enough to realize that natural resources provide us with inputs for product manufacturing, and, at the same time, products at the end of their life cycle constitute a supply of materials for further use. This way, we understand the circular economy and environmental sustainability.

In this Special Issue titled "Development and Modification of New or Recycled Materials and Technological Processes towards Sustainable Development", 16 papers were published that address specific tasks and problems to achieve long-term sustainability.

The five review papers address promising issues in various scientific fields. Fu et al. [1] describe advances in the development of layered double hydroxide-based materials for wastewater treatment applications. Takacova et al. [2] provide an overview of the current state of the field in spent portable lithium battery recycling at both the research and industrial scales. Another interesting review is offered by Pan et al. [3], focusing on the applications of luminescent silver nanoclusters embedded in zeolites for lighting, gas monitoring, and sensing. Mironovs et al. [4] concentrate their review on sustainable methods of recycling, using, or reusing perforated metal materials, and Rostami-Tapeh-Esmaeil and Rodrigue [5] provide a review based on the latest research on the morphological, physical, and mechanical properties of rubber foams, their applications, and future development.

The eleven research articles offer promising studies from various scientific fields with potential practical applications. Bottom zinc dross, as a valuable secondary material, is investigated in the paper by Pauerová et al. [6]. The recycling of cable waste and its application use are explored by Wędrychowicz et al. [7]. The improvement in electrocatalysts, based on a NiCoP surface for bifunctional water splitting, is examined by Sheng et al. [8]. The development of high-temperature in situ water sorbents, based on zeolite, dolomite, lanthanum oxide, and coke, is covered by Acha et al. [9]. The research by Pikna et al. [10] aims at optimizing the processing and recovery of chromium through the electrocoagulation method. The study by Mamrilla et al. [11] provides valuable insights into zinc-based alloys containing magnesium, calcium, and manganese, which can be used to

Citation: Liptai, P. Development and Modification of New or Recycled Materials and Technological Processes towards Sustainable Development. *Materials* **2024**, *17*, 3060. https://doi.org/10.3390/ma17133060

Received: 5 June 2024
Accepted: 14 June 2024
Published: 21 June 2024

Copyright: © 2024 by the author. Licensee MDPI, Basel, Switzerland. This article is an open access article distributed under the terms and conditions of the Creative Commons Attribution (CC BY) license (https://creativecommons.org/licenses/by/4.0/).

design alloys for specific biomedical applications. Research on h-BN and g-C3N4 nanoparticles as oil-based additives for significantly reducing wear rates and thereby extending product life is presented by Zhong et al. [12]. The authors of Takacova et al. [13] focus on optimizing the process of impurity removal to obtain high-purity ZnO. The optimization of process parameters in induction hardening is investigated by Dziatkiewicz et al. [14]. The potential usability of recycled titanium scrap as a substitute for primary sources is explored in the work by Judd et al. [15]. Finally, the paper by Hejazi et al. [16] significantly enhances knowledge of the processes of the restoration and conservation of Persian historical buildings.

Conflicts of Interest: The author declares no conflicts of interest.

References

1. Fu, Y.; Fu, X.; Song, W.; Li, Y.; Li, X.; Yan, L. Recent Progress of Layered Double Hydroxide-Based Materials in Wastewater Treatment. *Materials* **2023**, *16*, 5723. [CrossRef]
2. Takacova, Z.; Orac, D.; Klimko, J.; Miskufova, A. Current Trends in Spent Portable Lithium Battery Recycling. *Materials* **2023**, *16*, 4264. [CrossRef]
3. Pan, L.; Ye, S.; Xv, X.; Lin, P.; Huang, R.; Wang, D. Zeolite-Encaged Luminescent Silver Nanoclusters. *Materials* **2023**, *16*, 3736. [CrossRef]
4. Mironovs, V.; Kuzmina, J.; Serdjuks, D.; Usherenko, Y.; Lisicins, M. Sustainable Lifecycle of Perforated Metal Materials. *Materials* **2023**, *16*, 3012. [CrossRef]
5. Rostami-Tapeh-Esmaeil, E.; Rodrigue, D. Morphological, Mechanical and Thermal Properties of Rubber Foams: A Review Based on Recent Investigations. *Materials* **2023**, *16*, 1934. [CrossRef]
6. Pauerová, K.; Trpčevská, J.; Briančin, J.; Plešingerová, B. Theoretical and Practical Evaluation of the Feasibility of Zinc Evaporation from the Bottom Zinc Dross as a Valuable Secondary Material. *Materials* **2022**, *15*, 8843. [CrossRef]
7. Wędrychowicz, M.; Papacz, W.; Walkowiak, J.; Bydałek, A.; Piotrowicz, A.; Skrzekut, T.; Kurowiak, J.; Noga, P.; Kostrzewa, M. Determining the Mechanical Properties of Solid Plates Obtained from the Recycling of Cable Waste. *Materials* **2022**, *15*, 9019. [CrossRef]
8. Sheng, M.; Yang, Y.; Bin, X.; Que, W. One-Step Electrochemical Synthesis and Surface Reconstruction of NiCoP as an Electrocatalyst for Bifunctional Water Splitting. *Materials* **2023**, *16*, 1529. [CrossRef]
9. Acha, E.; Agirre, I.; Barrio, V.L. Development of High Temperature Water Sorbents Based on Zeolites, Dolomite, Lanthanum Oxide and Coke. *Materials* **2023**, *16*, 2933. [CrossRef]
10. Pikna, Ľ.; Heželová, M.; Remeteiová, D.; Ružičková, S.; Findorák, R.; Briančin, J. A Comprehensive View of the Optimization of Chromium (VI) Processing through the Application of Electrocoagulation Using a Pair of Steel Electrodes. *Materials* **2023**, *16*, 3027. [CrossRef]
11. Mamrilla, W.; Molčanová, Z.; Ballóková, B.; Džupon, M.; Džunda, R.; Csík, D.; Michalik, Š.; Lisnichuk, M.; Saksl, K. The Influence of Manganese Addition on the Properties of Biodegradable Zinc-Manganese-Calcium Alloys. *Materials* **2023**, *16*, 4655. [CrossRef]
12. Zhong, W.; Dong, J.; Chen, S.; Tong, Z. The Synergistic Lubrication Effects of h-BN and g-C3N4 Nanoparticles as Oil-Based Additives for Steel/Steel Contact. *Materials* **2023**, *16*, 4979. [CrossRef]
13. Takacova, Z.; Piroskova, J.; Miskufova, A.; Vindt, T.; Hezelova, M.; Orac, D. Removal of Impurities from EAFD Ammonium Carbonate Leachate and Upgrading the Purity of Prepared ZnO. *Materials* **2023**, *16*, 5004. [CrossRef]
14. Dziatkiewicz, G.; Kuska, K.; Popiel, R. Evolutionary Optimizing Process Parameters in the Induction Hardening of Rack Bar by Response Surface Methodology and Desirability Function Approach under Industrial Conditions. *Materials* **2023**, *16*, 5791. [CrossRef]
15. Judd, K.G.; Tsaknopoulos, K.; Sousa, B.C.; Pepi, M.; Cote, D.L. Comparative Evaluation of Titanium Feedstock Powder Derived from Recycled Battlefield Scrap vs. Virgin Powder for Cold Spray Processing. *Materials* **2024**, *17*, 1122. [CrossRef]
16. Hejazi, B.; Luz, C.; Grüner, F.; Frick, J.; Garrecht, H. Characterisation of Adobe and Mud–Straw for the Restoration and Rehabilitation of Persian Historical Adobe Buildings. *Materials* **2024**, *17*, 1764. [CrossRef]

Disclaimer/Publisher's Note: The statements, opinions and data contained in all publications are solely those of the individual author(s) and contributor(s) and not of MDPI and/or the editor(s). MDPI and/or the editor(s) disclaim responsibility for any injury to people or property resulting from any ideas, methods, instructions or products referred to in the content.

materials

MDPI

Review

Recent Progress of Layered Double Hydroxide-Based Materials in Wastewater Treatment

Yanli Fu †, Xiaoqian Fu †, Wen Song, Yanfei Li, Xuguang Li and Liangguo Yan *

School of Water Conservancy and Environment, University of Jinan, Jinan 250022, China; f18766162009@163.com (Y.F.); fuxiaoqian02@163.com (X.F.); stu_songw@ujn.edu.cn (W.S.); stu_liyf@ujn.edu.cn (Y.L.); stu_lixg@ujn.edu.cn (X.L.)
* Correspondence: yanyu-33@163.com or chm_yanlg@ujn.edu.cn
† These authors contributed equally to this work.

Abstract: Layered double hydroxides (LDHs) can be used as catalysts and adsorbents due to their high stability, safety, and reusability. The preparation of modified LDHs mainly includes coprecipitation, hydrothermal, ion exchange, calcination recovery, and sol–gel methods. LDH-based materials have high anion exchange capacity, good thermal stability, and a large specific surface area, which can effectively adsorb and remove heavy metal ions, inorganic anions, organic pollutants, and oil pollutants from wastewater. Additionally, they are heterogeneous catalysts and have excellent catalytic effect in the Fenton system, persulfate-based advanced oxidation processes, and electrocatalytic system. This review ends with a discussion of the challenges and future trends of the application of LDHs in wastewater treatment.

Keywords: LDHs; coprecipitation; hydrothermal; adsorbent; catalyst; advanced oxidation processes; electrocatalyst

Citation: Fu, Y.; Fu, X.; Song, W.; Li, Y.; Li, X.; Yan, L. Recent Progress of Layered Double Hydroxide-Based Materials in Wastewater Treatment. *Materials* **2023**, *16*, 5723. https://doi.org/10.3390/ma16165723

Academic Editor: Sabino De Gisi

Received: 18 July 2023
Revised: 17 August 2023
Accepted: 19 August 2023
Published: 21 August 2023

Copyright: © 2023 by the authors. Licensee MDPI, Basel, Switzerland. This article is an open access article distributed under the terms and conditions of the Creative Commons Attribution (CC BY) license (https://creativecommons.org/licenses/by/4.0/).

1. Introduction

Hydrotalcite, hydrotalcite-like compounds, and their intercalation materials are collectively referred to as layered double hydroxides (LDHs), which have the chemical formula of $[M^{2+}_{1-x}M^{3+}_{x}(OH)_2]^{x+}[(A^{n-})_{x/n}\cdot mH_2O]^{x-}$ (M^{2+} for divalent metal cations, such as Mg^{2+}, Co^{2+}, etc., M^{3+} for trivalent metal cations, such as Fe^{3+}, Al^{3+}, etc., and A^{n-} for inorganic anions, such as NO_3^-, CO_3^{2-}, etc.). Compared with other layered materials, the type and ratio of metals in LDHs and the type of interlayer anions can be regulated to a certain extent; thus, various LDHs with different structures and functions are prepared [1]. It is possible to modulate or modify the composition or structure of LDHs according to different requirements to obtain LDH-based materials that meet different needs. Due to the advantages of easy synthesis, low toxicity, and chemical stability, LDHs are widely used in the fields of adsorption [2], catalytic chemistry [3], electrochemistry [4], photochemistry [5,6], etc. For example, LDHs and LDH-based materials can be used as adsorbents to remove heavy metal ions [7], refractory organics [8], and other pollutants from water because of their "ion-exchangeability", "memory effect", and other properties.

Among various water and wastewater treatment methods, adsorption and catalytic methods are commonly used because of their lower operating cost and higher treatment efficiency [9,10]. In addition, LDHs also have a broad application prospect in the field of water pollution treatment. To date, several review articles have discussed the application of LDHs and their composites in water treatment, including LDHs for the treatment of dyes, heavy metals, and anionic wastewater [11–14]. In recent years, lots of work has been conducted on the application of LDH-based materials in adsorption and catalysis, and few comprehensive reviews have been published [15,16].

To further explore the research progress of LDHs in the wastewater treatment process, the preparation and improvement of LDHs and their applications as adsorbents and catalysts are reviewed, which can provide a reference for the application of LDHs in wastewater treatment. The objectives of this work include the following: (1) to summarize the synthesis methods of properties of LDHs and their composites; (2) to review the research progress of adsorption and catalytic performance of LDH-based materials for the removal of inorganic and organic pollutants from wastewater; (3) to focus on the interaction mechanisms between LDHs and pollutants using adsorption and catalytic technologies. The challenges and future trends of the application of LDHs in wastewater treatment are also discussed at the end of this review.

2. Preparation and Modification of LDHs

LDHs have attracted much attention since its discovery of brucite, and the limited amount and purity of natural LDHs have made synthetic LDHs a hot spot for various applications due to their metal type and interlayer space tunability [17]. LDH can be classified as binary, ternary, and tetradentate LDHs according to the composition and number of metal elements. A binary LDH consists of two cations; a ternary one involves three cations, and the quaternary LDH contains four cations [18]. Chen et al. [19] prepared a binary LiAl-LDH using a one-step co-precipitation method to investigate the effect of interlayer anion of SO_4^{2-} on the adsorption performance. Li et al. [20] synthesized a ternary $NiCo_2Mn$-LDH for arsenate adsorption based on the M-O octahedral structure of Ni(II), Co(II), and Mn(III). Enkhtuvshin et al. [21] prepared a tetradentate NiFeAlCo-LDH using a MOF-derived electrochemical activation method. The overpotential of NiFeAlCo-LDH was as low as 220 mV at a current of 100 mA cm^{-2}, which was caused by the oxygen intermediate that could synergize with neighboring metals.

LDH-based composites also have been studied in recent years. Zhang et al. [4] loaded CoAl-LDH onto biochar to remove heavy metals and PAHs by electrochemical activation of peroxymonosulfate (PMS). Li et al. [22] prepared flower-like nanosheets with Fe_2O_3/NiFe-LDH reaching a low overpotential of 220 mV at a high current density of 500 mA cm^{-2}, in that Fe_2O_3 and NiFe-LDH exhibited extremely high catalytic activity in concert. Li et al. [23] deposited LDHs on the ZIF-67 surface along the nanofibers to obtain nanoflower-branch composites (CoNi-LDH@CNFs) with a rich cavity structure supported by electrospun nanofibers, which have excellent cathodic performance in microbial fuel cells (MFC) with a maximum power density of 1390.37 mW/m^2.

In addition, LDHs are of great interest because of their adjustable metal ions, multiple metal combinations, simple preparation methods, and high functionality. By controlling the nucleation and growth rate of metal salts, the morphology and structure of LDHs can be controlled, and the preparation method of LDHs with excellent properties is the focus [24]. The common synthesis approaches of modified LDHs mainly include coprecipitation, hydrothermal, ion exchange, calcination recovery, and sol–gel methods.

2.1. Coprecipitation Method

The coprecipitation method is commonly used for the preparation of various LDHs and is also applied in the synthesis of LDH-based composites. Generally, the divalent and trivalent cations that constitute the LDHs laminate were mixed as a salt solution, and the target materials were added to the salt solution in a certain proportion. Then, the pH was adjusted to a constant value with Na_2CO_3 or NaOH solution under stirring at a certain temperature. Finally, the solids were washed and dried to obtain the LDH-based composites [25]. Lu et al. [26] formed a novel Fe_3O_4@MgAl-LDH on the surface of Fe_3O_4 microspheres. The dense negative charge on the surface of Fe_3O_4 microspheres and the electrostatic interaction between cations in the metal salt solution were used to form LDHs on the surface of Fe_3O_4. During the preparation process, the pH of the Fe_3O_4 suspension was adjusted to 10 with salt and alkaline solutions to make the surface of Fe_3O_4 microspheres with a partial negative charge. When adjusting the pH, vigorous stirring was required to make the Mg^{2+} and Al^{3+} cations adsorb quickly on the surface of Fe_3O_4 by electrostatic attraction. Mohamed et al. [27] prepared ZnFe-NO_3 LDH with n(Zn)/n(Fe) of 2.0 by coprecipitation method. Then, ZnFe-NO_3 LDH was mixed with chloroform containing pyrrole and stirred for 1 h to form a solution, and the oxidants, by mixing hydrochloric acid with potassium persulfate, were transferred to this solution to form a black–brown precipitate. The solids were washed and dried with ethanol and distilled water at room temperature to obtain PPy NF@ZnFe-LDH. The above process is shown in Figure 1A. The synthesized ZnFe-NO_3 LDH exhibited plate-like or sheet-like particles in the form of separated particles or agglomerated particles. After the formation of PPy NF@ZnFe-LDH nanocomposites, the ZnFe-NO_3 LDH particles were completely covered by a layer of fibrous PPy with a smooth surface and uniform diameter. The characterization image of polymer fibers reflected that the interlocking between fibers formed a porous network structure, giving the final product a higher surface area and pore volume and thus facilitating the performance as an adsorbent or catalyst carrier.

Yan's research group [28–33], prepared a series of LDHs and their composites using the coprecipitation method by changing the type of metal salts, the composition of alkali solution, and adding Fe_3O_4, biochar, and other substrates. Characterization results, such as X-ray diffraction (XRD), Fourier transform infrared (FTIR) spectroscopy, and scanning electron microscopy (SEM), showed that the as-prepared LDHs were well crystalline with a typical hexagonal layered structure [28]. The intercalated LDHs obtained by dissolving EDTA, L-cysteine, and other intercalating agents in an alkaline solution not only retained the crystal structure of LDHs but also brought in functional groups, which improved the ability of intercalated LDHs to remove pollutants from water [29,30]. The LDH-based hybrids with magnetic core-shell structures were also synthesized using Fe_3O_4 as a substrate. Fe_3O_4@MAl-LDHs (M = Mg^{2+}, Zn^{2+}, Ni^{2+}) possessed superparamagnetism as well as the crystal structure and properties of LDHs, which enabled the rapid solid–liquid separation of the material from the mixture after reaction and facilitated material recovery [29,31]. The biochar/MgAl layered double oxide (LDO) was prepared from vegetable biochar and calcination, and the schematic diagram is shown in Figure 1B. The biochar distributed as a contributing matrix in the LDH interlayer after pyrolysis significantly increased the maximum adsorption capacity of both heavy metals and anionic pollutants in the water [33].

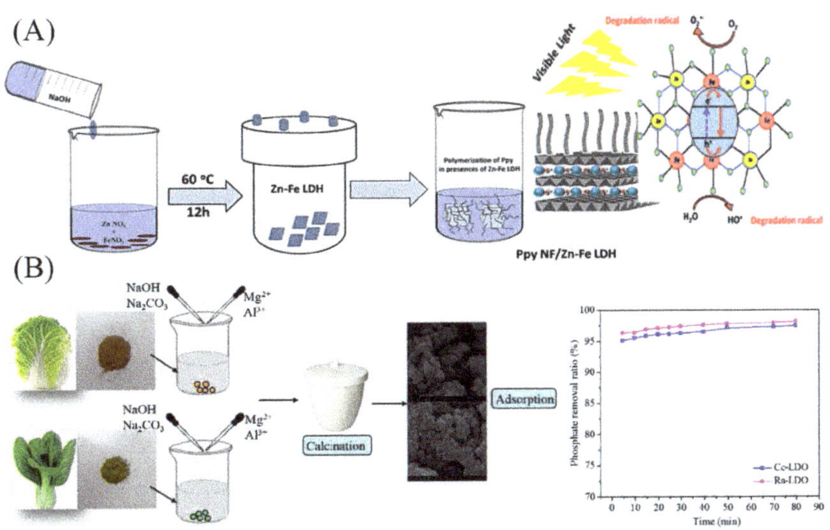

Figure 1. Schematic diagram of the synthesis of (**A**) PPy NF@ZnFe LDH [27], (**B**) Biochar/MgAl-LDO by coprecipitation method [33]. Reprinted with permission from Elsevier.

In order to obtain a good crystalline structure of the LDHs prepared by coprecipitation, the resulting products need to be aged for a long time. In addition, the particle size of LDHs obtained by the coprecipitation method tends to be unevenly distributed [34].

2.2. Hydrothermal Method

The hydrothermal method is a crystallization process under a higher temperature and pressure for a period of time; then, the solid is filtered, washed, and dried to produce LDHs. The LDHs prepared by this method have the characteristics of high purity, good dispersion, uniform particles, and controllable morphology than that of the coprecipitation method [35]. The high-temperature hydrothermal treatment allowed for the growth of LDH microcrystals with regular shapes and the controlled LDHs with different morphologies and sizes could be achieved by controlling the reaction temperature and reaction time [36]. Liang et al. [37] prepared the MgAl-LDHs with a stoichiometric ratio by reacting magnesium nitrate as well as aluminum nitrate salt solutions in a mixed solution of formamide and sodium hydroxide at a temperature of 60 °C for 24 h. During the hydrothermal process, the reagents with slow hydrolysis, such as urea, are often added to promote the crystallization in the 2D direction and the formation of stable ultrathin structure of LDHs [38]. Rathee et al. [39] synthesized NiFeTi-LDH in different proportions by mixing nickel nitrate, iron nitrate, titanium tetrachloride, and urea and hydrothermally aged in an autoclave at 160 °C for two days. The composites showed a distinct lamellar structure.

Some LDH-based composites were prepared using the hydrothermal method by changing the type of metal salts, the composition of the alkali solution, and the addition of Fe_3O_4 [40,41]. For example, the CaAl-LDH (CAL) was prepared by hydrothermal method and had typical LDH peaks, good crystallinity, and hexagonal layered structure, which could efficiently and rapidly remove hexavalent chromium (Cr(VI)) from the water [40]. To further increase the capacity, CaAl-LDH was combined with polypyrrole (PPy) to obtain (CAL-PPy) (Figure 2A) [40]. The characterization results showed that PPy was coated on the surface of hexagonal CAL. The addition of PPy introduced many amine functional groups, which improved the adsorption efficiency. To reduce the cost of composite materials and increase their reusability, MgAl-LDH was assembled onto graphene oxide and its magnetic product containing Fe_3O_4 [41]. The preparation process is shown in Figure 2B. The composites had abundant functional groups and large specific surface area and could effectively remove heavy metals under neutral and weak alkaline conditions. Moreover, magnetic nanomaterials could achieve simple and effective solid–liquid separation under a magnetic field. However, the preparation process of LDHs using the hydrothermal method is more energy-intensive and generally requires specific reaction vessels.

2.3. Ion Exchange Method

The ion exchange method is based on the characteristics of LDHs with anion exchange. Usually, synthetic LDHs are used as precursors, and the anions of the target product are exchanged with the interlayer anions to produce the desired LDHs. The ion exchange method allows for the synthesized LDHs with various interlayer anions, and the loading amounts need to be further improved. González et al. [42] synthesized MgAl-LDH-Cl with Cl^- as the interlayer anion by coprecipitation method, and then, the humic acid (HA) was added to synthesize LDH-HA adsorbent by ion exchange. It was found that the functional groups of HA had a high affinity for metal cations, and LDH-HA could remove Cu^{2+}, Pb^{2+}, and Cd^{2+} efficiently and rapidly. Li et al. [43] synthesized Fe_3O_4@(DS-LDH) by coprecipitation and ion exchange methods and used blue and methyl orange in water for the removal of methylene. The Fe_3O_4 powder was firstly prepared by solvothermal method; then, Fe_3O_4@MgAl-LDH with flaky core-shell structure was synthesized by coprecipitation method, and finally, Fe_3O_4@(DS-LDH) with SDS intercalation was prepared by ion exchange method (Figure 3A). The transmission electron microscopy image in Figure 3B showed that the LDH particles were hexagonal nanosheets of 50–100 nm. The shape of Fe_3O_4 particles was close to spherical. The morphology of DS-LDH particles was similar to that of Mg_3Al-LDH, and the size of DS-LDH particles was slightly increased. In addition, DS-LDH particles were significantly aggregated, which may be due to the enhanced hydrophobicity of the particles. The surfaces of Fe_3O_4@LDH and Fe_3O_4@(DS-LDH) were rougher than the pristine Fe_3O_4, and a thin layer of LDH (or DS-LDH) was found to cover Fe_3O_4, indicating that Fe_3O_4@LDH and Fe_3O_4@(DS-LDH) formed a core-shell structure.

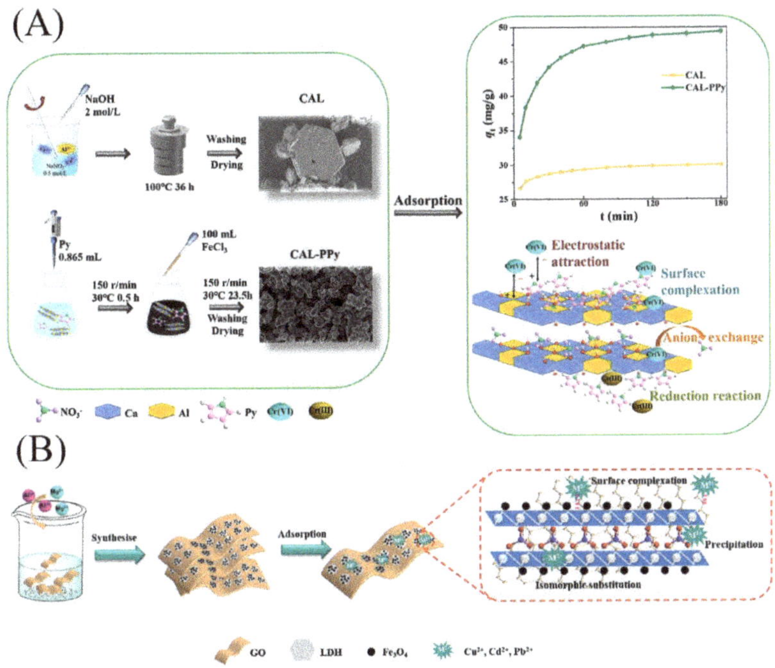

Figure 2. Schematic diagram of the synthesis of (**A**) CAL and CAL-PPy [40] and (**B**) magnetic MgAl-LDH/graphene oxide composite [41]. Reprinted with permission from Elsevier.

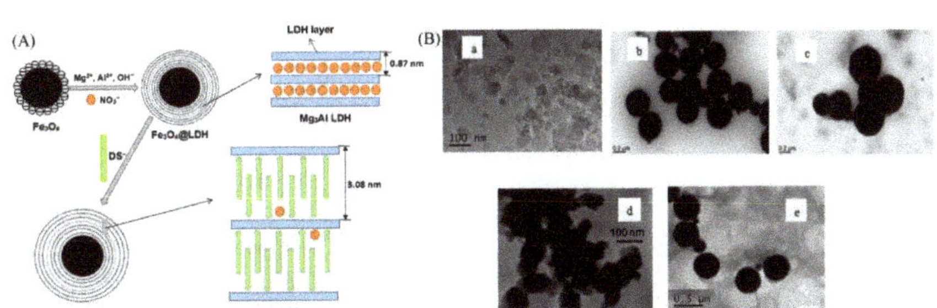

Figure 3. (**A**) Schematic diagram for the preparation of Fe_3O_4@(DS-LDH) [43]. (**B**) TEM photos of (**a**) Mg_3Al LDH, (**b**) Fe_3O_4, (**c**) Fe_3O_4@LDH, (**d**) DS-LDH, and (**e**) Fe_3O_4@ (DS-LDH) [43]. Reprinted with permission from Elsevier.

2.4. Calcination Recovery Method

The calcination recovery method is based on the special "memory effect" of LDHs; i.e., the LDHs precursors are firstly heat-treated at a certain temperature to obtain roasting products of LDOs, and then, it is added to a solution containing target anions to restore the original lamellar structure to obtain a new LDHs. LDHs are limited in their application as adsorbents for anionic pollutants because the interlayer carbonate anions are not easily exchanged with other anions, and most of the prepared classical LDHs are not hierarchically porous nanostructures. To solve this problem, Jia et al. [44] roasted MgAl-LDH at 500 °C for 4 h to obtain the precursor MgAl-LDO, reduced MgAl-LDO to obtain the borate intercalated MgAl-LDH (B-LDH), and finally, roasted at 800 °C for 4 h to obtain B-LDO (Figure 4A). The synthesized B-LDO had higher adsorption efficiency, good reusability, and stability and were more suitable for wastewater treatment. Li et al. [45] prepared ZnAl-LDH by coprecipitation method, and then, Fe_3O_4/LDH was synthesized by solvothermal method at 160 °C for 8 h. Finally, the obtained catalyst was calcined at 500 °C for 2 h to synthesize Fe_3O_4/LDO/BiOBr composite for efficient photoreduction in a high concentration of Cr(VI) (Figure 4B). A magnetic recycling experiment revealed that Fe_3O_4/LDO/BiOBr-1.5 can be easily reused for 4 cycles.

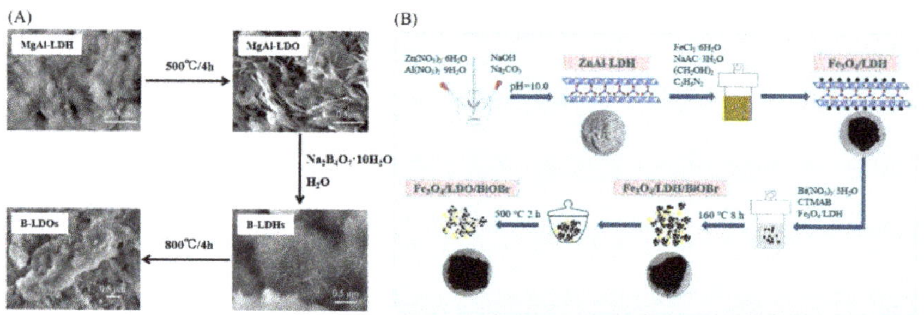

Figure 4. Schematic diagram of the preparation of (**A**) B-LDO [44] and (**B**) Fe_3O_4/LDO/BiOBr-x [45]. Reprinted with permission from Elsevier.

It is worth noting that the calcination recovery method requires higher temperatures and higher energy consumption and may produce harmful gases during the calcination process.

2.5. Sol–Gel Method

Sol–gel method uses organometallic salts through a sol–gel transformation process [46,47]. It can control the morphology and size of the particle well, and then, the phase separation can be controlled by the hydrolytic condensation reaction of the alcohol salt [48]. Chubar et al. [49] investigated the synthesis of MgAl-LDH via sol–gel method with and without alcohol salts. The adsorption effect of the alkoxide-free sol–gel (Mg-Al-AFSG) is better than the alkoxide sol–gel (Mg-Al-ASG). This is because the interlayer carbonate ions of Mg-Al-AFSG are involved in the removal of anions, while the tightly bound bidentate coordination of Mg-Al-ASG leads to poor adsorption without the involvement of carbonate. The commonly used chemicals in sol–gel method are metal alkoxides, while inorganic salts are used to achieve more economical and environmental-friendly materials. Valeikiene et al. [50] studied the reconstitution properties of $Mg_{2-x}M_x/Al_1$-LDH (M = Ca, Sr, Ba) prepared by sol–gel method. The results show that the microstructure of LDHs produced by sol–gel method is basically consistent with that of LDHs. Except for the adsorption performance, the stability of the adsorbent also plays an important role. Luo et al. [51] used the phase-separated sol–gel method to synthesize the nanocomposite of FeLiAl-LDH and FeOOH. The Fe(III) doping and FeOOH nanocomposite reduced the dissolution of LiAl-LDH in the granular nanocomposite and did not negatively affect the adsorption capacity.

3. Adsorptive Removal of Pollutants by LDH-Based Materials

There are many technical methods for wastewater treatment, among which the adsorption method has a good application prospect which has the advantages of simple operation, high efficiency, low cost, and reusability as an adsorbent. In recent years, lots of low-cost adsorbents, including natural clay minerals, industrial by-products, chitosan, organic composite materials, and biological adsorbents, have been prepared and used to remove different pollutants from water [52–54]. Among them, LDH-based materials are widely concerned as adsorbents because of their large specific surface area, low cost, easy synthesis, and reusability, and are expected to become promising new adsorptive materials.

3.1. Inorganic Pollutants

3.1.1. Heavy Metals

The application of LDHs in the adsorption of cationic heavy metals has received increasing attention [15,36]. Special functional materials can be obtained that have the advantages of both the interlayer anion and the main body of the LDHs [55]. Therefore, the adsorption capacity and selectivity of heavy metals can be improved by the functional ligands of the LDHs [55,56]. As can be seen from Table 1, a variety of modified LDH materials were synthesized to remove Cd^{2+}, Pb^{2+}, Cu^{2+}, and Cr^{6+} in water [28–30,40,41,57–61]. The adsorption performance of LDHs for heavy metal ions was investigated in detail by optimizing the experimental conditions, such as adsorbent dosage, oscillation time, and solution pH, and by studying the adsorption kinetics, isotherms, and recycling properties. The mechanisms were also conducted by integrating spectroscopic characterizations and solution chemistry. Zhang et al. [29] used a coprecipitation method to synthesize L-cysteine intercalated MgAl-LDH, thereby introducing complexing groups, such as carboxyl, sulfhydryl, and amino groups, and then achieving efficient removal of heavy metals of Cu^{2+}, Pb^{2+}, and Cd^{2+} through hydroxide or sulfide precipitation, complexation of surface-rich groups, and isomorphous replacement of LDH (Figure 5A). Lyu et al. [60] synthesized CS-LDH (chitosan-modified MgAl-LDH), which was rich in surface functional groups, such as amino, hydroxyl, and amide groups, and had a loose and porous laminar structure. The adsorption process for Pb^{2+} quickly reached an equilibrium state within 10 min, and that of Cd^{2+} was 60 min. The adsorption capacities of CS-LDH for Pb^{2+} and Cd^{2+} were significantly larger than those of CS and MgAl-LDH and reached 333.3 mg/g and 140.8 mg/g, respectively. The adsorption mechanism was chemical precipitation, surface complexation, and isomorphous replacement (Figure 5B). The above works indicated that the adsorption of cationic heavy metals was mainly through the formation of surface hydroxide precipitation, complexation with functional groups, and isomorphous replacement of LDHs.

Table 1. Adsorptive removal of heavy metal ions by LDHs.

Adsorbent	Heavy Metal Ions	Adsorption Conditions			Theoretical Adsorption Capacity (mg/g)	Adsorption Mechanism	Ref.
		Adsorbent Dosage	pH	Oscillation Time (min)			
MgAl-CO_3-LDH, Fe_3O_4/MgAl-CO_3-LDH	Cd^{2+} Cd^{2+}	0.08 g	4	60 300	61.40~70.20 45.60~54.70	precipitation, surface adsorption, surface complexation	[28]
MgAl-Cys-LDH	Cu^{2+} Pb^{2+} Cd^{2+}	0.05 g	4	90 180 10	58.07 186.20 93.11	precipitation, surface complexation, isomorphous replacement	[29]
Fe_3O_4/LDH-AM	Cd^{2+} Pb^{2+} Cu^{2+}	0.05 g	5~6	240 180 240	74.06 266.60 64.66	surface complexation, precipitation	[58]
Magnetic MgAl-LDO/carbon	Cd^{2+} Pb^{2+} Cu^{2+}		6	90 120 400	386.10 359.70 192.70	precipitation, surface complexation, electrostatic attraction	[59]
CS/MgAl-LDH	Pb^{2+} Cu^{2+}	0.05 g	6	10 60	333.30 140.80	precipitation, surface complexation, isomorphous replacement	[60]

Table 1. *Cont.*

Adsorbent	Heavy Metal Ions	Adsorption Conditions			Theoretical Adsorption Capacity (mg/g)	Adsorption Mechanism	Ref.
		Adsorbent Dosage	pH	Oscillation Time (min)			
CaAl-LDH	Cu^{2+} Cd^{2+}	0.01 g	5/5.8		381.90 1035.40	precipitation, isomorphous replacement, surface complexation	[57]
GO/LDH Fe_3O_4@GO/LDH	Cu^{2+} Cd^{2+} Pb^{2+}	20 mg		240	89.26~80.72 76.67~70.26 226.98~213.96	surface complexation, precipitation, isomorphous replacement	[41]
LDH-EDTA-AM	Cr^{6+}			30	48.47	electrostatic attraction, surface complexation, ion exchange, reduction	[30]
Fe_3O_4-ZnAl-LDH/TiO_2	Cr^{6+}	0.02 g/L	3	480		electrostatic attraction, ion exchange, photoreduction	[61]
CaAl-LDH CAL-PPy	Cr^{6+}	0.03 g			34.06 66.14	electrostatic attraction, surface complexation, anion exchange, reduction	[40]

For the heavy metals presented as anions, in addition to the above mechanisms, they could be adsorbed by LDHs via electrostatic attraction of positive surface charges and interlayer anions exchange [62]. As for Cr^{6+}, there was a simultaneous adsorption-reduction coupling mechanism in which the electron-giving groups of LDHs reduced the adsorbed Cr^{6+} to Cr^{3+} and then adsorbed Cr^{3+} by complexation and isomorphous replacement [30]. This depended on some factors, such as the divalent and trivalent laminate ions of the prepared LDHs, the type of interlayer anions, and the conditions of the adsorption experiments. Yang et al. [40] prepared CAL and CAL-Ppy by hydrothermal and in situ polymerization methods, respectively. The addition of Ppy introduced many amine functional groups, increased the number of active sites, and improved the efficiency of CAL in the removal of Cr^{6+}. The main removal mechanisms included electrostatic attraction, surface complexation, anion exchange, and reduction to less toxic Cr^{3+}. Poudel et al. [63] prepared CoAl-LDH containing hematite (α-Fe_2O_3) @3D porous carbon nanofibers (Co-Al-LDH@Fe_2O_3/3DPCNF) by hydrothermal method, which had a maximum adsorption capacity of 400.40 mg/g for Cr^{6+}. The increased specific surface area, ultra-high hydrophilicity, high affinity for metal ions, and three-dimensional cross-network nanostructure are the main reasons for the larger adsorption capacity. The mechanisms mainly included surface complexation, ion exchange, and reduction (Figure 5C). In addition, after 10 cycles, the composite was able to remove about 60% of the metal ions.

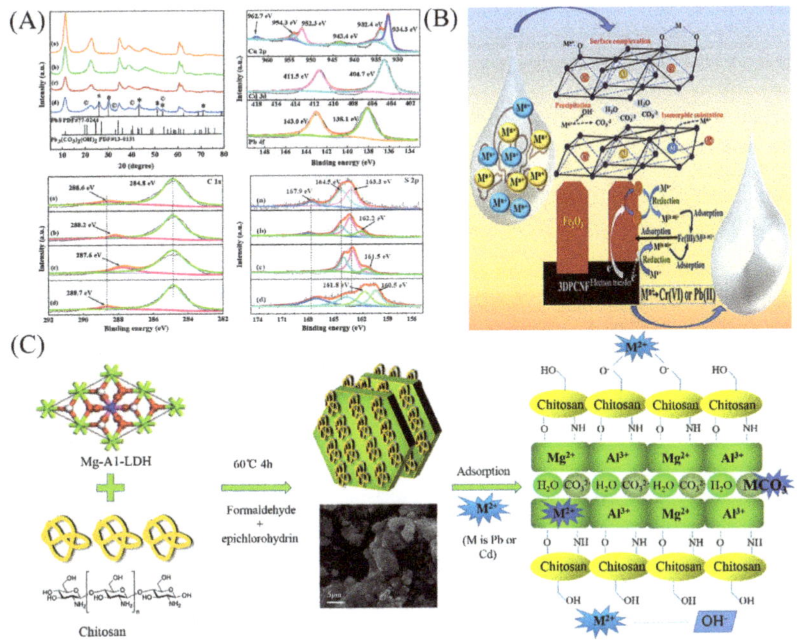

Figure 5. (**A**) XRD and XPS spectra of Pb 4f/Cd 3d/Cu 2p, C 1s and S 2p of MgAl-Cys-LDH before adsorption (a) and after adsorption of Cu(II) (b), Cd(II) (c) and Pb(II) (d) [29]. Schematic diagram of the mechanisms of heavy metal adsorption by (**B**) CS-LDH [60] and (**C**) Co-Al-LDH@Fe_2O_3/3DPCNF [63]. Reprinted with permission from Elsevier.

3.1.2. Inorganic Anions

Sulfate, phosphate, and nitrate ions in wastewater are common pollutants. Among them, sulfate ions are generally removed by adding solid calcium hydroxide ($Ca(OH)_2$) to precipitate gypsum, but the formed precipitate usually takes much time to settle, and the limited solubility of $Ca(OH)_2$ makes the removal of sulfate unsatisfactory. To improve the removal efficiency of sulfate ions, Maziarz et al. [64] investigated the effect of LDHs addition, $Ca(OH)_2$ addition, a mixture of $Ca(OH)_2$ and LDHs, and the use of LDHs after precipitation with $Ca(OH)_2$ on the removal of sulfate from water. It was found that the use of a mixture of $Ca(OH)_2$ and LDHs resulted in simultaneous precipitation and adsorption of sulfate with higher removal efficiency. In addition, the application of the mixture also resulted in a significant reduction in sludge volume compared to the experiments using only $Ca(OH)_2$. Zhang et al. [33] loaded MgAl-LDH onto cabbage and rape biomass by coprecipitation and followed by oxygen-limited calcination to obtain two biochar/LDO composites with large specific surface areas and abundant surface functional groups. At a dosage of only 0.05 g of biochar/LDOs, the removal process of 50 mg/L phosphate in the pH range of 2–10 was rapid, removing more than 92% of phosphate in 5 min. Based on FTIR, XPS, and zeta potential analysis, the adsorption mechanisms included the memory effect of LDHs, electrostatic attraction, surface complexation, and interlayer anion exchange.

Radioactive waste from nuclear power endangers the existence of living things. Adsorption is one of the main treatment methods for radioactive wastewater. Yang et al. [65] synthesized a composite of MgCo-LDHs and dendritic fibrous nanosilica (DFNS) via an interface-constrained strategy. The DFNS@Mg-Co LDH showed high uranium (U(VI)) uptake capacity (1143 mg g^{-1}) at pH = 3 and C_0 = 598.7 mg L^{-1}, which was about 4.8-fold higher than that of pristine DFNS. Yang et al. [66] prepared a MgAl-LDH intercalated with salicylaldoxime (SA-LDH), which exhibited excellent U(VI) capture performance. The tremendous maximum sorption capacity was 502 mg·g^{-1}, surpassing most known sorbents. The interaction mechanisms included complexation (UO_2^{2+} with SA^- and/or CO_3^{2-}), ion exchange, and precipitation.

3.2. Organic Pollutants

3.2.1. Dyes

Dyes are widely used in industry and can prevent sunlight and oxygen from entering the water when they are discharged into water bodies [67]. Because of the anion exchange capacity, memory effect, and large specific surface area, LDHs have shown excellent adsorption performance in the treatment of printing and dyeing wastewater. The adsorption mechanisms of LDHs for dyes can usually be divided into three main categories: (1) electrostatic attraction, whereby the hydrophilicity and hydrophobicity of LDH surface can be controlled and used to adsorb different types of dyes; (2) ion exchange, whereby the anions in the dye molecules are exchanged with the interlayer anions of LDH; (3) complexation, whereby the dye molecules complexed or chemical bonded with functional groups of LDHs to achieve efficiency removal of dye molecules from wastewater.

A series of functional LDH-based materials for removing dyes from water were synthesized [30,31]. To better investigate the removal of dyes from combined contaminants, Li et al. [30] prepared LDH-EDTA-AM (EDTA-intercalated MgAl-LDH crosslinked with acrylamide) composite. It was found that the adsorption of congo red (CR) by LDH-EDTA-AM did not affect by Cr^{6+}, while the removal of Cr^{6+} was promoted by CR. The main adsorption mechanisms include electrostatic attraction, surface complexation, and anion exchange. Soltani et al. [68] synthesized LDH/MOF composite for the treatment of a reactive dye Orange II. The higher specific surface area (1282 m^2 g^{-1}) improved the adsorption for orange II, and the maximum adsorption amount reached 1173 mg/g. Mallakpour et al. [69] prepared the nanocomposite beads containing chitosan (Chi), tannic acid (TA), LDH, and mixed metal oxides (MMO). The beads were applied for the simultaneous removal of three reactive dyes, and the maximum adsorption capacities of Chi/TA@LDH and Chi/TA@MMO beads were between 257 and 483 mg g^{-1}.

3.2.2. Oil

Oily wastewater typically contains dispersed oil, suspended particles, fats, greases, hydrocarbons, and petroleum fractions [70]. The oil forms an insoluble film on the surface of the water, preventing aeration and natural lighting of the water body. In addition, it can contaminate groundwater, aquifers, and drinking water due to the infiltration of contaminants, causing adverse effects and ecological problems [71].

Due to the persistence of oily wastewater in the aquatic environment, conventional biological processes cannot remove it quickly, which often hinders the treatment of industrial-scale systems [72]. With in situ feasibility, cost-effectiveness, and environmental friendliness, adsorption is one of the most promising technologies for efficient oil–water separation [73,74], and LDHs are considered to be potential adsorbents. Menino et al. [75] used sodium dodecyl sulfate (SDS) and Fe_3O_4 to change the properties of LDHs and prepared a hydrophobic magnetic adsorbent (LDH-MSDS) with a maximum adsorption capacity of 659.90 mg g^{-1} for oil. The results showed that the hydrophobic surface combined with the superwetting and strong adhesion of Fe_3O_4 were favorable for the selective adsorption of oil-containing wastewater.

As is well known, superhydrophobic/superoleophilic materials can be used to remove oil from oil–water mixtures. However, the filtration process often results in the blockage of membrane pores by oil droplets [76]. Then, the development of superhydrophilic/superoleophobic films with anti-fouling properties has received much attention [68,69]. Cui et al. [77] prepared the grass-like composite membranes (NiCo-LDH/PVDF) by growing NiCo-LDH on the surface of polydopamine-modified polyvinylidene fluoride (PVDF) membrane using the hydrothermal method. The membrane exhibited a superhydrophilic/superoleophobic surface with good antifouling properties and could be recycled over a long period. Xie et al. [78] prepared a new active antifouling carbon cloth@NiCo-LDH/Ag (CC@LDH/Ag) membrane. The results showed that the CC@LDH/Ag membrane had superhydrophilic/superoleophobic performance and high separation efficiency for different oil–water mixtures. In addition, CC@LDH/Ag membrane also exhibited good self-cleaning ability.

3.2.3. Persistent Organic Pollutants (POPs)

Contamination of groundwater by pesticides in modern agriculture systems is a matter of big concern and needs to be treated urgently. Lartey-Young and Ma [79] investigated the adsorption performance of CuZnFe-LDH and the composite with bamboo biochar (LDHBC) for atrazine removal from an aqueous solution. The removal of atrazine was up to 74.8%, and the adsorption mechanisms could be determined by π-π interactions occurring at the interfaces by hydrogen bonding and pore-filling effect. Mourid et al. [80] prepared the Zn_2Al-LDH by coprecipitation at a constant pH, which has been used to remove 2,4,5-trichlorophenoxyacetic acid (2,4,5-T) herbicide. The removal rate reached 96% for an optimal molar ratio 2,4,5-T/Cl = 0.5. The partial or total release of the herbicide was observed to depend on the composition of the desorbing solution. This suggested the possibility of LDH recycling and confirmed its effectiveness in eliminating this type of pollutant from water.

Per- and polyfluoroalkyl substances (PFAS) are widely used in a variety of industries and have a high chemical stability because of the high C-F bond energies, which makes them particularly persistent in environmental media [81]. Removal of PFAS from contaminated water by adsorption is considered to be an effective and simple method. Chen et al. [82] prepared an LDH with the metal composition of Cu(II)Mg(II)Fe(III) for the fast adsorption of perfluorooctane sulfonate (PFOS) and perfluorooctanoate (PFOA). A total of 84% of PFOS and 48% of PFOA were adsorbed in the first minutes when contacted with 0.1 g/L of suspended μm-sized LDH particles. Hydrophobic interactions were primarily responsible for the adsorption of these compounds in accordance with the different adsorption affinities of long-chain and short-chain perfluorinated carboxylic acids. It was previously demonstrated that the modification of minerals with organic functional groups enhanced the adsorption of organic pollutants [14]. Min et al. [83] introduced organic functional groups to the ZnAl-LDH by covalent bonding through a post-grafting process. The organically functionalized LDHs were efficient for PFOA adsorption. This resulted from the synergy of the positively charged structural layers of LDHs that provided strong electrostatic interactions and the modified organic functional groups that provided enhanced hydrophobic interactions to capture PFOA.

3.3. *Other Pollutants*

Except for the above-mentioned inorganic and organic pollutants, microplastics (MPs) are becoming an intractable environmental issue due to their non-biodegradable nature and wide spreading. Tiwari et al. [84] synthesized the nanostructured ZnAl-LDH via the co-precipitation method and used it for the adsorption of nano-scale plastic debris (NPDs) from an aqueous solution. The negatively charged NPDs were adsorbed on ZnAl-LDH by electrostatic interaction, and 96% of NPDs was removed in deionized water. The removal efficiency was found to be 100% at a pH of 4 and declined to 37% at a pH of 9 due to the increased competitive binding and destabilization of LDH under alkaline conditions.

To address the challenge of eliminating MPs in the case of acidic or alkaline wastewater, Peng et al. [85] engineered a high-performance capture agent for polystyrene (PS) in a wide pH range, i.e., three-dimensional graphene-like carbon-assembled LDO (G@LDO). In virtue of the mutual protection effect of graphene-like carbon and LDO, G@LDO featured the preeminent acid and base resistance for PS removal. The removal efficiency of PS was $\geq$80% at pH = 3–11, and nearly 60% of PS was removed at pH of 1 and 13. Furthermore, the removal mechanisms were hydrogen bond and complexation from LDH recovered from LDO, as well as π-π/p-π interactions of carbon-containing sulfur.

4. Application of LDHs in Advanced Oxidation Processes

The adsorption process only removes pollutants from the aqueous solution to the adsorbent surface, and the adsorbed contaminants are not degraded [18]. Therefore, there is an urgent need for an efficient way to degrade pollutants. In recent years, advanced oxidation processed (AOPs) have gained much attention for their ability to completely degrade pollutants to CO_2 and H_2O [86]. It mainly includes Fenton, persulfate, and electrochemical processes.

4.1. Fenton-like Reaction

Fenton reaction is a process that uses the catalytic reaction of H_2O_2 and Fe^{2+} to rapidly generate highly reactive $^{\bullet}OH$ radicals, which is widely used in the wastewater treatment [87]. The generation of $^{\bullet}OH$ in the Fenton reaction is closely related to iron ions, and it has been shown that the incorporation of FeS and Fe_3O_4 nanoparticles significantly enhanced the coupling of Fe^{2+} and S^- and the synergistic effect of FeS on the Fenton process accelerated the dechlorination and degradation of organochlorine compounds [88]. Furthermore, Huang et al. [87] obtained FeS@LDH and Fe_3O_4@LDH composites to catalyze the Fenton reaction for the degradation of methoxychlor. In the catalytic degradation process of Fe_3O_4@LDH, the reductive dechlorination, cleavage, and oxidation of methoxychlor occurred; while using FeS@LDH and Fe_3O_4@LDH hybrid catalysts, the hydrogenolysis and cleavage of methoxychlor occurred (Figure 6A).

Most of the variable-valence transition metals, such as Cu and Co ions, can replace Fe ions in the Fenton reaction [9]. Compared with Fe, Cu facilitates the interfacial electron transfer and, thus, accelerates the regeneration of the catalyst. Yan et al. [89] established a highly catalytic system for the degradation of ethylbenzene using Cu-based LDHs (CuMgFe-LDH) and H_2O_2. The degradation of ethylbenzene was achieved through the formation of active sites by surface-bound hydroxyl groups of LDHs and the generation of $^{\bullet}OH$ in the redox process of Cu(II)-Cu(III). In addition, due to the tunability of multi-metals in LDH laminates, LDHs with synergistic redox cycle of multiple transition metals can be designed to promote the regeneration of low-valent metal and improve the efficiency of the Fenton reaction. Wang et al. [90] developed a ternary CuNiFe-LDH to degrade phenol. Compared with NiFe-, CuFe-, and CuNiAl-LDHs, CuNiFe-LDH had better catalytic performance. In the conventional Fenton reaction process in which Fe^{3+} in the LDHs laminate was reduced to Fe^{2+} and then oxidized to Fe^{3+} again, Cu^+ in the laminate could also be oxidized to Cu^{2+} to catalyze the Fenton reaction. Moreover, the metal–oxygen–metal bridge and highly dispersed NiO_6 octahedra in LDHs allowed for a quick electron transfer and generated more $^{\bullet}OH$ radicles to degrade phenol to CO_2 and H_2O, and the mechanism of CuNiFe-LDH catalytic degradation of phenol is shown in Figure 6B.

Some substances, such as carbon-based materials, can also be loaded on LDHs to form stable and efficient Fenton catalysts. For example, the trimetallic LDH of CuMgAl-LDH combined with reduced graphene oxide (rGO) exhibited a triphasic synergistic effect with enhanced catalytic performance [91]. The structure of the catalyst improved the stability and electronic conductivity. Peng et al. [92] prepared CuAl-LDH for the degradation of ammonia nitrogen and found that CuAl-LDH aggregated easily. The addition of carbon fiber (CF) prevented the aggregation of the delaminated LDH nanosheets. The conductivity and porous structure of CF also accelerated the diffusion of ammonia nitrogen, thus improving the catalytic performance of CuAl/CF-LDH. The large specific surface area, abundant active sites, and strong chemical forces were involved in the ammonia nitrogen degradation (Figure 6C).

In the application of LDHs as catalysts in Fenton reactions, most studies focused on the effect of different cations in the lamellae of LDHs. Few studies considered the effects of interlayer anions, which also affect catalytic activity. Costa-Serge et al. [93] evaluated the effect of different interlayer anions in the boronic acid-based LDH (CuMgFe-$B(OH)_4$) on the degradation of 5-fluorouracil. It was found that the intermolecular B-O and B-OH can induce changes in the surface charge density of CuMgFe-$B(OH)_4$ and accelerate the reduction in Cu(II) and Fe(III), which favors the generation of reactive oxygen species. Figure 6D explains the possibility that the presence of boric acid leads to the partial reduction in iron and copper on the catalyst surface. The Lewis acid-base bond between LDH hydroxy-boron and oxygen was strengthened and improved the stabilization of the hydrogen bond between boric acid and CuMgFe-B$(OH)_4$ surface.

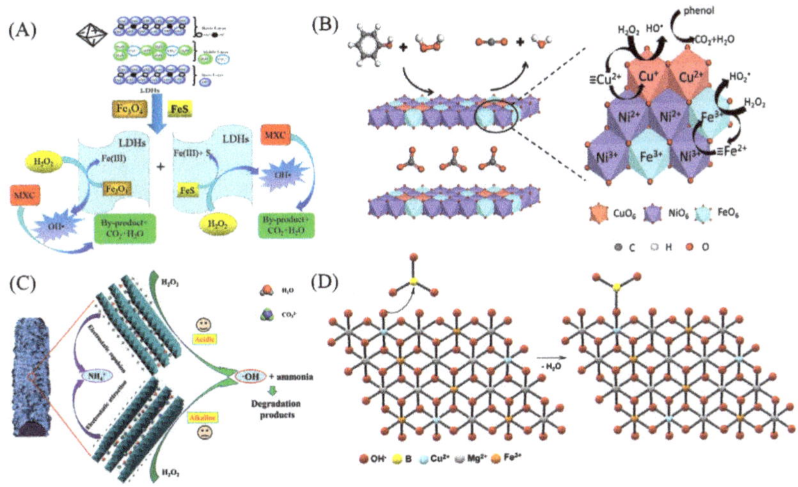

Figure 6. (**A**) Schematic diagram of Fenton reaction catalyzed by FeS@LDH and Fe_3O_4@LDH [87]. (**B**) Reaction mechanism of phenol degradation by CuNiFe-LDH [90]. (**C**) Illustration of the mechanism of ammonia nitrogen degradation by 2-CuAl/CF-LDH [92]. (**D**) Surface reaction between CuMgFe-B$(OH)_4$ and B$(OH)_3$ [93]. Reprinted with permission from Elsevier.

4.2. *Persulfate-Based AOPs*

Among different AOPs, $SO_4^{\bullet-}$ dominated AOPs not only overcome the limitations of conventional Fenton system in terms of narrow pH applicability, large use of H_2O_2, ferrous salts, and lots of sludge yield in practical applications but also have the advantages of strong oxidative capacity, long half-life of radicals, and strong mass transfer capacity [94]. It is noteworthy that LDHs have a great advantage in the effective activation of PMS due to their controllable lamellar structure [95].

Transition metals usually act as electron donors to activate PMS due to their advantages, such as low energy consumption and high efficiency [96]. Co^{2+}, Ni^{2+}, Fe^{2+}, Cu^{2+}, and Mn^{2+} have shown strong persulfate catalytic activity [97,98]. Gong et al. [99] prepared FeCo-LDH by coprecipitation method and used to degrade RhB by non-homogeneous activated PMS. In FeCo-LDH/PMS system, RhB was completely degraded within 10 min, and the removal efficiency reached more than 90%. As shown in Figure 7A, the redox process of Fe(II) and Co(II) generated electrons, activated PMS to produce reactive radicals, and the generated $SO_4^{\bullet-}$ and $^{\bullet}OH$ simultaneously rapidly degraded RhB. The efficient regeneration of Co(II) may be the reason for the significant activation of PMS and the enhanced degradation of RhB. Li et al. [100] prepared CoFeLa-LDH catalyst by coprecipitation method, which exhibited excellent performance to activate PMS for tetracycline (TC) elimination. The degradation efficiency of TC in the CoFeLa-LDH$_2$/PMS system was 90.1% within 10 min.

To solve the problem of metal leaching, it is proposed to prepare LDH-based composites by immobilizing LDHs onto a specific functional carrier. Carbon-based-LDHs composites with activated carbon (AC), GO, biochar, and various carbon polymer films as carriers can increase the degradation efficiency and stability compared to single LDHs catalysts [101–103]. Ma et al. [103] prepared the CoFe-LDH using the coprecipitation method, loaded it onto AC, and purified it with HCl to synthesize AC@CoFe-LDH. Compared with CoFe-LDH/PS system, the degradation efficiency of lomefloxacin by AC@CoFe-LDH was increased by 56% within 60 min, where $SO_4^{\bullet-}$ played a major role in the degradation process. The catalyst itself could be regenerated in the redox reaction between Fe(II)/Fe(III) and Co(II)/Co(III), and the reaction of AC components with H_2O could also restore the functional groups (Figure 7B). As dissolved organic matter (DOM) is of great interest as an important component of biochar, Ye et al. [102] obtained DOM-FeAl-LDH by incorporating DOM into FeAl-LDH and further used it to activate PMS to degrade BPA in water. Compared with FeAl-LDH/PMS system, the removal rate of BPA in DOM-FeAl-LDH/PMS system increased from 60% to 93% within 60 min. It was found that the catalyst activity and stability were improved after DOM incorporation. This was because the HA-like compounds in DOM could act as electron transfer mediators, which could accelerate the reduction in Fe(III). In addition, DOM itself could produce ROS, which reacts with dissolved oxygen and may lead to the production of $^1O_2$ (Figure 7C).

For LDH-based catalysts, some metal oxides or metal sulfides have been intensively utilized [104,105]. Ali et al. [104] dispersed Co-LDH in copper salt solution and calcined to obtain CuOx@Co-LDH and found that phenol was degraded within 40 min. The Cu^{2+} was detected in the solution, and no Co^{2+} was detected after the reaction. The degradation mechanism was mainly through the synergistic interaction between mixed valence metals to activate PS to produce $SO_4^{\bullet-}$ and $^{\bullet}OH$ for phenol degradation (Figure 7D). The strong synergistic interaction between the active components in the LDH structure was the reason for the superior stability and activity of the CuOx@Co-LDH catalyst. Zeng et al. [105] synthesized hollow flower-like and rich oxygen vacancies of CoAl-LDH@CoSx catalyst and found that 98.5% of sulfamethoxazole (40 μmol/L) could be removed in 4 min, and the mineralization rate reached 20.4% in 6 min. Both results of quenching and radical trapping experiments showed that the main active substance in CoAl-LDH@CoSx/PMS system was $^1O_2$.

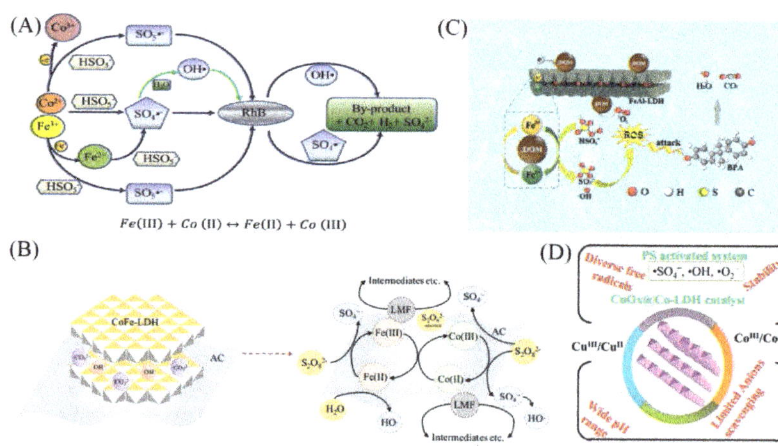

Figure 7. (**A**) Mechanism of RhB degradation in FeCo-LDH/PMS system [99]. (**B**) Activation of PS by AC@CoFe-LDH for lomefloxacin degradation [103]. (**C**) Removal mechanism of BPA in DOM-FeAl-LDH/PMS system [102]. (**D**) Advantages of CuOx@Co-LDH as an activator for PS [104]. Reprinted with permission from Elsevier.

4.3. Electrocatalytic System

The diversity of layer metal ions allows for the LDHs to have different physicochemical and electrical properties, which can be used as electrocatalysts in the wastewater treatment [106]. Currently, the most widely used oxygen evolution reaction (OER) catalysts in alkaline environments are commercial ruthenium and iridium oxides (RuO_2, IrO_2) [107], which cannot be used commercially on a large scale due to their high production costs and low electrochemical stability. It is well known that transition metals are abundant and have good catalytic properties for both OER and hydrogen evolution reactions (HER) [108]. Therefore, transition metal hydroxides are used to replace precious metal catalysts with great advantages, especially some composites containing multiple transition metals. Among them, Ni-based LDHs have become a subject of intensive materials in recent years [109]. Liu et al. [110] synthesized a series of Cr, Mn, and Co-doped NiFe-LDHs via the hydrothermal method, all of which enhanced the catalytic activity of OER in an alkaline medium. The improved catalytic performance could be attributed to the tuning of the electronic states of Ni and Fe atoms by the doped metal atoms. However, the stability of transition metal catalysts needs to be further improved compared to precious metals.

Non-precious metal catalysts are in full swing. For example, carbon nanotubes and graphene, which are rich in functional groups (−COOH, −OH, etc.) on their surfaces and have strong electrical conductivity, large specific surface area, and good thermal stability, and have attracted much attention in the field of electrocatalysis [111]. Xu et al. [112] prepared the three-dimensional nanocomposites by assembling NiAl-LDHs nanosheets on GO. The structure was unique, and it increased the specific surface area. The specific capacitance could reach 1300 F g^{-1} at higher current densities. After 500 cycles at 15.30 A g^{-1}, the specific capacitance remained at 91% compared to 74% for pure NiAl-LDH. Meanwhile, the good stability of non-metallic materials under acidic and alkaline conditions makes them the preferred materials for electrocatalysts under high currents. However, the lower electronic conductivity hinders the further development of LDHs as high-performance electrocatalysts. To solve this problem, Duan et al. [113] prepared the NiFe-LDH/CNTs composites with alternating stacks of NiFe-LDH and CNTs layers. This unique structure provided ample active sites and efficient charge/electron transfer channels for OER electrocatalysis. The optimized composite required only 234 mV to achieve a current density of 10 mA cm^{-2} and exhibited better OER activity than RuO_2. The excellent performance should be attributed to the synergistic effect of active NiFe-LDH with conductive CNTs and an intercalation structure with a highly accessible surface area. To meet the growing demand for electrochemical conversion systems, LDH electrocatalysts should focus more on compact design, low cost, and versatility.

5. Conclusions and Prospect

We reviewed the main research advances in LDH-based materials for wastewater treatment in recent years. The preparation approaches of modified LDHs mainly include coprecipitation, hydrothermal, ion exchange, calcination recovery, and sol–gel methods. Wastewater treatment methods using LDHs mainly involve adsorption, Fenton reaction, persulfate-based AOPs, and electrocatalytic method. The advantages of LDHs as adsorbents and catalysts, as well as the reaction mechanism, were mainly reviewed.

Although the modification of LDHs and the removal of contaminants from water are studied, some shortcomings still remain and need to be further investigated.

(1) In terms of the preparation of LDHs, the method of high-performance LDHs is relatively cumbersome and not conducive to industrial production, which limits the application. In addition, LDHs are mostly in powder form and are difficult to be recovered after the treatment process;
(2) In terms of wastewater treatment using LDH-based materials, most of the studies are still on the laboratory scale. LDHs are often used to remove a single pollutant, while various pollutants usually coexist in the water environment;
(3) In terms of mechanism research, there is a lack of in-depth methods, and the quantitative contribution of each mechanism to the total capacity needs to be further investigated.

In view of the shortcomings of LDHs, we can put forward the challenges to be faced in future work:

(1) Due to the urgent demand for high-efficient and low-cost materials for treating wastewater, there is a need to produce sustainable, low-cost, and industry-scale LDHs. For better recovery of LDHs and improvement of reusability during large-scale use, magnetic or non-powdered LDHs can be designed and synthesized for wastewater treatment;
(2) To make full use of the advantages of LDHs and other nanomaterials, high-performance LDH-based composites can be prepared via various methods. For example, LDHs with multivariate ions and the combination of LDHs with other high-performance materials may be one of the research directions to obtain high-performance catalysts;
(3) More work from laboratory to industrial production is urgently needed. There is also a need to strengthen the research on the performance of modified LDHs in removing specific pollutants selectively from multiple pollutants;
(4) Further in-depth studies on the mechanisms of LDHs for various pollutants can be carried out by means of spectroscopic characterization, interfacial chemical methods, and theoretical calculations. The contribution of each mechanism should be conducted to understand the interfacial process.

Author Contributions: Y.F.: conceptualization, methodology, investigation, formal analysis, and writing—original draft; L.Y.: writing—review and editing, methodology, funding acquisition, and resources; X.F.: formal analysis and writing—original draft; W.S., Y.L. and X.L.: conceptualization, methodology, investigation, and formal analysis. All authors have read and agreed to the published version of the manuscript.

Funding: National nature science foundation of China (21577048).

Institutional Review Board Statement: Not applicable.

Informed Consent Statement: Not applicable.

Data Availability Statement: Not applicable.

Conflicts of Interest: The authors declare that they have no known competing financial interests or personal relationships that could have appeared to influence the work reported in this paper.

References

1. Mishra, G.; Dash, B.; Pandey, S. Layered double hydroxides: A brief review from fundamentals to application as evolving biomaterials. *Appl. Clay Sci.* **2018**, *153*, 172–186. [CrossRef]
2. Ma, L.; Wang, Q.; Islam, S.M.; Liu, Y.; Ma, S.; Kanatzidis, M.G. Highly selective and efficient removal of heavy metals by layered double hydroxide intercalated with the MoS_4^{2-} ion. *J. Am. Chem. Soc.* **2016**, *138*, 2858–2866. [CrossRef]
3. Li, C.; Wei, M.; Evans, D.G.; Duan, X. Layered double hydroxide-based nanomaterials as highly efficient catalysts and adsorbents. *Small* **2014**, *10*, 4469–4486. [CrossRef]
4. Zhang, X.; Zhang, X.; An, C.; Wang, S. Electrochemistry-enhanced peroxymonosulfate activation by CoAl-LDH@biochar for simultaneous treatment of heavy metals and PAHs. *Sep. Purif. Technol.* **2023**, *311*, 123341. [CrossRef]
5. Mohapatra, L.; Parida, K. A review on the recent progress, challenges and perspective of layered double hydroxides as promising photocatalysts. *J. Mater. Chem. A* **2016**, *4*, 10744–10766. [CrossRef]
6. Boumeriame, H.; Da Silva, E.S.; Cherevan, A.S.; Chafik, T.; Faria, J.L.; Eder, D. Layered double hydroxide (LDH)-based materials: A mini-review on strategies to improve the performance for photocatalytic water splitting. *J. Energy Chem.* **2022**, *64*, 406–431. [CrossRef]
7. Guan, X.; Yuan, X.; Zhao, Y.; Bai, J.; Li, Y.; Cao, Y.; Chen, Y.; Xiong, T. Adsorption behaviors and mechanisms of Fe/Mg layered double hydroxide loaded on bentonite on Cd (II) and Pb (II) removal. *J. Colloid Interface Sci.* **2022**, *612*, 572–583. [CrossRef]
8. Zhang, G.; Zhang, X.; Meng, Y.; Pan, G.; Ni, Z.; Xia, S. Layered double hydroxides-based photocatalysts and visible-light driven photodegradation of organic pollutants: A review. *Chem. Eng. J.* **2020**, *392*, 123684. [CrossRef]
9. Nerud, F.; Baldrian, P.; Gabriel, J.; Ogbeifun, D. Decolorization of synthetic dyes by the Fenton reagent and the Cu/pyridine/H_2O_2 system. *Chemosphere* **2001**, *44*, 957–961. [CrossRef]
10. Pan, D.; Ge, S.; Tian, J.; Shao, Q.; Guo, L.; Liu, H.; Wu, S.; Ding, T.; Guo, Z. Research progress in the field of adsorption and catalytic degradation of sewage by hydrotalcite-derived materials. *Chem. Rec.* **2020**, *20*, 355–369. [CrossRef]
11. Dong, Y.; Kong, X.; Luo, X.; Wang, H. Adsorptive removal of heavy metal anions from water by layered double hydroxide: A review. *Chemosphere* **2022**, *303*, 134685. [CrossRef]
12. Keyikoglu, R.; Khataee, A.; Yoon, Y. Layered double hydroxides for removing and recovering phosphate: Recent advances and future directions. *Adv. Colloid Interface* **2022**, *300*, 102598. [CrossRef]

13. Dias, A.C.; Ferreira Fontes, M.P. Arsenic (V) removal from water using hydrotalcites as adsorbents: A critical review. *Appl. Clay Sci.* **2020**, *191*, 105615. [CrossRef]
14. Mittal, J. Recent progress in the synthesis of layered double hydroxides and their application for the adsorptive removal of dyes: A review. *J. Environ. Manag.* **2021**, *295*, 113017. [CrossRef]
15. He, X.; Qiu, X.; Hu, C.; Liu, Y. Treatment of heavy metal ions in wastewater using layered double hydroxides: A review. *J. Dispers. Sci. Technol.* **2018**, *39*, 792–801. [CrossRef]
16. Tang, Z.; Qiu, Z.; Lu, S.; Shi, X. Functionalized layered double hydroxide applied to heavy metal ions absorption: A review. *Nanotechnol. Rev.* **2020**, *9*, 800–819. [CrossRef]
17. Taoufik, N.; Sadiq, M.H.; Abdennouri, M.; Qourzal, S.; Khataee, A.; Sillanpaa, M.; Barka, N. Recent advances in the synthesis and environmental catalytic applications of layered double hydroxides-based materials for degradation of emerging pollutants through advanced oxidation processes. *Mater. Res. Bull.* **2022**, *154*, 111924. [CrossRef]
18. Xie, Z.; Zhou, H.; He, C.; Pan, Z.; Yao, G.; Lai, B. Synthesis, application and catalytic performance of layered double hydroxide based catalysts in advanced oxidation processes for wastewater decontamination: A review. *Chem. Eng. J.* **2021**, *414*, 128713. [CrossRef]
19. Chen, J.; Yuan, H.; Yu, J.; Yan, M.; Yang, Y.; Lin, S. Regulating lithium extraction based on intercalated SO_4^{2-} in Li/Al-LDHs. *J. Colloid Interface Sci.* **2023**, *649*, 694–702. [CrossRef]
20. Li, K.; Li, S.; Li, Q.; Liu, H.; Yao, W.; Wang, Q.; Chai, L. Design of a high-performance ternary LDHs containing Ni, Co and Mn for arsenate removal. *J. Hazard. Mater.* **2022**, *427*, 127865. [CrossRef]
21. Enkhtuvshin, E.; Kim, K.M.; Kim, Y.K.; Mihn, S.; Kim, S.J.; Jung, S.Y.; Thao, N.T.T.; Ali, G.; Akbar, M.; Chung, K.Y.; et al. Stabilizing oxygen intermediates on redox-flexible active sites in multimetallic Ni-Fe-Al-Co layered double hydroxide anodes for excellent alkaline and seawater electrolysis. *J. Mater. Chem. A* **2021**, *9*, 27332–27346. [CrossRef]
22. Li, C.F.; Xie, L.J.; Zhao, J.W.; Gu, L.F.; Wu, J.Q.; Li, G.R. Interfacial electronic modulation by Fe_2O_3/NiFe-LDHs heterostructures for efficient oxygen evolution at high current density. *Appl. Catal. B-Environ.* **2022**, *306*, 121097. [CrossRef]
23. Li, H.; Sun, Y.; Wang, J.; Liu, Y.; Li, C. Nanoflower-branch LDHs and CoNi alloy derived from electrospun carbon nanofibers for efficient oxygen electrocatalysis in microbial fuel cells. *Appl. Catal. B-Environ.* **2022**, *307*, 121136. [CrossRef]
24. Chen, Y.; Li, J.; Zhai, B.; Liang, Y. Enhanced photocatalytic degradation of RhB by two-dimensional composite photocatalyst. *Colloid Surf. A* **2019**, *568*, 429–435. [CrossRef]
25. Tang, S.; Yao, Y.; Chen, T.; Kong, D.; Shen, W.; Lee, H.K. Recent advances in the application of layered double hydroxides in analytical chemistry: A review. *Anal. Chim. Acta* **2020**, *1103*, 32–48. [CrossRef]
26. Lu, L.; Li, J.; Ng, D.H.L.; Yang, P.; Song, P.; Zuo, M. Synthesis of novel hierarchically porous Fe_3O_4@MgAl-LDH magnetic microspheres and its superb adsorption properties of dye from water. *J. Ind. Eng. Chem.* **2017**, *46*, 315–323. [CrossRef]
27. Mohamed, F.; Abukhadra, M.R.; Shaban, M. Removal of safranin dye from water using polypyrrole nanofiber/Zn-Fe layered double hydroxide nanocomposite (PPy NF/Zn-Fe LDH) of enhanced adsorption and photocatalytic properties. *Sci. Total Environ.* **2018**, *640*, 352–363. [CrossRef]
28. Shan, R.; Yan, L.; Yang, K.; Hao, Y.; Du, B. Adsorption of Cd(II) by Mg-Al-CO_3- and magnetic Fe_3O_4/Mg-Al-CO_3-layered double hydroxides: Kinetic, isothermal, thermodynamic and mechanistic studies. *J. Hazard. Mater.* **2015**, *299*, 42–49. [CrossRef]
29. Zhang, X.; Yan, L.; Li, J.; Yu, H. Adsorption of heavy metals by L-cysteine intercalated layered double hydroxide: Kinetic, isothermal and mechanistic studies. *J. Colloid Interface Sci.* **2020**, *562*, 149–158. [CrossRef]
30. Li, J.; Yu, H.; Zhang, X.; Zhu, R.; Yan, L. Crosslinking acrylamide with EDTA-intercalated layered double hydroxide for enhanced recovery of Cr(VI) and Congo red: Adsorptive and mechanistic study. *Front. Environ. Sci. Eng.* **2020**, *14*, 52. [CrossRef]
31. Shan, R.; Yan, L.; Yang, K.; Yu, S.; Hao, Y.; Yu, H.; Du, B. Magnetic Fe_3O_4/MgAl-LDH composite for effective removal of three red dyes from aqueous solution. *Chem. Eng. J.* **2014**, *252*, 38–46. [CrossRef]
32. Su, X.; Chen, Y.; Li, Y.; Li, J.; Song, W.; Li, X.; Yan, L. Enhanced adsorption of aqueous Pb(II) and Cu(II) by biochar loaded with layered double hydroxide: Crucial role of mineral precipitation. *J. Mol. Liq.* **2022**, *357*, 119083. [CrossRef]
33. Zhang, Z.; Yan, L.; Yu, H.; Yan, T.; Li, X. Adsorption of phosphate from aqueous solution by vegetable biochar/layered double oxides: Fast removal and mechanistic studies. *Bioresour. Technol.* **2019**, *284*, 65–71. [CrossRef]
34. Chaillot, D.; Bennici, S.; Brendle, J. Layered double hydroxides and LDH-derived materials in chosen environmental applications: A review. *Environ. Sci. Pollut. Res.* **2021**, *28*, 24375–24405. [CrossRef]
35. Kozerozhets, I.V.; Panasyuk, G.P.; Semenov, E.A.; Avdeeva, V.V.; Danchevskaya, M.N.; Simonenko, N.P.; Vasiliev, M.G.; Kozlova, L.O.; Ivakin, Y.D. Recrystallization of nanosized boehmite in an aqueous medium. *Powder Technol.* **2023**, *413*, 118030. [CrossRef]
36. Gu, P.; Zhang, S.; Li, X.; Wang, X.; Wen, T.; Jehan, R.; Alsaedi, A.; Hayat, T.; Wang, X. Recent advances in layered double hydroxide-based nanomaterials for the removal of radionuclides from aqueous solution. *Environ. Pollut.* **2018**, *240*, 493–505. [CrossRef]
37. Liang, D.; Yue, W.; Sun, G.; Zheng, D.; Ooi, K.; Yang, X. Direct Synthesis of unilamellar MgAl-LDH nanosheets and stacking in aqueous solution. *Langmuir* **2015**, *31*, 12464–12471. [CrossRef]
38. Omwoma, S.; Chen, W.; Tsunashima, R.; Song, Y. Recent advances on polyoxometalates intercalated layered double hydroxides: From synthetic approaches to functional material applications. *Coord. Chem. Rev.* **2014**, *258*, 58–71. [CrossRef]
39. Rathee, G.; Awasthi, A.; Sood, D.; Tomar, R.; Tomar, V.; Chandra, R. A new biocompatible ternary layered double hydroxide adsorbent for ultrafast removal of anionic organic dyes. *Sci. Rep.* **2019**, *9*, 16225. [CrossRef]

40. Yang, D.; Chen, Y.; Li, J.; Li, Y.; Song, W.; Li, X.; Yan, L. Synthesis of calcium-aluminum-layered double hydroxide and a polypyrrole decorated product for efficient removal of high concentrations of aqueous hexavalent chromium. *J. Colloid Interface Sci.* **2022**, *607*, 1963–1972. [CrossRef]
41. Li, J.; Huang, Q.; Yu, H.; Yan, L. Enhanced removal performance and mechanistic study of Cu^{2+}, Cd^{2+}, and Pb^{2+} by magnetic layered double hydroxide nanosheets assembled on graphene oxide. *J. Water Process Eng.* **2022**, *48*, 102893. [CrossRef]
42. Gonzalez, M.A.; Pavlovic, I.; Barriga, C. Cu(II), Pb(II) and Cd(II) sorption on different layered double hydroxides. A kinetic and thermodynamic study and competing factors. *Chem. Eng. J.* **2015**, *269*, 221–228. [CrossRef]
43. Li, Y.; Bi, H.; Liang, Y.; Mao, X.; Li, H. A magnetic core-shell dodecyl sulfate intercalated layered double hydroxide nanocomposite for the adsorption of cationic and anionic organic dyes. *Appl. Clay Sci.* **2019**, *183*, 105309. [CrossRef]
44. Jia, Y.; Liu, Z. Preparation of borate anions intercalated MgAl-LDHs microsphere and its calcinated product with superior adsorption performance for Congo red. *Colloid Surf. A* **2019**, *575*, 373–381. [CrossRef]
45. Li, J.; Yu, H.; Yan, T.; Sun, M.; Li, X.; Song, W.; Yan, L. Insight into the enhanced visible-light photoreduction of aqueous Cr(VI) by assembled Fe_3O_4/LDO/BiOBr composites. *Colloid Surf. A* **2022**, *634*, 128021. [CrossRef]
46. Feng, X.; Long, R.; Wang, L.; Liu, C.; Bai, Z.; Liu, X. A review on heavy metal ions adsorption from water by layered double hydroxide and its composites. *Sep. Purif. Technol.* **2022**, *284*, 120099. [CrossRef]
47. Kameliya, J.; Verma, A.; Dutta, P.; Arora, C.; Vyas, S.; Varma, R.S. Layered double hydroxide materials: A review on their preparation, characterization, and applications. *Inorganics* **2023**, *11*, 121. [CrossRef]
48. Danks, A.E.; Hall, S.R.; Schnepp, Z. The evolution of 'sol-gel' chemistry as a technique for materials synthesis. *Mater. Horiz.* **2016**, *3*, 91–112. [CrossRef]
49. Chubar, N.; Gerda, V.; Megantari, O.; Micusik, M.; Omastova, M.; Heister, K.; Man, P.; Fraissard, J. Applications versus properties of Mg-Al layered double hydroxides provided by their syntheses methods: Alkoxide and alkoxide-free sol-gel syntheses and hydrothermal precipitation. *Chem. Eng. J.* **2013**, *234*, 284–299. [CrossRef]
50. Valeikiene, L.; Roshchina, M.; Grigoraviciute-Puroniene, I.; Prozorovich, V.; Zarkov, A.; Ivanets, A.; Kareiva, A. On the reconstruction peculiarities of sol-gel derived Mg_2-xMx/Al-1(M = Ca, Sr, Ba) layered double hydroxides. *Crystals* **2020**, *10*, 470. [CrossRef]
51. Luo, Q.; Dong, M.; Li, Q.; Wu, Z.; Liu, Z.; Li, J. Improve the durability of lithium adsorbent Li/Al-LDHs by Fe^{3+} substitution and nanocomposite of FeOOH. *Miner. Eng.* **2022**, *185*, 107717. [CrossRef]
52. Zhang, Y.; Zhao, M.; Cheng, Q.; Wang, C.; Li, H.; Han, X.; Fan, Z.; Su, G.; Pan, D.; Li, Z. Research progress of adsorption and removal of heavy metals by chitosan and its derivatives: A review. *Chemosphere* **2021**, *279*, 130927. [CrossRef]
53. Wu, M.J.; Wu, J.Z.; Zhang, J.; Chen, H.; Zhou, J.Z.; Qian, G.R.; Xu, Z.P.; Du, Z.; Rao, Q.L. A review on fabricating heterostructures from layered double hydroxides for enhanced photocatalytic activities. *Catal. Sci. Technol.* **2018**, *8*, 1207–1228. [CrossRef]
54. El Ashmawy, A.A.; Tada, M.; Yoshimura, C. Weak dehydration enhances the adsorption capacity of boehmite for anionic dyes. *Colloid Surf. A* **2023**, *674*, 131954. [CrossRef]
55. Laipan, M.; Yu, J.; Zhu, R.; Zhu, J.; Smith, A.T.; He, H.; O'Hare, D.; Sun, L. Functionalized layered double hydroxides for innovative applications. *Mater. Horiz.* **2020**, *7*, 715–745. [CrossRef]
56. Ma, S.; Huang, L.; Ma, L.; Shim, Y.; Islam, S.M.; Wang, P.; Zhao, L.-D.; Wang, S.; Sun, G.; Yang, X.; et al. Efficient uranium capture by polysulfide/layered double hydroxide composites. *J. Am. Chem. Soc.* **2015**, *137*, 3670–3677. [CrossRef]
57. Zhang, S.; Chen, Y.; Li, J.; Li, Y.; Song, W.; Li, X.; Yan, L.; Yu, H. Highly efficient removal of aqueous Cu(II) and Cd(II) by hydrothermal synthesized CaAl-layered double hydroxide. *Colloid Surf. A* **2022**, *641*, 128584. [CrossRef]
58. Sun, J.; Chen, Y.; Yu, H.; Yan, L.; Du, B.; Pei, Z. Removal of Cu^{2+}, Cd^{2+} and Pb^{2+} from aqueous solutions by magnetic alginate microsphere based on Fe_3O_4/MgAl-layered double hydroxide. *J. Colloid Interface Sci.* **2018**, *532*, 474–484. [CrossRef]
59. Hou, T.; Yan, L.; Li, J.; Yang, Y.; Shan, L.; Meng, X.; Li, X.; Zhao, Y. Adsorption performance and mechanistic study of heavy metals by facile synthesized magnetic layered double oxide/carbon composite from spent adsorbent. *Chem. Eng. J.* **2020**, *384*, 123331. [CrossRef]
60. Lyu, F.; Yu, H.; Hou, T.; Yan, L.; Zhang, X.; Du, B. Efficient and fast removal of Pb^{2+} and Cd^{2+} from an aqueous solution using a chitosan/Mg-Al-layered double hydroxide nanocomposite. *J. Colloid Interface Sci.* **2019**, *539*, 184–193. [CrossRef]
61. Yang, Y.; Li, J.; Yan, T.; Zhu, R.; Yan, L.; Pei, Z. Adsorption and photocatalytic reduction of aqueous Cr(VI) by Fe_3O_4-ZnAl-layered double hydroxide/TiO_2 composites. *J. Colloid Interface Sci.* **2020**, *562*, 493–501. [CrossRef]
62. Hai, N.T.; Dong, T.N.; Giang, T.L.; Tomul, F.; Lima, E.C.; Woo, S.H.; Sarmah, A.K.; Hung, Q.N.; Phuong, T.N.; Dinh, D.N.; et al. Adsorption mechanism of hexavalent chromium onto layered double hydroxides-based adsorbents: A systematic in-depth review. *J. Hazard. Mater.* **2019**, *373*, 258–270. [CrossRef]
63. Poudel, M.B.; Awasthi, G.P.; Kim, H.J. Novel insight into the adsorption of Cr(VI) and Pb(II) ions by MOF derived Co-Al layered double hydroxide @hematite nanorods on 3D porous carbon nanofiber network. *Chem. Eng. J.* **2021**, *417*, 129312. [CrossRef]
64. Maziarz, P.; Matusik, J.; Leiviska, T. Mg/Al LDH enhances sulfate removal and clarification of AMD wastewater in precipitation processes. *Materials* **2019**, *12*, 2334. [CrossRef]
65. Yang, P.; Li, S.; Liu, C.; Liu, X. Interface-constrained layered double hydroxides for stable uranium capture in highly acidic industrial wastewater. *ACS Appl. Mater. Interfaces* **2021**, *13*, 17988–17997. [CrossRef]

66. Yang, L.; Qiao, B.; Zhang, S.; Yao, H.; Cai, Z.; Han, Y.; Li, C.; Li, Y.; Ma, S. Intercalation of salicylaldoxime into layered double hydroxide: Ultrafast and highly selective uptake of uranium from different water systems via versatile binding modes. *J. Colloid Interface Sci.* **2023**, *642*, 623–637. [CrossRef]
67. Mudhoo, A.; Paliya, S.; Goswami, P.; Singh, M.; Lofrano, G.; Carotenuto, M.; Carraturo, F.; Libralato, G.; Guida, M.; Usman, M.; et al. Fabrication, functionalization and performance of doped photocatalysts for dye degradation and mineralization: A review. *Environ. Chem. Lett.* **2020**, *18*, 1825–1903. [CrossRef]
68. Soltani, R.; Marjani, A.; Shirazian, S. A hierarchical LDH/MOF nanocomposite: Single, simultaneous and consecutive adsorption of a reactive dye and Cr(vi). *Dalton Trans.* **2020**, *49*, 5323–5335. [CrossRef]
69. Mallakpour, S.; Radfar, Z.; Feiz, M. Chitosan/tannic acid/ZnFe layered double hydroxides and mixed metal oxides nanocomposite for the adsorption of reactive dyes. *Carbohydr. Polym.* **2023**, *305*, 120528. [CrossRef]
70. Jamaly, S.; Giwa, A.; Hasan, S.W. Recent improvements in oily wastewater treatment: Progress, challenges, and future opportunities. *J. Environ. Sci.* **2015**, *37*, 15–30. [CrossRef]
71. Yaacob, N.; Sean, G.P.; Nazri, N.A.M.; Ismail, A.F.; Abidin, M.N.Z.; Subramaniam, M.N. Simultaneous oily wastewater adsorption and photodegradation by ZrO_2-TiO_2 heterojunction photocatalysts. *J. Water Process Eng.* **2021**, *39*, 101644. [CrossRef]
72. Ikhsan, S.N.W.; Yusof, N.; Aziz, F.; Ismail, A.F.; Jaafar, J.; Salleh, W.N.W.; Misdan, N. Superwetting materials for hydrophilic-oleophobic membrane in oily wastewater treatment. *J. Environ. Manag.* **2021**, *290*, 112565. [CrossRef]
73. Su, Y.; Li, Z.; Zhou, H.; Kang, S.; Zhang, Y.; Yu, C.; Wang, G. Ni/carbon aerogels derived from water induced self-assembly of Ni-MOF for adsorption and catalytic conversion of oily wastewater. *Chem. Eng. J.* **2020**, *402*, 126205. [CrossRef]
74. Wang, Y.; Wang, B.; Wang, J.; Ren, Y.; Xuan, C.; Liu, C.; Shen, C. Superhydrophobic and superoleophilic porous reduced graphene oxide/polycarbonate monoliths for high-efficiency oil/water separation. *J. Hazard. Mater.* **2018**, *344*, 849–856. [CrossRef]
75. Menino, N.M.; Salla, J.D.S.; do Nascimento, M.S.; Dallago, R.M.; Peralta, R.A.; Moreira, R.F.P.M. High-performance hydrophobic magnetic hydrotalcite for selective treatment of oily wastewater. *Environ. Technol.* **2021**, *44*, 2003441. [CrossRef]
76. Lu, Y.; Sathasivam, S.; Song, J.; Crick, C.R.; Carmalt, C.J.; Parkin, I.P. Robust self-cleaning surfaces that function when exposed to either air or oil. *Science* **2015**, *347*, 1132–1135. [CrossRef]
77. Cui, J.; Zhou, Z.; Xie, A.; Wang, Q.; Liu, S.; Lang, J.; Li, C.; Yan, Y.; Dai, J. Facile preparation of grass-like structured NiCo-LDH/PVDF composite membrane for efficient oil-water emulsion separation. *J. Membr. Sci.* **2019**, *573*, 226–233. [CrossRef]
78. Xie, A.; Cui, J.; Yang, J.; Li, C.; Wang, Y.; Dai, J. Active antifouling carbon cloth@Ni-Co LDH/Ag membrane for efficient oil/water separation. *Appl. Clay Sci.* **2021**, *211*, 106161. [CrossRef]
79. Lartey-Young, G.; Ma, L. Optimization, equilibrium, adsorption behaviour of Cu/Zn/Fe LDH and LDHBC composites towards atrazine reclamation in an aqueous environment. *Chemosphere* **2022**, *293*, 133526. [CrossRef]
80. Mourid, E.H.; Lakraimi, M.; Legrouri, A. Removal and release of the 2,4,5-trichlorophenoxyacetic acid herbicide from wastewater by layered double hydroxides. *J. Inorg. Organomet. Polym. Mater.* **2021**, *31*, 2116–2128. [CrossRef]
81. Gobelius, L.; Hedlund, J.; Duerig, W.; Troger, R.; Lilja, K.; Wiberg, K.; Ahrens, L. Per- and polyfluoroalkyl substances in swedish groundwater and surface water: Implications for environmental quality standards and drinking water guidelines. *Environ. Sci. Technol.* **2018**, *52*, 4340–4349. [CrossRef]
82. Chen, Y.; Georgi, A.; Zhang, W.; Kopinke, F.; Yan, J.; Saeidi, N.; Li, J.; Gu, M.; Chen, M. Mechanistic insights into fast adsorption of perfluoroalkyl substances on carbonate-layered double hydroxides. *J. Hazard. Mater.* **2021**, *408*, 124815. [CrossRef]
83. Min, X.; Huo, J.; Dong, Q.; Xu, S.; Wang, Y. Enhanced sorption of perfluorooctanoic acid with organically functionalized layered double hydroxide. *Chem. Eng. J.* **2022**, *446*, 137019. [CrossRef]
84. Tiwari, E.; Singh, N.; Khandelwal, N.; Monikh, F.A.; Darbha, G.K. Application of Zn/Al layered double hydroxides for the removal of nanoscale plastic debris from aqueous systems. *J. Hazard. Mater.* **2020**, *397*, 122769. [CrossRef]
85. Peng, G.; Xiang, M.; Wang, W.; Su, Z.; Liu, H.; Mao, Y.; Chen, Y.; Zhang, P. Engineering 3D graphene-like carbon-assembled layered double oxide for efficient microplastic removal in a wide pH range. *J. Hazard. Mater.* **2022**, *433*, 128672. [CrossRef]
86. Ge, L.; Shao, B.; Liang, Q.; Huang, D.; Liu, Z.; He, Q.; Wu, T.; Luo, S.; Pan, Y.; Zhao, C.; et al. Layered double hydroxide based materials applied in persulfate based advanced oxidation processes: Property, mechanism, application and perspectives. *J. Hazard. Mater.* **2022**, *424*, 127612. [CrossRef]
87. Huang, Y.; Yang, Y.; Wang, X.; Yuan, X.; Pi, N.; Yuan, H.; Liu, X.; Ni, C. Heterogeneous Fenton-like degradation of methoxychlor in water using two different FeS@hydrotalcites (LHDs) and Fe_3O_4@LHDs catalysts prepared via an in situ growth method. *Chem. Eng. J.* **2018**, *342*, 142–154. [CrossRef]
88. Ba, Y.; Hu, Z.; Bao, P.; Qiao, M.; Hua, J.; Wu, C.; Wu, J. Ferrous ions accelerate sulfide-induced abiotic dechlorination of DDT in waterlogged paddy soil and in soil solution. *J. Soil. Sediment* **2011**, *11*, 1209–1220. [CrossRef]
89. Yan, J.; Chen, Y.; Gao, W.; Chen, Y.; Qian, L.; Han, L.; Chen, M. Catalysis of hydrogen peroxide with Cu layered double hydrotalcite for the degradation of ethylbenzene. *Chemosphere* **2019**, *225*, 157–165. [CrossRef]
90. Wang, H.; Jing, M.; Wu, Y.; Chen, W.; Ran, Y. Effective degradation of phenol via Fenton reaction over CuNiFe layered double hydroxides. *J. Hazard. Mater.* **2018**, *353*, 53–61. [CrossRef]
91. Dou, L.; Zhang, H. Facile assembly of nanosheet array-like CuMgAl-layered double hydroxide/rGO nanohybrids for highly efficient reduction of 4-nitrophenol. *J. Mater. Chem. A* **2016**, *4*, 18990–19002. [CrossRef]
92. Peng, X.; Wang, M.; Hu, F.; Qiu, F.; Zhang, T.; Dai, H.; Cao, Z. Multipath fabrication of hierarchical CuAl layered double hydroxide/carbon fiber composites for the degradation of ammonia nitrogen. *J. Environ. Manag.* **2018**, *220*, 173–182. [CrossRef]

93. Costa-Serge, N.D.M.; Goncalves, R.G.L.; Ramirez-Ubillus, M.A.; Li, C.; Hammer, P.; Chiron, S.; Nogueira, R.F.P. Effect of the interlamellar anion on CuMgFe-LDH in solar photo-Fenton and Fenton-like degradation of the anticancer drug 5-fluorouracil. *Appl. Catal. B-Environ.* **2022**, *315*, 121537. [CrossRef]
94. Dhakshinamoorthy, A.; Navalon, S.; Alvaro, M.; Garcia, H. Metal nanoparticles as heterogeneous Fenton catalysts. *ChemSusChem* **2012**, *5*, 46–64. [CrossRef]
95. Wang, A.; Chen, Y.; Zheng, Z.; Wang, H.; Li, X.; Yang, Z.; Qiu, R.; Yan, K. In situ N-doped carbon-coated mulberry-like cobalt manganese oxide boosting for visible light driving photocatalytic degradation of pharmaceutical pollutants. *Chem. Eng. J.* **2021**, *411*, 128497. [CrossRef]
96. Yang, Z.; Li, X.; Huang, Y.; Chen, Y.; Wang, A.; Wang, Y.; Li, C.; Hu, Z.; Yan, K. Facile synthesis of cobalt-iron layered double hydroxides nanosheets for direct activation of peroxymonosulfate (PMS) during degradation of fluoroquinolones antibiotics. *J. Clean. Prod.* **2021**, *310*, 127584. [CrossRef]
97. Ren, Y.; Lin, L.; Ma, J.; Yang, J.; Feng, J.; Fan, Z. Sulfate radicals induced from peroxymonosulfate by magnetic ferrospinel MFe_2O_4 (M = Co, Cu, Mn, and Zn) as heterogeneous catalysts in the water. *Appl. Catal. B-Environ.* **2015**, *165*, 572–578. [CrossRef]
98. Vigneshwaran, S.; Jun, B.M.; Prabhu, S.M.; Elanchezhiyan, S.S.; Ok, Y.S.; Meenakshi, S.; Park, C.M. Enhanced sonophotocatalytic degradation of bisphenol a using bimetal sulfide-intercalated MXenes, 2D/2D nanocomposite. *Sep. Purif. Technol.* **2020**, *250*, 117178. [CrossRef]
99. Gong, C.; Chen, F.; Yang, Q.; Luo, K.; Yao, F.; Wang, S.; Wang, X.; Wu, J.; Li, X.; Wang, D.; et al. Heterogeneous activation of peroxymonosulfate by Fe-Co layered doubled hydroxide for efficient catalytic degradation of Rhoadmine B. *Chem. Eng. J.* **2017**, *321*, 222–232. [CrossRef]
100. Li, X.; Hou, T.; Yan, L.; Shan, L.; Meng, X.; Zhao, Y. Efficient degradation of tetracycline by CoFeLa-layered double hydroxides catalyzed peroxymonosulfate: Synergistic effect of radical and nonradical pathways. *J. Hazard. Mater.* **2020**, *398*, 122884. [CrossRef]
101. Deng, J.; Xiao, L.; Yuan, S.; Wang, W.; Zhan, X.; Hu, Z.-H. Activation of peroxymonosulfate by CoFeNi layered double hydroxide/graphene oxide (LDH/GO) for the degradation of gatifloxacin. *Sep. Purif. Technol.* **2021**, *255*, 117685. [CrossRef]
102. Ye, Q.; Wu, J.; Wu, P.; Wang, J.; Niu, W.; Yang, S.; Chen, M.; Rehman, S.; Zhu, N. Enhancing peroxymonosulfate activation of Fe-Al layered double hydroxide by dissolved organic matter: Performance and mechanism. *Water Res.* **2020**, *185*, 116246. [CrossRef]
103. Ma, Q.; Nengzi, L.C.; Li, B.; Wang, Z.; Liu, L.; Cheng, X. Heterogeneously catalyzed persulfate with activated carbon coated with CoFe layered double hydroxide (AC@CoFe-LDH) for the degradation of lomefloxacin. *Sep. Purif. Technol.* **2020**, *235*, 116204. [CrossRef]
104. Jawad, A.; Lang, J.; Liao, Z.; Khan, A.; Ifthikar, J.; Lv, Z.; Long, S.; Chen, Z.; Chen, Z. Activation of persulfate by CuOx@Co-LDH: A novel heterogeneous system for contaminant degradation with broad pH window and controlled leaching. *Chem. Eng. J.* **2018**, *335*, 548–559. [CrossRef]
105. Zeng, H.; Deng, L.; Zhang, H.; Zhou, C.; Shi, Z. Development of oxygen vacancies enriched CoAl hydroxide@hydroxysulfide hollow flowers for peroxymonosulfate activation: A highly efficient singlet oxygen-dominated oxidation process for sulfamethoxazole degradation. *J. Hazard. Mater.* **2020**, *400*, 123297. [CrossRef]
106. Song, A.; Yang, W.; Yang, W.; Sun, G.; Yin, X.; Gao, L.; Wang, Y.; Qin, X.; Shao, G. Facile synthesis of cobalt nanoparticles entirely encapsulated in slim nitrogen-doped carbon nanotubes as oxygen reduction catalyst. *ACS Sustain. Chem. Eng.* **2017**, *5*, 3973–3981. [CrossRef]
107. Pfeifer, V.; Jones, T.E.; Velez, J.J.V.; Arrigo, R.; Piccinin, S.; Haevecker, M.; Knop-Gericke, A.; Schloegl, R. In situ observation of reactive oxygen species forming on oxygen-evolving iridium surfaces. *Chem. Sci.* **2017**, *8*, 2143–2149. [CrossRef]
108. Zheng, R.; Zhao, C.; Xiong, J.; Teng, X.; Chen, W.; Hu, Z.; Chen, Z. Construction of a hierarchically structured, NiCo-Cu-based trifunctional electrocatalyst for efficient overall water splitting and 5-hydroxymethylfurfural oxidation. *Sustain. Energy Fuels* **2021**, *5*, 4023–4031. [CrossRef]
109. Yang, L.; Liu, Z.; Zhu, S.; Feng, L.; Xing, W. Ni-based layered double hydroxide catalysts for oxygen evolution reaction. *Mater. Today Phys.* **2021**, *16*, 100292. [CrossRef]
110. Liu, X.; Fan, X.; Huang, H.; Lin, H.; Gao, J. Electronic modulation of oxygen evolution on metal doped NiFe layered double hydroxides. *J. Colloid Interface Sci.* **2021**, *587*, 385–392. [CrossRef]
111. Ansari, S.M.; Khan, M.Z.; Anwar, H.; Ikram, M.; Sarfraz, Z.; Alam, N.; Khan, Y. Tungsten oxide-reduced graphene oxide composites for photoelectrochemical water splitting. *Arab. J. Sci. Eng.* **2021**, *46*, 813–825. [CrossRef]
112. Xu, J.; Gai, S.; He, F.; Niu, N.; Gao, P.; Chen, Y.; Yang, P. A sandwich-type three-dimensional layered double hydroxide nanosheet array/graphene composite: Fabrication and high supercapacitor performance. *J. Mater. Chem. A* **2014**, *2*, 1022–1031. [CrossRef]
113. Duan, M.; Qiu, M.; Sun, S.; Guo, X.; Liu, Y.; Zheng, X.; Cao, F.; Kong, Q.; Zhang, J. Intercalating assembly of NiFe LDH nanosheets/CNTs composite as high-performance electrocatalyst for oxygen evolution reaction. *Appl. Clay Sci.* **2022**, *216*, 106360. [CrossRef]

Disclaimer/Publisher's Note: The statements, opinions and data contained in all publications are solely those of the individual author(s) and contributor(s) and not of MDPI and/or the editor(s). MDPI and/or the editor(s) disclaim responsibility for any injury to people or property resulting from any ideas, methods, instructions or products referred to in the content.

materials

MDPI

Review

Current Trends in Spent Portable Lithium Battery Recycling

Zita Takacova *, Dusan Orac, Jakub Klimko and Andrea Miskufova

Institute of Recycling Technologies, Faculty of Materials, Metallurgy and Recycling, Technical University of Kosice, Letna 9, 04200 Kosice, Slovakia; dusan.orac@tuke.sk (D.O.); jakub.klimko@tuke.sk (J.K.); andrea.miskufova@tuke.sk (A.M.)
* Correspondence: zita.takacova@tuke.sk; Tel.: +421-556022493

Abstract: This paper provides an overview of the current state of the field in spent portable lithium battery recycling at both the research and industrial scales. The possibilities of spent portable lithium battery processing involving pre-treatment (manual dismantling, discharging, thermal and mechanical-physical pre-treatment), pyrometallurgical processes (smelting, roasting), hydrometallurgical processes (leaching followed by recovery of metals from the leachates) and a combination of the above are described. The main metal-bearing component of interest is the active mass or cathode active material that is released and concentrated by mechanical-physical pre-treatment procedures. The metals of interest contained in the active mass include cobalt, lithium, manganese and nickel. In addition to these metals, aluminum, iron and other non-metallic materials, especially carbon, can also be obtained from the spent portable lithium batteries. The work describes a detailed analysis of the current state of research on spent lithium battery recycling. The paper presents the conditions, procedures, advantages and disadvantages of the techniques being developed. Moreover, a summary of existing industrial plants that are focused on spent lithium battery recycling is included in this paper.

Keywords: spent portable lithium battery; recycling; pyrometallurgy; hydrometallurgy

Citation: Takacova, Z.; Orac, D.; Klimko, J.; Miskufova, A. Current Trends in Spent Portable Lithium Battery Recycling. *Materials* **2023**, *16*, 4264. https://doi.org/10.3390/ma16124264

Academic Editor: Elza Bontempi

Received: 5 May 2023
Revised: 31 May 2023
Accepted: 7 June 2023
Published: 8 June 2023

Copyright: © 2023 by the authors. Licensee MDPI, Basel, Switzerland. This article is an open access article distributed under the terms and conditions of the Creative Commons Attribution (CC BY) license (https://creativecommons.org/licenses/by/4.0/).

1. Introduction

In the European Union (EU), the production of metals from primary raw materials is confronted with a number of problems, including the depletion or complete absence of primary raw material resources and the associated raw material dependency, high extraction costs, low metal content of primary raw materials, etc. As a consequence, EU raw materials policy has started to focus on promoting the supply of own-sourced raw materials and increasing resource efficiency and recycling rates [1]. The main objective is to reduce the dependence of EU countries on imports of raw materials. In 2010, the EU established a list of critical raw materials in terms of their supply risk and economic importance, which is reviewed every three years [1,2]. The current list of critical raw materials from the year 2020 includes 30 critical raw materials instead of the original 14 from the year 2010 [3]. Among them are also the components present in lithium batteries, such as cobalt, lithium and graphite. Material recycling of spent lithium cells is essential in the effort:

- to obtain a valuable secondary source of the present materials, mainly metals, whose content in spent lithium cells often exceeds their content in primary raw materials;
- to save natural resources and energy;
- obtain saleable products with high added value;
- to achieve self-sufficiency in raw materials, etc.

Material recycling is also important in the development of a circular economy. Implementing the circular economy principle will close the life cycle of spent lithium cells, which consists of several phases, including production, consumption and recycling of spent lithium batteries.

The amount of spent lithium batteries (LiBs) is growing every year as a result of the boom in IT and telecommunication technologies and the expansion of electromobility supported by EU. The current increase in the consumption of lithium cells will be reflected in a rapid increase in the amount of spent lithium cells intended for recycling.

By 2030, the amount of spent LiBs in the European waste market is expected to increase 5–6 times compared with that in 2022 [1]. Currently, mainly portable and industrial LiBs are available on the waste market, but LiBs from electric vehicles (EVs) and hybrid electric vehicles (HEVs) are expected to start increasing rapidly in the near future. However, these batteries have further uses as energy storage sources, which will delay the need for their material recycling for some time. In the EU, a new regulation is in the process of being approved which will replace the existing Directive 2006/66/EC. The new regulation will, for example, increase the minimum limits for the collection and recycling efficiency of spent electrochemical cells and introduce limits for the material recovery of specific metals contained in them, such as cobalt and lithium.

Currently in the EU, the recycling rate of metals present in spent portable LiBs is 22% for cobalt and less than 1% for Li [3], indicating a large potential for increasing. The price of these metals and their content in portable LiBs also play in favor of recycling. While the cobalt content in primary raw materials tends to be very low (0.06–0.7%), the cobalt content in the active mass of portable LiBs can be up to about 20 wt.%. The lithium content in ores and in portable LiBs is approximately the same (1–5%) [4].

Lithium batteries can be divided into portable, automotive and industrial batteries depending on the point of application. According to Directive 2006/66/EC, portable batteries are defined as airtight, hermetically closed cells that can be hand-carried, except for automotive and industrial batteries [5]. The new regulation will introduce a weight limit of 5 kg in order to differentiate portable batteries and accumulators from industrial ones [2]. In general, the recycling of spent lithium batteries can be divided into pre-treatment processes, pyrometallurgical processes and hydrometallurgical processes. These processes are most often combined in order to achieve comprehensive treatment, economic and environmental requirements and maximum recovery of the present components.

Among the pre-treatment procedures can be included: manual dismantling, discharge of residual voltages, mechanical-physical processing and thermal processing.

Pyrometallurgical processing methods include roasting and melting in different types of furnaces.

Hydrometallurgical processes are an alternative or a complement to another type of treatment for spent portable LiBs. The first step consists of leaching, followed by refining of the leachate and finally extraction of metals from the leachate. Spent portable LiBs enter the leaching process after mechanical-physical or thermal pre-treatment. Hydrometallurgical treatment is most often carried out on the active mass of spent portable LiBs. Active mass is a mixture of cathode and anode active material containing the main metals of interest such as cobalt; lithium; to a minor extent, nickel; manganese; and also containing significant amounts of graphite from the anode material. The active mass usually contains minor amounts of copper, aluminum and iron, which represent electrode residues and which could not be removed by available mechanical-physical methods.

For the extraction of cobalt and lithium from the active mass, acid leaching is mainly applied with the following leaching reagents: sulfuric acid, hydrochloric acid, nitric acid, various organic acids, etc., most commonly in the temperature range 25–100 °C. At the same time, nickel and manganese are mostly leached. In the case of cobalt, reductive leaching is appropriate.

Refining of the obtained leachate is carried out if the leachate has also received the accompanying metals and other impurities. The valuable metals, especially cobalt and lithium, and to a minor extent nickel and manganese, may be extracted from the solution:

- by precipitation in the form of sparingly soluble compounds;
- by cementation in the form of a cementation precipitate;
- by solvent extraction;
- by electrowinning.

Ion exchange is also an option. Electrochemical processes are another method for obtaining metals from active mass. The chosen technology is mainly based on the composition of the leachate, the content of the interested metals and the presence of accompanying metals and other impurities in the leachate.

The aim of this paper is to summarize the available information and to provide an overview of current technologies and trends in the recycling of spent portable LiBs in the research and at a practical scale, to describe the advantages and disadvantages of proposed or preferred approaches.

2. Characterization of Portable LiBs

Portable LiBs can be divided into primary non-rechargeable and secondary rechargeable batteries. Compared with portable zinc or alkaline batteries, portable non-rechargeable LiBs have higher specific energy, high nominal voltage, high current density, long discharge cycle, light weight, etc. They are used in various electronic applications, e.g., in medical devices such as hearing aids [6–8].

Primary LiBs consist of a lithium metal anode and a cathode, which can be composed of several materials, e.g., MnO_2, $SOCl_2$ and FeS_2. Li-MnO_2 batteries are the most common type of portable non-rechargeable LiBs. They are produced in cylindrical shapes in different sizes and as buttons or coin cells. Portable Li-$SOCl_2$ batteries contain a liquid cathode, which consists of a porous carbon current collector filled with thionyl chloride [6]. Furthermore, primary LiBs contain iron, aluminum and plastics coming from the cover, organic electrolytes containing lithium salts, polypropylene-based separators, carbon as a catalyst and possibly others. The composition of portable primary LiBs of cylindrical and button shape from selected producers is given in Table 1.

Table 1. Composition of portable primary LiBs.

Producer	Saft [9]	BiPOWER [10]	Duracell [11]	Ultralife [12]
Type	Li–$SOCl_2$ cylindrical	Li–$SOCl_2$ cylindrical, prismatic	Li–MnO_2 cylindrical	Li–MnO_2 cylindrical, button
Component	Content [%]			
Li	3.5–5	≤5	1–5	3–4
$SOCl_2$ or MnO_2	40–46	≤47	15–45	40–45
$AlCl_3$, other Li salts	1–5	≤5	0–5	1
C	3–4	≤5	0–5	
Ethylene glycol dimethyl ether			5–10	3–4
Propylene Carbonate (PC), Ethylene carbonate (EC)			0–10	4–5
Tetrahydrofuran				5–9

Recently, primary lithium batteries often replace alkaline batteries in cameras, children's toys, watches and other electronic devices due to their advantageous properties. Their use is mainly limited by their higher cost. The disadvantage of primary LiBs is the risk of explosion and fire in case of improper handling, dismantling and recycling, which is mainly related to the high reactivity of metallic lithium and the content of flammable organic electrolytes.

Portable rechargeable LiBs have a wider application than non-rechargeable ones and are used in mobile phones, laptops, tablets, electronic toys, etc. The advantages of portable rechargeable LiBs can be listed as long life and charging cycle, high specific energy and energy density, wide operating temperature range, no memory effect, long shelf life and more. The main disadvantages include relatively high cost, cell degradation at high temperatures, cell overcharging in improper use, which can lead to overheating and degradation of the components, followed by combustion and even explosion [7,8].

Portable secondary LiBs contain lithium, mostly in the chemical compounds. The cathode active material in portable secondary LiBs tends to be mainly $LiMeO_2$-based metal oxides deposited on an aluminum foil cathode, where Me = Co, Ni, Mn and their combination. The anode active material consists of carbon or graphite deposited on a copper foil-anode. Between the anode and the cathode there is a porous separator, most often made of polypropylene, which serves as an electric insulator [8].

The most widely used cathode active material for portable secondary LiBs so far is $LiCoO_2$ (LCO), due to its wide application temperature range (~10–100 °C), high specific and bulk capacitance, low self-discharge and the longest lifetime among currently available materials [13,14]. $LiMn_2O_4$ (LMO) represents a cheaper alternative to cathode active material; it is used in a variety of applications such as drills and other power tools, medical devices, e-bikes and also EVs [15,16].

Recently, the use of $LiNiMnCoO_2$ (NMC) with reduced Co content compared with LCO, which partially replaces Ni and Mn, has been increasing. NMC is one of the most successful formulas for cathode material and is becoming a popular material for electric tools, e-bikes and EVs as well [17]. Furthermore, $LiMePO_4$-based compounds, where Me = Fe, Co, Mn, Ni, are also used as cathode active material. $LiFePO_4$ (LFP) is mainly used due to its favorable price, good thermal stability and large specific capacity. However, the disadvantage is its low conductivity [18–20].

The electrolytes include ethylene carbonate, propylene carbonate, dimethyl carbonate and dimethyl sulfoxide, similar to primary LiBs. Li salts are dissolved in organic solvents, e.g., $LiClO_4$, LiF and $LiPF_6$ [17,21]. To improve the conductivity, carbon or acetylene black is added to the cathode material [18,22]. The material composition of portable secondary LiBs is shown in Table 2.

Table 2. Material composition of portable secondary LiBs [8].

Component of Portable Secondary LiBs	Material	Content [%]
Cover	Steel, aluminum	20–25
Cathode active material	LCO, NMC, NCA, LFP, LMO	25–35
Anode active material	graphite	14–19
Electrolyte	$LiPF_6$ dissolved in PC, EC, dimethyl carbonate, diethyl carbonate	10–15
Cathode	Aluminum	5–7
Anode	Copper	5–9
Separator	PP, polyethylene (PE)	1–4
Additives	Carbon black, silicone, etc.	unspecified

3. Possibilities of Spent Portable LiB Recycling

Recycling processes include pre-treatment procedures, pyrometallurgical, hydrometallurgical and combined processes.

3.1. Pre-Treatment

Pre-treatment processes for spent portable LiB recycling include manual dismantling, residual voltage discharge, mechanical-physical treatment and thermal treatment to isolate and concentrate the material suitable for further processing.

Figure 1 shows a principle scheme for the pre-treatment of spent portable LiBs.

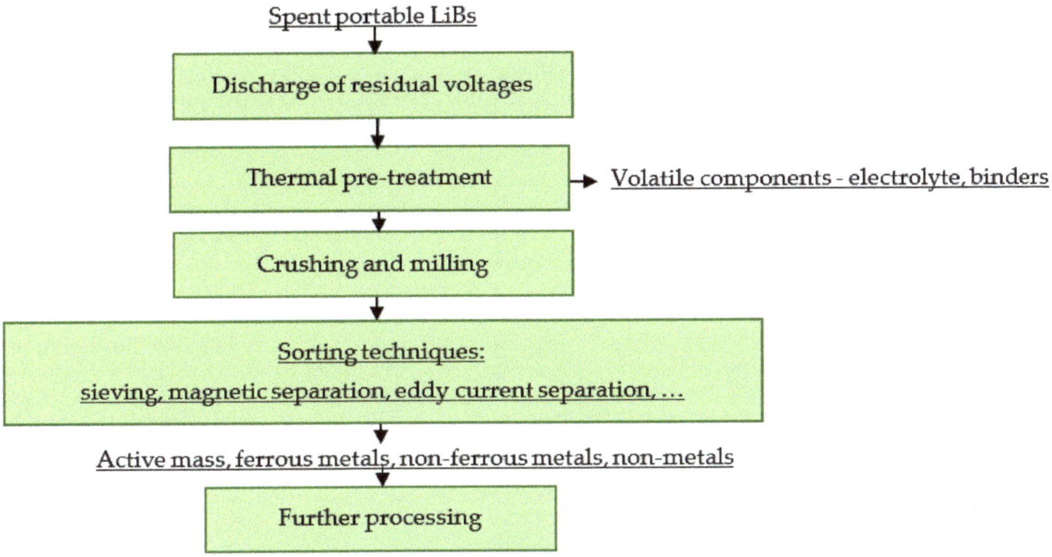

Figure 1. Principle scheme for pre-treatment of spent portable LiBs.

Manual dismantling is carried out in order to release and isolate the individual components. In industry, it is particularly applicable for the processing of oversized lithium cells, e.g., from EVs. The discharge of the residual voltage of spent portable LiBs may be carried out by thermal, wet or cryogenic processes, in particular to avoid short circuits and subsequent burning of the present electrolytes and binders. The aim is also to reduce the reactivity of the metallic lithium present in the primary LiBs. The mechanical-physical processing of spent portable LiBs mainly involves crushing, grinding and subsequent sorting. The output of the mechanical-physical processing is an active mass, metallic electrodes, aluminum packaging in the non-ferrous fraction, steel packaging in the ferrous fraction and separators and plastic packaging in the non-metallic fraction.

In current research, various techniques and combinations of crushing and grinding are being developed and validated in order to efficiently disintegrate the spent portable LiBs and release the maximum amount of the individual components, especially the active mass. The aim is also to avoid undesirable phenomena during crushing or grinding, e.g., encapsulation of the active mass inside the shell [23]. An example of the efficient recovery of active mass from spent portable LiBs is the mechanical-physical pretreatment procedure using a double-rotor crusher followed by a hammer crusher with a gain of active mass in the <0.71 mm fraction, which accounts for 40–57% of the input sample weight [24,25]. The active mass from portable LCO-based LiBs can contain more than 20% of Co, about 4% of Li and a minor fraction of Mn, Ni, Al, Fe and Cu (<1%). Al, Fe and Cu represent impurities in the active mass as residual metal electrodes and packaging. Graphite contributes to ca. 50% of the active mass [24].

In addition to mechanical-physical pre-treatment, thermal pre-treatment is frequently used, e.g., heating to evaporate volatile components, pyrolysis, calcination and others. Potentially hazardous gases released during thermal treatment have to be captured and cleaned in flue gas cleaning installations. The captured evaporated electrolyte can be recycled theoretically.

The aim of pyrolysis of the spent LiBs is to remove separators, binders and other organic components or changes in the structure in order to obtain material with a suitable composition for further processing. Calcination is applied for a similar purpose to pyrolysis or to modify the composition of the final product.

One way to effectively separate the cathode material from the aluminum cathode is to use a variety of organic reagents that dissolve the present binders. Organic solvents for dissolving binders are usually applied after manual dismantling or mechanical-physical treatment. Perspective organic solvents are n-methylpyrrolidone, dimethylacetamide, acetone and trifluoroacetate (TFA) [26–30]. A PVDF-based binder can also be dissolved using dimethylformamide; however, a polytetrafluoroethylene (PTFE)-based binder does not dissolve in the above medium [31].

The cathode active material can be separated from the Al cathode via leaching with NaOH as well [28,32] or by wet ultrasonic separation combined with crushing. Carbon can also be separated from the copper anode in this way [33,34].

3.2. *Pyrometallurgical Processing*

Pyrometallurgical treatment methods for spent portable LiBs include roasting and melting in different types of furnaces, e.g., in a shaft furnace, rotary kiln or converter.

Figure 2 shows a principle scheme for the pyrometallurgical processing of spent portable LiBs.

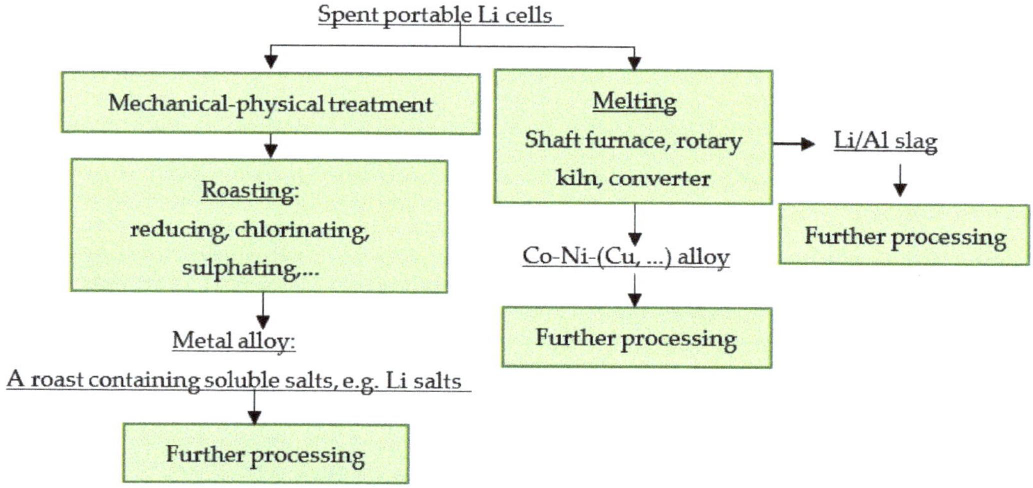

Figure 2. Principle scheme for pyrometallurgical processing of spent portable LiBs.

3.2.1. Roasting

During roasting of spent portable LiBs, various insoluble metal compounds convert into products that are soluble, e.g., even in water. In this way, the use of the aggressive leaching reagents required in the conventional hydrometallurgical process is eliminated. Prior to roasting, it is necessary to carry out a mechanical-physical pre-treatment. Depending on the used reagent, the roasting processes can be divided into chlorination roasting, sulfating roasting, nitration roasting, reduction roasting and possibly others. Chlorination roasting involves heating the feed material together with a chlorinating agent such as $HCl_{(g)}$, NH_4Cl, NaCl or $Cl_{2(g)}$ to produce readily soluble chlorides of the cobalt and lithium [35]. In sulfating roasting, the following sulfating reagents—$SO_{2(g)}$, $MgSO_4$, NH_4SO_4, $NaHSO_4.H_2O$ and Na_2SO_4—have been confirmed to produce readily soluble Li_2SO_4 [36].

In reduction roasting, the active mass is heated with a reducing agent, which can be carbon, charcoal or coke. The carbon or graphite present in spent LiBs is often used for this purpose [37]. New phases such as Li_2CO_3, CoO and NiO are formed during roasting at temperatures of 600–900 °C. However, the use of higher temperatures (ca. 700 °C) results in graphite burning out, thus reducing the reduction conditions [38]. The use of a vacuum at higher temperatures (ca. 900 °C) is preferred, where the carbothermal reduction of the present oxides to the metallic form occurs. The carbothermal reduction during roasting of the cathode material containing $LiCoO_2$, the decomposition of $LiCoO_2$ and the formation of Li_2CO_3 proceeds according to reactions (1)–(3) [39].

$$LiCoO_2 + 3C = Li_2CO_3 + 4Co + CO_2, \quad (1)$$

$$LiCoO_2 = Li_2O + CoO + O_2, \quad (2)$$

$$Li_2O + CO_2 = Li_2CO_3. \quad (3)$$

Waste biomass can be used as an alternative reducing agent for the carbothermal reduction of metals during roasting [40].

3.2.2. Melting

The melting of spent portable LiBs consists of two steps. First, they are heated to a lower temperature in a furnace in order to discharge residual voltages and evaporate the electrolyte. In the second step, metal alloys are formed at high temperatures in a reducing carbon atmosphere and other organic materials such as plastics and separators are burned off. Carbon from the anode and aluminum from the cathode act as reducing agents. In the pyrometallurgical process of spent LiBs, cobalt and nickel from the cathode active material and copper from the anode are recovered most efficiently. The obtained alloy usually proceeds to hydrometallurgical processing. Lithium passes to the slag, from where it can be recovered with additional processing [41].

During melting of spent LiBs, the high viscosity of the slag and high melting temperature are often a problem, which causes metal particles to drift into the slag during the casting process, resulting in losses. An essential tool for reducing the losses and for the efficient transfer of lithium and aluminum into the slag is an optimal slag system and multi-stage slag casting. CaO and SiO_2 have proven to be useful as slag-forming additives. Al_2O_3, which comes from spent portable LiBs as an oxidized cathode, is also a suitable slag-forming additive. The resulting alloy mainly consists of a solid solution of Fe-Co-Cu-Ni, including a small amount of matte [42].

3.3. Hydrometallurgical Processing

Hydrometallurgical processing is another alternative or complement to the treatment of spent portable LiBs. They enter the leaching process after mechanical-physical or even thermal pre-treatment. Hydrometallurgical treatment of spent portable LiBs is most often carried out on the active mass. In the case of leaching of the active mass to recover cobalt, acid leaching with the addition of a reducing agent predominates. The interesting metals, which are mainly cobalt and lithium and, to a minor extent, nickel and manganese, can be extracted from solution via precipitation, cementation, solvent extraction, electrowinning, ion exchange and their combination.

The type of technology is based mainly on the composition of the leachate, the content of the suitable metals and the presence of accompanying metals and other impurities in the leachate.

The principle scheme for hydrometallurgical processing of spent portable LiBs is shown in Figure 3.

3.3.1. Leaching

In the research of leaching valuable metal from active mass or cathode active material, various leaching systems are studied, searching for optimal parameters (temperature, concentration of reagents, L:S ratio, leaching time, etc.) in order to achieve the maximum extraction of the present metals and to explain the theoretical basis of the processes. Sulfuric acid, as the most common leaching agent in terms of availability and cost, can be considered as a suitable leaching agent for leaching the metals from the active mass of spent LiBs.

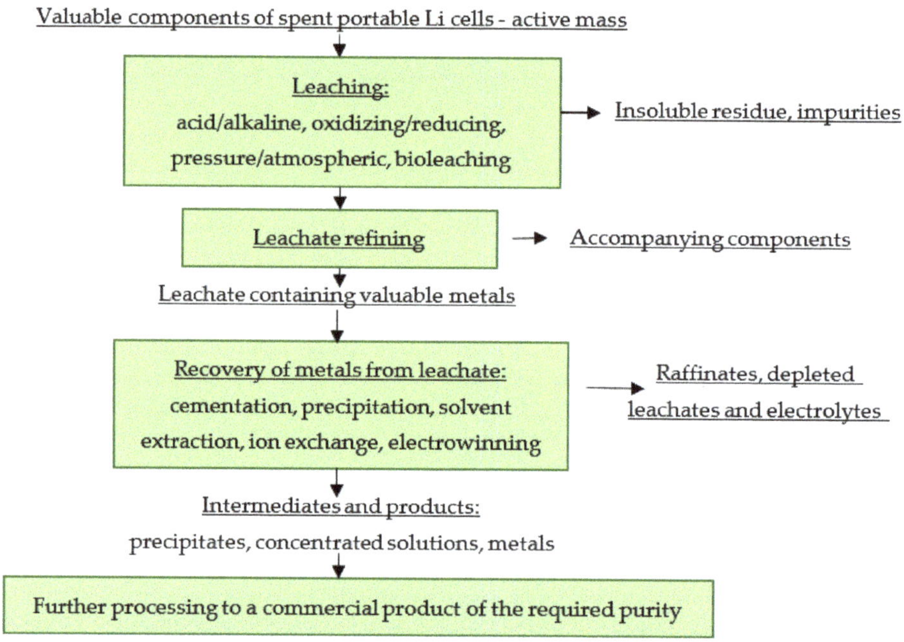

Figure 3. Principle scheme for hydrometallurgical processing of spent portable LiBs.

For efficient leaching of the metals, hydrogen peroxide is often added to the sulfuric acid in varying volumes.

During leaching of a $LiCoO_2$-based active mass in sulfuric acid without and with the addition of hydrogen peroxide, the following reactions can be expected (4) and (5) [24,43]

$$LiCoO_2 + 1.5\ H_2SO_4 = CoSO_4 + 0.5\ Li_2SO_4 + 0.25\ O_2 + 1.5\ H_2O,\ \Delta G^0{}_{293} = -612.2\ kJ, \quad (4)$$

$$LiCoO_2 + 1.5\ H_2SO_4 + 1.5\ H_2O_2 = CoSO_4 + 0.5\ Li_2SO_4 + O_2 + 3\ H_2O,\ \Delta G^0{}_{293} = -617.3\ kJ. \quad (5)$$

In both cases, the value of $\Delta G^0{}_{293}$ is negative, indicating that the progress of both reactions is thermodynamically possible. Hydrogen peroxide acts as a reducing agent to reduce Co^{3+} from the $LiCoO_2$ structure to Co^{2+}, which is not so resistant to leaching [43].

Using sulfuric acid without the addition of hydrogen peroxide, maximum cobalt yields of 50–60% can be achieved from the active mass of the spent portable LiBs even at higher acid concentrations and at high temperatures (80–100 °C). In the $H_2SO_4 + H_2O_2$, cobalt yields can reach up to 100%, depending on the concentration of the leaching agent, the amount of addition of reducing agent, the temperature, the slurry density, the leaching time, etc. In addition, lithium (95–100%) and other metals present, such as nickel, manganese, copper, aluminum and others, are also leached [44–49].

Sulfuric acid with hydrogen peroxide can also be used for metal leaching from the reduction-roasted active mass of spent LiBs with high efficiency [38]. However, the addition of hydrogen peroxide has some limitations. The problem is leaching at higher temperatures (80–100 °C), where cobalt yields decrease [50]. The decrease in cobalt yield with increasing temperature may be due to the decomposition of hydrogen peroxide, which theoretically proceeds according to reaction (6) [4,50].

$$2H_2O_2 = O_2 + 2H_2O,\ \Delta G^0{}_{293} = -116.41\ kJ. \quad (6)$$

For these reasons, a temperature between 40 °C and 60 °C can be considered as the optimum temperature for leaching Co with sulfuric acid and hydrogen peroxide.

Another reducing agent that has confirmed its potential is metallic copper, e.g., in the ratio Cu: $LiCoO_2$ = 1:1. Copper also serves as a catalyst. It is also suitable to add chloride ions to this system to stabilize Cu^+ in the form of complexes. The dissolution of Cu occurs through an intermediate step in which solid CuCl precipitates on the Cu surface, potentially inhibiting the leaching process of the valuable metals [51].

Iron scrap can also be considered as a suitable, economical and efficient reducing agent for leaching Co, Ni and Li from the active mass in sulfuric acid. The reduction is provided by the release of Fe^{2+} from its dissolution. The use of ferrous scrap significantly increases the yields of cobalt and nickel, while it has a negligible effect on the leaching of lithium. The added iron is subsequently most efficiently removed from the leachate via precipitation at pH = 4, without co-precipitation of other metals. After iron removal, cobalt and nickel can be recovered from the leachate via cementation with zinc scrap [52].

Sulfuric acid can also be used as a leaching agent for leaching lithium slag from the melting of spent LiBs. However, in addition to lithium, the use of sulfuric acid also results in the simultaneous leaching of silica, which causes the formation of gels, making it difficult to filter and recover the lithium from the leachate. An alternative in slag processing is dry digestion using concentrated sulfuric acid, which avoids the formation of gels while maintaining a high lithium yield [53].

A promising approach for the extraction of metals from the active mass of spent LiBs is the leaching and simultaneous precipitation of the present metals using oxalic acid. Under suitable conditions (e.g., 0.25 M oxalic acid, 10% slurry density, addition of 0.5 vol.% H_2O_2, temperature 80 °C, leaching time 90 min.), 100% of copper and lithium can be selectively converted to a leachate. At the same time, oxalates of cobalt, nickel and manganese are formed. Diffusion of oxalic acid to $LiCoO_2$ occurs through a layer of solid reaction product which forms on the surface; this step is probably rate limiting. The formed oxalates can be subsequently leached in, e.g., sulfuric acid or directly used as precursors for the synthesis of new cathode material [54–56].

Another suitable leaching agent for metals extraction from the active mass of the spent LiBs is hydrochloric acid [43,57–59]. The leaching of Co and Li from LCO-based active mass using HCl is represented by the following chemical reaction (7) [43]:

$$2LiCoO_2 + 8HCl = 2LiCl + 2CoCl_2 + 2H_2O + Cl_2,\ \Delta G^0{}_{293} = -337.16\ kJ. \quad (7)$$

Practically, almost 100% of Co and Li can be leached from the active mass using 2–4 M HCl at 60–80 °C in a relatively short time (0.5–1 h) [24,58,59]. A comparison of the effect of H_2SO_4 and HCl shows that maximum yields of Co and Li can be achieved using HCl without the addition of any reducing agents, whereas a reducing agent is required for 100% leaching of the valuable metals using sulfuric acid. The cobalt extraction in both H_2SO_4 and HCl takes place in two time periods. In the case of the leaching in H_2SO_4, in the first period, i.e., 15–20 min from the start of leaching, the process is controlled by the rate of the chemical reaction, $Ea_{(Co)}$ = 43–48 $kJ \cdot mol^{-1}$. In the second time period, the process changes to a diffusion-controlled rate, as demonstrated by the apparent activation energy, $Ea_{(Co)}$ = 3–3.5 $kJ \cdot mol^{-1}$. In the case of hydrochloric acid, the Co extraction in the first time period is controlled by the rate of the chemical reaction, $Ea_{(Co)}$ = 40–44 $kJ \cdot mol^{-1}$. In the second time period, the process has a mixed mechanism, $Ea_{(Co)}$ = 20–26 $kJ \cdot mol^{-1}$. The lithium extraction is diffusion controlled in both time periods, or proceeds in a mixed mode, $Ea_{(Li)}$ = 2–20 $kJ \cdot mol^{-1}$, in both leaching agents. The extraction of cobalt and lithium is influenced by the internal structure of the LCO-based active mass, and the extraction of cobalt depends on the extraction of lithium from the $LiCoO_2$ structure [24].

Other leaching reagents such as Na_2CO_3, ammonia reagents and H_3PO_4 are also suitable for leaching the active mass from the spent LiBs. Na_2CO_3 has proven to be a suitable leaching agent for the leaching of nearly 99% of Li from the roasted active mass of the spent LiBs. Ammonia leaching of roasted active mass (e.g., $NH_3.H_2O + (NH_4)_2CO_3 + Na_2SO_3$ leaching system) can leach metals such as cobalt and nickel in nearly 100% yields after lithium leaching, leaving manganese in the insoluble residue [60]. Ammonia leaching can also be carried out at higher pressure using various reducing agents—$Na_2S_2O_3$, Na_2HPO_3 and $(NH_4)_2SO_3$. Using $(NH_4)_2SO_3$, up to 100% recovery of Co, 98% recovery of Ni and more than 90% recovery of Li can be achieved [61]. H_3PO_4 with hydrogen peroxide can also provide a suitable leaching medium for Co and Li leaching from the active mass at high temperatures (90 °C) [62].

Among the intensification methods for metals extraction from the active mass of the spent LiBs can be included ultrasonic leaching, which aims to achieve high metal yields in a short time. Ultrasonic leaching (360 W) of the cathode active material using H_2SO_4 with H_2O_2 can achieve more than 90% of Co and Li yields in a short time of 30 min. Ultrasonic leaching provides a higher leaching rate and increased process efficiency compared with conventional leaching under otherwise identical experimental conditions [63].

In recent years, several organic acids such as citric acid [64,65], tartaric acid [66], lactic acid [67], adipic acid [68] and DL-malic acid [69] have been studied as leaching agents for Co and Li extraction in the research. The use of organic reducing agents, e.g., ascorbic acid [70,71], glucose and fructose [57,72] is also being investigated. The advantage of using reducing saccharides as reducing agents is that wastes from the sugar industry or agriculture can be used as saccharide sources [57].

The use of such leaching systems is mainly aimed at achieving higher process selectivity, lower toxicity and lower environmental impact. The disadvantage is the high cost compared with conventional mineral acids. High metal yields can be achieved in these systems, but the disadvantage, for instance in the case of adipic acid and glycine, is the long leaching times (3–6 h) [69,73].

In addition to the interesting metals, a significant part of the active matter consists of graphite, which remains in the insoluble residue during leaching. There are also studies describing the possibility of recovering graphite from the anode during the processing of LiBs. The most effective seems to be its separation during leaching, when it is concentrated in the insoluble residue [74].

3.3.2. Refining and Recovery of Metals from Leachates

The leachates from spent portable LiBs intended for further processing mainly contain cobalt—approximately five times more cobalt than lithium, magnesium and nickel. The other accompanying metals (iron, aluminum, copper) are only present in minor concentrations. The processes used to extract the valuable metals from the leachate include cementation, solvent extraction, precipitation, electrochemical processes, ion exchange and combinations of these.

Cementation

During the cementation of valuable metals, Al, Zn and Fe can be used as a cementing agent [75]. In practice, zinc powder is most commonly used. The disadvantage is that the obtained cementation precipitate is a mixture of the several present metals. In addition to cobalt, the present nickel, copper and other metals are also cemented. Subsequently, such a cementation precipitate has to be refined. In contrast, cementation as a method for recovering metals from leachates cannot be used for recovering lithium because of its high electronegativity, which also provides the possibility of mutual separation of, e.g., cobalt and lithium.

Precipitation

Using precipitation to produce sparingly/slightly soluble or insoluble compounds, almost all present metals can be recovered from the leachate after leaching of the active mass of spent LiBs, depending on the precipitating agent and other parameters. Examples of compounds of the valuable metals, the Ksp values and their solubility in water at 25 °C are given in Table 3.

Table 3. Ksp and solubility of compounds of the valuable metals at 25 °C [76,77].

Valuable Metal	Compounds	K_{sp}	Solubility in H_2O [g/L]	
Co	$Co_3(PO_4)_2$	2.05×10^{-35}	1.66×10^{-5}	Insoluble *
	$Co(OH)_2$	5.95×10^{-15}	1.06×10^{-3}	Insoluble
	$CoCO_3$	1.00×10^{-10}	1.19×10^{-3}	Insoluble
	$Co_2C_2O_4 \cdot 2H_2O$ (20 °C)	2.70×10^{-9}	2.12×10^{-1}	Insoluble
Li	Li_3PO_4	2.37×10^{-11}	9.94×10^{-1}	Slightly soluble *
	Li_2CO_3	8.15×10^{-4}	4.35	Slightly soluble
Mn	$Mn(OH)_2$	2.00×10^{-13}	3.28×10^{-3}	Insoluble
	$MnCO_3$	2.24×10^{-11}	5.44×10^{-4}	Insoluble
Ni	$Ni(OH)_2$	5.48×10^{-16}	4.78×10^{-4}	Insoluble
	$NiCO_3$	1.42×10^{-7}	4.47×10^{-2}	Insoluble

* insoluble $\leq$ 0.1 g/L, slightly soluble = 1–10 g/L [78].

The difficulty with precipitation is that, in addition to the required metal, other present metals also pass into the precipitate. High process selectivity can be achieved via multi-step precipitation with an appropriate choice of precipitating reagents and a suitably chosen pH. Cobalt can be efficiently precipitated from sulfate leachates after leaching of the active mass by several proven reagents such as NaOH, KOH, Na_2CO_3, oxalic acid and ammonium oxalate [79–81]. The behavior of the metals during precipitation can be predicted through speciation diagrams [44].

Oxalic acid has been confirmed as a suitable precipitating agent for cobalt, but it also precipitates copper, nickel and possibly other metals [79]. Direct use of oxalic acid is appropriate if the leachate contains no or minimal amounts of copper [82]. If copper is present in the leachate, it is preferable to precipitate it first. Sodium sulfide is suitable for precipitating copper with high efficiency (up to 99.9%) to form copper sulfide at a molar ratio of $Na_2S:Cu^{2+}$ = 3:1, at 25 °C, with a precipitation time of 30 min [44].

In the case of a targeted preparation of a mixed precipitate, oxalic acid may be the sole used reagent. The advantage in such a recycling process is the absence of complicated metal separation and low processing costs. The mixed precipitate of cobalt, manganese and nickel oxalates can serve as a precursor in the production of new cathode materials [83].

Dimethylglyoxime may be considered as a suitable nickel precipitating agent, and nickel precipitation may be included at the beginning of the process or preceded by manganese precipitation using $KMnO_4$. Nickel precipitation by dimethylglyoxime is represented by chemical reaction (8) [84–86]

$$Ni^{2+} + 2C_4H_8N_2O_2 + 2OH^- = [Ni\text{-}(C_4H_7N_2O_2)_2] + 2H_2O. \quad (8)$$

Nickel and manganese can also be effectively precipitated using Na_2CO_3 at a suitable pH. Manganese is selectively precipitated as $MnCO_3$ at pH = 7.5 and nickel as $NiCO_3$ at pH = 9. At the highest pH = 12–14, lithium is precipitated as Li_2CO_3 by Na_2CO_3, but often only after the solution has been concentrated to the required Li concentration (ca. 10 g·L^{-1}) [81]. Lithium is most often precipitated from the leachate as a last step. With a molar ratio of Li:Na_2CO_3 = 0.7, pH = 12 and temperature = 100 °C, a lithium precipitation efficiency of almost 99% can be achieved in a time of 40 min [84]. Na_3PO_4 is also appropriate for lithium precipitation from the acid leachate with high efficiency (more than 90%), producing Li_3PO_4 [87].

Solvent extraction

In addition to cementation and precipitation, solvent extraction (SX) can be used in the recovery of metals from these types of leachates using acid extraction reagents [88–90]. Solvent extraction is characterized by several parameters, such as distribution ratio, partition coefficient, separation factor and extraction efficiency. In particular, the extraction efficiency and selectivity of the process are influenced by the type of organic extraction reagent and its concentration, the type of stripping reagent and its concentration, the pH, the extraction time, the ratio of the A:O = inorganic (aqueous) phase to the organic phase during extraction, the ratio of the O:A = organic phase to the aqueous phase during stripping and possibly others. Operating pressure and temperature have negligible or only a small influence [91]. Among the reagents used in the SX of valuable metals can be included Cyanex 272, D2EHPA, ALIQUAT 336, VERSATIC, TBP—tributyl phosphate, etc.

Cyanex 272 has proven to be an excellent extractant for the mutual separation of cobalt and nickel, where the cobalt passes into the organic phase with high efficiency and the nickel remains in the leachate [92]. The cobalt extraction process using Cyanex 272 can be written by Equation (9) [92].

$$Co^{2+(aq)} + (HA)_{2(org)} \Leftrightarrow CoA_{2(org)} + 2H^{+(aq)}. \tag{9}$$

From the sulfate leachate, 100% of Co extraction efficiency can be achieved at pH = 5.5–8; however, other metals present—Ni and Mn—can be co-extracted along with Co. The optimum pH can therefore be considered to be pH = 5.5–7, where 90–100% of Co passes into the organic phase and nickel remains in the leachate.

If manganese is also passed into the enriched solution, it is advisable to pre-treat the cobalt recovery with selective manganese extraction, e.g., with D2EHPA at pH = 4 [25,92]. The extraction reagent D2EHPA at a concentration of 12% v/v is suitable for the extraction of copper (90–100%) and aluminum (80%) at pH = 1.5–3, depending on the type of leachate, in addition to the extraction of manganese [64,87]. At higher reagent concentrations (about 20 vol.%), it can also be used for the extraction of lithium at pH = 5.5, O:A = 4 with 94% efficiency [64].

Other reagents for the selective extraction of cobalt include Mextral®272P, using which Co is extracted but nickel and lithium remain in the leachate [93]. The use of a two-stage countercurrent system is advantageous, where nearly 100% of cobalt can be obtained under certain conditions [84,94]. If copper is present in the leachate, it can be selectively extracted using Mextral®5640H with efficiencies up to 100% at pH~2 using two countercurrent extraction stages [93]. LIX 84-IC can also be used to extract copper at pH = 3–6, while nickel is extracted as well [95]. Another useful extraction reagent for the extraction of cobalt from leachates after acid leaching is PC-88A. Together with cobalt, manganese is co-extracted at pH = 5, but Ni and Li remain in the leachate [96].

Recently, the use of ionic liquids Cyphos IL-102, Cyphos IL-101 [97,98] and deep eutectic solvents (DESs) [99] in various combinations has also been investigated for the solvent extraction of cobalt. Cyphos IL-102 has confirmed its potential for the selective quantitative extraction of cobalt at a concentration of 0.2 M using countercurrent extraction at a phase ratio of A:O = 1, with metals such as Mn and Li not being extracted [97]. Similarly, Cyphos IL-101 can be described as a suitable extraction reagent for cobalt with excellent stability and a high extraction capacity of about 35 $g \cdot L^{-1}$ [98]. Among the DESs group, the combination of choline chloride with phenylacetic acid in a 1:2 ratio can be considered among the most effective [99].

Ion exchange

Ion exchange rarely occurs in the leachate treatment process after leaching of the active mass of spent LiBs. It is generally more suitable for dilute solutions, e.g., wastewater, where metal concentrations are much lower than those in leachates. Potentially suitable resins for cobalt recovery include a chelating polyamide oxime resin, which has confirmed its potential in recovering cobalt from the leachate after the leaching of LCO-based cathode material in HCl. High adsorption of cobalt (97%) can be achieved at pH = 5.5 with a resin content of 0.04 $g \cdot mL^{-1}$ after 15 min of processing. The advantage of using this resin is that alkali metals and rare earth metals do not form complexes, that is, lithium remains in the leachate [100]. Dowex M4195 can also be described as an resin suitable for the selective recovery of cobalt and nickel from leachates, with an efficiency of almost 99% for both metals [101].

3.3.3. Electrochemical Methods

Electrowinning as a method for metals' recovery from leachate after leaching of the active mass of spent LiBs can be effectively used for the recovery of cobalt and nickel or manganese. However, refined solutions with a high content of the valuable metal have to be fed to the process. The problem in the electrowinning of cobalt is the presence of nickel. The small difference in the standard E^0 potentials of these metals ($E^0{}_{Co2+/Co} = -0.28$ V, $E^0{}_{Ni2+/Ni} = -0.25$ V) makes it almost impossible to selectively electrolytically recover them when they are in solution together. For this reason, it is preferable if the electrowinning is preceded by their mutual separation, e.g., by solvent extraction. Cobalt electrowinning is also sensitive to the content of other metals such as Fe, Cr, Cu, Pb, Zn, etc., because cobalt has a negative electrochemical potential, which means that metals with a positive potential are preferentially deposited at the cathode [102].

Among the perspective electrochemical procedures can be included the recycling of LCO-active masses via a neutral aqueous electrolytic cell. During this process, an electrochemical pH gradient is generated via the electrolysis of water, which allows the leaching of metals at low pH in the anode unit and the precipitation of solid $Co(OH)_2$ at high pH in the cathode unit. In addition, lithium in the form of Li_2CO_3 is obtained [103]. The above method can also be used in the recycling of LMO-based active mass [104]. The LCO-active mass can also be electrochemically treated by simultaneous leaching and metal precipitation from the solution. In the first step, the metals are leached using H_2SO_4, which is formed in the reactor during the electrodeposition of copper from the copper sulfate electrolyte. Extracted Co^{2+} ions are recovered electrolytically as metallic cobalt with simultaneous regeneration of sulfuric acid. Lithium is recovered from the leachate as Li_2CO_3 [105].

4. Industrial Processing of Spent Portable LiBs

In the industry, pyrometallurgical and combined processing of spent portable LiBs is currently predominant. In combined processing, the pyrometallurgical step is usually followed by a hydrometallurgical refining step. The purpose of these hydrometallurgical processes is the efficient separation of the valuable metal from the mixed material as an alloy, a matte or a slag—and its extraction in the form of commercially interesting compounds. These are mostly "in-house" developed methods, where the recycling procedures are developed by the companies themselves [106–110]. A general scheme for the processing of spent portable LiBs in the industrial sector is shown in Figure 4.

4.1. Accurec Recycling

Accurec Recycling is a German company that deals with the recycling of almost all types of spent LiBs. The processing procedure consists of three basic steps—residual voltage discharge, thermal pre-treatment and sorting. Thermal pre-treatment is carried out in a rotary kiln where the temperature does not exceed 600 °C in order to avoid oxidation of metals such as aluminum. Subsequently, the cells are crushed and the obtained fractions are multi-stage sorted using magnetic separation and air sorting. The output of this process is an iron fraction, a Cu/Al fraction and a Co-Ni active mass. The metal fractions are sold for further processing. From an active mass with a particle size < 0.2 mm, pellets are produced. The pellets are treated with a carbothermal reduction at 800 °C to produce a cobalt alloy. The lithium is passed into the dust and slag. Lithium from slag can be recovered by subsequent hydrometallurgical steps [109,110]. A flowsheet of the described process is shown in Figure 5.

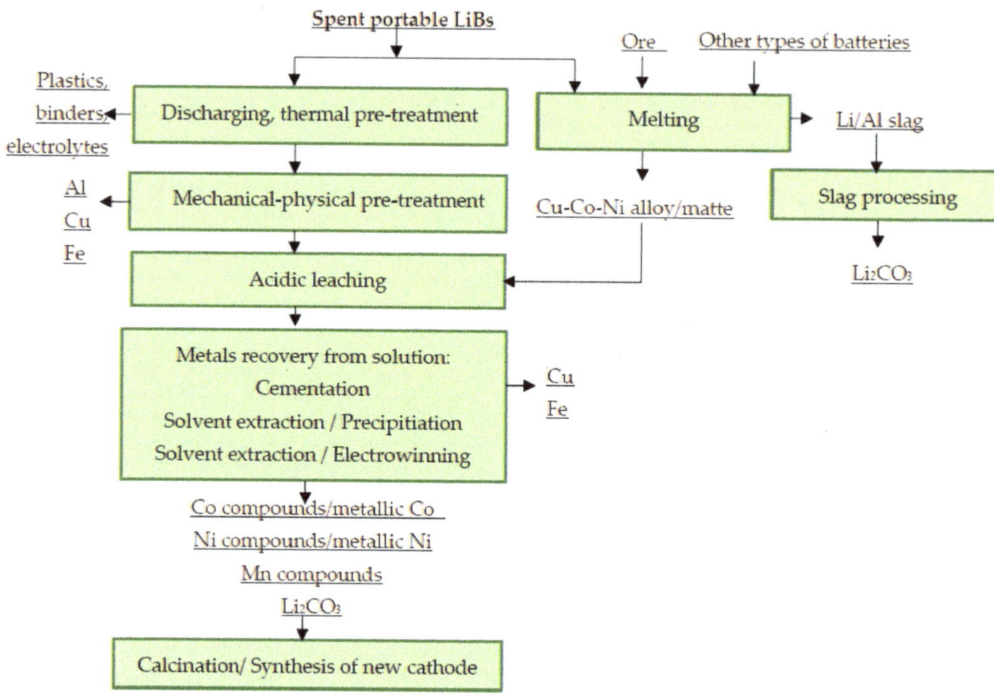

Figure 4. Principle scheme for industrial processing of spent LiBs.

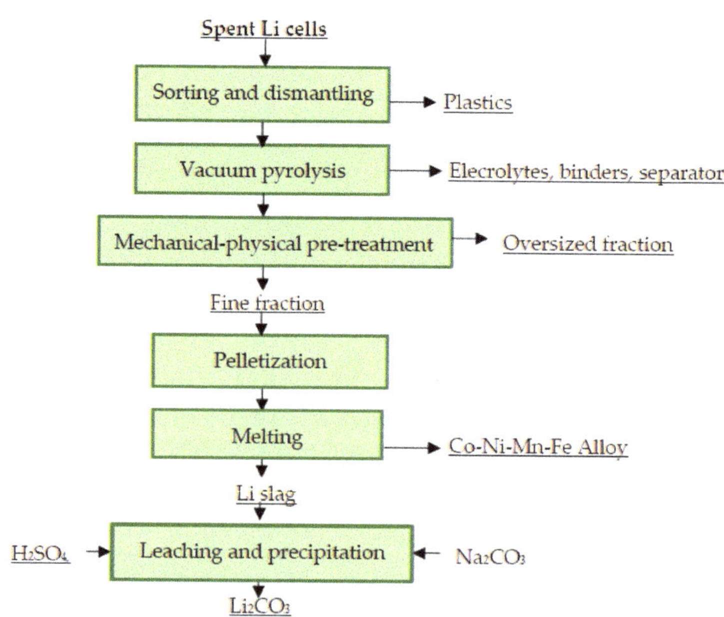

Figure 5. Processing of spent Li cells at Accurec Recycling.

4.2. Akkuser

AkkuSer in Finland is involved in the mechanical and physical pre-treatment of several types of spent electrochemical cells using dry technology to obtain enriched fractions for further processing. The pre-treatment process for sorted spent LiBs consists of crushing using two types of crushers. The first crusher operates at a temperature of 40–50 °C, at 100–400 rpm, reducing the cell size to 1.25–2.5 cm. The crushed material is then transferred to the secondary crusher, which operates at 1000–1200 rpm. This crusher reduces the material to a size of <6 mm. Iron is removed from this fraction via magnetic separation. The resulting fraction with Co-Cu content is a saleable product [110,111]. A flowsheet of AkkuSer's spent LiB recycling process is shown in Figure 6.

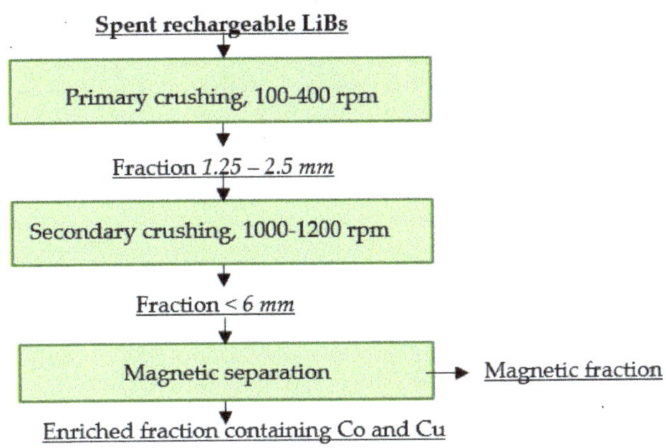

Figure 6. Processing of spent rechargeable LiBs at Akkuser.

4.3. Batrec AG

Batrec AG's facility in Switzerland processes all types of electrochemical cells [110]. At the company, spent LiBs are subjected to mechanical pre-treatment by crushing them in a protective CO_2 atmosphere to minimize possible reactions. Subsequently, the cells are exposed to humid air. The fractions obtained by the crushing and subsequent sorting are sold for further processing [7]. A flowsheet of the described process is shown in Figure 7.

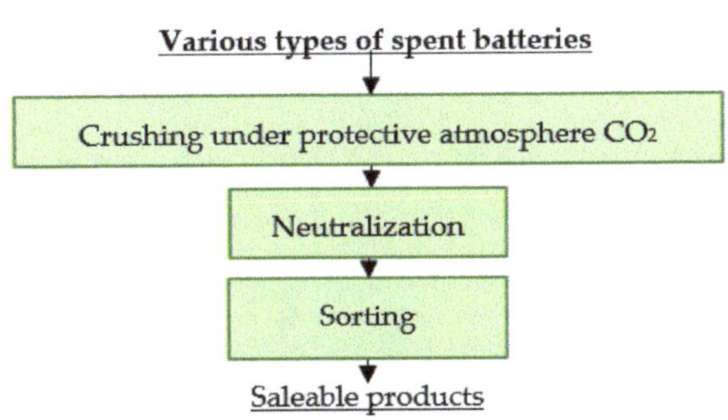

Figure 7. Processing of spent batteries at Batrec AG.

4.4. Glencore

Canadian Glencore, which previously operated as XSTRATA, uses a combined processing of spent LiBs. Spent LiBs are injected to a rotary kiln or the material is melted in a converter. The temperature in the melt reaches 1300 °C. A formed matte is subsequently processed by a hydrometallurgical route in order to complete cobalt extraction. The organic components are burned in the melting process and the resulting gas is treated via additional combustion. The disadvantage is the loss of lithium in the slag [106,110,112]. A flowsheet of Glencore's recycling process is shown in Figure 8.

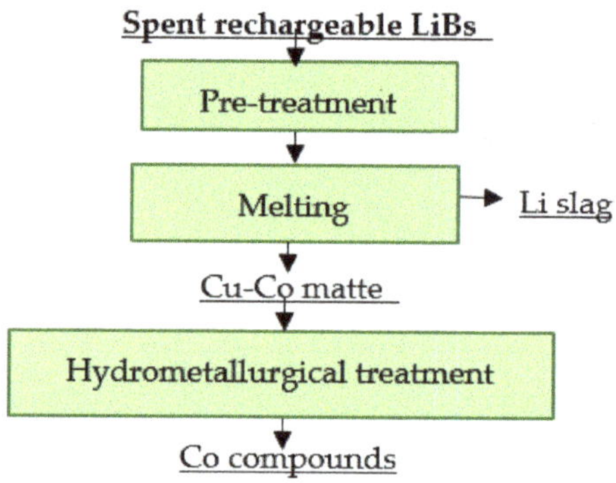

Figure 8. Processing of spent LiBs at Glencore.

4.5. Inmetco

At the American company INMETCO, spent LiBs represent a secondary input to a pyrometallurgical process that was originally designed for steelmaking waste treatment. Recycling of spent LiBs involves mechanical pre-treatment and reduction melting. Pre-treatment consists of disassembly, electrolyte evaporation and subsequent crushing. Crushed lithium cells are melted together with the pellets from Ni/Cd waste. The process is carried out at 1260 °C for 20 min. The organic components and the present carbon are combusted or used as a reducing agent in the carbothermal reduction of the metals. A submerged electric arc furnace is used to refine the melt, yielding Co, Ni and Fe in alloy form, and lithium is passed to the slag [109,110]. A flowsheet of the recycling process is shown in Figure 9.

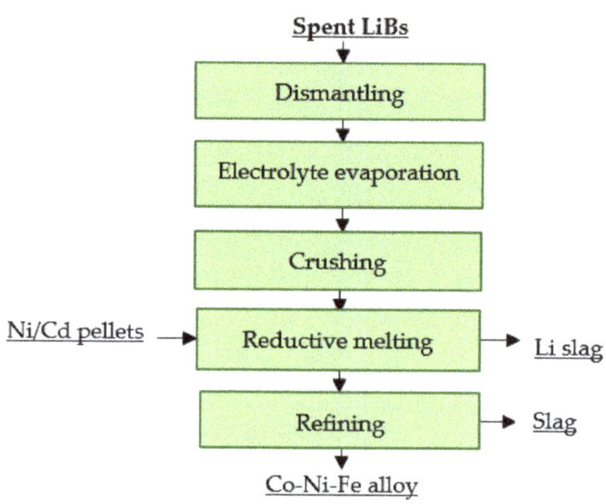

Figure 9. Processing of spent LiBs at Inmetco.

4.6. JX Nippon Mining and Metals

Japanese company JX Nippon Mining and Metals operates a pyro-hydrometallurgical method for spent LiB recycling. The spent LiBs are injected into a stationary furnace, where they are subjected to thermal pre-treatment in order to obtain the cathode material and the other usable components separately. The organic electrolyte is evaporated and the evaporated fluorine is recovered via precipitation. The material obtained from the thermal pre-treatment is crushed and subsequently sieved and separated into several grain size fractions. The fine fraction, which mainly consists of cathode material, proceeds to leaching. The metals are recovered from the leachate via multiple solvent extraction followed by precipitation in the case of Mn recovery and electrowinning in the case of Co and Ni recovery. The main products of the process are electrolytic Co and Ni. The by-products are Li_2CO_3 and $MnCO_3$ [113]. A flowsheet of the described process is shown in Figure 10.

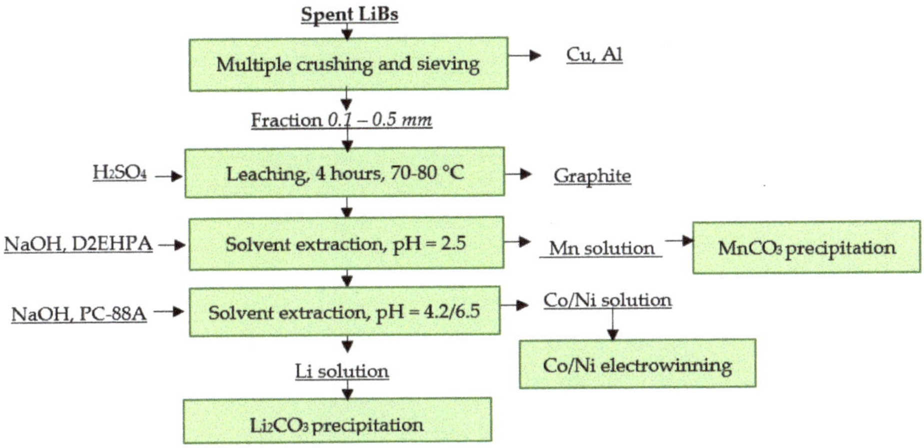

Figure 10. Processing of spent LiBs at JX Nippon Mining and Metals.

4.7. Recupyl Valibat

The French company Recupyl treats spent LiBs in the Valibat recycling process. At first, spent LiBs are subjected to a two-stage crushing process. The first stage of crushing takes place in a rotary knife crusher in an argon or CO_2 atmosphere, where the metallic Li present in the non-rechargeable LiBs is passivated to Li_2CO_3. In the second stage of crushing, an impact crusher is used at 90 rpm in order to reduce the particle size below 3 mm. The material is then separated into two fractions using a vibrating screen. The oversized fraction is subjected to magnetic separation to remove the iron present. The non-magnetic fraction is then processed on a densimetric bench separator. This results in a high density fraction containing Cu and Al and a low density fraction consisting of paper and plastics. The fraction <3 mm is further sorted using 500 µm sieves, where most of the Cu is removed. The fine fraction <500 µm, rich in metal oxides, is then mixed with water and pH is adjusted to 12. The lithium salts present are dissolved and the metal oxides and graphite are concentrated in the insoluble residue. Lithium is recovered from the solution by the addition of CO_2. The insoluble residue is leached with H_2SO_4 at 80 °C in order to obtain Co. Residual copper is recovered from leachate via cementation. Any remaining lithium may be precipitated as Li_3PO_4 at this point by the addition of H_3PO_4. Finally, cobalt is precipitated from a solution with NaClO as $Co(OH)_2$, or is obtained electrolytically [114–116]. A flowsheet of the described recycling process is shown in Figure 11.

4.8. Retriev Technologies

Retriev Technologies' forerunner was TOXCO, which processed spent portable non-rechargeable LiBs. Today, the Canadian company also recycles spent rechargeable LiBs. The process involves spent LiB discharging with a brine solution and crushing. If spent non-rechargeable LiBs are processed, cryogenic crushing is used. The crushed cells are passed in a hammer mill and then the metallic fractions are removed by sieving. Active mass is removed from the aluminum and plastic fractions on a vibrating table. Active mass is leached in water, concentrating the lithium in the leachate. Lithium from solution is obtained with the Na_2CO_3 as Li_2CO_3. Retriev Technologies also has a patented process for obtaining high quality cathode active material [110,112,117]. A flowsheet of the described recycling process is shown in Figure 12.

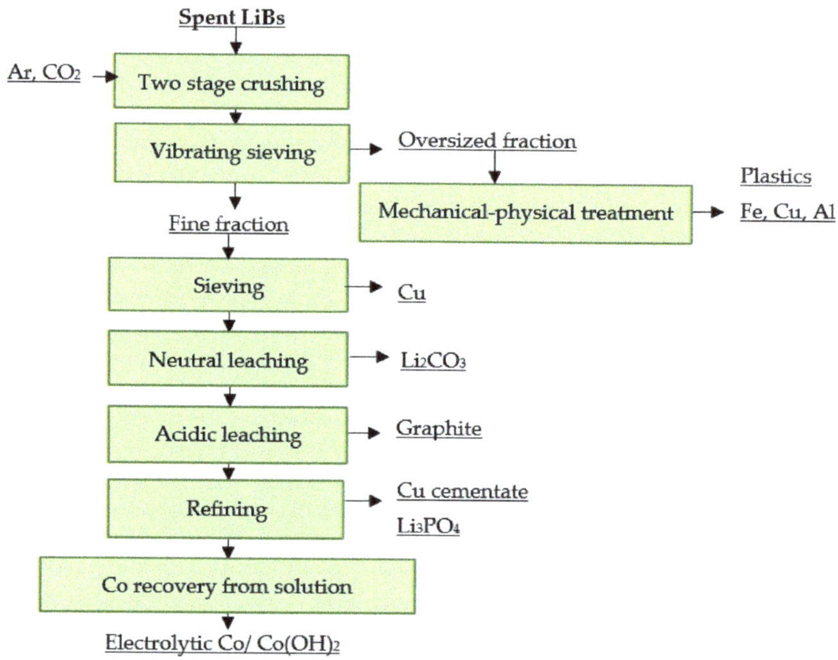

Figure 11. Processing of spent LiBs at Recupyl Valibat.

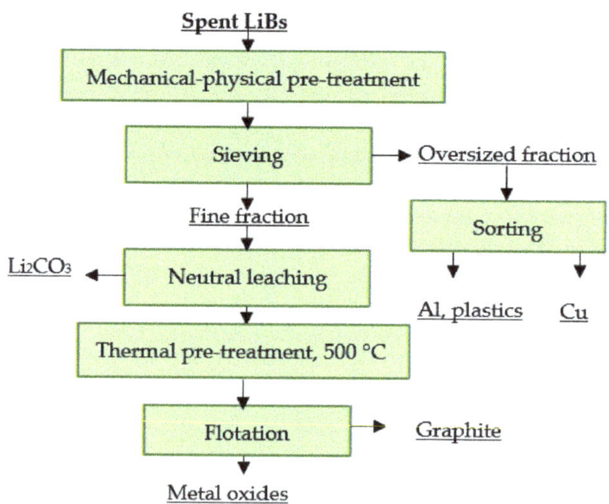

Figure 12. Processing of spent LiBs at Retriev Technologies.

4.9. SNAM

French company SNAM focuses on recycling almost all types of electrochemical cells, including spent portable and industrial LiBs. In the first step, spent LiBs are subjected to a discharge process. This is followed by pyrolysis in order to remove separators and electrolytes. Subsequent sieving produces fractions containing Fe, Al, Cu and a fine fraction, which is processed hydrometallurgically with cobalt and lithium, obtaining saleable products [118,119]. A flowsheet of the described recycling process is shown in Figure 13.

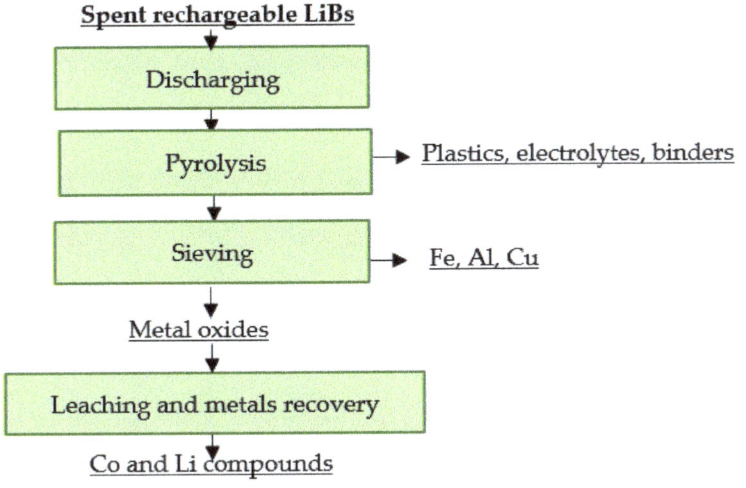

Figure 13. Processing of spent rechargeable LiBs at SNAM.

4.10. *Sumitomo Metal Mining*

Sumitomo Metal Mining (SMM) uses a combined method of recycling spent LiBs. It is also designed to process spent LiBs from EVs. The smelting process produces a Cu-Co-Ni alloy and Li slag. Both intermediates are processed using hydrometallurgical refining. The output of the process is a new cathode active material and electrolytic copper [120,121]. A flowsheet of the described recycling process is shown in Figure 14.

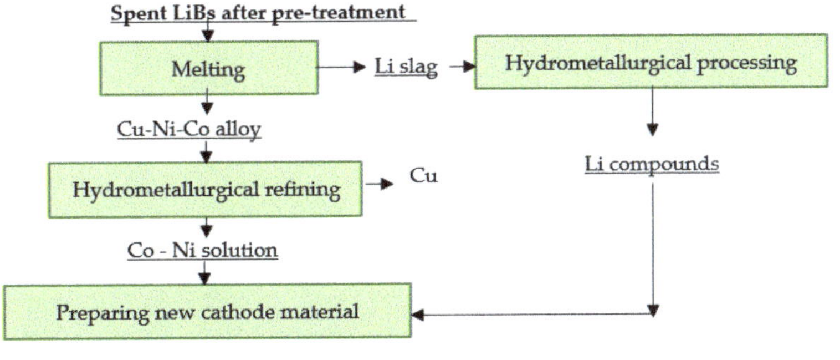

Figure 14. Processing of spent LiBs at SMM.

4.11. *Umicore Battery Recycling*

Umicore Battery Recycling in Belgium processes all types of spent LiBs—from portable to industrial—using reduction melting in a vertical shaft furnace [122]. The feed consisting of spent LiBs, NiMH batteries and slag-forming additives—SiO_2, $CaCO_3$ and coke—is supplied to the preheating zone of the shaft furnace with a temperature of up to 300 °C, where the present electrolyte is evaporated. The feed is then passed to the pyrolysis zone of the furnace where, at a temperature of 700 °C, all the organic components are completely decomposed, which also serves as fuel. This partially reduces the required coke content. The melting of the input material begins in the third zone of the furnace, where temperatures of 1200 to 1450 °C can be reached by injecting oxygen. The melting process results in three intermediate products: a Cu-Fe-Ni-Co alloy, slag and blast furnace gas. The alloy proceeds to leaching with HCl and the subsequent extraction of metals from the leachate via solvent extraction. The obtained products are Cu, Fe, $CoCl_2$ and $Ni(OH)_2$. The lithium is passed to the slag, where it is recovered by a complex hydrometallurgical process. Subsequently, $LiCoO_2$ is produced via synthesis with $CoCl_2$ [122–124]. A flowsheet of the described recycling process is shown in Figure 15.

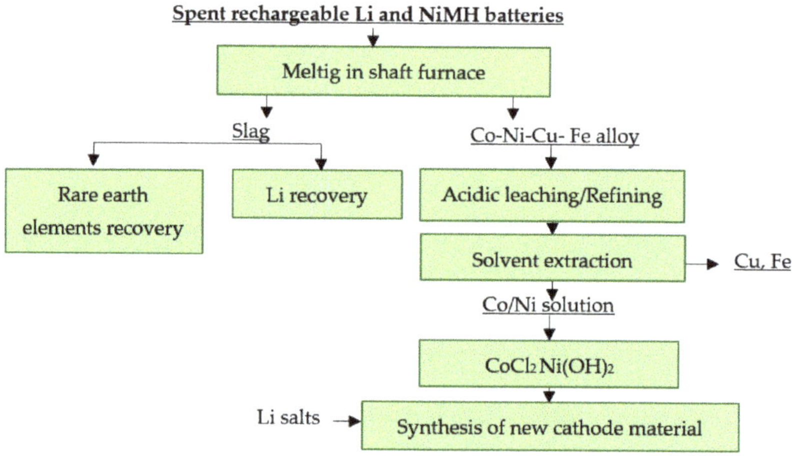

Figure 15. Processing of spent LiBs at Umicore Battery Recycling.

4.12. Pilot Plant

4.12.1. Battery Resources Recycling Process

The process called the Battery Resources Recycling Process (BRRP) is a combination of mechanical and hydrometallurgical processing of spent LiBs using a thermal step in the refining stage. The process consists of hammer crushing followed by magnetic separation. The non-magnetic fraction is leached in NaOH to extract Al as $NaAlO_2$. The insoluble residue is separated into two fractions by sieving. The oversized fraction is sorted on the basis of different densities and a copper-rich fraction is obtained. The fine fraction is passed to a four-stage hydrometallurgical process. In the first stage, the material is leached in H_2SO_4 with the addition of H_2O_2 at 65–70 °C. The undissolved graphite is separated and the lithium is recovered from solution as $LiFePO_4$. The leachate containing Co, Ni, Mn, Li, Al and Cu is subjected to pH adjustment to pH = 6.5 by the addition of NaOH in order to precipitate the remaining Al. Gaseous N_2 is also added at this stage to limit the oxidation of Mn^{2+} ions. In the third leaching stage, after adjusting the leach composition with pure compounds to achieve the required Co:Ni:Mn = 1:1:1 ratio, the pH of the solution is raised to 11 with NaOH, resulting in the precipitation of Co, Mn and Ni hydroxides. In the last step, Na_2CO_3 is added to the leachate at 40 °C to precipitate the remaining lithium. From the obtained $Co(OH)_2$, $Mn(OH)_2$ and $Ni(OH)_2$ together with precipitated and pure Li_2CO_3, the pellets are prepared for the production of a new cathode material [110,124]. A flowsheet of the described recycling process is shown in Figure 16.

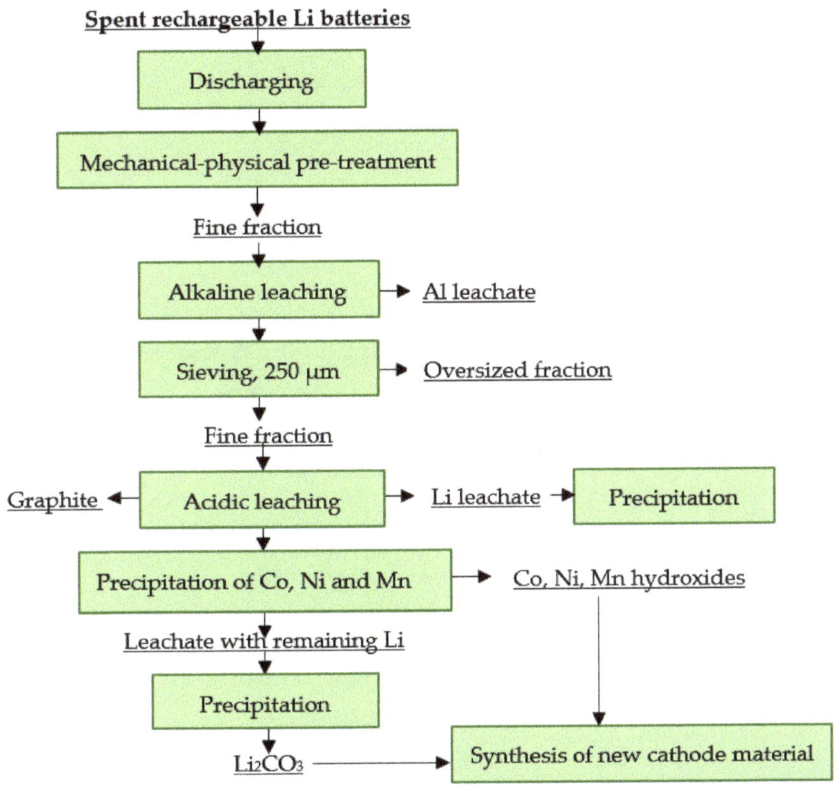

Figure 16. Processing of spent LiBs at BRRP.

4.12.2. LithoRec

The LithoRec process focuses on obtaining high-quality cathode material, mainly by processing spent LiBs from EVs. These cells have a higher content of plastics compared with portable LiBs, which also require a more complex recycling process, especially in the pre-treatment area. LithoRec focuses on mechanical-physical pretreatment and subsequent hydrometallurgical operations. This process produces metals such as Co, Ni and Mn in oxide form and lithium in the form of LiOH and/or Li_2CO_3. The obtained intermediates are processed via calcination to form a precursor material suitable for the production of a new cathode material [7,110,123,125]. A flowsheet of the process is shown in Figure 17.

4.12.3. OnTo Process

Additionally, the OnTo pilot process focuses on spent LiBs from EVs. Discharging is followed by disassembly and the removal of oversized components from the input material. The treated cells proceed to a perforation. Then, they are exposed to CO_2 under supercritical conditions (pressure 7.4 MPa and temperature 31 °C). These conditions are designed to allow extraction and recovery of the organic solvent and electrolyte. The remaining cell is stabilized at atmospheric pressure and temperature. This is followed by crushing and sorting. The obtained fractions are refined hydrometallurgically [110,126]. A flowsheet of the described recycling process is shown in Figure 18.

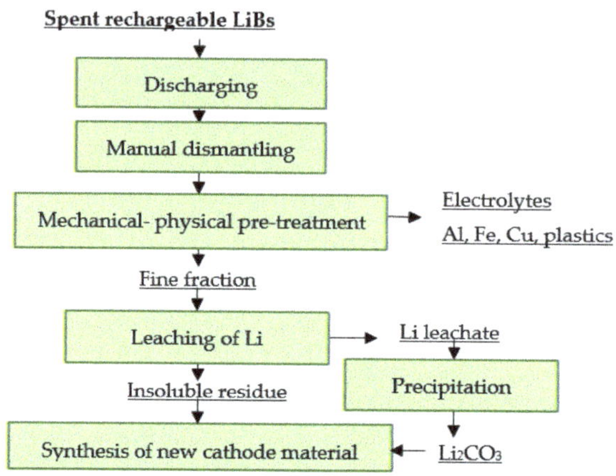

Figure 17. Processing of spent rechargeable LiBs at LithoRec.

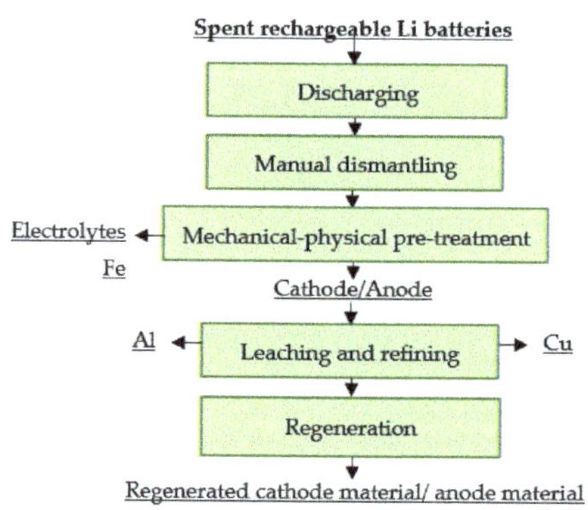

Figure 18. Processing of spent rechargeable LiBs in the OnTo process.

An overview of the companies and their technologies applied in the recycling of spent LiBs is given in Table 4.

Table 4. Companies and their technologies for spent LiB recycling [7,109–127].

Company	Input	Processing	Output	Capacity [Tons/Year]
Accurec Recycling	All types of spent LiBs	Combined	Co-Ni-Mn-Fe alloy; Li_2CO_3 from slag	3000
AkkuSer	All types of spent batteries	Mechanical-physical pre-treatment	From spent LiBs—fraction of Co, Cu	4000
Batrec AG	All types of spent batteries	Mechanical-physical pre-treatment	Metals fractions	200
Glencore	Spent rechargeable LiBs	Combined	Co compounds Li slag	7000
INMETCO	Spent rechargeable LiBs	Mechanical-physical pre-treatment Pyrometallurgy	Co-Ni-Fe alloy	6000

Table 4. *Cont.*

Company	Input	Processing	Output	Capacity [Tons/Year]
JX Nippon Mining and Metals	Waste from cathode active material production	Mechanical-physical pre-treatment Hydrometallurgy	Electrolytic Co, Ni $MnCO_3$, Li_2CO_3	5000
Recupyl Valibat	Spent rechargeable LiBs	Mechanical-physical pre-treatment Hydrometallurgy	Electrolytic Co $Co(OH)_2$ Li_3PO_4	110
Retriev Technologies	Spent rechargeable and non-rechargeable LiBs	Combined	$LiMeO_2$—new cathode active material Li_2CO_3, graphite	4500
SNAM	Spent rechargeable LiBs	Thermal pre-treatment Hydrometallurgy	Saleable products of Co and Li	300
SMM	Spent rechargeable LiBs	Combined	Cathode active material	n.d. **
Umicore Battery Recycling	Spent rechargeable LiBS, Ni-Cd, NiMH batteries	Combined	$LiCoO_2$ $Ni(OH)_2$	7000
BRRP *	Spent rechargeable LiBs	Mechanical-physical pre-treatment Hydrometallurgy	Cathode active material based on Co, Li, Mn, Ni	n.d. **
LithoRec *	Spent rechargeable LiBs from EV	Mechanical-physical pre-treatment Hydrometallurgy	Cathode active material based on Co, Li, Mn, Ni	2000
OnTo *	Spent rechargeable LiBs from EV	Mechanical-physical pre-treatment Hydrometallurgy	Cathode and anode active material	n.d.

* pilot plant. ** not defined.

5. Discussion

5.1. Pre-Treatment

Pretreatment is an essential part of effective recycling of spent LiBs. An important step is the discharging; in industrial applications, it is frequently carried out via heating or with brine solutions (companies). The aim is to prevent burning and explosion. Manual dismantling as the subsequent pre-treatment step has been successfully applied in the industry, mainly for LiBs from EVs (e.g., Retriev Technologies, INMETCO, pilot plant LithoRec, OnTO company). It is not applicable for portable LiBs, mainly due to the small size of the cells.

The mechanical-physical pre-treatment of spent portable LiBs is a key to efficiently obtaining the individual components and fractions that can be successfully recycled in the own recycling process. Several mechanical pre-treatment processes have proven their suitability to obtain active mass, packaging materials, metal electrodes and other fractions with high efficiency and purity. Using appropriately selected steps, a major proportion of the active mass can be released and obtained with high purity; this means with a minimum of impurities such as residual electrodes, separators and packaging. There are a number of plants in the industry that only deal with mechanical-physical recycling (Batreg AG, Akkuser, Acurec) and sell the recovered fractions for further use.

The most common practices for the mechanical-physical treatment of portable LiBs include crushing and grinding, which are essential for the release of active mass. In order to make the process more efficient, multiple crushing is preferable, which is also used in industry (e.g., Akkuser, JX Nippon Minning and Metals, Recupyl Valibat). Then, sorting techniques are applied, often magnetic separation, where iron is removed from the cover, air sorting, vibrating sorting, sieving, etc. In this way, it is possible to separately recover the iron from the casing, the aluminum from the cathode, the copper from the anode and the metal oxides in the active mass. This is a sophisticated method of separating individual metals, with only minimal metal losses compared with those of pyrometallurgy, where the input tends to be whole cells without pre-treatment.

5.2. Pyrometallurgical Treatment

Currently, pyrometallurgical or combined processing of spent LiBs is the technology that is most widely used in the industry. The resulting metal alloy, or matte from melting, proceeds to hydrometallurgical processing. The main advantages of pyrometallurgical processes are as follows:

- high input variability; spent LiBs can be processed with other types of lithium cells, such as EVs batteries, as well as with other types of electrochemical cells, e.g., NiMH batteries;
- the relative simplicity of the process and the fact that no discharge and no mechanical-physical pre-treatment is required; spent portable LiBs can be embedded as a whole in the furnace, e.g., at Umicore Battery Recycling. Only proportionally larger pieces of industrial LiBs are dismantled;
- the possible use of existing plants for the treatment of other feedstock and/or wastes;
- the use of carbon from the anode active material, plastic packaging and separators as a partial substitute for fuel, thus improving the energy balance; in addition, the carbon also serves as a reducing agent;
- high capacities.

However, there are more disadvantages, mainly:

- high investment and operating costs;
- high energy consumption;
- risk of explosion;
- the necessity of refining the obtained products;
- the formation of CO_2 and other gases and their necessary capture and purification;
- loss of metals and other materials in the process; for example, aluminum and lithium are transferred to the slag. Currently, there are also technologies for lithium recovery from slag, but this increases the economic costs of the recycling process.

The graphite from the anode burns off in the melting process, which improves the energy balance. On the other hand, its material potential is lost, which can be considered as a disadvantage.

Research in this field is mainly focused on the optimization of the slag system in order to avoid metal losses during slag casting and for effective slagging of lithium. A promising process being investigated is roasting using temperatures up to 900 °C to suitably modify the material composition for further processing. The disadvantages of roasting are the energy consumption, the addition of chemical reagents and their cost.

A separate area is so-called direct recycling, which is a new and efficient concept for recycling spent LiBs, especially from EVs. The principle is efficient direct recycling based on targeted healing. Specifically, in the case of recycling $LiFePO_4$ cathode material, this involves combining low-temperature aqueous solution relithiation and rapid post-annealing. In this way, ready-to-use recycled cathode materials can be obtained that match the electrochemical properties of the new materials. This method can significantly reduce energy consumption and greenhouse gas emissions, leading to significant economic and environmental advantages compared with current hydrometallurgical and pyrometallurgical methods [128,129].

5.3. Hydrometallurgical Treatment

Compared with pyrometallurgy, hydrometallurgical processes offer several advantages, for example:

- lower investment and operating costs;
- low production of gaseous emissions;

- higher yields and lower metal losses in the process.

The environmental impact of hydrometallurgical methods is generally lower than pyrometallurgical methods, and energy consumption also tends to be lower [117].

The main disadvantage is the production of large amounts of wastewater and solutions. However, they can be successfully recovered and reused in the process in many cases.

The input for hydrometallurgical processing is usually the active mass or cathode active material of spent portable or industrial LiBs. Currently, LCO-based mass is chemically predominant. NMC- and NCA-based materials are available for recycling to a lesser extent; however, like LCO, they contain cobalt in their structure. There are a number of different leaching systems that are suitable for leaching metals from active mass, e.g., sulfuric acid with added hydrogen peroxide, hydrochloric acid, phosphoric acid with added hydrogen peroxide and others.

The mechanism of leaching metals, especially cobalt and lithium, using sulfuric acid with and without hydrogen peroxide or using hydrochloric acid has been described quite successfully over the years. In both media, quantitative conversions of metals to leachates can be achieved with suitably selected conditions. However, the disadvantage is the low selectivity. Today, research is largely focused on various alternatives for leaching systems, mainly with the aim of reducing costs, increasing selectivity and improving environmental acceptability. The addition of alternative reducing agents to sulfuric acid, such as copper and iron scrap, or the use of leaching agents from the group of organic acids that are generally considered to be more environmentally acceptable, e.g., citric acid, lactic acid and others, are being investigated. Under certain conditions, high yields of the metals of interest can also be achieved in these media. A promising approach is the simultaneous leaching and precipitation of metals using oxalic acid.

The graphite from the anode, which is part of the active mass, is not leached in the above media, but is part of the insoluble residue, from where it can be carefully extracted, e.g., by flotation. However, its re-use in the production of new anode material has a number of limitations, as it is often of insufficient quality. Its structure is very damaged and disordered and the main problem in recycling is to restore the degree of graphitization. Theoretically, spent graphite has three possible uses: regeneration into battery materials, low-value utilization (as reductant, adsorbent) and high-value conversion. At present, its recycling is not carried out in industry, but the possibilities are being investigated at a laboratory scale [74].

The combination of precipitation and solvent extraction in varying order has proven effective for the efficient extraction of individual metals from obtained leachate. When precipitation is used as a method to recover cobalt from the leachate, oxalic acid is one of the effective precipitating agents. The advantages of using oxalic acid are the high precipitation efficiencies and the fact that no additional metal ion is added to the leachate, as is the case, for example, with NaOH precipitation. The disadvantage is the co-precipitation of other present metals, with the exception of lithium. Therefore, it is suitable to precipitate manganese with Na_2CO_3 or $KMnO_4$ and nickel with dimethylglyoxime prior to cobalt precipitation. Lithium is preferably recovered from solution by precipitation with Na_2CO_3 at pH ~12–14.

In the case of solvent extraction, Cyanex 272, Mextral®272P and PC-88A are considered as suitable extraction agents for cobalt. The application of ionic liquids and DESs has been a specific recent development, and they have confirmed their potential in the extraction of cobalt with high efficiency. D2EHPA is suitable for the extraction of manganese and lithium; Mextral®5640H is suitable for the copper extraction. Under appropriately selected conditions such as suitable pH, phase ratio, extraction time, suitable stripping agent, etc., high efficiencies can be achieved using solvent extraction. The problem with solvent extraction is the requirement to repeat the individual operations in a multi-step process in order to achieve the required efficiency, which makes the extraction process longer and more expensive. The disadvantage of solvent extraction is also the cost of extraction and stripping reagents and the necessity of their regeneration. However, regeneration can achieve cost savings and increase environmental acceptability as well.

Electrochemical processes for spent LiB recycling are a sophisticated method which can include the processes of electrowinning, electrooxidation and electrodeposition of the valuable metal. In addition, electrochemical processes using selective membranes that only transmit one type of ion can also be applied. In this way, the selected metals, e.g., cobalt from lithium, are mutually separated. During electrochemical processes, high metal yields and selectivity can be achieved without the use of aggressive leaching reagents. However, their wider application may be inhibited by relatively high process costs and limited operational capacity.

In the industry, companies combining mechanical-physical pre-treatment and hydrometallurgical processing can be considered as more sophisticated, environmentally friendly and energy efficient. They have usually a lower processing capacity (150–300 tons/year) compared with pyrometallurgical processes. However, the lower processing capacity can partly be considered as an advantage due to the high process flexibility.

Recycling plants that have successfully introduced lithium battery recycling are mainly pyrometallurgical plants that have built on years of experience in processing other types of feedstock and that have adapted existing technologies to the new input materials as spent LiBs, e.g., Umicore. The problem is the safety of the process due to the hazardous nature of spent LiBs and the high energy consumption. Hydrometallurgical plants whose processes are based on "in house" developed methods using sophisticated and highly flexible steps and appropriate mechanical pre-treatment can also be considered successful. A separate group are companies that only use mechanical-physical treatment in their processing, with the benefit of saleable fractions obtained at relatively low input and operating costs. However, in this case it is appropriate if all types of electrochemical cells are the input. For the future, the most promising for portable LiB recycling seems to be hydrometallurgical plants operating today as pilot plants (e.g., OnTO, LithoRec), which provide complex recycling of LiBs with recovery of all recoverable metals and components. These are mainly a combination of leaching with solvent extraction followed by metal electrowinning or precipitation. A major advantage of solvent extraction is the possibility of recovering the organic agent.

6. Conclusions

Material recycling of used portable lithium cells has been carried out around the world for several years. However, in the context of their increasing consumption, it is necessary to increase capacities, streamline processes and adapt technologies to other types of spent LiBs, e.g., LiBs from EVs. A comprehensive approach is preferred with the aim of maximizing the recovery of present components. The main challenge in recycling processes is to achieve high selectivity and efficiency of individual processes. The demand for high selectivity results from the heterogeneity of the composition of the spent portable LiBs, which represent a wide mixture of materials such as metal oxides present in the active mass, graphite, copper and aluminum electrodes and a polypropylene separator. Almost all of these components can be recycled in their own cycle after appropriate mechanical-physical pre-treatment. Metals present in the active mass, such as cobalt, magnesium and nickel, pass through the treatment process into shared intermediates with high efficiency, from where they can be selectively extracted. A combination of the following steps occurs in almost every proposed process for complex processing of spent lithium cells: mechanical-physical pre-treatment, thermal pre-treatment, leaching (also multi-stage), precipitation and solvent extraction in several stages and variations and preparation of the final product, e.g., via calcination. EU legislation defines a minimum recycling efficiency of 50% for spent lithium cells and up to 65% in 2025. In the case of pyrometallurgy, where the graphite and separator are combusted and some metals, e.g., lithium and aluminum are lost in the slag, achieving the required recycling efficiency is questionable. From this point of view, hydrometallurgical treatment processes can be considered more appropriate.

Hydrometallurgy achieves high metal yields at low losses, and the processes are more flexible and energy efficient. However, the disadvantages include the generation of waste water and waste solutions, as well as the requirement to pre-treat the cells in order to obtain material suitable for leaching.

Author Contributions: Conceptualization, Z.T., D.O. and J.K.; methodology, Z.T. and A.M.; resources, Z.T.; writing—original draft preparation, Z.T.; writing—review and editing, Z.T., D.O., A.M. and J.K.; project administration, Z.T. All authors have read and agreed to the published version of the manuscript.

Funding: This work was funded by the Ministry of Education of the Slovak Republic under grant VEGA 1/0678/23.

Institutional Review Board Statement: Not applicable.

Informed Consent Statement: Not applicable.

Data Availability Statement: Not applicable.

Conflicts of Interest: The authors declare no conflict of interest.

References

1. Colthorpe, A. Energy Storage, Is There Life after Death for Europe's Lithium-Ion Batteries? Available online: https://www.energy-storage.news/is-there-life-after-death-for-europes-lithium-ion-batteries (accessed on 14 November 2022).
2. Council of the European Union: Proposal for A Regulation of the European Parliament and of the Council Concerning Batteries and Waste Batteries, Repealing Directive 2006/66/EC and Amending Regulation (EU) No 2019/1020. Available online: https://data.consilium.europa.eu/doc/document/ST-5469-2023-INIT/en/pdf (accessed on 26 January 2023).
3. Talens Peiro, L.; Nuss, P.; Mathieux, F.; Blengini, G. *Towards Recycling Indicators Based on EU Flows and Raw Materials System Analysis Data*; EUR 29435 EN; Publications Office of the European Union: Luxembourg, 2018; ISBN 978-92-79-97247-8. [CrossRef]
4. Takacova, Z. Material Recycling of Spent Lithium Accumulators (in Slovak Language). Ph.D. Thesis, Technical University of Kosice, Košice, Slovakia, 2014.
5. Directive 2006/66/EC of the European Parliament and of the Council of 6 September 2006 on Batteries and Accumulators and Waste Batteries and Accumulators and Repealing Directive 91/157/EEC. Available online: http://data.europa.eu/eli/dir/2006/66/oj (accessed on 26 January 2023).
6. Reddy, T.B. *Linden's Handbook of Batteries*, 4th ed.; McGraw-Hill Companies: New York, NY, USA, 2011; ISBN 978-0-07-162419-0.
7. HELIS Project: High Energy Lithium Sulphur Cells and Batteries. Available online: https://cordis.europa.eu/project/id/666221 (accessed on 28 January 2023).
8. Sojka, R.; Pan, Q.; Billmann, L. Comparative Study of Li-Ion Battery Recycling Processes. ACCUREC Recycling GmbH. Available online: https://accurec.de/wp-content/uploads/2021/04/Accurec-Comparative-study.pdf (accessed on 29 May 2023).
9. Saft: Material/Product Safety Data Sheet (MSDS-PSDS). Available online: https://www.battery-kutter.de/main/media/content/downloads/MSDS-Li-SOCl2-English.pdf (accessed on 16 November 2022).
10. BiPOWER: (Material) Safety Data Sheets (MSDSs/SDSs). Available online: http://www.bipowerusa.com/documents/bp-msds-li-socl2.pdf (accessed on 16 November 2022).
11. Duracell: Safety Data Sheets. Available online: https://docs.rs-online.com/9392/0900766b80c4677b.pdf (accessed on 16 November 2022).
12. Ultralife: Material Safety Data Sheet. Available online: https://emergencybeaconcorp.com/wp-content/uploads/2014/11/msds-gs-46-battery.pdf (accessed on 16 November 2022).
13. Nitta, N.; Wu, F.; Lee, J.T.; Yushin, G. Li-Ion Battery Materials: Present and Future. *Mater. Today* **2015**, *18*, 252–264. [CrossRef]
14. Chang, S.-K.; Kweon, H.-J.; Kim, B.-K.; Jung, D.-Y.; Kwon, Y.-U. Syntheses of $LiCoO_2$ for Cathode Materials of Secondary Batteries from Reflux Reactions at 130–200 °C. *J. Power Sources* **2002**, *104*, 125–131. [CrossRef]
15. LMO Batteries: Properties and Usage. Available online: https://sinovoltaics.com/learning-center/storage/lmo-batteries (accessed on 26 January 2023).
16. George, G.; Posada-Pérez, S.; Poater, A.; Solà, M. Density Functional Investigation of the Interaction of H_2O with Spinel $Li_{1-X}Mn_2O_4$ Surfaces: Implications for Aqueous Li-Ion Batteries. *Appl. Surf. Sci.* **2023**, *612*, 155822. [CrossRef]
17. NMC Battery Material (LiNiMnCoO2). Available online: https://www.targray.com/li-ion-battery/cathode-materials/nmc (accessed on 23 January 2023).
18. Gerold, E.; Lerchbammer, R.; Antrekowitsch, H. Parameter Study on the Recycling of LFP Cathode Material Using Hydrometallurgical Methods. *Metals* **2022**, *12*, 1706. [CrossRef]
19. Tolganbek, N.; Zhalgas, N.; Kadyrov, Y.; Umirov, N.; Bakenov, Z.; Mentbayeva, A. Facile Deposition of the LiFePO4 Cathode by the Electrophoresis Method. *ACS Omega* **2023**, *8*, 8045–8051. [CrossRef]

20. Neumann, J.; Petranikova, M.; Meeus, M.; Gamarra, J.; Younesi, R.; Winter, M.; Nowak, S. Recycling of Lithium-Ion Batteries-Current State of the Art, Circular Economy, and Next Generation Recycling. *Adv. Energy Mater.* **2022**, *12*, 2102917. [CrossRef]
21. Li, Q.; Chen, J.; Fan, L.; Kong, X.; Lu, Y. Progress in Electrolytes for Rechargeable Li-Based Batteries and Beyond. *Green. Energy Environ.* **2016**, *1*, 18–42. [CrossRef]
22. Orion Engineered Carbons: Premium Carbon Black Grades for Lithium-Ion Batteries. Available online: https://www.innovationnewsnetwork.com/premium-carbon-black-grades-for-lithium-ion-batteries/26365 (accessed on 23 January 2023).
23. Takahashi, V.C.I.; Botelho Junior, A.B.; Espinosa, D.C.R.; Tenório, J.A.S. Enhancing Cobalt Recovery from Li-Ion Batteries Using Grinding Treatment Prior to the Leaching and Solvent Extraction Process. *J. Environ. Chem. Eng.* **2020**, *8*, 103801. [CrossRef]
24. Takacova, Z.; Havlik, T.; Kukurugya, F.; Orac, D. Cobalt and Lithium Recovery from Active Mass of Spent Li-Ion Batteries: Theoretical and Experimental Approach. *Hydrometallurgy* **2016**, *163*, 9–17. [CrossRef]
25. Granata, G.; Pagnanelli, F.; Moscardini, E.; Takacova, Z.; Havlik, T.; Toro, L. Simultaneous Recycling of Nickel Metal Hydride, Lithium Ion and Primary Lithium Batteries: Accomplishment of European Guidelines by Optimizing Mechanical Pre-Treatment and Solvent Extraction Operations. *J. Power Sources* **2012**, *212*, 205–211. [CrossRef]
26. He, L.-P.; Sun, S.-Y.; Yu, J.-G. Performance of $LiNi_{1/3}Co_{1/3}Mn_{1/3}O_2$ Prepared from Spent Lithium-Ion Batteries by a Carbonate Co-Precipitation Method. *Ceram. Int.* **2018**, *44*, 351–357. [CrossRef]
27. Yang, L.; Xi, G.; Xi, Y. Recovery of Co, Mn, Ni, and Li from Spent Lithium Ion Batteries for the Preparation of $LiNiCo_yMn_zO_2$ Cathode Materials. *Ceram. Int.* **2015**, *41*, 11498–11503. [CrossRef]
28. Weng, Y.; Xu, S.; Huang, G.; Jiang, C. Synthesis and Performance of $Li[(Ni_{1/3}Co_{1/3}Mn_{1/3})_{1-x}Mg_x]O_2$ Prepared from Spent Lithium Ion Batteries. *J. Hazard. Mater.* **2013**, *246–247*, 163–172. [CrossRef] [PubMed]
29. Chandran, V.; Ghosh, A.; Patil, C.K.; Mohanavel, V.; Priya, A.K.; Rahim, R.; Madavan, R.; Muthuraman, U.; Karthick, A. Comprehensive Review on Recycling of Spent Lithium-Ion Batteries. *Mater. Today Proc.* **2021**, *47*, 167–180. [CrossRef]
30. Zhang, X.; Xie, Y.; Cao, H.; Nawaz, F.; Zhang, Y. A Novel Process for Recycling and Resynthesizing $LiNi_{1/3}Co_{1/3}Mn_{1/3}O_2$ from the Cathode Scraps Intended for Lithium-Ion Batteries. *Waste Manag.* **2014**, *34*, 1715–1724. [CrossRef]
31. Zhou, X.; He, W.Z.; Li, G.M.; Zhang, X.J.; Zhu, S.G.; Huang, J.W. Recycling of electrode materials from spent lithium-ion batteries. In Proceedings of the 4th International Conference on Bioinformatics and Biomedical Engineering, Chengdu, China, 18–20 June 2010; pp. 1–4.
32. Nan, J.; Han, D.; Zuo, X. Recovery of Metal Values from Spent Lithium-Ion Batteries with Chemical Deposition and Solvent Extraction. *J. Power Sources* **2005**, *152*, 278–284. [CrossRef]
33. Li, J.; Shi, P.; Wang, Z.; Chen, Y.; Chang, C.-C. A Combined Recovery Process of Metals in Spent Lithium-Ion Batteries. *Chemosphere* **2009**, *77*, 1132–1136. [CrossRef]
34. Yu, D.; Huang, Z.; Makuza, B.; Guo, X.; Tian, Q. Pretreatment Options for the Recycling of Spent Lithium-Ion Batteries: A Comprehensive Review. *Miner. Eng.* **2021**, *173*, 107218. [CrossRef]
35. Fan, E.; Li, L.; Lin, J.; Wu, J.; Yang, J.; Wu, F.; Chen, R. Low-temperature molten-salt-assisted recovery of valuable metals from spent lithium-ion batteries. *ACS Sustain. Chem. Eng.* **2019**, *7*, 16144–16150. [CrossRef]
36. Lin, J.; Li, L.; Fan, E.; Liu, C.; Zhang, X.; Cao, H. Conversion mechanisms of selective extraction of lithium from spent lithium-ion batteries by sulfation roasting. *ACS Appl. Mater. Interface* **2020**, *12*, 18482–18489. [CrossRef]
37. Makuza, B.; Tian, Q.; Guo, X.; Chattopadhyay, K.; Yu, D. Pyrometallurgical Options for Recycling Spent Lithium-Ion Batteries: A Comprehensive Review. *J. Power Sources* **2021**, *491*, 229622. [CrossRef]
38. Vieceli, N.; Nogueira, C.A.; Guimarães, C.; Pereira, M.F.C.; Durão, F.O.; Margarido, F. Hydrometallurgical Recycling of Lithium-Ion Batteries by Reductive Leaching with Sodium Metabisulphite. *Waste Manag.* **2018**, *71*, 350–361. [CrossRef] [PubMed]
39. Rostami, T.; Khoshandam, B.; Maroufi, S. Recovery of Lithium, Cobalt, Nickel, and Manganese from Spent Lithium-Ion Batteries through a Wet-Thermal Process. *Mater. Res. Bull.* **2022**, *153*, 111897. [CrossRef]
40. Chen, X.; Wang, Y.; Li, S.; Jiang, Y.; Cao, Y.; Ma, X. Selective Recycling of Valuable Metals from Waste $LiCoO_2$ Cathode Material of Spent Lithium-Ion Batteries through Low-Temperature Thermochemistry. *Chem. Eng. J.* **2022**, *434*, 134542. [CrossRef]
41. Peng, C.; Liu, F.; Wang, Z.; Wilson, B.P.; Lundström, M. Selective Extraction of Lithium (Li) and Preparation of Battery Grade Lithium Carbonate (Li_2CO_3) from Spent Li-Ion Batteries in Nitrate System. *J. Power Sources* **2019**, *415*, 179–188. [CrossRef]
42. Songwen, X.; Ren, G.; Xie, M.; Pan, B.; Fan, Y.; Wang, F.; Xia, X. Recovery of Valuable Metals from Spent Lithium-Ion Batteries by Smelting Reduction Process Based on $MnO–SiO_2–Al_2O_3$ Slag System. *J. Sustain. Met.* **2017**, *3*, 211–218. [CrossRef]
43. Takacova, Z.; Havlik, T. Thermodynamic aspects of the extraction of Co and Li from active mass of LiBs. *Metall* **2015**, *69*, 339–343.
44. Kang, J.; Sohn, J.; Chang, H.; Senanayake, G.; Shin, S.M. Preparation of Cobalt Oxide from Concentrated Cathode Material of Spent Lithium Ion Batteries by Hydrometallurgical Method. *Adv. Powder Technol.* **2010**, *21*, 175–179. [CrossRef]
45. Kang, J.; Senanayake, G.; Sohn, J.; Shin, S.M. Recovery of Cobalt Sulfate from Spent Lithium Ion Batteries by Reductive Leaching and Solvent Extraction with Cyanex 272. *Hydrometallurgy* **2010**, *100*, 168–171. [CrossRef]
46. Ferreira, D.A.; Prados, L.M.Z.; Majuste, D.; Mansur, M.B. Hydrometallurgical Separation of Aluminium, Cobalt, Copper and Lithium from Spent Li-Ion Batteries. *J. Power Sources* **2009**, *187*, 238–246. [CrossRef]
47. Chen, L.; Tang, X.; Zhang, Y.; Li, L.; Zeng, Z.; Zhang, Y. Process for the Recovery of Cobalt Oxalate from Spent Lithium-Ion Batteries. *Hydrometallurgy* **2011**, *108*, 80–86. [CrossRef]
48. Paulino, J.F.; Busnardo, N.G.; Afonso, J.C. Recovery of Valuable Elements from Spent Li-Batteries. *J. Hazard. Mat.* **2008**, *150*, 843–849. [CrossRef] [PubMed]

49. Tian, G.; Yuan, G.; Aleksandrov, A.; Zhang, T.; Li, Z.; Fathollahi-Fard, A.M.; Ivanov, M. Recycling of Spent Lithium-Ion Batteries: A Comprehensive Review for Identification of Main Challenges and Future Research Trends. *Sustain. Energy Technol. Assess.* **2022**, *53*, 102447. [CrossRef]
50. Takacova, Z.; Jajcisinova, J.; Havlik, T. Hydrometallurgical processing of spent portable Li-ion accumulators. In Proceedings of the Kammel's Quo Vadis Hydrometallurgy 6, 6th International Conference, Kosice, Slovakia, 4–7 June 2012.
51. Partinen, J.; Halli, P.; Helin, S.; Wilson, B.P.; Lundström, M. Utilizing Cu^+ as Catalyst in Reductive Leaching of Lithium-Ion Battery Cathode Materials in H_2SO_4–NaCl Solutions. *Hydrometallurgy* **2022**, *208*, 105808. [CrossRef]
52. Ghassa, S.; Farzanegan, A.; Gharabaghi, M.; Abdollahi, H. Iron Scrap, a Sustainable Reducing Agent for Waste Lithium Ions Batteries Leaching: An Environmentally Friendly Method to Treating Waste with Waste. *Resour. Conserv. Recycl.* **2021**, *166*, 105348. [CrossRef]
53. Klimko, J.; Oráč, D.; Miškufová, A.; Vonderstein, C.; Dertmann, C.; Sommerfeld, M.; Friedrich, B.; Havlík, T. A Combined Pyro- and Hydrometallurgical Approach to Recycle Pyrolyzed Lithium-Ion Battery Black Mass Part 2: Lithium Recovery from Li Enriched Slag—Thermodynamic Study, Kinetic Study, and Dry Digestion. *Metals* **2020**, *10*, 1558. [CrossRef]
54. Chabhadiya, K.; Srivastava, R.R.; Pathak, P. Two-Step Leaching Process and Kinetics for an Eco-Friendly Recycling of Critical Metals from Spent Li-Ion Batteries. *J. Environ. Chem. Eng.* **2021**, *9*, 105232. [CrossRef]
55. Verma, A.; Corbin, D.R.; Shiflett, M.B. Lithium and Cobalt Recovery for Lithium-Ion Battery Recycle Using an Improved Oxalate Process with Hydrogen Peroxide. *Hydrometallurgy* **2021**, *203*, 105694. [CrossRef]
56. Sun, L.; Qiu, K. Vacuum Pyrolysis and Hydrometallurgical Process for the Recovery of Valuable Metals from Spent Lithium-Ion Batteries. *J. Hazard. Mat.* **2011**, *194*, 378–384. [CrossRef]
57. Granata, G.; Moscardini, E.; Pagnanelli, F.; Trabucco, F.; Toro, L. Product Recovery from Li-Ion Battery Wastes Coming from an Industrial Pre-Treatment Plant: Lab Scale Tests and Process Simulations. *J. Power Sources* **2012**, *206*, 393–401. [CrossRef]
58. Guo, Y.; Li, F.; Zhu, H.; Li, G.; Huang, J.; He, W. Leaching lithium from the anode electrode materials of spent lithium-ion batteries by hydrochloric acid (HCl). *Waste Manag.* **2016**, *51*, 227–233. [CrossRef] [PubMed]
59. Wang, R.-C.; Lin, Y.-C.; Wu, S.-H. A Novel Recovery Process of Metal Values from the Cathode Active Materials of the Lithium-Ion Secondary Batteries. *Hydrometallurgy* **2009**, *99*, 194–201. [CrossRef]
60. Cao, N.; Zhang, Y.; Chen, L.; Jia, Y.; Huang, Y. Priority Recovery of Lithium and Effective Leaching of Nickel and Cobalt from Spent Lithium-Ion Battery. *Trans. Nonferrous Met. Soc. China* **2022**, *32*, 1677–1690. [CrossRef]
61. Wang, S.; Wang, C.; Lai, F.; Yan, F.; Zhang, Z. Reduction-Ammoniacal Leaching to Recycle Lithium, Cobalt, and Nickel from Spent Lithium-Ion Batteries with a Hydrothermal Method: Effect of Reductants and Ammonium Salts. *Waste Manag.* **2020**, *102*, 122–130. [CrossRef] [PubMed]
62. Pinna, E.G.; Ruiz, M.C.; Ojeda, M.W.; Rodriguez, M.H. Cathodes of Spent Li-Ion Batteries: Dissolution with Phosphoric Acid and Recovery of Lithium and Cobalt from Leach Liquors. *Hydrometallurgy* **2017**, *167*, 66–71. [CrossRef]
63. Jiang, F.; Chen, Y.; Ju, S.; Zhu, Q.; Zhang, L.; Peng, J.; Wang, X.; Miller, J.D. Ultrasound-Assisted Leaching of Cobalt and Lithium from Spent Lithium-Ion Batteries. *Ultrason. Sonoch.* **2018**, *48*, 88–95. [CrossRef]
64. Punt, T.; Akdogan, G.; Bradshaw, S.; van Wyk, P. Development of a Novel Solvent Extraction Process Using Citric Acid for Lithium-Ion Battery Recycling. *Miner. Eng.* **2021**, *173*, 107204. [CrossRef]
65. Li, L.; Ge, J.; Wu, F.; Chen, R.; Chen, S.; Wu, B. Recovery of Cobalt and Lithium from Spent Lithium Ion Batteries Using Organic Citric Acid as Leachant. *J. Hazard. Mater.* **2010**, *176*, 288–293. [CrossRef]
66. Nayaka, G.P.; Zhang, Y.; Dong, P.; Wang, D.; Pai, K.V.; Manjanna, J.; Santhosh, G.; Duan, J.; Zhou, Z.; Xiao, J. Effective and Environmentally Friendly Recycling Process Designed for $LiCoO_2$ Cathode Powders of Spent Li-Ion Batteries Using Mixture of Mild Organic Acids. *Waste Manag.* **2018**, *78*, 51–57. [CrossRef]
67. Santhosh, G.; Nayaka, G.P. Cobalt Recovery from Spent Li-Ion Batteries Using Lactic Acid as Dissolution Agent. *Clean. Eng. Technol.* **2021**, *3*, 100122. [CrossRef]
68. Nayaka, G.P.; Zhang, Y.; Dong, P.; Wang, D.; Zhou, Z.; Duan, J.; Li, X.; Lin, Y.; Meng, Q.; Pai, K.V.; et al. An Environmental Friendly Attempt to Recycle the Spent Li-Ion Battery Cathode through Organic Acid Leaching. *J. Environ. Chem. Eng.* **2019**, *7*, 102854. [CrossRef]
69. Li, L.; Ge, J.; Chen, R.; Wu, F.; Chen, S.; Zhang, X. Environmental Friendly Leaching Reagent for Cobalt and Lithium Recovery from Spent Lithium-Ion Batteries. *Waste Manag.* **2010**, *30*, 2615–2621. [CrossRef] [PubMed]
70. Peng, C.; Hamuyuni, J.; Wilson, B.P.; Lundström, M. Selective Reductive Leaching of Cobalt and Lithium from Industrially Crushed Waste Li-Ion Batteries in Sulfuric Acid System. *Waste Manag.* **2018**, *76*, 582–590. [CrossRef] [PubMed]
71. Chen, H.; Gu, S.; Guo, Y.; Dai, X.; Zeng, L.; Wang, K.; He, C.; Dodbiba, G.; Wei, Y.; Fujita, T. Leaching of Cathode Materials from Spent Lithium-Ion Batteries by Using a Mixture of Ascorbic Acid and HNO_3. *Hydrometallurgy* **2021**, *205*, 105746. [CrossRef]
72. Takacova, Z.; Ivankova, T.; Havlik, T.; Vindt, T. Acidic leaching of active mass from spent lithium ion batteries with addition of saccharides. *Metall* **2015**, *69*, 31–35.
73. Nayaka, G.P.; Pai, K.V.; Santhosh, G.; Manjanna, J. Recovery of Cobalt as Cobalt Oxalate from Spent Lithium Ion Batteries by Using Glycine as Leaching Agent. *J. Environ. Chem. Eng.* **2016**, *4*, 2378–2383. [CrossRef]
74. Liu, J.; Shi, H.; Hu, X.; Geng, Y.; Yang, L.; Shao, P.; Luo, X. Critical Strategies for Recycling Process of Graphite from Spent Lithium-Ion Batteries: A Review. *Sci. Total Environ.* **2022**, *816*, 151621. [CrossRef]

75. Standard Electrode Potentials. Available online: http://hyperphysics.phy-astr.gsu.edu/hbase/Chemical/electrode.html (accessed on 27 October 2022).
76. Solubility-Product Constants (Ksp) for Compounds at 25 °C. Available online: https://saylordotorg.github.io/text_general-chemistry-principles-patterns-and-applications-v1.0/s30-appendix-b-solubility-product-.html (accessed on 26 October 2022).
77. Solubility Product Constants. Available online: https://www.princetonschools.net/site/handlers/filedownload.ashx?moduleinstanceid=714&dataid=1472&FileName=Solubility_Product_Constants.pdf (accessed on 26 October 2022).
78. Levels of Solubility. Available online: https://www.solubilityofthings.com/levels-of-solubility (accessed on 30 May 2023).
79. Takacova, Z.; Dzuro, V.; Havlik, T. Cobalt precipitation from leachate originated from leaching of spent li-ion batteries active mass-characterization of inputs, intermediates and outputs. *World Metall. Erzmetall* **2017**, *70*, 336–344.
80. Shu, S.; He, W.; Li, G.; Zhou, X.; Zhang, X.; Huang, J. Recovery of Co and Li from Spent Lithium-Ion Batteries by Combination Method of Acid Leaching and Chemical Precipitation. *Trans. Nonferrous Met. Soc. China* **2012**, *22*, 2274–2281. [CrossRef]
81. Barbieri, E.M.S.; Lima, E.P.C.; Lelis, M.F.F.; Freitas, M.B.J.G. Recycling of Cobalt from Spent Li-Ion Batteries as β-$Co(OH)_2$ and the Application of Co_3O_4 as a Pseudocapacitor. *J. Power Sources* **2014**, *270*, 158–165. [CrossRef]
82. Meshram, P.; Pandey, B.D.; Mankhand, T.R. Hydrometallurgical Processing of Spent Lithium Ion Batteries (LIBs) in the Presence of a Reducing Agent with Emphasis on Kinetics of Leaching. *Chem. Eng. J.* **2015**, *281*, 418–427. [CrossRef]
83. Gao, R.; Sun, C.; Xu, L.; Zhou, T.; Zhuang, L.; Xie, H. Recycling LiNi0.5Co0.2Mn0.3O2 Material from Spent Lithium-Ion Batteries by Oxalate Co-Precipitation. *Vacuum* **2020**, *173*, 109181. [CrossRef]
84. Asadi Dalini, E.; Karimi, G.; Zandevakili, S. Treatment of Valuable Metals from Leaching Solution of Spent Lithium-Ion Batteries. *Miner. Eng.* **2021**, *173*, 107226. [CrossRef]
85. Sattar, R.; Ilyas, S.; Bhatti, H.N.; Ghaffar, A. Resource Recovery of Critically-Rare Metals by Hydrometallurgical Recycling of Spent Lithium Ion Batteries. *Sep. Purif. Technol.* **2019**, *209*, 725–733. [CrossRef]
86. Chen, X.; Chen, Y.; Zhou, T.; Liu, D.; Hu, H.; Fan, S. Hydrometallurgical Recovery of Metal Values from Sulfuric Acid Leaching Liquor of Spent Lithium-Ion Batteries. *Waste Manag.* **2015**, *38*, 349–356. [CrossRef]
87. Ma, L.; Nie, Z.; Xi, X.; Han, X. Cobalt Recovery from Cobalt-Bearing Waste in Sulphuric and Citric Acid Systems. *Hydrometallurgy* **2013**, *136*, 1–7. [CrossRef]
88. SX Kinetics. Available online: http://www.sxkinetics.com/sxprocess.htm (accessed on 27 October 2022).
89. Metallurgy and Materials Technology. Available online: http://www.hydrometallurgy.metal.ntua.gr/Solvent%20Extraction.htm (accessed on 27 October 2022).
90. Extraction Theory. Available online: https://chem.libretexts.org/Bookshelves/Organic_Chemistry/Organic_Chemitry_Lab_Techniques_(Nichols)/04%3A_Extraction/4.05%3A_Extraction_Theory (accessed on 27 October 2022).
91. Optimize Liquid-Liquid Extraction. Available online: http://www.cheresources.com/content/articles/separationtechnology/optimize-liquid-liquid-extraction?pg=3 (accessed on 27 October 2022).
92. Takacova, Z.; Marcisova, S.; Havlik, T. Solvent extraction of cobalt from leach liquor after leaching of the active mass of spent lithium accumulators. *Metall* **2013**, *67*, 450–454.
93. Chen, X.; Xu, B.; Zhou, T.; Liu, D.; Hu, H.; Fan, S. Separation and Recovery of Metal Values from Leaching Liquor of Mixed-Type of Spent Lithium-Ion Batteries. *Sep. Purif. Technol.* **2015**, *144*, 197–205. [CrossRef]
94. Swain, B.; Jeong, J.; Lee, J.; Lee, G.-H.; Sohn, J.-S. Hydrometallurgical Process for Recovery of Cobalt from Waste Cathodic Active Material Generated during Manufacturing of Lithium Ion Batteries. *J. Power Sources* **2007**, *167*, 536–544. [CrossRef]
95. Choubey, P.K.; Dinkar, O.S.; Panda, R.; Kumari, A.; Jha, M.K.; Pathak, D.D. Selective Extraction and Separation of Li, Co and Mn from Leach Liquor of Discarded Lithium Ion Batteries (LIBs). *Waste Manag.* **2021**, *121*, 452–457. [CrossRef]
96. Yang, Y.; Lei, S.; Song, S.; Sun, W.; Wang, L. Stepwise Recycling of Valuable Metals from Ni-Rich Cathode Material of Spent Lithium-Ion Batteries. *Waste Manag.* **2020**, *102*, 131–138. [CrossRef] [PubMed]
97. Dhiman, S.; Gupta, B. Partition Studies on Cobalt and Recycling of Valuable Metals from Waste Li-Ion Batteries via Solvent Extraction and Chemical Precipitation. *J. Clean. Prod.* **2019**, *225*, 820–832. [CrossRef]
98. Xu, L.; Chen, C.; Fu, M.-L. Separation of Cobalt and Lithium from Spent Lithium-Ion Battery Leach Liquors by Ionic Liquid Extraction Using Cyphos IL-101. *Hydrometallurgy* **2020**, *197*, 105439. [CrossRef]
99. Lukomska, A.; Wisniewska, A.; Dabrowski, Z.; Kolasa, D.; Luchcinska, S.; Domanska, U. Separation of cobalt, lithium and nickel from the "black mass" of waste Li-ion batteries by ionic liquids, DESs and organophosphorous-based acids extraction. *J. Mol. Liq.* **2021**, *343*, 117694. [CrossRef]
100. Badawy, S.M.; Nayl, A.A.; El Khashab, R.A.; El-Khateeb, M.A. Cobalt separation from waste mobile phone batteries using selective precipitation and chelating resin. *J. Mater. Cycles Waste* **2014**, *16*, 739–746. [CrossRef]
101. Strauss, M.L.; Diaz, L.A.; McNally, J.; Klaehn, J.; Lister, T.E. Separation of Cobalt, Nickel, and Manganese in Leach Solutions of Waste Lithium-Ion Batteries Using Dowex M4195 Ion Exchange Resin. *Hydrometallurgy* **2021**, *206*, 105757. [CrossRef]
102. Basics of Cobalt Electrowinning in under 3 Minutes. Available online: https://blog.emew.com/basics-of-cobalt-eletrowinning-in-under-3-minutes (accessed on 31 October 2022).
103. Ni, J.; Zhou, J.; Bing, J.; Guan, X. Recycling the Cathode Materials of Spent Li-Ion Batteries in a H-Shaped Neutral Water Electrolysis Cell. *Sep. Purif. Technol.* **2021**, *278*, 119485. [CrossRef]
104. Zhou, J.; Bing, J.; Ni, J.; Wang, X.; Guan, X. Recycling the Waste LiMn2O4 of Spent Li-Ion Batteries by PH Gradient in Neutral Water Electrolyser. *Mater. Today Sustain.* **2022**, *20*, 100205. [CrossRef]

105. Li, S.; Wu, X.; Jiang, Y.; Zhou, T.; Zhao, Y.; Chen, X. Novel Electrochemically Driven and Internal Circulation Process for Valuable Metals Recycling from Spent Lithium-Ion Batteries. *Waste Manag.* **2021**, *136*, 18–27. [CrossRef]
106. Chen, M.; Ma, X.; Chen, B.; Arsenault, R.; Karlson, P.; Simon, N.; Wang, Y. Recycling End-of-Life Electric Vehicle Lithium-Ion Batteries. *Joule* **2019**, *3*, 2622–2646. [CrossRef]
107. Pinegar, H.; Smith, Y.R. Recycling of End-of-Life Lithium Ion Batteries, Part I: Commercial Processes. *J. Sustain. Metall.* **2019**, *5*, 402–416. [CrossRef]
108. Peters, L.; Friedrich, B. Proven Methods for Recovery of Lithium from Spent Batteries. In Proceedings of the DERA Workshop Lithium, Aachen, Germany, 27 June 2017. [CrossRef]
109. Georgi-Maschler, T.; Friedrich, B.; Weyhe, R.; Heegn, H.; Rutz, M. Development of a Recycling Process for Li-Ion Batteries. *J. Power Sources* **2012**, *207*, 173–182. [CrossRef]
110. Velazquez-Martinez, O.; Valio, J.; Santasalo-Aarnio, A.; Reuter, M.; Serna-Guerrero, R. A Critical Review of Lithium-Ion Battery Recycling Processes from a Circular Economy Perspective. *Batteries* **2019**, *5*, 68. [CrossRef]
111. Pudas, J.; Erkkila, A.; Viljamaa, J. Battery Recycling Method. U.S. Patent 8,979,006 B2, 17 March 2015.
112. CEC. Environmentally Sound Management of End-of-Life Batteries from Electric-Drive Vehicles in North America, Canada: Commission for Environmental Cooperation. 2015, p. 103. Available online: http://www.cec.org/files/documents/publications/11637-environmentally-sound-management-end-life-batteries-from-electric-drive-vehicles-en.pdf (accessed on 24 June 2022).
113. Haga, Y.; Saito, K.; Hatano, K. Waste lithium-ion battery recycling in JX Nippon mining & metals corporation. In *The Materials Processing Fundamentals 2018, Proceedings of the TMS Annual Meeting & Exhibition, Phoenix, AZ, USA, 11–15 March 2018*; Lambotte, G., Lee, J., Allanore, A., Wagstaff, S., Eds.; Springer International Publishing: Cham, Switzerland, 2018; pp. 143–147.
114. Botelho, B., Jr.; Stopic, S.; Friedrich, B.; Soares Tenorio, J.A.; Espinosa, D.C.R. Cobalt Recovery from Li-Ion Battery Recycling: A Critical Review. *Metals* **2021**, *11*, 1999. [CrossRef]
115. Vezzini, A. Manufacturers, Materials and Recycling Technologies. In *Lithium-Ion. Batteries*; Pistoia, G., Ed.; Elsevier: Amsterdam, The Netherlands, 2014; pp. 529–551. ISBN 978-0-444-59513-3.
116. Tedjar, F.; Foudraz, J.C. Method for the Mixed Recycling of Lithium-Based Anode Batteries and Cells. U.S. Patent 7,820,317 B2, 26 October 2010.
117. Mossali, E.; Picone, N.; Rodriguez, O.; Perez, J.M.; Colledan, M. Lithium-Ion Batteries towards Circular Economy: A Literature Review of Opportunities and Issues of Recycling Treatments. *J. Environ. Manag.* **2020**, *264*, 110500. [CrossRef] [PubMed]
118. SNAM Recycling. Available online: https://www.snam.com/activites/recycling-snam.php (accessed on 24 June 2022).
119. Liu, C.; Lin, J.; Cao, H.; Zhang, Y.; Sun, Z. Recycling of spent lithium-ion batteries in view of lithium recovery: A critical review. *J. Clean. Prod.* **2019**, *228*, 801–813. [CrossRef]
120. SMM: Strengthening of Battery Recycling. Available online: https://www.smm.co.jp/en/sustainability/activity_highlights/article_15/ (accessed on 13 January 2023).
121. Sumitomo Metal Mining Co. Develops Recycling Process for Lithium-Ion Batteries. Available online: https://www.recyclingtoday.com/news/japanese-company-develops-lithium-ion-battery-recycling-process/ (accessed on 13 January 2023).
122. Santen, S. Obtaining Nickel or Cobalt by Dry Processes with Formation of Ferro-Nickel or Ferro-Cobalt. European Patent EP1589121B1. Available online: https://patents.google.com/patent/EP1589121B1 (accessed on 27 June 2022).
123. Lv, W.; Wang, Z.; Cao, H.; Sun, Y.; Zhang, Y.; Sun, Z. A critical review and analysis on the recycling of spent lithium-ion batteries. *ACS Sustain. Chem. Eng.* **2018**, *6*, 1504–1521. [CrossRef]
124. SBIR Phase II: A Closed Loop Process for the Recycle of End-of-Life Li-Ion Batteries. Available online: https://www.nsf.gov/awardsearch/showAward?AWD_ID=1738027&HistoricalAw (accessed on 27 June 2022).
125. Diekmann, J.; Hanisch, C.; Frobose, L.; Schalicke, G.; Loellhoeffel, T.; Folster, A.-S.; Kwade, A. Ecological Recycling of Lithium-Ion Batteries from Electric Vehicles with Focus on Mechanical Processes. *J. Electrochem. Soc.* **2017**, *164*, A6184–A6191. [CrossRef]
126. OnTo Technology's Advanced Battery Recycling. Available online: https://www.onto-technology.com/ (accessed on 27 June 2022).
127. Ali, H.; Khan, H.A.; Pecht, M.G. Circular Economy of Li Batteries: Technologies and Trends. *J. Energy Storage* **2021**, *40*, 102690. [CrossRef]
128. Xu, P.; Dai, Q.; Gao, H.; Liu, H.; Zhang, M.; Li, M.; Chen, Y.; An, K.; Meng, Y.S.; Liu, P.; et al. Efficient Direct Recycling of Lithium-Ion Battery Cathodes by Targeted Healing. *Joule* **2020**, *4*, 2609–2626. [CrossRef]
129. Shahjalal, M.; Roy, P.K.; Shams, T.; Fly, A.; Chowdhury, J.I.; Ahmed, M.d.R.; Liu, K. A Review on Second-Life of Li-Ion Batteries: Prospects, Challenges, and Issues. *Energy* **2022**, *241*, 122881. [CrossRef]

Disclaimer/Publisher's Note: The statements, opinions and data contained in all publications are solely those of the individual author(s) and contributor(s) and not of MDPI and/or the editor(s). MDPI and/or the editor(s) disclaim responsibility for any injury to people or property resulting from any ideas, methods, instructions or products referred to in the content.

materials

MDPI

Review

Zeolite-Encaged Luminescent Silver Nanoclusters

Ling Pan, Song Ye *, Xinling Xv, Peixuan Lin, Ruihao Huang and Deping Wang

School of Materials Science and Engineering, Tongji University, Shanghai 201804, China
* Correspondence: yesong@tongji.edu.cn

Abstract: Silver nanoclusters (Ag NCs) are nanoscale aggregates that possess molecular-like discrete energy levels, resulting in electronic configuration-dependent tunable luminescence spanning the entire visible range. Benefiting from the efficient ion exchange capacity, nanometer dimensional cages, and high thermal and chemical stabilities, zeolites have been employed as desirable inorganic matrices to disperse and stabilize Ag NCs. This paper reviewed the recent research progresses on the luminescence properties, spectral manipulation, as well as the theoretical modelling of electronic structure and optical transition of Ag NCs confined inside various zeolites with different topology structures. Furthermore, potential applications of the zeolite-encaged luminescent Ag NCs in lighting, gas monitoring and sensing were presented. This review concludes with a brief comment on the possible future directions in the study of zeolite-encaged luminescent Ag NCs.

Keywords: zeolite; silver nanoclusters; luminescence properties; spectral manipulation; luminescence mechanism; applications

Citation: Pan, L.; Ye, S.; Xv, X.; Lin, P.; Huang, R.; Wang, D. Zeolite-Encaged Luminescent Silver Nanoclusters. *Materials* **2023**, *16*, 3736. https://doi.org/10.3390/ma16103736

Academic Editors: Pavol Liptai, Jaroslav Briančin and Maroš Halama

Received: 5 April 2023
Revised: 10 May 2023
Accepted: 10 May 2023
Published: 15 May 2023

Copyright: © 2023 by the authors. Licensee MDPI, Basel, Switzerland. This article is an open access article distributed under the terms and conditions of the Creative Commons Attribution (CC BY) license (https://creativecommons.org/licenses/by/4.0/).

1. Introduction

Noble metal nanoclusters are new types of ultra-small luminescent centers that have different electronic structures compared with their bulk counterparts and exhibit size-related luminescence properties as quantum dots. Ag NCs are one of the most researched high luminescent metal nanoclusters, which can be effectively excited by UV light, and the resulting emission is tunable throughout the visible region with a quantum yield approaching that of commercial phosphors. The extraordinary luminescence properties of Ag NCs have attracted considerable research interest in many applications, such as photocatalysis, imaging, illumination, detection and biosensors [1–5]. The emissive character of Ag NCs is closely related to their constitution, geometric structure and chemical state. So far, the explicitly reported luminescent Ag NCs consist of only a few Ag^+/Ag^0, for example, the $Ag_2{}^+$ dimer [6], the $Ag_3{}^{n+}$ trimer [7], the $Ag_4{}^{2+}$ tetramer [8], and the $Ag_6{}^{2+}$ hexamer [9].

Due to the high surface energy, Ag NCs are very easily agglomerated into larger silver nanoparticles (Ag NPs) with surface plasma resonance (SPR) absorption, resulting in the quenching of radiative emission. In order to obtain bright and tunable emission, suitable scaffolds are required to stabilize the isolated metal nanoclusters by limiting their overgrowth and, meanwhile, offer proper crystal fields and site symmetry [10–12]. The incorporations of Ag NCs with organic molecules, such as DNA/RNA, proteins, and polymers, result in satisfied biocompatibility, brightness, and photostability, making them applicable as fluorescent biomarkers; however, the loss of stability over time may limit their applications in devices [13–16]. Inorganic scaffolds for Ag NCs include glasses [17–20], metal-organic frameworks (MOFs) [21–23], and zeolites [7,8,24]. The rigid structure of glasses can limit the aggregation of Ag NCs due to immobility, but it is harder to limit their growth precisely. The weaker thermal and chemical stability of MOFs compared with inorganics and the introduction of organic templates may hinder the control of Ag NCs growth and meanwhile bring the risk of environmental contamination.

Zeolites are desirable scaffolds for the stabilization of Ag NCs due to their efficient ion exchange capacity that allows the efficient intake of Ag^+ and molecular-sized cages for the confinement of domain-limited functional Ag NCs. The basic units of the zeolite framework are silicon-oxygen tetrahedra $[SiO_4]^{4-}$ and aluminum-oxygen tetrahedra $[AlO_4]^{5-}$. Due to the presence of $[AlO_4]^{5-}$, the zeolite framework is negatively charged, and therefore some extra framework cations, such as Na^+, K^+, Li^+, Ca^{2+}, are usually introduced for charge balance [25]. The luminescence properties of Ag NCs are not only sensitive to the topology of the zeolite but can also be well manipulated by adjusting the Si/Al ratio and the extra framework cations of the host zeolite, the silver loading degree, as well as the external energy activation condition. Therefore, the diverse topology structures and multiple modification methods of zeolites enable easier and more flexible control of the luminescence properties of Ag NCs. Additionally, the hydration state and adsorption of noxious fume also influence the luminescence performance of Ag NCs that are confined in some types of zeolites or formatted under mild external energy activation, indicating a new sensor and detector application vista besides lighting and display.

This short review started with the preparation and structural characterization of various zeolites-encaged luminescent Ag NCs, followed by the recent studies on the luminescence properties, spectral manipulation, luminescence mechanism and quantum chemistry-based computational study of Ag NCs in zeolites. The applications of zeolite-encaged Ag NCs in white LEDs, tunable luminescence, and moisture and gas monitoring were also presented. This review is finalized by a short comment on the particularly interesting area of future research, including the spectral modification methods, the understanding of the luminescence mechanism, and the possible applications of the high bright zeolite-encaged Ag NCs.

2. Radiative Emission and Spectral Manipulation of Ag NCs in Zeolites

2.1. Building and Characterization of Zeolite-Encaged Ag NCs

Research on the luminescent Ag NCs-loaded zeolites started decades ago when researchers first discovered the radiative emission of Ag NCs in LTA and FAU zeolites under UV excitation [26–28]. To obtain a bright emission of Ag NCs with zeolites as scaffolds, a low Si/Al ratio of the host zeolite is essential as it is beneficial for the efficient intake of Ag^+ through the replacement of extra framework cations, such as Li^+, Na^+, and NH^{4+}, during liquid ion-exchange process. Subsequently, the application of external energies, such as thermal treatment [29,30], electron beam [31], and X-rays irradiations [32,33], may promote the reduction of Ag^+ into Ag^0 by trapping electrons from the O^{2-} that belonged to zeolite framework or adsorbed water, and induce the aggregation of Ag^+ and Ag^0 for the formation of Ag NCs in zeolites. Recent research indicated that Ag^+ exchange efficiency and the external energy driving obviously influence the luminescence behavior of Ag NCs, as it controls the formation efficiency and distribution of Ag NCs inside zeolites [34].

Benefiting from the advanced characterization techniques, the zeolite-encaged Ag NCs can now be observed directly [35–38]. One example is the study on the correlation between the atomic-scale structure and the luminescence properties of Ag NCs in FAU zeolite, in which complete three-dimensional structural characterizations of Ag NCs were performed by the combination of XRD Rietveld analysis and high-angle annular dark field scanning transmission electron microscopy (HAADF-STEM), as shown in Figure 1a,b. In this study, two different luminescent clusters were identified after thermal treatment, one is the $[Ag_4]$ cluster in Ag-FAUX zeolite, and the other one is the $[Ag_3]$ cluster in Ag-FAUY zeolite. It was reported that the $[Ag_4]$ clusters associated with yellow emission are constituted by the Ag^+/Ag^0 that are mainly populated at sites I, I′, II and II′, as shown in Figure 1c. The linear $[Ag_3]$ clusters associated with green emission are located along the hexagonal prism connectors and appear to interact weakly with Ag^0 located at sites I, I′ and II, as shown in Figure 1d. This work laid the foundation for the luminescence mechanism investigations in terms of structure [36].

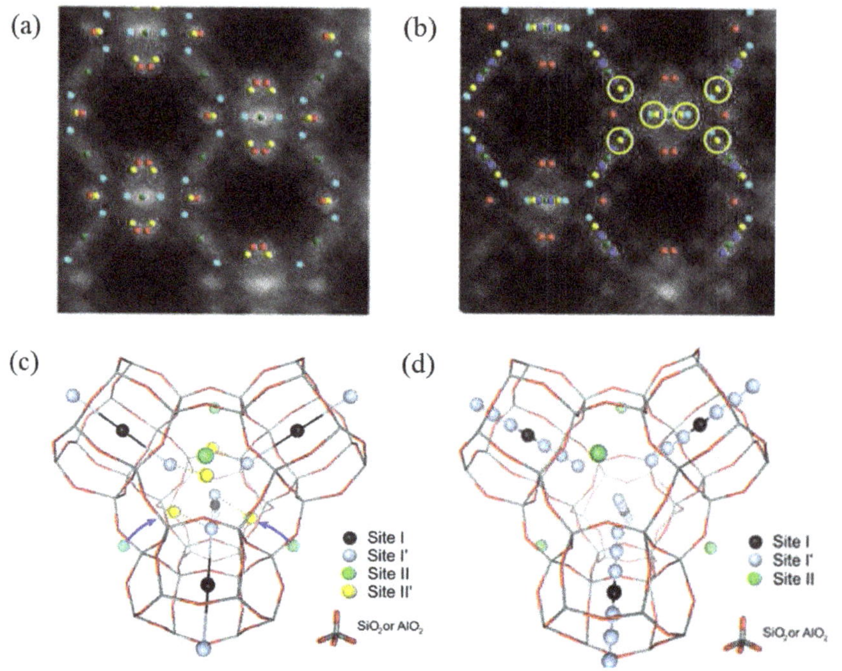

Figure 1. Overlay between the averaged HAADF-STEM images and the silver atomic positions obtained from XRD along the [110] zone axis for Ag-FAUX (**a**) and FAUY (**b**) zeolites in luminescent states, and three-dimensional structural models for Ag-FAUX (**c**) and FAUY (**d**) [36].

The electron spin resonance (ESR) characterization provided structural information of the confined Ag NCs [7,39–41]. According to the ESR analysis, the 690 nm emission in LTA zeolites is associated with Ag_6^+ with a doublet electronic ground state, while the 550 nm emitter is related to Ag_3^+ [40]. Additionally, the more powerful extended X-ray absorption fine structure (EXAFS) and the X-ray excited optical luminescence-EXAFS (XEOL-EXAFS) was also employed to obtain the direct structural information of Ag NCs formatted inside LTA, FAUY, and FAUX zeolites, such as Ag-Ag, Ag-O, and Ag-Li distances [9,42–44]. In recent years, a series of Ag NCs models have been built by taking the sodalite cavity of LTA zeolite as framework [8,45,46], aimed to calculate the energy level structure and possible optical transitions between the discrete energy levels of Ag NCs with different structural configurations and valence states. The combination of experimental characterization and theoretical calculation is helpful for the systematical investigation on the luminescence mechanism of Ag NCs confined inside zeolites with different topologies from the perspectives of constitution, configuration, energy level structure, and luminescence kinetics, providing a strong support for achieving bright, efficient, and tunable emission of zeolite-encaged Ag NCs.

2.2. Luminescence Properties of Ag NCs Confined Inside Zeolites

Low Si/Al ratio zeolites, such as FAU, LTA and SOD, have been widely employed as ideal scaffolds for Ag NCs. These zeolites all consist of sodalite cages, the differences in structure are that the sodalite cages are connected by a double six-membered ring (D6r) in FAU, a double four-membered ring (D4r) in LTA, while SOD consists entirely of the sodalite cages with a homogeneous structure [47]. Recent studies indicated that the luminescence quantum yields of Ag NCs are strongly dependent on the topology structure of the host zeolites, which can reach around 97% in FAU zeolite while below 83% in LTA zeolite [48,49].

FAU zeolites can be subdivided into FAUX ($1 < Si/Al < 1.5$) and FAUY ($1.5 < Si/Al < 3$) types according to the Si/Al ratio. Generally, the Ag NCs confined inside FAUY zeolite exhibit a blue-shifted emission compared with those confined inside FAUX zeolite due to the reduced lattice parameter. It was reported that the strong green visible emission centered at 540–550 nm and around 530 nm can be obtained after thermal treatment of the Ag^+-exchanged FAUX and FAUY zeolites, respectively [24,43,50,51]. According to emission spectra, XPS studies and extended Hückel molecular orbital calculation, the optimum geometry of Ag_3^{n+} is nearly linearly coordinated inside the D6r cage of FAU zeolites [7,50].

The correlation between the radiative emission and the structural configuration of Ag NCs confined inside LTA zeolites has been investigated with the help of ESR and EXAFS measurements. It was reported that the Ag NCs formed in LTA zeolites tend to be located in the sodalite cages, in which the Ag_5^{3+} cluster formed in Ag_{12}-LTA zeolite exhibits a yellow emission around 577 nm [52], the Ag_6^+ cluster gives off a red emission [35,41], the Ag_3^{n+} and Ag_4^{n+} clusters show a yellow-green emission around 550 nm [7,28,40,41], respectively. The effect of the silver loading degree on the luminescence properties of Ag NCs was also extensively explored in Li-LTA zeolites, and it was indicated that the emission wavelength of Ag NCs under UV excitation is strongly dependent on silver loading degree: Ag_{1-2}Li-LTA zeolite emits green light, Ag_{3-4}Li-LTA zeolite emits yellow light, and Ag_{5-12}Li-LTA zeolite emits red light, respectively [34].

The tunable emission of Ag NCs was also reported in SOD zeolites. For example, by substituting the extra framework cations of Na^+ with Cs^+ and adjusting silver loading degree, the emission color of Ag NCs can be turned from green through yellow to red. The reported tunable emission of Ag NCs is mainly due to the lattice shrinkage or expansion of the host zeolites, which conversely applies the structural feedback to the luminescence property of Ag NCs [25,53,54].

2.3. Luminescence Tailoring through Structure Modification of Host Zeolites

Many recent studies revealed that the extra framework cations influence not only the Ag^+ exchange capacity but also the formation efficiency and the local crystal fields of Ag NCs. Generally, small radius extra framework cations lead to reduced lattice parameters and, therefore, blue-shifted emission, and the Ag NCs showed the strongest emission intensity in Li-zeolites among a series of M-zeolites ($M = Na^+, K^+, Cs^+, Ca^{2+}$) [29,53,55,56]. Taking FAU zeolite as an example, as shown in Figure 2a, site I is located inside the D6r cage, sites II and III inside the sodalite cage, and sites IV and V inside the super cage, respectively. All the sites from I to V can be occupied by extra framework cations with different ionic radii, such as Li^+, Na^+, K^+, NH^{4+}, H^+, and Ca^{2+} [55,57]. The researchers studied the influence of extra framework cations on the luminescence properties of Ag NCs in the Na-FAUY zeolite and Li^+, Ca^{2+} partially exchanged zeolites. It was suggested that Ag NCs show stronger emission in Li-FAUY zeolites, and meanwhile, the emission peak is tunable between 462 and 530 nm, which drastically blue-shifted in Li-FAUY zeolites while slightly red-shifted in Ca-FAUY zeolites compared with in Na-FAUY zeolites, as shown in Figure 2b [29]. It was also reported that the peak emission wavelength of Ag NCs can also be well turned from 470 to 610 nm when confined in the Li^+, Na^+, K^+, Cs^+, and Mg^{2+} partially exchanged SOD zeolites [25,53,54].

Furthermore, the effect of extra framework cations on the luminescence performance of Ag NCs was systematically studied with LTA as host zeolite by increasing the substitution ratio of Na^+ by Li^+. According to XEOL-EXAFS and Transmission-EXAFS (Tr-EXAFS) measurements, the Ag_c-Ag_c equilibrium distance in the ground state is about 2.68 Å, while the Ag_c-Ag_c equilibrium distance in the excited state is Li^+ content dependent, which decreases from 2.75 Å in Ag_1-exchanged Li_0Na_{12}-LTA to 2.70 Å in Ag_1-exchanged $Li_{12}Na_0$-LTA. The decreased difference in Ag_c-Ag_c distances between the excited state and ground state of Ag NCs from 0.07 to 0.02 Å revealed a stronger guest–host–guest interaction in the LTA zeolites with increasing Li^+ as extra framework cations. It was indicated that the emission peak of Ag NCs shows a drastic blue shift from 605 to 510 nm with increasing Li^+

exchange amount in Li_xNa_{12-x}-LTA zeolites because of the reduced lattice parameters of the host zeolites, as shown in Figure 3 [44].

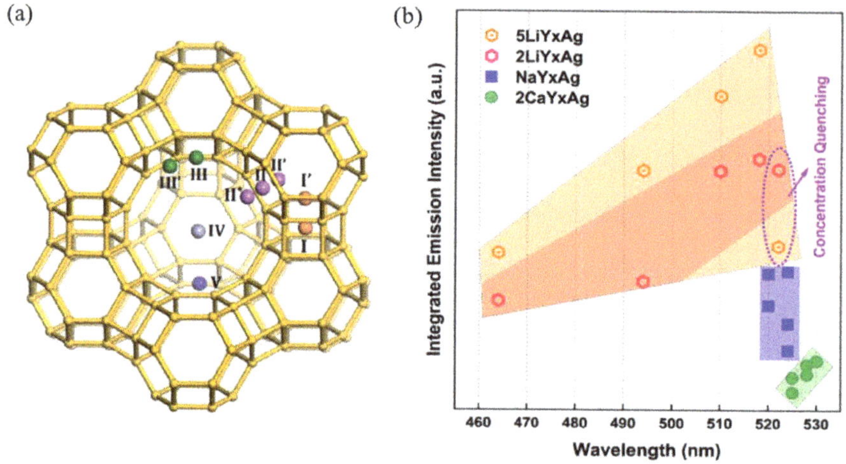

Figure 2. Typical extra framework cation sites I(I′), II(II′, II*), III(III′), IV, and V in the structure of FAU zeolites (**a**) [57], and summary of the integrated emission intensity and peak position of MYxAg (M = 5Li, 2Li, Na, and 2Ca; x = 5, 10, 20, 30, and 50) (**b**) [29].

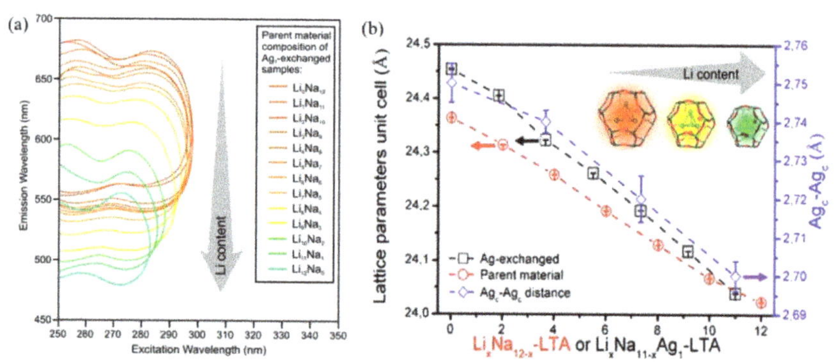

Figure 3. Two-dimensional excited emission spectra of Ag-LTA zeolites with various concentrations of Li^+ exchange (**a**), variations in lattice parameter of unit cell of Li_xNa_{12-x}-LTA and $Li_xNa_{11-x}Ag_1-$LTA zeolites and Ag_c-Ag_c distances of the tetrahedral Ag NCs in $Li_xNa_{11-x}Ag_1$-LTA zeolites (**b**) [44].

In addition to the extra framework cations substitution, the luminescence properties of Ag NCs can also be manipulated through framework modification by using desilication and dealumination approaches. During the desilication and dealumination processes, the Si/Al ratio of the zeolite framework can be modified, thus introducing a new method for regulating and optimizing the luminescence performance of Ag NCs in zeolites [58–60]. It was presented that the Ag NCs showed red-shifted emission in the desilicated FAUY zeolites and blue-shifted emission in the dealuminated FAUY zeolites, and a tunable emission in the wavelength range of 482–528 nm was obtained, as shown in Figure 4. This tunable luminescence property of Ag NCs is the result of controlling the local crystal field and the coupling between the host lattice and the luminescent center. Compared with adjusting the silver loading degree, the framework modification can tailor the emission of Ag NCs in a wider wavelength region and improve luminescence intensity [58].

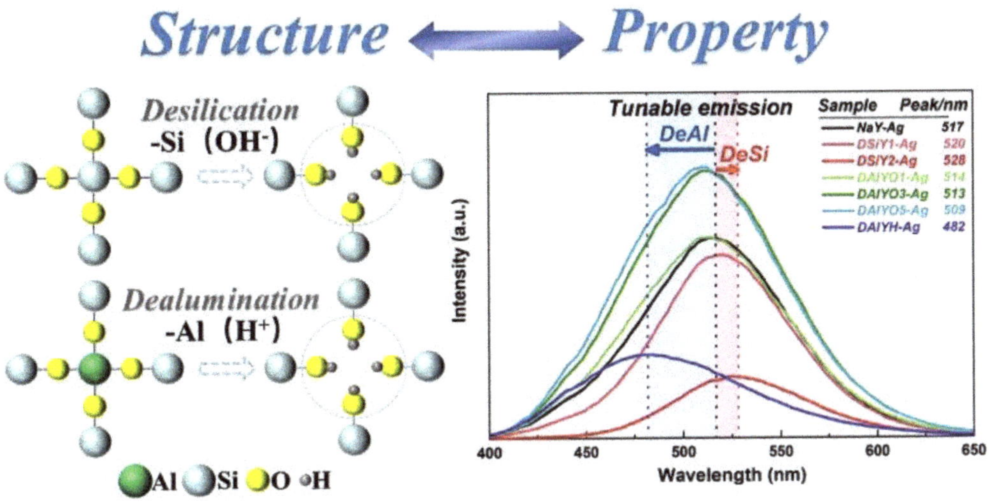

Figure 4. Schematic description of desilication and dealumination processes, and the emission spectra of NaY−Ag, DSiY1−Ag, DSiY2−Ag, DAlYO1−Ag, DAlYO3−Ag, DAlYO5−Ag, and DAlYH−Ag [58].

2.4. Water Manipulated Ag NCs Emission

The three-dimensional eight-ring diffusion channels present in LTA zeolites allow for rapid adsorption/desorption of H_2O; therefore, the luminescence of Ag NCs confined inside LTA zeolites is closely associated with hydration and dehydration states of the host zeolites [61,62]. Related studies suggested that the reversible luminescence response in the hydration and dehydration cycle is related to the structural dynamics of Ag NCs confined in the sodalite cages of LTA zeolites [9,34,63,64].

In the hydrated states, the Ag_4 cluster $[Ag_4(H_2O)_x, 0 \leq x \leq 8]$ is responsible for the green-yellowish emission in the range of 520–570 nm under UV excitation. According to the ESR and EXAFS investigations, the luminescent $[Ag_4(H_2O)_4]^{2+}$ clusters are distributed along the six-ring axis of the sodalite cage [34,63,64]. Upon dehydration, $[Ag_4(H_2O)_x]$ clusters are transformed into the non-luminescent $[Ag_6(O_F)_{14}]^{2+}$ clusters that appear to interact strongly with zeolite framework oxygen (O_F) [9]. This reversible $Ag_4{}^{2+} \leftrightarrow Ag_6{}^{2+}$ switching highlights the importance of the Ag NCs charge and valence electrons in the modulation of their optical properties.

The effect of hydration degree on the luminescence properties of Ag NCs has also been extensively investigated at low silver loading degrees. As shown in Figure 5, the emission of Ag_1Li–LTA zeolites shows red-shifted emission from blue to green-yellow with increasing hydration levels [34]. This hydration/dehydration-induced emission wavelength change of Ag NCs can also be observed when confined inside K-, Na- and Ca-LTA zeolites; however, the response is not as fast as in Li-LTA zeolite, which may be due to the lower mobility of those extra framework cations than Li^+ [9]. It was also noted that the external quantum efficiency (EQE) of Ag NCs in K-, Na- and Ca-LTA zeolites is below 16%, while Ag_1Li-LTA zeolite has a high EQE of up to 62%, due to the pronounced dynamic change in the emissive color of the Ag_1Li-LTA zeolite with the degree of hydration and the high EQE, which can be potentially applied as a humidity sensor.

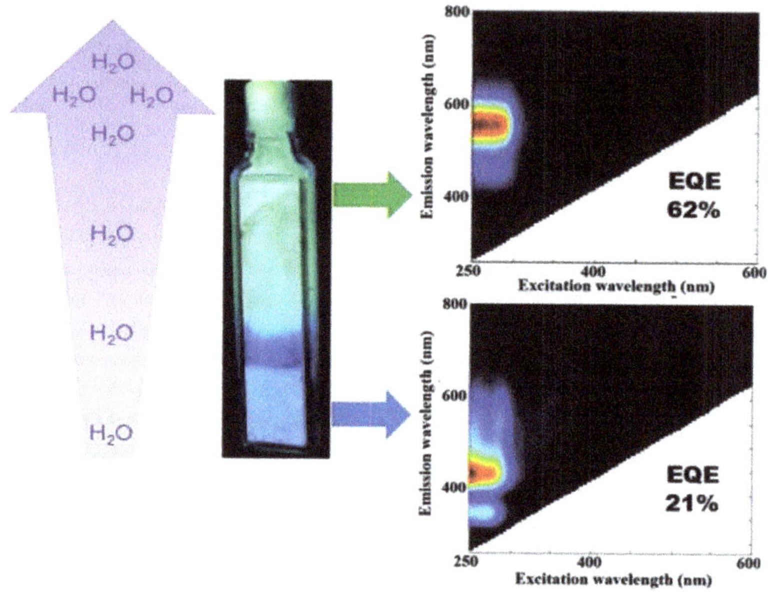

Figure 5. Water dependence of the emission color observed in Ag_1Li-LTA zeolite. In fully hydrated samples (19% water content) a green emission was found, whereas in partially hydrated samples a (2% water content) blue emission was recorded [34].

2.5. Emission of Ag NCs in Rare Earth Ions Exchanged Zeolites

In recent years, the rare earth (RE) ions and Ag NCs co-loaded zeolites have been developed for the generation of a bright visible emission, in which the Ag NCs act as the sensitizer for RE ions and the RE ions regulate the formation efficiency and luminescence properties of Ag NCs [50,65]. In the Ag^+-RE ions dual exchanged zeolites, Eu^{3+} may promote the formation of Ag NCs in Na-FAUX zeolites during the thermal treatment process, which benefits the reduction of Ag^+ [65], while the existence of Tb^{3+} can hinder the supply of electrons required for the reduction of Ag^+ to Ag^0 and thus hinder the effective formation of Ag NCs [50]. Moreover, the XPS measurement indicated that the prior Tb^{3+} exchange might lead to a reduced Si/Al ratio and thus result in blue-shifted emission and stronger ionic properties of Ag NCs. Due to the tunable broadband emission that can fully overlap the excitation band of RE ions, Ag NCs are expected as effective sensitizers for RE ions in zeolites. It was reported that the energy transfer efficiency from Ag NCs to Eu^{3+} in FAUX reaches around 73% with optimized Ag^+ exchange amount and thermal treatment condition, as shown in Figure 6 [51]. Benefiting from the efficient energy transfer from $[Ag_2]^{n+}$ pairs to Eu^{3+} and the combination of the red emission of Eu^{3+} and the green emission of Ag NCs, the warm white light emission was obtained in SOD zeolite [66].

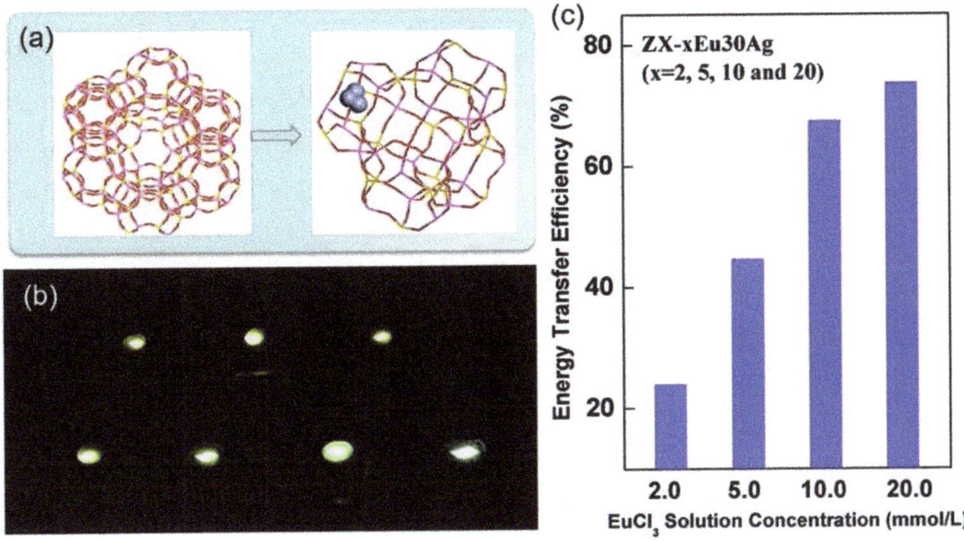

Figure 6. Three-dimensional structure of Na-FAUX zeolite viewed along the [111] direction and Ag NCs located in the D6r cage (**a**), physical diagram of samples with different ratios of Ag^+/Eu^{3+} exchange concentration (**b**) and energy transfer efficiency from Ag NCs to Eu^{3+} (**c**) [51].

3. Studies on the Luminescence Mechanism of Ag NCs in Zeolites

The mechanism behind the tunable luminescence of Ag NCs in zeolites attracts great research interest and has been investigated from the aspects of experimental characterization and theoretical calculation. As early as 1962, researchers discovered that Ag NCs loaded Na-LTA zeolites are colorless in a hydrated state, while activating to yellow when highly vacuumed at room temperature and turning brownish red when heated [26]. This hydration state dependence of reversible color change was attributed to electronic charge transfer from the oxygen lone pairs of the zeolite framework to the empty 5 s orbital of Ag^+ ions, and known as the ligand-to-metal charge transfer (LMCT) process [67]. Recent studies suggested that the electron transition from the neighboring Ag^+-Ag^+ metal-to-metal charge transfer (MMCT) state to the ground state may be responsible for the green- and red-dominated emissions in Ag NCs loaded LTA zeolites, as shown in Figure 7a. A similar luminescence mechanism was also proposed for the Ag NCs that are confined inside SOD zeolites, as shown in Figure 7b [54]. The mechanism of Ag NCs in Li^+ partially exchanged Na-LTA zeolites was also investigated with the help of temperature-dependent luminescence. It was suggested that the emissions located around 595 (298–233 K), 495 (233–153 K), and 395 nm (153–83 K) are due to the radiative transitions from the triplet states to the singlet states that resulted from the confinement effect of Ag NCs, and the difference between the lowest multiple minima in the excited state potential energy surface lead to different emissions, as shown in Figure 7c [68].

In order to acquire more information on the relationship between the structural configuration and the luminescence performance of Ag NCs, the DFT and TD-DFT calculations were carried out to compute the energy level structure, the HOMO and LUMO states, and the absorption spectra. In these computational studies, the unit sodalite cavity of $[Si_{24}H_{24}O_{36}]$ in LTA zeolite was used to encage Ag NCs [8,45,46,69]. One of the successful examples is the investigation of the origin of the bright-green emission of Ag NCs confined inside LTA zeolites. According to XEOL, EXAFS, TD-DFT calculation and time-resolved spectroscopy, a mixture of tetrahedral $[Ag_4(H_2O)_x]^{2+}$ (x = 2 and 4) clusters occupies the center of a fraction of the sodalite cages, as shown in Figure 8a, and the radiative emission is originated from a confined two-electron superatom quantum system with hybridized Ag and water O orbitals delocalized over the cluster. Based on the calculated energy level in Figure 8b, the modeled absorption spectra peaked at 343 and 320 nm for $[Ag_4(H_2O)_2]^{2+}$, and $[Ag_4(H_2O)_4]^{2+}$ isomers are in excellent agreement with the experimentally measured absorption and excitation spectra, as shown in Figure 8c,d. It was estimated that the Ag_3K_9-LTA zeolite consists of about 40% of $Ag_4(H_2O)_4$ and 60% of $Ag_4(H_2O)_2$ clusters [8].

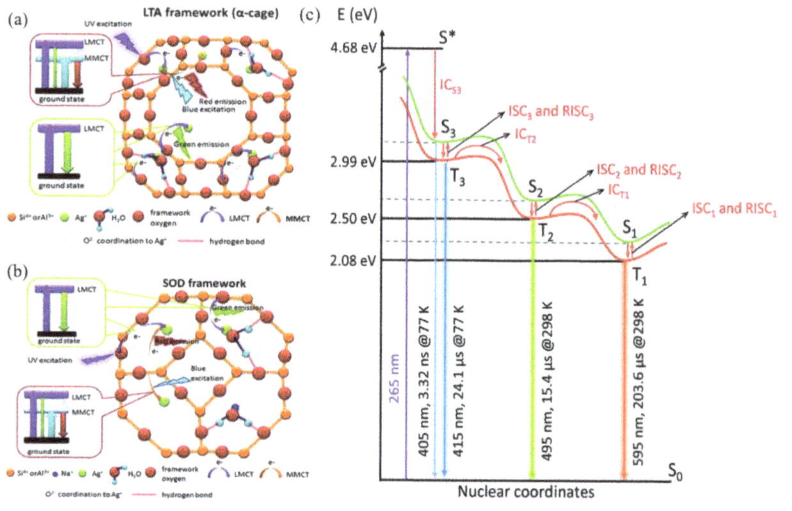

Figure 7. Proposed emission mechanisms in Ag^+ exchanged LTA (**a**) and SOD (**b**) zeolites [54]; simplified kinetic electron relaxation scheme for the processes observed upon photoexcitation of $Ag_1Li_6Na_5$−LTA zeolite (λ_{ex} = 265 nm), S* is Franck–Condon excited state (**c**) [68].

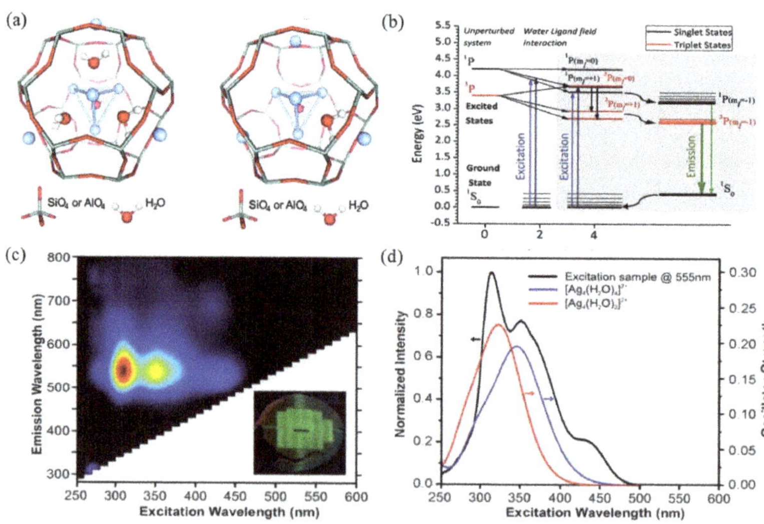

Figure 8. Distribution of $[Ag_4(H_2O)_2]^{2+}$ and $[Ag_4(H_2O)_4]^{2+}$ in LTA zeolite sodalite cage (**a**), energy level structure diagram (**b**), two–dimensional excitation–emission plot, inset: the picture of an X-ray–irradiated sample under 366 nm illumination (**c**), and excitation spectrum of Ag_3K_9−LTA zeolite (λ_{em} = 555 nm) and modeled absorption spectra for the $[Ag_4(H_2O)_2]^{2+}$ and $[Ag_4(H_2O)_4]^{2+}$ isomers (**d**) [8].

In recent research, the all Si zeolite model was further improved by taking framework Al into consideration; the researchers designed a sodalite cavity with different Si/Al ratios of 11, 5 and 1, respectively [69]. TD-DFT calculations show that the maximum positions of calculated absorption spectra are between 3 and 4 eV, which is consistent with the experimentally observed absorption peak positions. When the framework Si is replaced by Al, the triplet state energy is increased, and this asymmetric local structure is more favorable to the intersystem crossing of different excited spin states.

4. Applications of Zeolite-Encaged Ag NCs

The Ag NCs show bright and tunable emission when confined inside different zeolites, and their luminescence stability is host zeolite and preparation condition dependent. For the zeolite-encaged Ag NCs with stable and tunable emission, their potential application as LED phosphors has been demonstrated, while for those exhibiting sensitive luminescence properties to environmental variations, they can be developed as sensors for toxic and hazardous gases.

4.1. Phosphors for LEDs

White LEDs are replacing conventional lighting devices due to the advantages of being efficient, mercury-free and environmentally friendly. Recent research indicated that the LEDs fabricated with light-emitting metal-cluster-loaded zeolites provide not only light quality but also high stability and long life [70,71].

It was reported that the high thermal treatment temperature of 950 °C led to the structure collapses of Ag-loaded FAUY zeolite, and the formatted $[Ag_2]^+$ and $[Ag^+]_2$ clusters in the glassy host give off intense white light emission under UV excitation. Combining this white emitting phosphor with a UV-COB, the white LED with a CCT of 5986 K and a high Ra of 92.3 was successfully fabricated, as shown in Figure 9a,b [6]. The Ag NCs confined in Li-LTA zeolite exhibit bright green emission under 340 nm excitation, based on which the prototype green-emissive LED was fabricated using a 340 nm chip, as shown in Figure 9c, which shows uniform green emission from the phosphor layer with an EQE of 83% [48]. Alternatively, the introduction of luminescent Ag NCs-loaded LTA zeolites as emitters in a conductive polymer matrix has led to the emergence of a new type of LEDs named ZEOLEDs. By adjusting the initial silver-loading degree in zeolites, it is possible to achieve tunable emission from blue to red and even near-white [72].

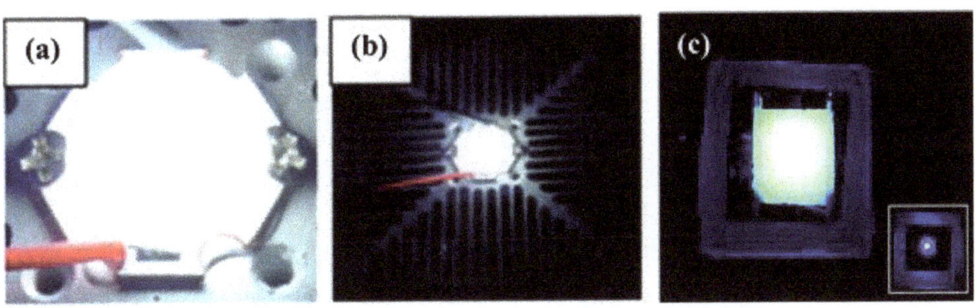

Figure 9. White-emitting block on a COB device driven by current (**a**,**b**) [6], remote phosphor LED prototype using a thin layer of the green luminescent dehydrated Ag_1Li_{11}–LTA zeolite (**c**) [48].

4.2. Detection and Sensor

The optical response of zeolite-encaged Ag NCs to external stimuli can also be applied to monitoring and sensing. For example, the reversible luminescence of Ag NCs in LTA zeolites is closely related to the hydration–dehydration process, providing new ideas for humidity sensors [9,34,63,73]. Recent research indicated that the emission wavelength of Ag NCs formatted in low silver-loaded Li-LTA zeolite is significantly related to the hydration state. The fully hydrated Ag_1Li-LTA composites (19% water content) show yellow emission, the partially hydrated samples (2% water content) show green and blue emission, and samples with less than 1% water content show blue emission, respectively, as demonstrated in Figure 10 [34]. It was suggested that this luminescent LTA(Li)-Ag composite with dramatic hydration-state-dependent color change could be used as a moisture sensor.

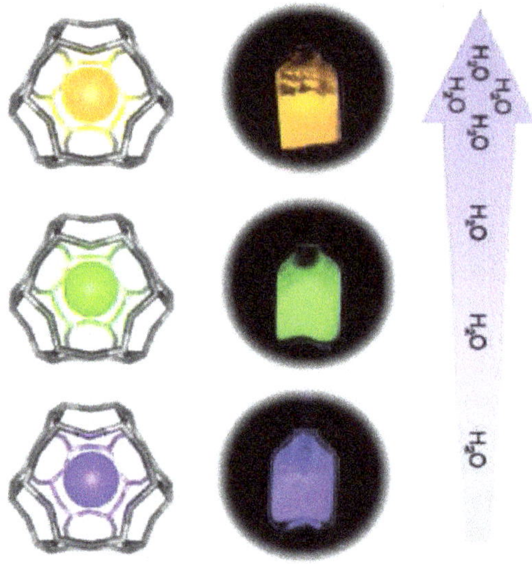

Figure 10. A scheme displaying the color change of real samples (Ag_1Li–LTA zeolite) with respect to the water content is depicted in the low-right panel [34].

Formaldehyde is the main indoor pollutant emitted by buildings and decorative materials and is harmful to human health [74,75]. Recent studies indicated that the luminescence properties of FAUY zeolite-encaged Ag NCs show a rapid and sensitive response to formaldehyde gas [76,77]. Figure 11a shows the PL spectra of Ag NCs-loaded FAUY zeolites in normal and increasing concentrations of formaldehyde atmospheres from 80 to 2560 ppb. A good linear relationship between the emission intensity of Ag NCs and formaldehyde concentration in the logarithmic range was obtained, as shown in Figure 11b [76].

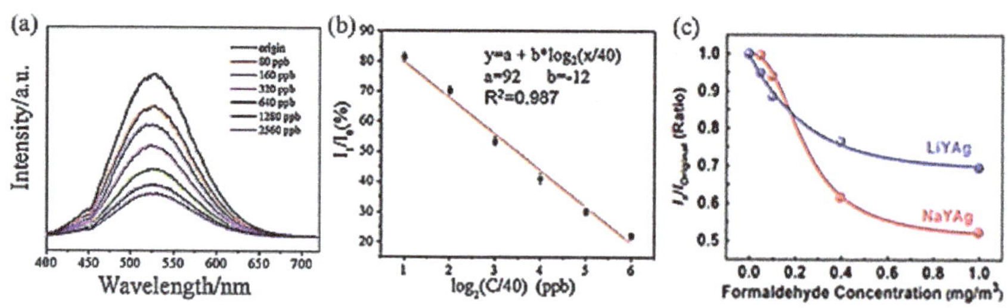

Figure 11. Emission spectra of Ag–FAUY under different formaldehyde concentration (**a**), and emission intensity of Ag–FAUY at different formaldehyde concentration (I_0: original, I_1:at different formaldehyde concentration) (**b**) [76], and relative emission intensity of Ag NCs in Li–FAUY and Na–FAUY zeolites (**c**) [77].

Further research indicated that the Ag NCs show different spectra response sensitivity in Na- and Li-FAU zeolites. Figure 11c shows relationships between formaldehyde concentration (x) and the emission intensity variation ($y = I_x/I_{original}$) of Ag CLs in Li-FAUY and Na-FAUY zeolites, which can be linearly fitted with Equations (1) and (2), respectively.

$$y = A_1 \times e^{\frac{-x}{t_0}} + A_2 \cdot \; (A_1 = 0.30,\ A_2 = 0.69,\ t_0 = 0.25,\ R_2 = 0.991) \quad (1)$$

$$y = \frac{B_2 - B_1}{\left[1 + \left(\frac{x}{x_0}\right)^p\right]} + B_2 \cdot \; (B_1 = 1.00,\ B_2 = 0.51,\ x_0 = 0.25,\ p = 2.43,\ R_2 = 0.998) \quad (2)$$

As can be observed from Figure 11c, AgLi-FAUY zeolite can more sensitively detect low-content formaldehyde than AgNa-FAUY zeolite, which is due to the high number of Ag^+/Ag^0 redox pairs in AgLi-FAUY zeolite that can promote the adsorption of formaldehyde. At higher formaldehyde concentrations, the significantly decreased emission intensity of Ag NCs in Na-FAUY zeolite is due to its relatively large specific surface area, which enables the adsorption of more formaldehyde gas. The detection limits of formaldehyde by Na- and Li-FAUY zeolite-encaged Ag NCs meet the requirements of WHO, OSHA and China's standards for indoor air quality.

5. Conclusions

Zeolites are promising scaffolds for the assembly of highly luminescent Ag NCs, as they allow for the flexible regulation of the luminescence properties of confined Ag NCs due to their abundant topology structures and exchangeable extra-framework cations. So far, the tunable emission of Ag NCs nearly across the visible region has been achieved by controlling the local crystal field and guest–host–guest interaction through host zeolite structure modification. Nevertheless, there is still a necessity to explore further methods to more precisely control the emission wavelength and improve the emission intensity of Ag NCs, for example, introducing non-luminescent RE ions as extra-framework cations or making use of intergrowth zeolites as scaffolds. On the other hand, how the geometric and electronic structure of Ag NCs influences their luminescence properties should be systematically studied with the assistance of quantum-chemistry-based computation, in which the framework Al and extra-framework cations should be taken into consideration in the building of sodalite or D6r cavities. Understanding the exact luminescent and spectral control mechanisms, as well as how the various parameters governing the luminescence tunability with the help of more advanced characterization techniques and theoretical calculations, is of great importance for the design of zeolite-encaged luminescent Ag NCs for efficient lighting phosphors and smart sensor devices applications, as well as other applications in adsorption and catalysis.

Funding: This work was financially supported by the National Natural Science Foundation of China (Nos. 51872200, 51772210), and the Natural Science Foundation of Shanghai (No. 18ZR1441900) and the National Key Research and Development Projects (No. 2018YFC1106302).

Conflicts of Interest: The authors declare no conflict of interest.

References

1. Liu, L.; Lopez-Haro, M.; Lopes, C.W.; Li, C.; Concepcion, P.; Simonelli, L.; Calvino, J.J.; Corma, A. Regioselective generation and reactivity control of subnanometric platinum clusters in zeolites for high-temperature catalysis. *Nat. Mater.* **2019**, *18*, 866–873. [CrossRef]
2. Ferreira, L.; Fonseca, A.M.; Botelho, G.; Almeida-Aguiar, C.; Neves, I.C. Antimicrobial activity of faujasite zeolites doped with silver. *Microporous Mesoporous Mater.* **2012**, *160*, 126–132. [CrossRef]
3. Goswami, N.; Yao, Q.; Luo, Z.; Li, J.; Chen, T.; Xie, J. Luminescent Metal Nanoclusters with Aggregation-Induced Emission. *J. Phys. Chem. Lett.* **2016**, *7*, 962–975. [CrossRef]
4. Vosch, T.; Antoku, Y.; Hsiang, J.-C.; Richards, C.I.; Gonzalez, J.I.; Dickson, R.M. Strongly emissive individual DNA-encapsulated Ag nanoclusters as single-molecule fluorophores. *Proc. Natl. Acad. Sci. USA* **2007**, *104*, 12616–12621. [CrossRef]
5. Devaraj, V.; Lee, I.H.; Kim, M.; Nguyen, T.M.; Son, J.P.; Lee, J.-M.; Lee, D.; Kim, K.H.; Oh, J.-W. Unveiling facet effects in metallic nanoparticles to design an efficient plasmonic nanostructure. *Curr. Appl. Phys.* **2022**, *44*, 22–28. [CrossRef]

6. Yao, D.; Xu, S.; Wang, Y.; Li, H. White-emitting phosphors with high color-rendering index based on silver cluster-loaded zeolites and their application to near-UV LED-based white LEDs. *Mater. Chem. Front.* **2019**, *3*, 1080–1084. [CrossRef]
7. De Cremer, G.; Coutino-Gonzalez, E.; Roeffaers, M.B.; De Vos, D.E.; Hofkens, J.; Vosch, T.; Sels, B.F. In situ observation of the emission characteristics of zeolite-hosted silver species during heat treatment. *Chemphyschem* **2010**, *11*, 1627–1631. [CrossRef]
8. Grandjean, D.; Coutino-Gonzalez, E.; Cuong, N.T.; Fron, E.; Baekelant, W.; Aghakhani, S.; Schlexer, P.; D'Acapito, F.; Banerjee, D.; Roeffaers, M.B.J.; et al. Origin of the bright photoluminescence of few-atom silver clusters confined in LTA zeolites. *Science* **2018**, *361*, 686–690. [CrossRef]
9. Aghakhani, S.; Grandjean, D.; Baekelant, W.; Coutino-Gonzalez, E.; Fron, E.; Kvashnina, K.; Roeffaers, M.B.J.; Hofkens, J.; Sels, B.F.; Lievens, P. Atomic scale reversible opto-structural switching of few atom luminescent silver clusters confined in LTA zeolites. *Nanoscale* **2018**, *10*, 11467–11476. [CrossRef]
10. Quintanilla, M.; Liz-Marzan, L.M. Caged clusters shine brighter. *Science* **2018**, *361*, 645. [CrossRef]
11. Diez, I.; Ras, R.H. Fluorescent silver nanoclusters. *Nanoscale* **2011**, *3*, 1963–1970. [CrossRef] [PubMed]
12. Romolini, G.; Steele, J.A.; Hofkens, J.; Roeffaers, M.B.J.; Coutino-Gonzalez, E. Tunable Luminescence from Stable Silver Nanoclusters Confined in Microporous Zeolites. *Adv. Opt. Mater.* **2021**, *9*, 2100526. [CrossRef]
13. Gwinn, E.; Schultz, D.; Copp, S.M.; Swasey, S. DNA-Protected Silver Clusters for Nanophotonics. *Nanomaterials* **2015**, *5*, 180–207. [CrossRef] [PubMed]
14. Kondo, J.; Tada, Y.; Dairaku, T.; Hattori, Y.; Saneyoshi, H.; Ono, A.; Tanaka, Y. A metallo-DNA nanowire with uninterrupted one-dimensional silver array. *Nat. Chem.* **2017**, *9*, 956–960. [CrossRef] [PubMed]
15. Yang, X.; Sun, J.-K.; Kitta, M.; Pang, H.; Xu, Q. Encapsulating highly catalytically active metal nanoclusters inside porous organic cages. *Nat. Catal.* **2018**, *1*, 214–220. [CrossRef]
16. Yonesato, K.; Ito, H.; Itakura, H.; Yokogawa, D.; Kikuchi, T.; Mizuno, N.; Yamaguchi, K.; Suzuki, K. Controlled Assembly Synthesis of Atomically Precise Ultrastable Silver Nanoclusters with Polyoxometalates. *J. Am. Chem. Soc.* **2019**, *141*, 19550–19554. [CrossRef] [PubMed]
17. Tikhomirov, V.K.; Rodríguez, V.D.; Kuznetsov, A.; Kirilenko, D.; Van Tendeloo, G.; Moshchalkov, V.V. Preparation and luminescence of bulk oxyfluoride glasses doped with Ag nanoclusters. *Opt. Express* **2010**, *18*, 22032–22040. [CrossRef]
18. Cuong, N.T.; Tikhomirov, V.K.; Chibotaru, L.F.; Stesmans, A.; Rodriguez, V.D.; Nguyen, M.T.; Moshchalkov, V.V. Experiment and theoretical modeling of the luminescence of silver nanoclusters dispersed in oxyfluoride glass. *J. Chem. Phys.* **2012**, *136*, 174108. [CrossRef]
19. Kuznetsov, A.S.; Cuong, N.T.; Tikhomirov, V.K.; Jivanescu, M.; Stesmans, A.; Chibotaru, L.F.; Velázquez, J.J.; Rodríguez, V.D.; Kirilenko, D.; Van Tendeloo, G.; et al. Effect of heat-treatment on luminescence and structure of Ag nanoclusters doped oxyfluoride glasses and implication for fiber drawing. *Opt. Mater.* **2012**, *34*, 616–621. [CrossRef]
20. Shestakov, M.V.; Cuong, N.T.; Tikhomirov, V.K.; Nguyen, M.T.; Chibotaru, L.F.; Moshchalkov, V.V. Quantum Chemistry Modeling of Luminescence Kinetics of Ag Nanoclusters Dispersed in Glass Host. *J. Phys. Chem. C* **2013**, *117*, 7796–7800. [CrossRef]
21. Houk, R.J.T.; Jacobs, B.W.; El Gabaly, F.; Chang, N.N.; Talin, A.A.; Graham, D.D.; House, S.D.; Robertson, I.M.; Allendorf, M.D. Silver Cluster Formation, Dynamics, and Chemistry in Metal-Organic Frameworks. *Nano Lett.* **2009**, *9*, 3413–3418. [CrossRef] [PubMed]
22. Sun, N.; Yan, B. Ag(+)-induced photoluminescence enhancement in lanthanide post-functionalized MOFs and Ag(+) sensing. *Phys. Chem. Chem. Phys.* **2017**, *19*, 9174–9180. [CrossRef]
23. Ge, K.-M.; Wang, D.; Xu, Z.-J.; Chu, R.-Q. A luminescent Eu(III)-MOF for selective sensing of Ag+ in aqueous solution. *J. Mol. Struct.* **2020**, *1208*, 127862. [CrossRef]
24. Taiji, K.; Iso, Y.; Isobe, T. Fluorescent Ag^+-exchanged zeolite nanoparticles with improved photoluminescence properties via X-ray irradiation. *J. Lumin.* **2018**, *196*, 214–220. [CrossRef]
25. Tian, X.; Li, Q.; Yao, D.; Li, P.; Li, H.; Wang, Y. Tunable luminescence of silver-exchanged SOD zeolite thermally treated under mild conditions. *J. Mater. Chem. C* **2022**, *10*, 1666–1671. [CrossRef]
26. Rálek, M.; Jírů, P.; Grubner, O.; Beyer, H. Molekularsiebe mit farbiger Indizierung des Wassergehaltes. *Collect. Czechoslov. Chem. Commun.* **1962**, *27*, 142–146. [CrossRef]
27. Tsutsumi, K.; Takahashi, H. The Formation of Metallic Silver in Silver-Form Zeolites. *Bull. Chem. Soc. Jpn.* **1972**, *45*, 2332–2337. [CrossRef]
28. Gellens, L.R.; Mortier, W.J.; Uytterhoeven, J.B. On the nature of the charged silver clusters in zeolites of type A, X and Y. *Zeolites* **1981**, *1*, 11–18. [CrossRef]
29. Yu, J.; Ye, S.; Liao, H.; Xv, X.; Wang, D. Luminescence Control of Silver Nanoclusters by Tailoring Extra-Framework Cations in FAU-Y Zeolite: Implications for Tunable Emission. *ACS Appl. Nano Mater.* **2021**, *4*, 13692–13699. [CrossRef]
30. Yu, J.; Ye, S.; Shi, Y.; Liao, H.; Wang, D. Thermally formation and luminescent performance of silver nanoclusters confined within LTA zeolites. *J. Alloys Compd.* **2021**, *857*, 157614. [CrossRef]
31. Mayoral, A.; Anderson, P.A. Production of bimetallic nanowires through electron beam irradiation of copper- and silver-containing zeolite A. *Nanotechnology* **2007**, *18*, 165708. [CrossRef]
32. Coutino-Gonzalez, E.; Grandjean, D.; Roeffaers, M.; Kvashnina, K.; Fron, E.; Dieu, B.; De Cremer, G.; Lievens, P.; Sels, B.; Hofkens, J. X-ray irradiation-induced formation of luminescent silver clusters in nanoporous matrices. *Chem. Commun.* **2014**, *50*, 1350–1352. [CrossRef] [PubMed]

33. Fenwick, O.; Coutino-Gonzalez, E.; Richard, F.; Bonacchi, S.; Baekelant, W.; de Vos, D.; Roeffaers, M.B.J.; Hofkens, J.; Samori, P. X-ray-Induced Growth Dynamics of Luminescent Silver Clusters in Zeolites. *Small* **2020**, *16*, e2002063. [CrossRef]
34. Coutino-Gonzalez, E.; Baekelant, W.; Grandjean, D.; Roeffaers, M.B.J.; Fron, E.; Aghakhani, M.S.; Bovet, N.; Van der Auweraer, M.; Lievens, P.; Vosch, T.; et al. Thermally activated LTA(Li)–Ag zeolites with water-responsive photoluminescence properties. *J. Mater. Chem. C* **2015**, *3*, 11857–11867. [CrossRef]
35. Mayoral, A.; Carey, T.; Anderson, P.A.; Lubk, A.; Diaz, I. Atomic resolution analysis of silver ion-exchanged zeolite A. *Angew. Chem. Int. Ed. Engl.* **2011**, *50*, 11230–11233. [CrossRef] [PubMed]
36. Altantzis, T.; Coutino-Gonzalez, E.; Baekelant, W.; Martinez, G.T.; Abakumov, A.M.; Tendeloo, G.V.; Roeffaers, M.B.; Bals, S.; Hofkens, J. Direct Observation of Luminescent Silver Clusters Confined in Faujasite Zeolites. *ACS Nano* **2016**, *10*, 7604–7611. [CrossRef]
37. Mayoral, A.; Carey, T.; Anderson, P.A.; Diaz, I. Atomic resolution analysis of porous solids: A detailed study of silver ion-exchanged zeolite A. *Microporous Mesoporous Mater.* **2013**, *166*, 117–122. [CrossRef]
38. Hiyoshi, N.; Ikeda, T.; Hasegawa, Y.; Sato, K. Observation of La-exchanged NaY zeolite using aberration-corrected scanning transmission electron microscopy. *Microporous Mesoporous Mater.* **2021**, *311*, 110711. [CrossRef]
39. Michalik, J.; Sadlo, J.; Kodaira, T.; Shimomura, S.; Yamada, H. ESR and optical studies of cationic silver clusters in zeolite rho. *J. Radioanal. Nucl. Chem.* **1998**, *232*, 135–138. [CrossRef]
40. De Cremer, G.; Coutino-Gonzalez, E.; Roeffaers, M.B.; Moens, B.; Ollevier, J.; Van der Auweraer, M.; Schoonheydt, R.; Jacobs, P.A.; De Schryver, F.C.; Hofkens, J.; et al. Characterization of fluorescence in heat-treated silver-exchanged zeolites. *J. Am. Chem. Soc.* **2009**, *131*, 3049–3056. [CrossRef]
41. Sun, T.; Seff, K. Silver Clusters and Chemistry in Zeolites. *Chem. Rev.* **1994**, *94*, 857–870. [CrossRef]
42. Lamberti, C.; Prestipino, C.; Bordiga, S.; Fitch, A.N.; Marra, G.L. Characterization of isolated Ag cations in homoionic Ag-Y zeolites: A combined anomalous XRPD and EXAFS study. *Nucl. Instrum. Methods Phys. Res. Sect. B Beam Interact. Mater. At.* **2003**, *200*, 155–159. [CrossRef]
43. Miyanaga, T.; Suzuki, Y.; Narita, S.; Nakamura, R. Local structure change of luminescent Ag zeolite-A and -X studied by in situ XAFS and IR spectroscopy. *J. Synchrotron Radiat.* **2020**, *27*, 1640–1647. [CrossRef] [PubMed]
44. Baekelant, W.; Aghakhani, S.; Coutino-Gonzalez, E.; Kennes, K.; D'Acapito, F.; Grandjean, D.; Van der Auweraer, M.; Lievens, P.; Roeffaers, M.B.J.; Hofkens, J.; et al. Shaping the Optical Properties of Silver Clusters Inside Zeolite A via Guest-Host-Guest Interactions. *J. Phys. Chem. Lett.* **2018**, *9*, 5344–5350. [CrossRef]
45. Cuong, N.T.; Nguyen, H.M.; Nguyen, M.T. Theoretical modeling of optical properties of Ag8 and Ag14 silver clusters embedded in an LTA sodalite zeolite cavity. *Phys. Chem. Chem. Phys.* **2013**, *15*, 15404–15415. [CrossRef]
46. Cuong, N.T.; Nguyen, H.M.; Pham-Ho, M.P.; Nguyen, M.T. Optical properties of the hydrated charged silver tetramer and silver hexamer encapsulated inside the sodalite cavity of an LTA-type zeolite. *Phys. Chem. Chem. Phys.* **2016**, *18*, 18128–18136. [CrossRef]
47. Selvam, T.; Warmuth, F.; Klumpp, M.; Warnick, K.G.; Lodes, M.A.; Korner, C.; Schwieger, W. Fabrication and pressure drop behavior of novel monolithic structures with zeolitic architectures. *Chem. Eng. J.* **2016**, *288*, 223–227. [CrossRef]
48. Baekelant, W.; Aghakhani, S.; Fron, E.; Martin, C.; Woong-Kim, C.; Steele, J.A.; De Baerdemaeker, T.; d'Acapito, F.; Chernysov, D.; van der Auweraer, M.; et al. Luminescent silver–lithium-zeolite phosphors for near-ultraviolet LED applications. *J. Mater. Chem. C* **2019**, *7*, 14366–14374. [CrossRef]
49. Fenwick, O.; Coutino-Gonzalez, E.; Grandjean, D.; Baekelant, W.; Richard, F.; Bonacchi, S.; De Vos, D.; Lievens, P.; Roeffaers, M.; Hofkens, J.; et al. Tuning the energetics and tailoring the optical properties of silver clusters confined in zeolites. *Nat. Mater.* **2016**, *15*, 1017–1022. [CrossRef]
50. Yu, J.; Ye, S.; Shi, Y.; Liao, H.; Xv, X.; Wang, D. Confinement of Highly Luminescent Silver Nanoclusters in FAUY Zeolite: A Study on the Formation Effects of Silver Nanoclusters and the Sensitization of Tb^{3+}. *ACS Appl. Nano Mater.* **2021**, *4*, 6290–6298. [CrossRef]
51. Shi, Y.L.; Ye, S.; Liao, H.Z.; Liu, J.F.; Wang, D.P. Formation of luminescent silver-clusters and efficient energy transfer to Eu^{3+} in faujasite NaX zeolite. *J. Solid. State Chem.* **2020**, *285*, 121227. [CrossRef]
52. Hoshino, H.; Sannohe, Y.; Suzuki, Y.; Azuhata, T.; Miyanaga, T.; Yaginuma, K.; Itoh, M.; Shigeno, T.; Osawa, Y.; Kimura, Y. Photoluminescence of the Dehydrated Ag-type Zeolite A Packed under Air. *J. Phys. Soc. Jpn.* **2008**, *77*, 064712. [CrossRef]
53. Yao, D.; Wang, Y.; Li, P.; Li, H. Luminescent $Ag(^{+})$ exchanged SOD zeolites with their potential applications in white LEDs. *Dalton Trans.* **2020**, *49*, 8179–8185. [CrossRef] [PubMed]
54. Lin, H.; Imakita, K.; Fujii, M.; Prokof'ev, V.Y.; Gordina, N.E.; Said, B.; Galarneau, A. Visible emission from $Ag(^{+})$ exchanged SOD zeolites. *Nanoscale* **2015**, *7*, 15665–15671. [CrossRef]
55. Frising, T.; Leflaive, P. Extraframework cation distributions in X and Y faujasite zeolites: A review. *Microporous Mesoporous Mater.* **2008**, *114*, 27–63. [CrossRef]
56. Aono, H.; Yahara, K.; Johan, E.; Itagaki, Y. Effect of coexisting lithium content on fluorescent properties of silver ion-exchanged LTA zeolite. *J. Ceram. Soc. Jpn.* **2020**, *128*, 670–676. [CrossRef]
57. Li, W.; Chai, Y.; Wu, G.; Li, L. Stable and Uniform Extraframework Cations in Faujasite Zeolites. *J. Phys. Chem. Lett.* **2022**, *13*, 11419. [CrossRef]
58. Xv, X.; Ye, S.; Pan, L.; Lin, P.; Liao, H.; Wang, D. Tailoring the Luminescence Properties of Silver Clusters Confined in Faujasite Zeolite through Framework Modification. *Materials* **2022**, *15*, 7431. [CrossRef]

59. Qin, Z.; Shen, W.; Zhou, S.; Shen, Y.; Li, C.; Zeng, P.; Shen, B. Defect-assisted mesopore formation during Y zeolite dealumination: The types of defect matter. *Microporous Mesoporous Mater.* **2020**, *303*, 110248. [CrossRef]
60. Groen, J.C.; Moulijn, J.A.; Pérez-Ramírez, J. Decoupling mesoporosity formation and acidity modification in ZSM-5 zeolites by sequential desilication–dealumination. *Microporous Mesoporous Mater.* **2005**, *87*, 153–161. [CrossRef]
61. Seifert, R.; Rytz, R.; Calzaferri, G. Colors of Ag^+-Exchanged Zeolite A. *J. Phys. Chem. A* **2000**, *104*, 7473–7483. [CrossRef]
62. Seifert, R.; Kunzmann, A.; Calzaferri, G. The Yellow Color of Silver-Containing Zeolite A. *Angew. Chem. Int. Ed.* **1998**, *37*, 1521–1524. [CrossRef]
63. Lin, H.; Imakita, K.; Fujii, M. Reversible emission evolution from Ag activated zeolite Na-A upon dehydration/hydration. *Appl. Phys. Lett.* **2014**, *105*, 211903. [CrossRef]
64. Fron, E.; Aghakhani, S.; Baekelant, W.; Grandjean, D.; Coutino-Gonzalez, E.; Van der Auweraer, M.; Roeffaers, M.B.J.; Lievens, P.; Hofkens, J. Structural and Photophysical Characterization of Ag Clusters in LTA Zeolites. *J. Phys. Chem. C* **2019**, *123*, 10630–10638. [CrossRef]
65. Shi, Y.; Ye, S.; Yu, J.; Liao, H.; Liu, J.; Wang, D. Simultaneous energy transfer from molecular-like silver nanoclusters to Sm^{3+}/Ln^{3+} (Ln = Eu or Tb) in glass under UV excitation. *Opt. Express* **2019**, *27*, 38159–38167. [CrossRef]
66. Yao, D.; Yang, J.; Xie, Y.; Wang, Y.; Wang, Y.; Li, H. Warm white-light phosphor based on a single-phase of $Ag^+/Eu^{3+}/Zn^{2+}$ loading SOD zeolites with application to white LEDs. *J. Alloys Compd.* **2020**, *823*, 153778. [CrossRef]
67. Calzaferri, G.; Leiggener, C.; Glaus, S.; Schurch, D.; Kuge, K. The electronic structure of Cu^+, Ag^+, and Au^+ zeolites. *Chem. Soc. Rev.* **2003**, *32*, 29–37. [CrossRef]
68. Sun, L.; Keshavarz, M.; Romolini, G.; Dieu, B.; Hofkens, J.; de Jong, F.; Fron, E.; Roeffaers, M.B.J.; Van der Auweraer, M. Origin of the polychromatic photoluminescence of zeolite confined Ag clusters: Temperature- and co-cation-dependent luminescence. *Chem. Sci.* **2022**, *13*, 11560–11569. [CrossRef]
69. Chan, B. Re-examining the electronic structure of fluorescent tetra-silver clusters in zeolites. *Phys. Chem. Chem. Phys.* **2021**, *23*, 1984–1993. [CrossRef]
70. Baekelant, W.; Coutino-Gonzalez, E.; Steele, J.A.; Roeffaers, M.B.J.; Hofkens, J. Form Follows Function: Warming White LEDs Using Metal Cluster-Loaded Zeolites as Phosphors. *ACS Energy Lett.* **2017**, *2*, 2491–2497. [CrossRef]
71. Zhou, X.; Ren, T.; Yin, J.; Zhang, X.; Li, P.; Li, H. A full visible LED enabled by a broadband yellow emission nepheline phosphor derived from a europium doped SOD zeolite. *J. Mater. Chem. C* **2022**, *10*, 15613–15619. [CrossRef]
72. Kennes, K.; Coutino-Gonzalez, E.; Martin, C.; Baekelant, W.; Roeffaers, M.B.J.; Van der Auweraer, M. Silver Zeolite Composites-Based LEDs: A Novel Solid-State Lighting Approach. *Adv. Funct. Mater.* **2017**, *27*, 1606411. [CrossRef]
73. Alcantara, G.P.; Ribeiro, L.E.B.; Alves, A.F.; Andrade, C.M.G.; Fruett, F. Humidity sensor based on zeolite for application under environmental conditions. *Microporous Mesoporous Mater.* **2017**, *247*, 38–45. [CrossRef]
74. Zhang, L.; Xie, Y.; Jiang, Y.; Li, Y.; Wang, C.; Han, S.; Luan, H.; Meng, X.; Xiao, F.-S. Mn-promoted Ag supported on pure siliceous Beta zeolite (Ag/Beta-Si) for catalytic combustion of formaldehyde. *Appl. Catal. B Environ.* **2020**, *268*, 118461. [CrossRef]
75. Wang, Y.; Dai, C.; Chen, B.; Wang, Y.; Shi, C.; Guo, X. Nanoscale HZSM-5 supported PtAg bimetallic catalysts for simultaneous removal of formaldehyde and benzene. *Catal. Today* **2015**, *258*, 616–626. [CrossRef]
76. Yao, D.; Wang, Y.; Li, H. Silver clusters based sensor for Low content formaldehyde detection in colorimetric and fluorometric dual Mode. *Sens. Actuators B Chem.* **2020**, *305*, 127451. [CrossRef]
77. Yu, J.; Ye, S.; Xv, X.; Pan, L.; Lin, P.; Liao, H.; Wang, D. Thermal-Driven Formation of Silver Clusters Inside Na/Li FAUY Zeolites for Formaldehyde Detection. *Nanomaterials* **2022**, *12*, 3215. [CrossRef]

Disclaimer/Publisher's Note: The statements, opinions and data contained in all publications are solely those of the individual author(s) and contributor(s) and not of MDPI and/or the editor(s). MDPI and/or the editor(s) disclaim responsibility for any injury to people or property resulting from any ideas, methods, instructions or products referred to in the content.

materials

MDPI

Review

Sustainable Lifecycle of Perforated Metal Materials

Viktors Mironovs [1], Jekaterina Kuzmina [1,*], Dmitrijs Serdjuks [2], Yulia Usherenko [1] and Mihails Lisicins [1]

[1] Scientific Laboratory of Powder Materials, Faculty of Mechanical Engineering, Transport and Aeronautics, Riga Technical University, 6B Kipsalas Street, LV-1048 Riga, Latvia; viktors.mironovs@rtu.lv (V.M.); yuliausherenko@gmail.com (Y.U.); mihails.lisicins@gmail.com (M.L.)

[2] Institute of Structural Engineering, Riga Technical University, 6A Kipsalas Street, LV-1048 Riga, Latvia; dmitrijs.serdjuks@rtu.lv

* Correspondence: ekaterina.kuzmina@inbox.lv; Tel.: +371-29146944

Abstract: In an era of rapidly growing consumer demand and the subsequent development of production, light materials and structures with a wide range of applications are becoming increasingly important in the field of construction and mechanical engineering, including aerospace engineering. At the same time, one of the trends is the use of perforated metal materials (PMMs). They are used as finishing, decorative and structural building materials. The main feature of PMMs is the presence of through holes of a given shape and size, which makes it possible to have low specific gravity; however, their tensile strength and rigidity can vary widely depending on the source material. In addition, PMMs have several properties that cannot be achieved with solid materials; for example, they can provide considerable noise reduction and partial light absorption, significantly reducing the weight of structures. They are also used for damping dynamic forces, filtering liquids and gases and shielding electromagnetic fields. For the perforation of strips and sheets, cold stamping methods are usually used, carried out on stamping presses, particularly using wide-tape production lines. Other methods of manufacturing PMMs are rapidly developing, for example, using liquid and laser cutting. An urgent but relatively new and little-studied problem is the recycling and further efficient use of PMMs, primarily such materials as stainless and high-strength steels, titanium, and aluminum alloys. The life cycle of PMMs can be prolonged because they can be repurposed for various applications such as constructing new buildings, designing elements, and producing additional products, making them more environmentally friendly. This work aimed to overview sustainable ways of PMM recycling, use or reuse, proposing different ecological methods and applications considering the types and properties of PMM technological waste. Moreover, the review is accompanied by graphical illustrations of real examples. PMM waste recycling methods that can prolong their lifecycle include construction technologies, powder metallurgy, permeable structures, etc. Several new technologies have been proposed and described for the sustainable application of products and structures based on perforated steel strips and profiles obtained from waste products during stamping. With more developers aiming for sustainability and buildings achieving higher levels of environmental performance, PMM provides significant environmental and aesthetic advantages.

Keywords: perforated metal materials; PMM technological waste recycling; structural members; sustainable

Citation: Mironovs, V.; Kuzmina, J.; Serdjuks, D.; Usherenko, Y.; Lisicins, M. Sustainable Lifecycle of Perforated Metal Materials. *Materials* **2023**, *16*, 3012. https://doi.org/10.3390/ma16083012

Academic Editors: Pavol Liptai, Jaroslav Briančin and Maroš Halama

Received: 10 March 2023
Revised: 27 March 2023
Accepted: 28 March 2023
Published: 11 April 2023

Copyright: © 2023 by the authors. Licensee MDPI, Basel, Switzerland. This article is an open access article distributed under the terms and conditions of the Creative Commons Attribution (CC BY) license (https://creativecommons.org/licenses/by/4.0/).

1. Introduction

Worldwide annual resource use reached almost 90 billion metric tones in 2017 and could more than double by 2050 [1]. By 2030, European Raw Materials Alliance (ERMA) activities will increase the production of raw and advanced materials and address the circular economy by encouraging the recovered processing of essential raw materials [2]. In an era of rapidly growing consumer demand and the subsequent development of production, lightweight materials and structures with a wide range of applications are becoming increasingly important in the field of construction and mechanical engineering. Unfortunately, metal recycling tends to be at a low level globally. In 2010, the International Resource Group, hosted by the United Nations Environment Programme, issued reports on the metal stocks that exist in society and their recycling rates [3].

Perforated metal is a valuable source of secondary raw materials to be processed in metallurgical enterprises. The secondary use of perforated metal makes the process less expensive than conventional remelting methods, as well as significantly reducing emissions of harmful substances into the atmosphere and making it more sustainable as well as prolonging the lifecycle. There is an urgent need to change how we use non-renewable resources, especially metals, to maintain their sustainable availability and minimize the negative impacts associated with their production and use [4].

Products made of ferrous and non-ferrous metals or alloys are a valuable source of secondary raw materials to be processed in metallurgical enterprises. The use of scrap metal makes it possible to make the technological process of smelting metals less energy-intensive compared to obtaining products from ore and significantly reduces emissions of harmful substances into the atmosphere. Metals also can be returned to the production process without impairment. The steel sector will indeed be a key contributor to Europe achieving its 2050 targets. Thanks to their unique properties, ferrous and non-ferrous metals can be recycled indefinitely. The strong recycling performance is widely recognized and long-lasting. Moreover, the industry's business model is circular by nature as the input of steel scrap is a necessary component for making new steel at any one of more than 500 steel plants in Europe. The more quality scrap that can be used in new steel production, the less raw materials and energy are needed and in turn, this reduces emissions. Steel, for instance, is circular by nature, because it is recycled repeatedly without loss of quality. Currently, of all steel packaging put on the market in Europe, 84% is being recycled into new steel products [5].

Solid waste generation is increasing in the world today. Zero-waste strategy is a smart solution to minimizing the increasing solid waste. To keep to a minimum solid waste, even more work is needed in the long run [6]. Reducing the resource intensity of products and emissions into the environment and improving their socio-economic performance throughout the life cycle of a material or product is an important task. Its solution can significantly facilitate the links between economic, social, and environmental aspects along the entire value chain of a product. At every stage of the life cycle, there is potential to reduce resource consumption and increase productivity.

Metals can be recycled back into the manufacturing process without compromising the quality of the subsequent product. In fact, recycling represents an advantage because it takes less energy to melt the metal. The recycling quota for scrap metal sent for recycling is 90% [7].

Modern architecture uses building materials in new and innovative ways. The fascination comes from the materials themselves and how they are processed and used. The emphasis is not so much on the material to be matched to the building, but on the effect of the particular materials created for the building with their material and visual qualities [5].

One of the trends in the manufacture of products and structures is the use of perforated metal materials (PMMs). These are used as finishing, decorative and structural building materials [8]. The growing interest is due to the fact that they significantly reduce the weight of the structure [9]. In addition, PMMs have a number of properties that cannot be achieved with solid materials; for example, they provide significant noise reduction [10], and light scattering [11]. They can also be used for damping dynamic force [12], filtering liquids and gases [13], and shielding electromagnetic fields [14]. PMMs are the simplest in terms of structure and implementation. They seem to be one of the most promising for wide production and application. The main feature of PMMs is the presence of through holes of a given shape and size. PMMs have low specific gravity, but their tensile strength and rigidity can vary widely depending on the source material [12]. This allows them to find a stable application, in particular in construction (Figure 1).

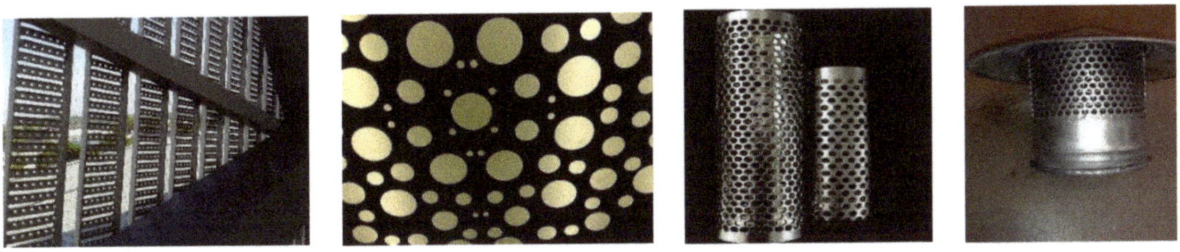

(a) (b) (c) (d)

Figure 1. The use of PMMs: to reduce the level of illumination inside the room (**a**), increase the uniformity of the lighting of the elevator cabin (**b**), mute the sound (**c**) and ventilate the premises (**d**).

There are many examples of the effective use of PMMs in the creation of enclosing structures for buildings and mechanical structures [15], for the manufacture of facades and decorative elements, as well as in the production of fittings for pipes and panels. PMMs are increasingly used in various interior solutions for premises and urban architecture [16,17].

In addition, PMMs have become popular and common materials for construction and installation works. Perforated steel strip components are widely used to connect and reinforce individual parts of building structures [16,17]. Most often, these are joints of load-bearing and enclosing structures made of drywall, as well as wood and concrete. In mechanical engineering, PMM products have found application in the manufacture of filters [18,19], separator assemblies, dryers, and other equipment [20] (Figure 2).

(a) (b) (c) (d)

Figure 2. Structural units (**a**,**b**) and elements of building and garden furniture (**c**,**d**) from PMM.

For the perforation of strips and sheets, the cold stamping method is usually used, which is implemented on stamping presses, as well as by using broad tape production lines [21]. Other methods for the production of PMMs are rapidly developing, for example, using liquid [22,23] and laser cutting [24]. The volumes of production and use of thin-sheet stamping products in the world are constantly growing and the biggest sector is the expanding automotive market (Figure 3) [25].

The most common PMMs are steel, aluminum and copper. The main feature of PMM is the presence of through holes of a given shape and size (Figure 4).

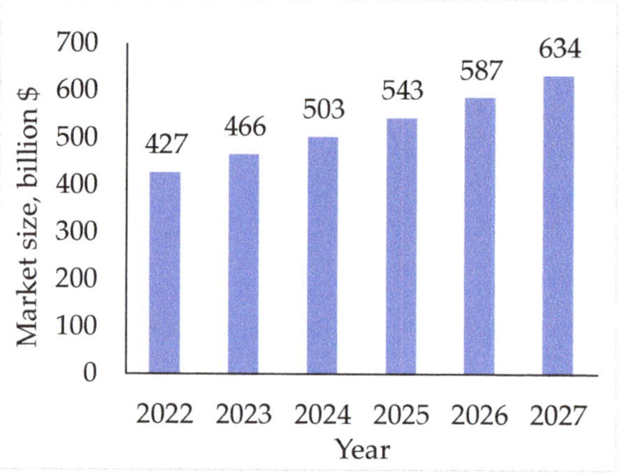

Figure 3. The estimated growth of the global stamped metal market [25].

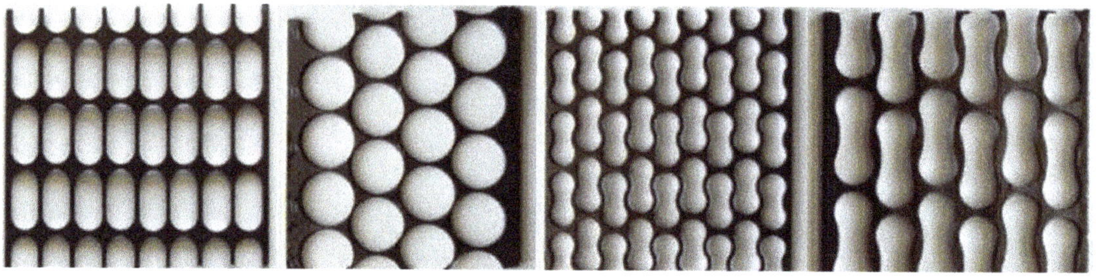

Figure 4. Some types of PMM residues with different shapes and sizes of hole samples.

In practice, the following physical and mechanical properties of PMMs are of the greatest interest: density, strength, ductility, weldability, corrosion resistance, hardness and microhardness and surface roughness [26,27]. When choosing materials for further use, an important point is the deformation state and the presence of corrosion phenomena. This is especially important when they are used in aggressive environments, such as marine, space, or aviation [28,29]. Experiments have shown that the tensile strength of steel perforated bands achieved as waste after punching is in the range of 100–250 MPa according to the type of perforation, width and thickness of the band and base material [30]. The percentage of perforation (%) of perforated bands, achieved after punching, varied from 66% to 76% when the thickness is in the range of 1.00 to 1.75 mm [30]. Studies have also shown that when perforations are made in the boundary layers, an increase in the microhardness of the material is observed and cracks may form. In particular, for steel, when stamping in the area of a punched hole (0.2–0.3 mm from the edge), the microhardness values almost doubled, but at a distance of 0.7–0.9 mm, it already reached the usual values for this material [31].

Punching or cutting perforation openings directly affects the mechanical and physical properties of the perforated metallic material. In many cases, the perforation process causes an increase in the surface hardness of the material, or forging [32]. During piercing or cutting, plastic deformation of the material occurs. Its degree on the edges of the perforation holes is very large and the grains of the metal microstructure are stretched in the direction of the applied load; therefore, their boundary surface can be invisible. In the process of forging the material, the specific volume of the metal increases and the density decreases, the ultimate strength, microhardness and brittleness increase while the plasticity and viscosity decrease. The stress concentration around the perforation opening increases.

Even changes in the magnetic properties of the material are possible. Gradually moving away from the edges of the openings, the degree of metal deformation decreases. For high carbon steels, the zone of obvious deformation may be 0.3 to 0.4 mm from the edge of the perforation slots.

2. Technological Waste PMM and Methods for Their Further Processing

PMM waste is generated during the reconstruction or demolition of buildings and structures in which they were used [33]. Another significant type of PMM technological waste is waste after stamping parts from a sheet or from a strip (cutting) (Figure 5).

Figure 5. PMM waste: rolls of steel tape (**a**); after rough grinding (**b**); in the form of pistons (**c**).

2.1. PMM Waste in Construction Technologies

Due to its diverse geometric parameters, physical and mechanical properties, PMM technological waste, primarily from steel and aluminum, can be effectively used in construction in the manufacture of enclosing structures, elements of facades and interiors, lighting devices, and the manufacture of some construction and finishing tools [34,35]. PMM can be used for producing several types of load-carrying, insulating and decorative structural members [17,36].

The main functions of building envelopes can be called thermal insulation and protection from negative external factors. Enclosing structures can consist of either one layer or several, for example, the outer cladding and the inner part, as well as insulation between them [37]. When we use the overlay method of perforated sheets or strips, we can obtain perforated frames. The imposition of PMMs can be multi-layered, which provides a more stable structure while maintaining the overall dimensions and creating a new ornament (Figure 6).

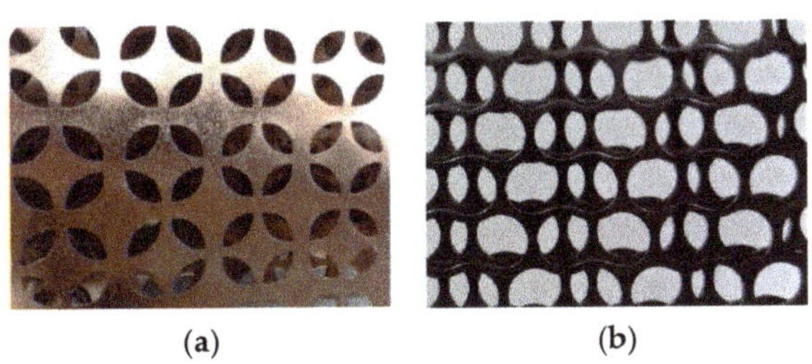

Figure 6. Fragments of panels: from two perforated steel strips laid with a copper shift (**a**), from steel several strips laid in parallel (**b**).

The connection of perforated sheets or strips is easily achieved by welding [38], soldering, bolting, or gluing [39]. The ability to overlap sheets in both parallel and perpendicular directions, adjust the mutual displacement of the perforation, and combine their sizes, allows us to significantly diversify the shapes of the resulting panels and their properties.

PMM-made perforated steel profiles can also be used for wall frames or structures as stiffening elements. Currently, self-bearing partitions and external wall finishes are often used for non-perforated metal profiles. One of the ways of creating the framework and providing the necessary stiffness for insulation wall panels is the use of perforated double-T profiles. In this way, the wall frame is considerably lightened [40]. The profile cavities are filled with insulation or soundproofing material inserted into the profiles, and the structure is insulated from the outside with plasterboard sheets or other finishing material. The perforation of the profiles provides opportunity for the plasterboard to connect it by fast-setting adhesive, such as gypsum-based glue. A very simple and effective wall structure, which contains heat and sound insulation, and the supporting frame of the wall is made of PMM in the form of a corner or channel (Figure 7).

Figure 7. The solution for the partition wall construction using perforated steel profiles: (**a**) structural solution of self-bearing insulation panel with the framework of PMM-made perforated steel profiles, (**b**) plasterboard and perforated steel tape connecting with gypsum-based glue [41].

The jointing of plasterboard sheets with perforated profiles is simple and has high productivity, which is why this method is recommended for self-bearing, quick-to-install construction. Plasterboard partition walls, using PMM and gypsum-based adhesive as the bonding element can be 20-27 cheaper than the traditional fixing option with tin profiles and screws [41].

PMM can also be used as a raw material for metal or composite sandwich panels where PMM forms the core of the panel and can also be used for upper or lower flanges. The principle of sandwich panels is similar to that of a double-T beam. In bending, the panel's top flange is compressed while the bottom flange is tensioned. The cell structure core analogs the beam wall, accommodates shear stresses, increases the stiffness of the structure and provides a certain distance between the flanges [15]. Compared to a beam wall, the panel core has the advantage of supporting the flanges over the entire surface area, forming a uniform reinforced structure. The separation of the flanges increases the moment of inertia of the panel cross-section by a slight increase in weight. This results in an efficient structure with good bending strength and stability. The use of perforated sheets also provides additional options for the fixing of the panels, depending on the structural solution [42,43].

Using PMM, it is possible to produce various honeycomb structures. Several sandwich panel solutions based on perforated steel strip cores are shown in Figure 8.

The panel constructions [44] shown in Figure 8 differ mainly in profile type and their arrangement. The profiles can be based on perforated steel of different widths and steel tapes of different thicknesses. The proportion of perforation may also vary [41]. The profiling of the tapes may differ, depending on the structural design of the panel and the expected load. The profiles may be arranged in both planes about the flange (Figure 9b) or upright (Figure 9a,c,d). All of these parameters are chosen considering the mechanical properties of the profile materials and the final product [45].

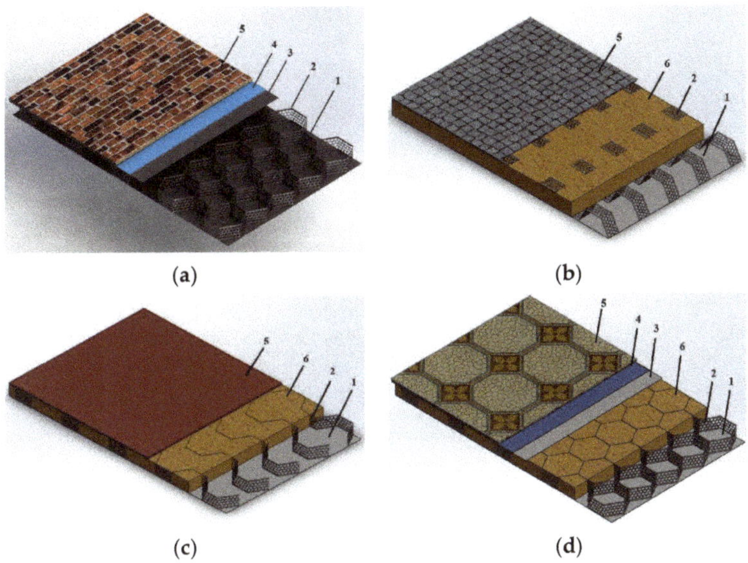

Figure 8. Sandwich panels with the PMM core, (**a**) external panel without heat insulation: (**b**,**c**) internal panel with heat or sound insulation; (**d**) ceiling panel; 1—bottom flange; 2—perforated profile; 3—load distribution layer; 4—damping layer or adhesive layer; 5—finishing layer; 6—heat or sound insulation [44].

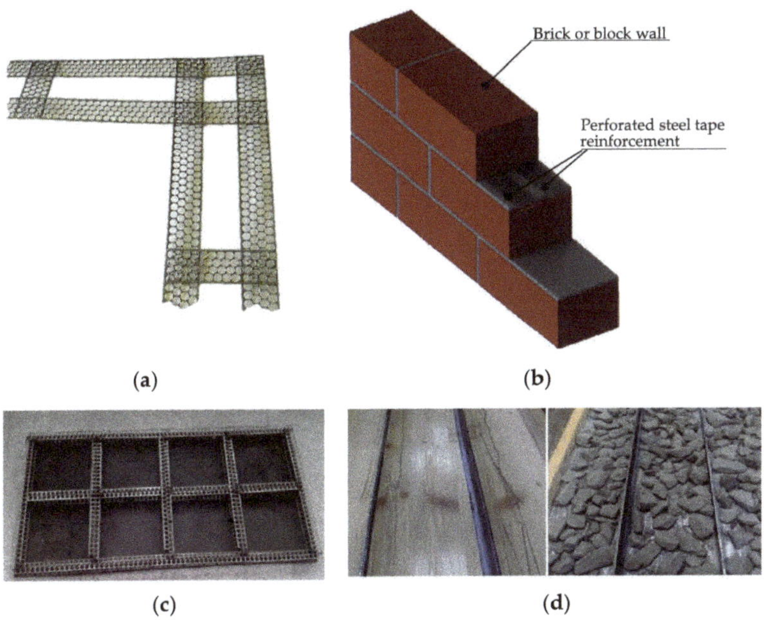

Figure 9. PMMs use different structural technologies: (**a**,**b**) PMM in masonry, (**c**) stiffeners in the mould members, (**d**) members providing timber-to-concrete connections in timber–concrete floor panels.

PMMs can also be effectively used as auxiliary elements in various construction technology processes, such as concrete pouring. There are some research and practical applications of PMMs in cement composite thin wall structures, where their use is of structural and operational efficiency but the labor intensity of such structures can also be a challenge that needs to be addressed. [43,46].

The application of PMM for timber-to-concrete connections in floor panels enables us to increase by 19% its load-carrying capacity panels. It has been stated that the existence of the holes has a significant influence on the load-carrying capacity of timber–concrete floor panels. The increased sustainability is provided by the 20% replacement of the metal materials by the PMM waste. [40]. It was stated, that existence of the holes has a significant influence on the load-carrying capacity of the timber–concrete floor. The use of steel PMM makes it possible to improve the quality of the brickwork by obtaining a thinner cement joint in the masonry, to increase the adhesion of metal to cement mortar. (Figure 9).

Another one of the possible directions for the use of tape PMMs is the production of armor-reinforced products for hydrotechnical construction [44].

2.2. Use of PMM Waste in Powder Metallurgy

With large volumes of PMM waste, it is advisable to consider the issue of their use as a raw material to produce powders. As an experiment, we carried out research on the grinding into powder of belt waste generated during the manufacture of drive chain elements [12,20]. To carry out research on the possibility of using PMM waste in powder metallurgy, a perforated tape made of steel 50 was used, which is a technological waste after stamping elements of drive chains. Grinding was carried out in several stages: cutting a sheet roll into fragments, coarse and fine crushing, and grinding the material into powder (Figure 10).

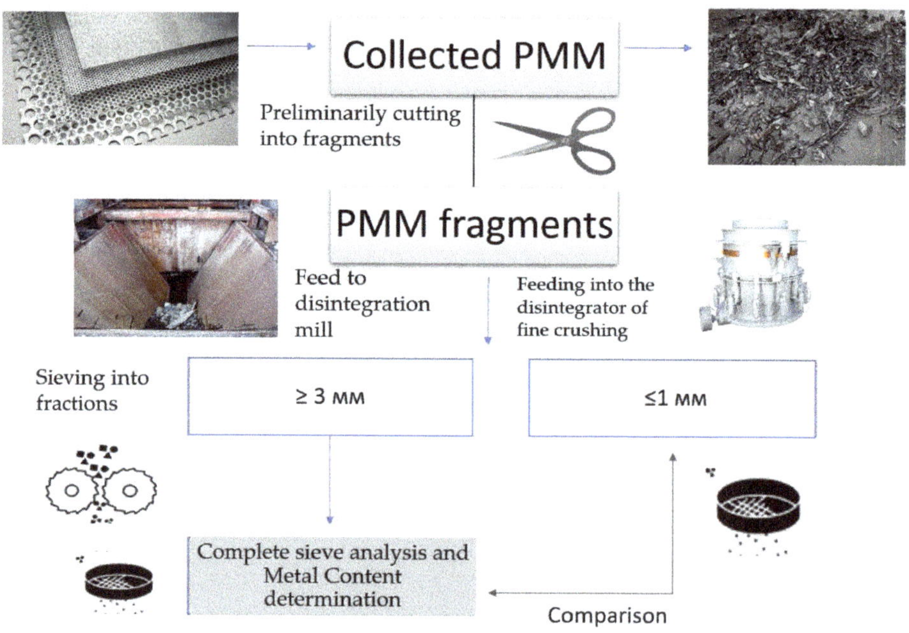

Figure 10. Scheme of processing PMM waste into powder by mechanical grinding.

The disadvantage of the method is increased energy consumption and an increased percentage of pollution (Figure 11a).

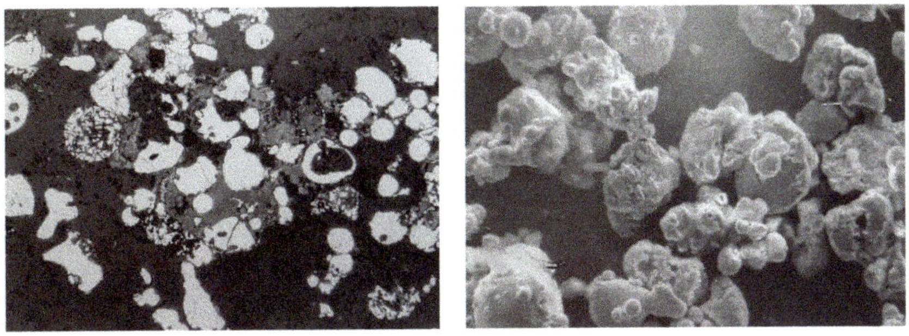

(**a**) (**b**)

Figure 11. Microstructure of powder from steel 50 after mechanical grinding (**a**); obtained by melt spraying (**b**).

Another method we used—obtaining a powder by melt spraying—allows for obtaining powders of higher quality (Figure 11b).

2.3. PMM Waste for the Manufacture of Frames of Technological Equipment and Devices

For the manufacture of frames, waste tape from PMM is most suitable, from which box-shaped profiles (PP) are pre-made. They have a number of advantages: low weight, ease of fastening sheet sheathing, etc. The example of a prismatic frame made of angle-type steel PP to obtain the device with steel sheet cladding Is shown in Figure 12.

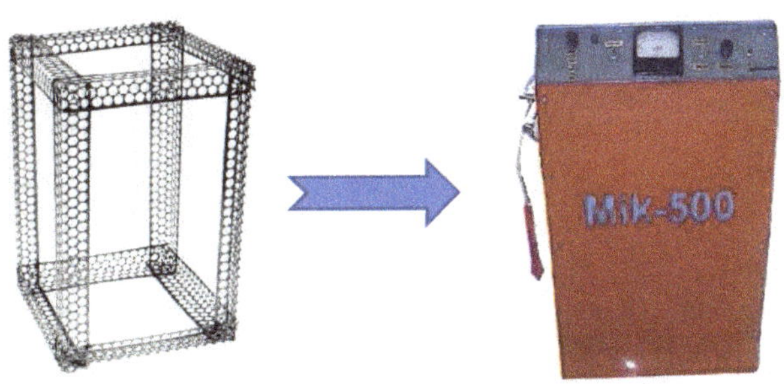

(**a**) (**b**)

Figure 12. Prismatic frame made of angle-type steel PP (**a**), the device with steel sheet cladding (**b**) [44].

Used materials, as a rule, have surface damage–corrosion areas, non-linearity and other defects (Figure 13).

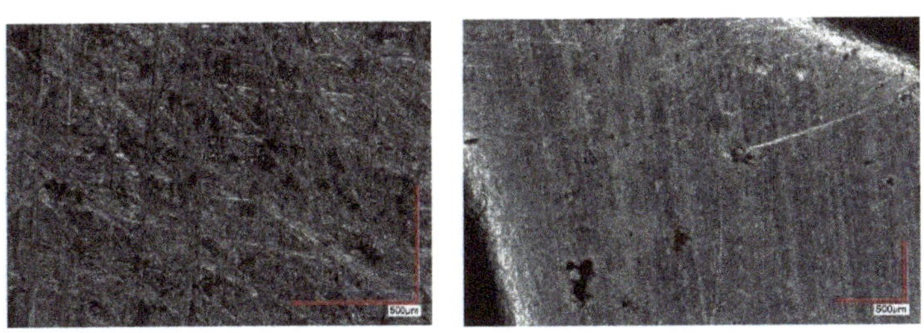

(**a**) (**b**)

Figure 13. PMM microstructure with traces of corrosion: they can be superficial (**a**) and deeper (**b**).

The growth of emerging markets increases the importance of a local presence [47]. That is why the reuse of perforated metals for architecture and design finds actuality in the market. As an example, decorative permeable and non-permeable lightweight grids, shields and wall structures can be developed. The grids can be partly covered with a polymeric material–rubber or plastic to produce reinforced decorative elements [45], which can simultaneously serve as delimiting and separating shields.

A protective coating can be applied to the fabricated grid, and the grid is then installed in a wooden or metal frame. In this way, the design becomes more attractive. Such types of elements can be both walls or hatches, delimiting different rooms, but allowing light to flow through. PMM can be successfully used in the design of lighting devices, in particular, in the manufacture of lamps for outdoor lighting. In this case, the decorative body is made from perforated steel strips obtained from stamping waste. Corps is running in the form of a sphere, cylinder or cone. To achieve this, you can use the processes of profiling and welding PMM The proposed solution allows us to reduce the weight of the lighting. device, reduce the cost of its production, as well as expand the scope of application by increasing and diversifying the luminous flux [48].

2.4. Permeable Structures from PMM Waste

A promising area of application of PMM frames is the creation of panels for absorbing mechanical energy. This refers to mechanical collision with moving bodies or vibration caused by the operation of the equipment. The core of the panels in this case should be of sufficient strength and rigidity, but at the same time sufficiently easily compressible. Its profile should be flat in relation to the facing sheets. Cellular hexagonal and lattice structures were used as cores for such panels in (Figure 14). This is achieved by layer-by-layer laying of PMM profiles. In this case, additional materials from polymers can be used.

(a) (b)

Figure 14. Lattice sandwich panel (**a**), using several perforated profiles (**b**).

The amount of absorbed energy depends to a large extent on the mechanical properties of the facing sheet materials, the material and the thickness of the panel core. Studies have shown that it is possible to form a panel from several profile layers and intermediate cladding layers. The thickness of the material naturally increases in this case. Therefore, the required energy absorption capacity of the panels must be carefully evaluated. In addition, increasing the thickness also increases material consumption and overall costs [44].

Various elements with a permeable structure can be made from waste sheet and tape materials, which are effectively used in the construction of various types of ventilation and filtering equipment, as well as in sound-absorbing devices [49]. Such structures made from PMM waste can be single-layer or multi-layer. The size and location of the perforations can also vary.

Figure 15a shows an example of a structure consisting of several cylinders of different diameters and perforations of the same shape with the possibility of displacement of one cylinder relative to another. Such a design, made from waste steel perforated tape, can be effectively used for drying materials and products [15]. Active ventilation is provided by air supply through the perforation. Drying channels made of PMM can be made in the form of semi-cylinders. In this case, they are easy to install and place in the factory yard or warehouse and easy to assemble and transport. The spatial structure can be made from tape PM using bending and welding processes (Figure 15b).

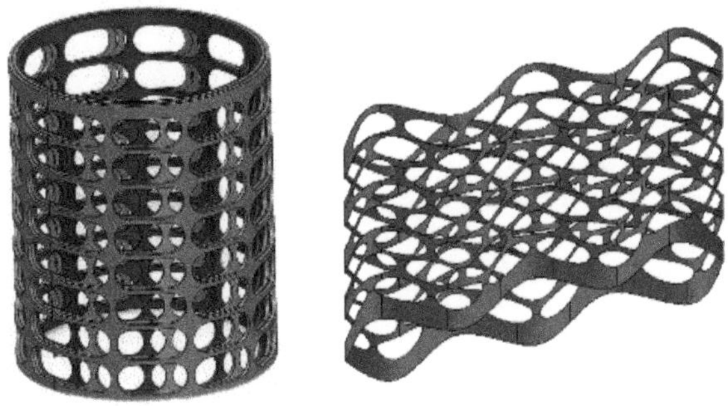

(a) (b)

Figure 15. Products with a permeable structure: cylindrical shape(s), obtained by folding a tape or sheet and welding to the height of the cylinder(s) (**a**); with complex spatial configuration, obtained by profiling tapes and connecting them by spot welding (**b**).

It is expedient to use cellular structures from PMM for the manufacture of deflectors, elements of refrigeration equipment, and climatic chambers. In [50], we proposed to fabricate metal permeable structures by layer-by-layer coating. The length and direction of the flow channels are adjusted by mutual displacement of the perforated sheets (Figure 16). Hexagonal mesh deflectors provide high flow efficiency, eliminate turbulence, fouling and moisture transfer, and reduce noise and energy consumption [10].

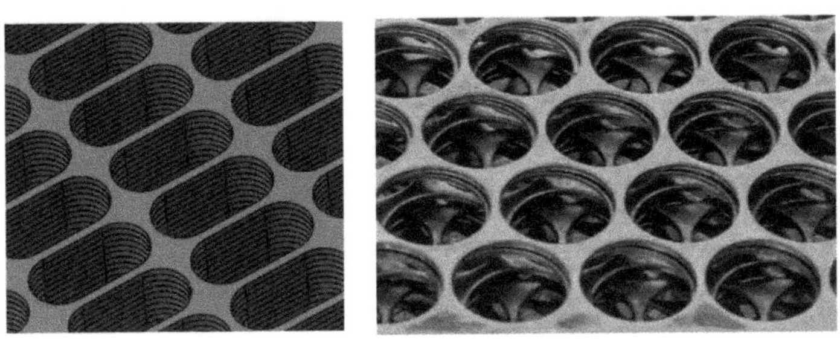

(a) (b)

Figure 16. Formation of flow channels in permeable structures: by the layer-by-layer laying of the tape without mutual displacement of layers (**a**); with mutual displacement of cell structure laminar airflow (**b**).

The permeable design allows the flow of liquid or gas to be controlled over time. This is achieved by mutual displacement of the perforated elements and by filling the created channels with filter material.

Perforated elements can be fixed in specially made cases or frames (Figure 17). In this case, PMM is used both as the filter material itself (usually this is the first stage of the filter), and as a frame for the internal filter.

Figure 17. Mesh filter in a perforated steel frame (**a**) and wood frame (**b**) [44].

2.5. Metal Screens from the Waste of Perforated Sheets

A number of works have shown that perforated metal screens (PMM made from tape or sheet waste can be successfully used to protect against light, sound, and electromagnetic radiation [51,52].

The presence of perforations allows some of the sound waves to penetrate inside and activate the absorption mechanisms. In the case of PMMs, there are two. The first one is the damping of sound energy during the passage of a wave through perforated holes due to frictional losses of oscillating air molecules interacting with the walls of the hole. Micro-perforated PMMs, in which the perforations are reduced to millimeter or sub-millimeter sizes, provide very effective sound absorption without any additional classic absorbing material [53]. Micro-perforated sheets are of particular interest to the aerospace and aircraft industries, as well as in various industrial cases [50]. The second mechanism is the Helmholtz resonator. In this case, a rigid wall reflecting sound waves is placed behind the PMM sheet at a distance L (Figure 18). Absorption occurs due to oscillations of the air column enclosed between the wall and the hole. To give the barriers maximum sound insulation, additional absorbers made of porous or woven materials with a thickness of 2–10 cm are provided in the design.

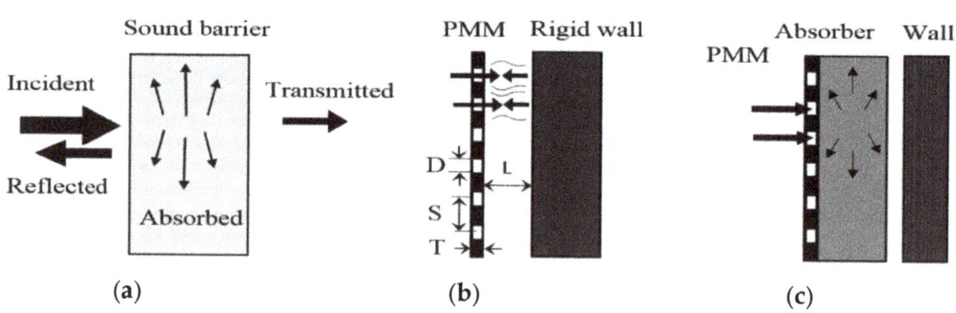

Figure 18. General schemes of sound absorption PMM: sound transformation scheme (**a**), absorption scheme using a protective perforated wall (**b**). also, with the addition of soundproofing material(s) (**c**) (D—diameter of holes, S—distance between hole centers, T—screen thickness, L—distance from screen to object surface).

Screens for protection against electromagnetic fields can be made from PMM waste. This experiment was considered in the work [54]. An experimental analysis of the screening efficiency of protective screens made of perforated steel tapes was carried out (Figure 19). It has been established that perforated screens reduce the intensity of the pulsed electromagnetic field (frequency 30 kHz) to 20–30%. However, the effectiveness of shielding materials depends on many parameters and must be evaluated for each specific application.

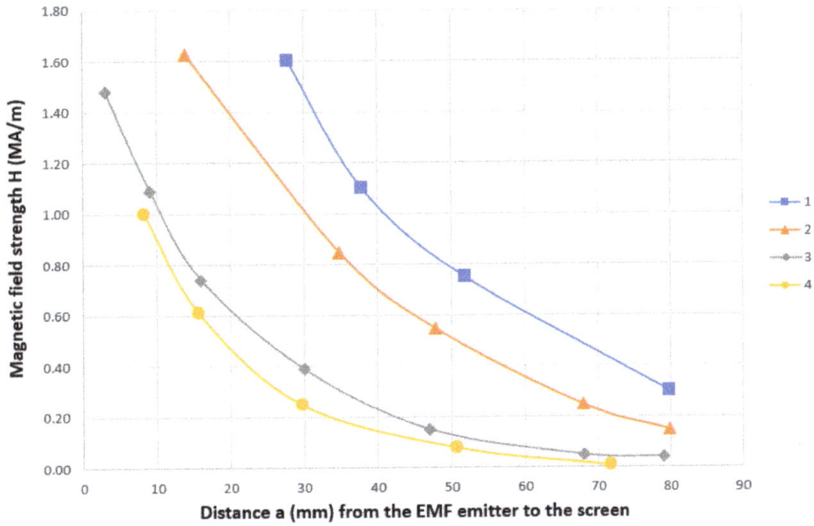

Figure 19. Damping of pulsed EMF in the axial and transversal position of the screen shielding: 1—axially to magnetic flux without shielding screen; 2—axially to magnetic flux shielding screen; 3—transversally to magnetic flux without shielding screen, 4—transversally to magnetic flux shielding screen.

Research in the field of passive protection of spacecraft structures has shown a growing demand for lightweight and economical materials that are needed to counteract the damaging effects of micrometeorite orbital debris [55].

The speed of collision of space vehicles with meteoroid bodies and space debris is in the range of 1–50 km s^{-1}. At such impact velocities, intense energy release occurs in a limited volume of matter, accompanied by the formation of shock waves, followed by mechanical destruction, melting, evaporation, and thermal ionization of the resulting vapors [56].

At the above-mentioned impact velocities, particles with a cross-section size of more than 0.5–1 cm can create penetrations in spacecraft walls and lead to catastrophic destructions. Impacts of small particles on the surface of a spacecraft cause the formation of craters and scratches on the surface, and in the case of a large number of impacts—noticeable erosion of the surface. Various optical elements suffer the most from small particle impacts: portholes, lenses, safety glasses, mirrors, etc. Plasma emissions and light flashes resulting from strong heating of the substance in the high-velocity impact zone can have a negative effect on sensors of scientific equipment and some units of spacecraft electrical and radio equipment [57].

A variety of structural materials, both homogeneous and composite, can be used to produce shields. The shield design can be either single- or multi-layer. Hypervelocity impact research led to the development of numerous shield configurations, such as the Whipple, Stuffed Whipple, metallic foam sandwich Whipple, Multi-Shock, and Mesh Double Bumper shields (Figure 20).

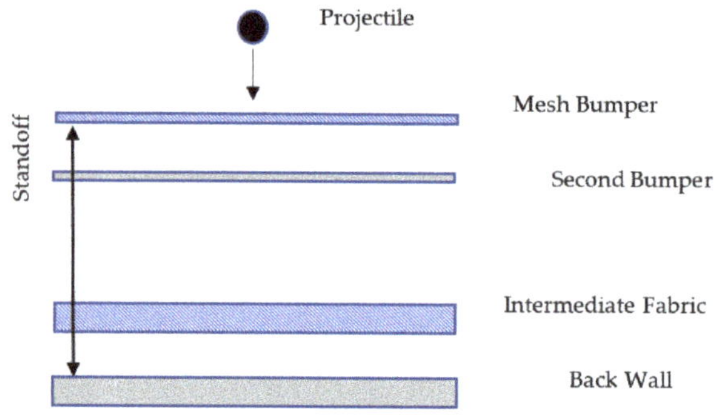

Figure 20. Mesh Double Bumper shield.

The study of metal mesh shields was carried out in [58,59]. It showed that they possessed higher efficiency than solid shields. It was found in the research [60–62] that the resistance of such screens can be increased due to various geometric features of their components (mutual arrangement of layers, angle of inclination, etc.), as well as from the materials of the layers and their order.

It is promising to use strip perforated steel materials in the form of cutting comb elements for the manufacture of machining tools. This proposal refers to the field of mechanical engineering, but can also be used in the construction industry, for example, in the manufacture of tools for finishing work (Figure 21). The shape and design of the cutting elements may vary depending on the intended use. Thus, for example, Figure 21a shows a cutting element consisting of a single perforated steel strip with cut and sharpened teeth. It is designed for cutting in the horizontal direction. In the second case (Figure 18b), the tool is designed to cut materials in the vertical direction [48].

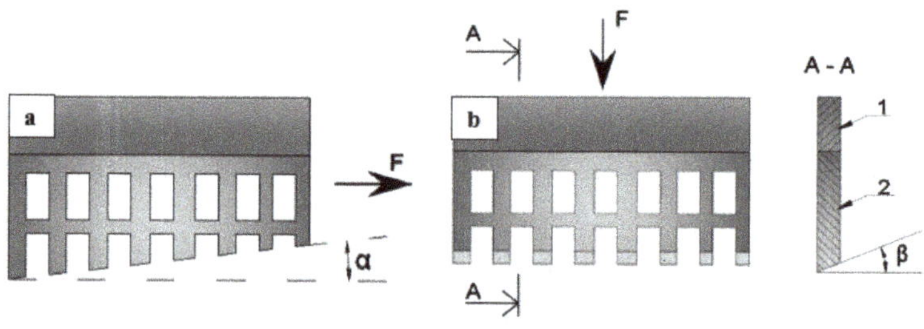

Figure 21. Single-layer cutting element with inclined teeth made of perforated steel strip: for horizontal (**a**) and vertical cutting (**b**) (F—pulling force, 1—gripper, 2—cutting blade).

3. Conclusions

The work showed sustainable ways to recycle or reuse perforated metal materials. Various new technologies are proposed and described for the sustainable application of PMMs. Reviewed methods are more environmentally friendly and they demonstrate a clear possibility of prolonging the lifecycle of PMMs through their reuse in numerous applications, thereby conserving resources and energy. The zero-waste orientation of PMMs makes them a smart way to minimize waste and attain the European Union's sustainable goals of becoming climate-neutral. PMMs provide significant environmental and aesthetic advantages, which align with the circular economy's principles of a closed-loop system and waste minimization.

The types and properties of PMM wastes were considered and investigated. It was stated that for small volumes, PMM waste can be used in load-carrying, insulating and decorative structural elements, for producing sound-protective screens, as well as frameworks and stiffeners for the casings of various devices. Moreover, the previous use of PMM waste in producing instrument frames, protective screens, and building fragments suggests that further investigation is warranted. Using strengthening elements made of PMM wastes increases the load-carrying capacity, stiffness and safety of timber, concrete, and masonry structural members. For example, the application of PMM for timber-to-concrete connections in floor panels enables its load-carrying capacity to be increased by 19%.

With large volumes of waste in the form of die-cutting, a sustainable lifecycle can be achieved by processing it into metal powder for the purposes of metallurgy. An important role is played by the processes of cutting, welding, bending and profiling of sheet and strip waste. Particular attention should be paid to assessing the general deformation state and especially the corrosion of perforated materials.

Author Contributions: Conceptualization: V.M., J.K. and Y.U.; formal analysis: J.K., V.M., Y.U., D.S. and M.L.; investigation: V.M., J.K., D.S. and M.L. All authors have read and agreed to the published version of the manuscript.

Funding: This study was funded with the support of the European Regional Development Fund within Activity 1.1.1.2 "Post-doctoral Research Aid" of the Specific Aid Objective 1.1.1 "To increase the research and innovative capacity of scientific institutions of Latvia and the ability to attract external financing, investing in human resources and infrastructure" of the Operational Program "Growth and Employment" (No. 1.1.1.2/VIAA/2/18/335). RTU project ID 3886.

Data Availability Statement: Data supporting the results presented can be provided upon request to the respective author.

Conflicts of Interest: The authors declare no conflict of interest.

References

1. Olivetti, E.A.; Cullen, J.M. Toward a Sustainable Materials System. *Science* **2018**, *360*, 1396–1398. [CrossRef] [PubMed]
2. European Raw Materials Alliance. Available online: https://single-market-economy.ec.europa.eu/industry/strategy/industrial-alliances/european-raw-materials-alliance_en (accessed on 27 March 2023).
3. Graedel, T.E.; Allwood, J.; Birat, J.-P.; Buchert, M.; Hagelüken, C.; Reck, B.K.; Sibley, S.F.; Sonnemann, G. *Recycling Rates of Metals: A Status Report*; United Nations Environment Programme: Nairobi, Kenya, 2011; ISBN 978-92-807-3161-3.
4. Weiser, A.; Bickel, M.W.; Kümmerer, K.; Lang, D.J. Towards a More Sustainable Metal Use—Lessons Learned from National Strategy Documents. *Resour. Policy* **2020**, *68*, 101770. [CrossRef]
5. Suharjanto, G.; Salim, K.A.; Mariana, Y.; Wijaksono, S. Perforated Metal Made from Recycled Material in the Application of Building Façade. *IOP Conf. Ser. Earth Environ. Sci.* **2020**, *426*, 012091. [CrossRef]
6. Song, Q.; Li, J.; Zeng, X. Minimizing the Increasing Solid Waste through Zero Waste Strategy. *J. Clean. Prod.* **2015**, *104*, 199–210. [CrossRef]
7. Hegger, M.; Auch-Schwelk, V.; Fuchs, M.; Rosenkranz, T. *Construction Materials Manual*, 1st ed.; Birkhäuser: Basel, Switzerland, 2006; ISBN 978-3-7643-7570-6.
8. Mironovs, V.; Boiko, I.; Koppel, T.; Lisicins, M. Cellular Tubular Structures from Perforated Metallic Tape and Its Application. *Agron. Res.* **2016**, *14*, 1053–1062.
9. Bogojavenskij, K.; Neubauer, A.; Ris, V. *Technologie Der Fertigung von Liechtbauprofilen*; DDR, Leipzig VEB Deutscher Verlag für Grundstoffindustrie: Leipzig, Germany, 1978.
10. Schultz, T.J. *Acoustical Uses for Perforated Metal: Principles and Applications*; Industrial Perforators Association: Milwaukee, WI, USA, 1986.
11. Mainini, A.G.; Poli, T.; Zinzi, M.; Speroni, A. Spectral Light Transmission Measure of Metal Screens for Glass Façades and Assessment of Their Shading Potential. *Energy Procedia* **2014**, *48*, 1292–1301. [CrossRef]
12. Mironovs, V.; Lisicins, M. Cellular Structures from Perforated Metal Bands and Its Uses. *Agron. Res.* **2011**, *12*, 279–284.
13. Fujita, K.; Inoue, T.; Emoto, H. Improvement of Permeability and Magnetic Shielding Effect of Pure Iron Magnetic Shield Materials. *JFE Tech. Rep.* **2005**, *6*, 24–28.
14. Koppel, T.; Tasa, T.; Tint, P. Electromagnetic Fields in Contemporary Office Workplaces. *Agron. Res.* **2013**, *11*, 421–434.
15. Mironovs, V.; Tatarinov, A.; Gorbacova, S. Expanding Application of Perforated Metal Materials in Construction and Architecture. *IOP Conf. Ser. Mater. Sci. Eng.* **2017**, *251*, 012027. [CrossRef]
16. Wadley, H.N.G. Multifunctional periodic cellular metals. *Phil. Trans. R. Soc.* **2006**, *364*, 31–68. [CrossRef] [PubMed]
17. Industrial Perforators Association. Perforated Handbook for Designers—Marco Specialty Steel. 1993. Available online: https://www.marcospecialtysteel.com/content/uploads/2017/01/Perforated-Handbook-for-designers-min.pdf (accessed on 27 March 2023).

18. Kim, T.J.; Thio, T.; Ebbesen, T.W.; Grupp, D.E.; Lezec, H.J. Control of Optical Transmission through Metals Perforated with Subwavelength Hole Arrays. *Opt. Lett.* **1999**, *24*, 256. [CrossRef]
19. Kawano, S.; Yamamoto, S.; Aikawa, S. Design and Evaluation of Space Filters Consisting Perforated Metal Plate and Dielectric Materials. *IEEJ Trans. Electron. Inf. Syst.* **2021**, *141*, 123–124. [CrossRef]
20. Mironovs, V.; Lisicins, M.; Boiko, I.; Zemchenkovs, V. Manufacturing of Cellular Structures of the Perforated Steel Tape. In Proceedings of the International Conference of DAAAM Baltic "Industrial Engineering", Zadar, Croatia, 24–27 October 2012.
21. Zaba, K.; Muzykiewicz, W.; Nowak, S. Analysis of the Perforation Process of Steel Strips Used in Automotive Industry. *Arch. Civ. Mech. Eng.* **2008**, *8*, 145–154. [CrossRef]
22. Adler, D.P.; Hii, W.W.-S.; Michalek, D.J.; Sutherland, J.W. Examing the Role of Cutting Fluids in Machining and Efforts to Address Associated Environmental Health Concerns. *Mach. Sci. Technol.* **2006**, *10*, 23–58. [CrossRef]
23. Baradie, M.A.E. Cutting Fluids: Part I. Characterisation. *J. Mater. Process. Technol.* **1996**, *56*, 786–797. [CrossRef]
24. Thomas, D.J. Laser Cut Hole Matrices in Novel Armour Plate Steel for Appliqué Battlefield Vehicle Protection. *Def. Technol.* **2016**, *12*, 351–359. [CrossRef]
25. The Business Research Company. *Stamped Metal Global Market Report (2023 edition)*; ReportLinker: Lyon, France, 2023; p. 175.
26. Karakaya, C. Numerical Investigation on Perforated Sheet Metals under Tension Loading. *Open Chem.* **2022**, *20*, 244–253. [CrossRef]
27. Sayed, A.M. Numerical Analysis of the Perforated Steel Sheets Under Uni-Axial Tensile Force. *Metals* **2019**, *9*, 632. [CrossRef]
28. Feng, Z.; Li, J.; Ma, J.; Su, Y.; Zheng, X.; Mao, Y.; Zhao, Z. EBSD Characterization of 7075 Aluminum Alloy and Its Corrosion Behaviors in SRB Marine Environment. *J. Mar. Sci. Eng.* **2022**, *10*, 740. [CrossRef]
29. Li, J.; Li, T.; Zeng, Y.; Chen, C.; Guo, H.; Lei, B.; Zhang, P.; Feng, Z.; Meng, G. A Novel Sol-Gel Coating via Catechol/Lysine Polymerization for Long-Lasting Corrosion Protection of Mg Alloy AZ31. *Colloids Surf. A Physicochem. Eng. Asp.* **2023**, *656*, 130361. [CrossRef]
30. Mironovs, V.; Boiko, I.; Lisicins, M. Methods of Profiling of the Metallic Materials. *J. Trends Dev. Mach. Assoc. Technol.* **2013**, 41–44.
31. Lisicins, M.; Mironovs, V.; Kalva, L. Analysis of Perforated Steel Tape Usage Possibility in Construction. In Proceedings of the Civil Engineering '11—3rd International Scientific Conference, Jelgava, Latvia, 12–13 May 2011; Volume 3, pp. 95–102.
32. Smith, W.F.; Hashemi, J. *Foundations of Materials Science and Engineering*, 4th ed.; McGraw-Hill Series in Materials Science; McGraw-Hill: Boston, MA, USA, 2005; ISBN 978-0-07-295358-9.
33. European Environment Agency. *Total Recycled Quantities for Six Key Material Types in the EU27 Including Exports for Recycling (11.1) (SCP 015)*; European Environment Agency: Kopenhagen, Denmark, 2013. Available online: http://www.eea.europa.eu/data-and-maps/indicators/11.1-total-recycling-amounts-of/assessment-1 (accessed on 27 March 2023).
34. Mironov, V.; Filippov, O.; Boiko, I. Al-W-B Powder Materials. *Est. J. Eng.* **2010**, *16*, 142. [CrossRef]
35. Hoornweg, D.; Bhada-Tata, P. *What a Waste: A Global Review of Solid Waste Management*; Urban Development Series; World Bank Publications: Washington, DC, USA, 2012; pp. 16–21.
36. Transforming Our World: The 2030 Agenda for Sustainable Development | Department of Economic and Social Affairs. Available online: https://sdgs.un.org/publications/transforming-our-world-2030-agenda-sustainable-development-17981 (accessed on 5 March 2023).
37. Mironov, V.; Boyko, I.; Serdjuks, D. Recycling and Application of Perforated Steel Band and Profiles. In Proceedings of the International Conference of DAAAM Baltic "Industrial Engineering", Vienna, Austria, 8–11 November 2006; pp. 285–288.
38. Spena, P.R.; Maddis, M.D.; D'Antonio, G.; Lombardi, F. Weldability and Monitoring of Resistance Spot Welding of Q&P and TRIP Steels. *Metals* **2016**, *6*, 270. [CrossRef]
39. Sinnes, K. *Welding and Cutting Science and Technology*, 10th ed.; American Welding Society: Miami, FL, USA, 2018; Volume 1.
40. Serdjuks, D.; Buka-Vaivade, K.; Goremikins, V.; Arturs, A.Y.M.S. Steel and Composite Tapes in Timber to Concrete Joint. In *Sustainable Energy Systems: Innovative Perspectives*; Lecture Notes in Civil Engineering; Sinitsyn, A., Ed.; Springer International Publishing: Cham, Switzerland, 2021; pp. 68–69. [CrossRef]
41. Mironovs, V.; Lapsa, V.A. Framework for Brick Masonry. LR patents LV 13429 B, 20 June 2006. Int. Cl. E04C5/01.
42. Mironovs, V.; Lisicins, M.; Boiko, I.; Zemčenkovs, V. Method of Manufacturing of Cellular Structures from Sheet Metal. In Proceedings of the 8th International DAAAM Baltic Conference, Tallinn, Estonia, 19–21 April 2012.
43. Mironovs, V.; Lisicins, M. Sandwich Panel with Cellular Structure. LR Patents LV 14825 B, 2014.
44. Mironovs, V.; Lisicins, M. *Perforated Metal Materials and Their Uses*; RTU Publishing House: Riga, Latvia, 2015; p. 159. ISBN 978-9934-10-694-1.
45. Baik, S.C.; Oh, K.H.; Lee, D.N. Analysis of the Deformation of a Perforated Sheet under Uniaxial Tension. *J. Mater. Process. Technol.* **1996**, *58*, 139–144. [CrossRef]
46. Li, C.; Li, G.; Zheng, L.; Liu, X.; Li, S.; Wang, X.; Soloveva, V.Y.; Dalerjon, H.; Fan, Z.; Qin, P. Study on the Mechanical Properties of Perforated Steel Plate Reinforced Concrete. *Materials* **2022**, *15*, 6944. [CrossRef]
47. Lehmhus, D. Advances in Metal Casting Technology: A Review of State of the Art, Challenges and Trends—Part I: Changing Markets, Changing Products. *Metals* **2022**, *12*, 1959. [CrossRef]
48. Mironovs, V.; Lisicins, M.; Zemčenkovs, V.; Urbahs, A. Comb-like Tool and Method for Its Manufacture. LV 14988 B; Riga, Latvia, 2015. Available online: https://worldwide.espacenet.com/publicationDetails/originalDocument?FT=D&date=20150920&DB=&locale=en_EP&CC=LV&NR=14988B&KC=B&ND=4 (accessed on 5 April 2023).

49. Takahashi, D. A New Method for Predicting the Sound Absorption of Perforated Absorber Systems. *Appl. Acoust.* **1997**, *51*, 71–84. [CrossRef]
50. Wetzig, A.; Herwig, P.; Hauptmann, J.; Baumann, R.; Rauscher, P.; Schlosser, M.; Pinder, T.; Leyens, C. Fast Laser Cutting of Thin Metal. *Procedia Manuf.* **2019**, *29*, 369–374. [CrossRef]
51. Mironovs, V.; Lisicins, M.; Boiko, I.; Koppel, T.; Zemchenkovs, V.; Lapkovskis, V.; Shishkin, A. Cellular Structures from Perforated Metallic Tape and Its Application for Electromagnetic Shielding Solutions. *Agron. Res.* **2014**, *12*, 279–284.
52. Morari, C.; Balan, I.; Pintea, J.; Chitanu, E.; Iordache, I. Electrical Conductivity and Electromagnetic Shielding Effectiveness of Silicone Rubber Filled with Ferrite and Graphite Powders. *Prog. Electromagn. Res. M* **2011**, *21*, 93–104. [CrossRef]
53. Chapter 2 Microperforation Methods. In *Membrane Science and Technology*; Elsevier: Amsterdam, The Netherlands, 2004; Volume 10, pp. 25–36. ISBN 978-0-444-51489-9.
54. Lapkovskis, V.; Mironovs, V.; Jevmenov, I.; Kasperovich, A.; Myadelets, V. Multilayer Material for Electromagnetic Field Shielding and EMI Pollution Prevention. *Agron. Res.* **2017**, *15*, 1067–1071.
55. Cherniaev, A.; Telichev, I. Weight-Efficiency of Conventional Shielding Systems in Protecting Unmanned Spacecraft from Orbital Debris. *J. Spacecr. Rocket.* **2017**, *54*, 75–89. [CrossRef]
56. Wozniakiewicz, P.J.; Burchell, M.J. Space Dust and Debris near the Earth. *Astron. Geophys.* **2019**, *60*, 3.38–3.42. [CrossRef]
57. Nuttall, A.; Close, S. A Thermodynamic Analysis of Hypervelocity Impacts on Metals. *Int. J. Impact Eng.* **2020**, *144*, 103645. [CrossRef]
58. Gerasimov, A.V.; Dobritsa, D.B.; Pashkov, S.V.; Khristenko, Y.F. Theoretical and Experimental Study of a Method for the Protection of Spacecraft from High–Speed Particles. *Cosmic Res.* **2016**, *54*, 118–126. [CrossRef]
59. Dobritsa, D.B.; Pashkov, S.V.; Khristenko, I.F. Protective Properties of Pleated Wire Mesh Shields for Spacecraft Protection against Meteoroids and Space Debris. *AIP Conf. Proc.* **2021**, *1*, 140006.
60. Dobritsa, D.B.; Pashkov, S.V.; Khristenko, Y.F. Study of the Efficiency of Corrugated Mesh Shields for Spacecraft Protection against Meteoroids and Manmade Space Debris. *Cosmic Res.* **2020**, *58*, 105–110. [CrossRef]
61. Development of a Ground-Based Method for Testing the Interaction of Materials with High-Speed Dust Particles—IAF Digital Library. Available online: https://dl.iafastro.directory/event/GLEX-2021/paper/62307/ (accessed on 5 March 2023).
62. Usherenko, Y.; Mironovs, V.; Pakrastiņš, L. Impact of the High-Speed Flow of Powder Particles on the Structure of Polymer Materials and Metal-Polymer Composites. *ETR* **2021**, *3*, 363–367. [CrossRef]

Disclaimer/Publisher's Note: The statements, opinions and data contained in all publications are solely those of the individual author(s) and contributor(s) and not of MDPI and/or the editor(s). MDPI and/or the editor(s) disclaim responsibility for any injury to people or property resulting from any ideas, methods, instructions or products referred to in the content.

materials

MDPI

Review

Morphological, Mechanical and Thermal Properties of Rubber Foams: A Review Based on Recent Investigations

Ehsan Rostami-Tapeh-Esmaeil and Denis Rodrigue *

Department of Chemical Engineering, Laval University, Quebec, QC G1V 0A6, Canada
* Correspondence: denis.rodrigue@gch.ulaval.ca; Tel.: +1-418-656-2903

Abstract: During recent decades, rubber foams have found their way into several areas of the modern world because these materials have interesting properties such as high flexibility, elasticity, deformability (especially at low temperature), resistance to abrasion and energy absorption (damping properties). Therefore, they are widely used in automobiles, aeronautics, packaging, medicine, construction, etc. In general, the mechanical, physical and thermal properties are related to the foam's structural features, including porosity, cell size, cell shape and cell density. To control these morphological properties, several parameters related to the formulation and processing conditions are important, including foaming agents, matrix, nanofillers, temperature and pressure. In this review, the morphological, physical and mechanical properties of rubber foams are discussed and compared based on recent studies to present a basic overview of these materials depending on their final application. Openings for future developments are also presented.

Keywords: rubber; foam; foaming agent; morphology; mechanical properties

Citation: Rostami-Tapeh-Esmaeil, E.; Rodrigue, D. Morphological, Mechanical and Thermal Properties of Rubber Foams: A Review Based on Recent Investigations. *Materials* **2023**, *16*, 1934. https://doi.org/10.3390/ma16051934

Academic Editors: Pavol Liptai, Jaroslav Briančin and Maroš Halama

Received: 31 December 2022
Revised: 9 February 2023
Accepted: 22 February 2023
Published: 26 February 2023

Copyright: © 2023 by the authors. Licensee MDPI, Basel, Switzerland. This article is an open access article distributed under the terms and conditions of the Creative Commons Attribution (CC BY) license (https://creativecommons.org/licenses/by/4.0/).

1. Introduction

Rubber foams are composite materials composed of a dispersed gaseous phase surrounded by an elastomeric matrix [1]. Rubber foams have high resilience, flexibility, tear resistance, deformability (especially at low temperature), abrasion resistance, buoyancy and lower weight, making them important materials in our daily lives [2,3]. In addition, their high elasticity, strength-to-weight ratio, impact resistance and energy absorption provide them with a wide range of various applications, including in automobiles, transportation, aeronautics, building, construction and cushioning [4]. However, more applications will expand to a larger variety of conditions with more technological development, different temperatures, pressures, mechanical solicitation, radiation or environmental conditions [5].

The first cellular polymer to enter the commercial market was a rubber foam developed in 1914 [6]. It was prepared by combining a natural rubber (NR) latex with gas-producing compounds (foaming agents), such as sodium and ammonium carbonate or sodium polysulfide. The objective is to give a broad insight related to important parameters including formulation (foaming agent, matrix and nanofiller) and foaming (temperature, pressure and time) parameters on the morphological, mechanical, physical and thermal properties of rubber foams.

2. Structural Features

2.1. Porosity: Open or Closed Cell

The expansion of a gaseous phase dispersed throughout a rubber matrix is necessary to produce a rubber foam. According to the porosity and cell hole, cellular materials can be typically classified as open cells (direct connection between the gas cells/broken walls) or closed cells (gas cell are distinct/unbroken walls) [7]. Foams with an open cell structure are more prevalent in less rigid materials, while closed cells represent more rigid ones [8]. The final application will depend on structural parameters. The high porosity of open cell foams lets them be good candidates for fuel tank inserts, filtration media, medical devices, fluid management and loudspeakers, due to their lower weight and lower cost. Because of their interconnections, the cells can allow fluids (gas or liquid) to travel across, which is beneficial for applications such as filtration and separation operations. Closed cell foams are generally more expensive, having lower porosity leading to better resistance and reduced adsorption/absorption which are interesting for thermal, sound and electrical insulators. In both cases, a chemical (CFA) and/or physical (PFA) foaming agent can be used to create closed or open cell structures [9].Vahidifar et al. prepared a closed cell NR/nanoclay (NC) foam using azodicarbonamide (ADC, 4 phr) as a CFA [10]. They found that the foaming of NR/NC (0–10 phr) at 155 °C for 30 min led to a closed cell structure. Another work stated that the addition of carbon nanotube (CNT = 1–3 phr) and nanoclay (NC = 2–6 phr) with different geometry and content led to a different closed/open cell ratio inside EPDM foams [11]. The tubular structure of CNT generated closed cells at low content, while higher concentration led to higher open cell content. Nevertheless, open cell formation was restricted by the presence of organo-modified NC, which had a plate-like structure leading to lower open cell content with increasing NC content.

Xie et al. reported on a "sea-island" structure resulting in a complex network inside NR by using expanded polymer microspheres as a foaming agent [3]. By adding 2.04 vol% CNT, they produced low density NR foams (0.50 g/cm^3) with good thermal (0.048 W/m.K) and electrical (36.6 S/m) conductivities. They also achieved NR foams with the lowest thermal conductivity of 0.02 W/m.K. Zonta et al. compared the salt leaching technique (Figure 1a) and traditional foaming process (Micropearl and Expancel) in terms of morphology and thermal conductivity for EPDM foams [12]. The salt leaching method led to an open cell structure, while the foaming agents led to closed cells with diameters ranging between 60 and 80 µm, as reported in Figure 1b–d. They also measured the thermal conductivity of their foams and observed that the value of NaCl/EPDM foam was 36.4% and 18.2% higher than for foams using Micropearl and Expancel, respectively. Furthermore, the compression set of conventional foaming methods was 160% and 392.5% higher than for NaCl/EPDM.

2.2. Cell Size and Cell Density Effect

According to the size of the cell, rubber foams can be divided into four categories: nanocellular (0.1–100 nm), ultramicrocellular (0.1–1 µm), microcellular (1–100 µm) and macrocellular (above 100 µm) [13]. In general, nanocellular foams are described as having cells less than 100 micron with a cell density higher than 10^{12} cells/cm^3. Nanocellular foams are expected to have several qualities superior to those of conventional foams because of their distinctive architectures (higher surface area), including a better strength-to-weight ratio, excellent filtration (separation) ability, lower dielectric constant and high thermal and sound insulation behavior [14]. Several studies have shown that the cells' dimensions directly affect the final foam's properties. For example, Rostami and coworkers prepared graded polyolefin elastomer (POE) foams with a range of cell size (98–337 µm) and cell density (204–820 cells/mm^3) across the thickness [15]. Their findings confirmed that the side of the foam with lower cell sizes/higher cell density provided improved thermal resistance. Tang et al. used supercritical carbon dioxide (scCO_2) foaming to create silicon rubber (SR) foams with various cell sizes ranging from 5 to 266 µm, leading to different cell shapes from spherical to polygonal, as presented in Figure 2 [16]. It was concluded that SR foams may be used in various applications depending on their cellular microstructures: foams with small cells (approximately 5 µm) had high compression strength, while large cells (above 50 µm) had good recovery performance and medium cell size (5–50 µm) showed good tensile properties. Syahrin et al. studied the effect of sodium bicarbonate ($NaHCO_3$) concentration (3 to 12 phr) on the cell size and compression properties of NR latex foam (NRLF) [17]. They reported that the average cell diameter of NRLF decreased by 15% as the CFA content increased from 3 to 12 phr, while the compression strength decreased from 0.54 to 0.08 MPa at 3 phr compared to an increase up to 0.19 MPa at 13 phr.

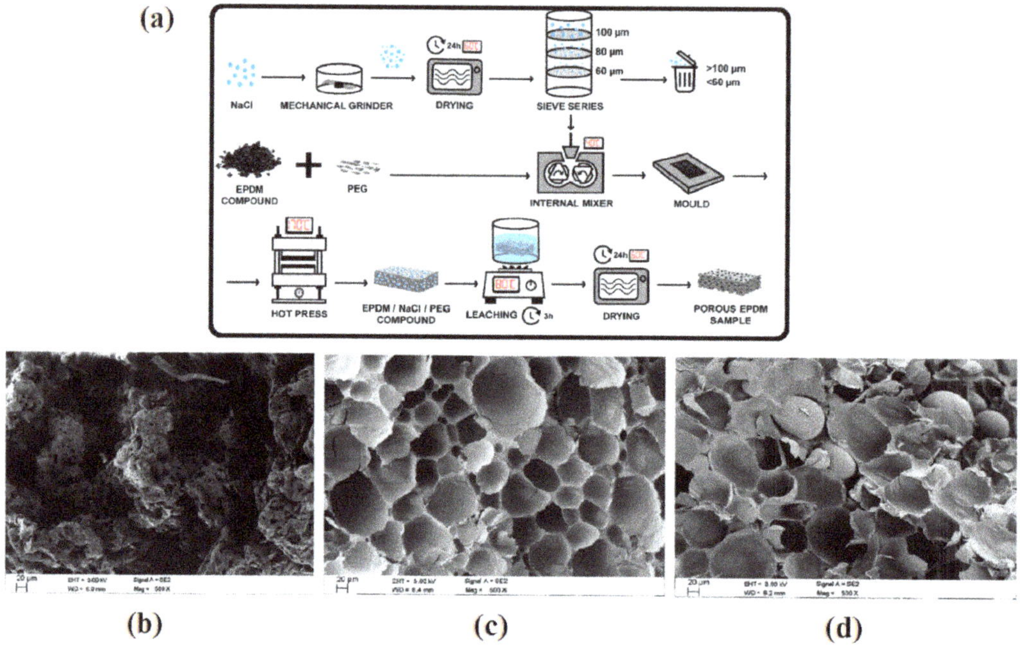

Figure 1. (**a**) The preparation steps for EPDM foams by the salt leaching technique and (**b**) SEM image of the resulting foam compared to foams based on (**c**) Micropearl and (**d**) Expancel [12].

Jia et al. studied the effect of expansion ratios on SR/nanographene (NG) composite foams' morphology [18]. It was found that these foams had both open and closed cells morphologies. Moreover, the cell density was increased (2.13–7.15 × 10^5/cm^3), while the cell diameter decreased (191–79.4 μm) as the foaming expansion ratio increased from 2 to 5. For composites at low ER, the large cells (>150 μm) dominated (73%), while smaller cells (<150 μm) dominated (93%) at higher ER. They also observed that a higher number of smaller cells enhanced the tunneling effects, leading to lower dielectric constants. Finally, higher cell density improved the NG distribution (more uniform), which also improved the dielectric constant.

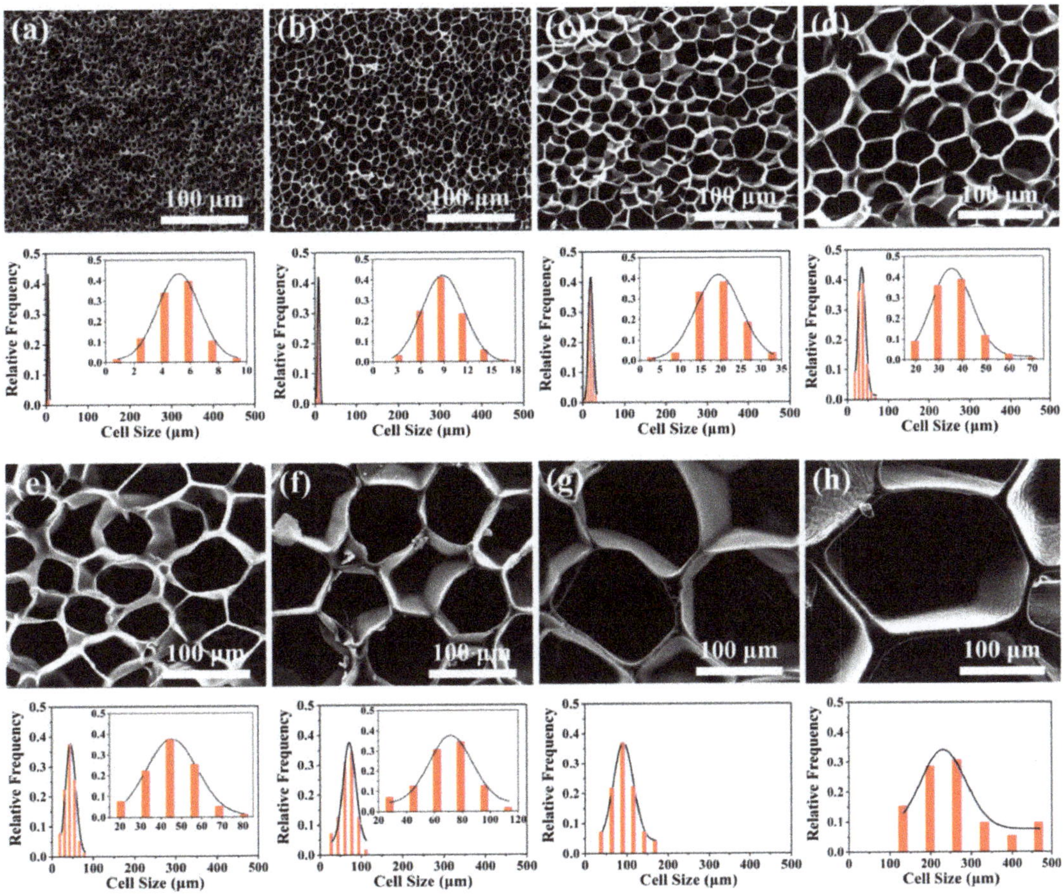

Figure 2. SEM images and cell morphology of SR foams designed by controlling the $scCO_2$ foaming conditions with different cell sizes (μm): (**a**) 5, (**b**) 9, (**c**) 20, (**d**) 34, (**e**) 45, (**f**) 71, (**g**) 107, (**h**) 266 [16].

3. Foaming Process

The foaming and crosslinking reactions are two opposing reactions that play a significant role in controlling the final foam properties by influencing the cell morphology during the rubber foam production. While the curing reaction between rubber molecules stabilizes the morphology as well as increases the viscosity, the gas generated via CFA decomposition leads to foaming. Gas molecules inside the rubber matrix decrease the curing characteristics (e.g., curing rate, torque, scorch time, etc.). They act as a solvent and aggregate between the macromolecule chains of the matrix, leading to higher free volume and chain mobility. On the other hand, the presence of foaming agents causes the system to absorb more heat, which inhibits the crosslinking process. During the decomposition of the blowing agent, the gas expands and decreases the crosslink density.

The foaming and curing processes can be effectively managed by selecting appropriate processing parameters, including temperature and formulation (components and concentration). To manage the final foam cellular structure (cell size and cell density), the kinetics of these two reactions must be balanced. When the pressure is released or the foaming agent begins to decompose and releases its gas, if the rubber has slow curing and is not adequately pre-cured, the gas will either escape or cell collapse will occur due to a lack of material strength (cell wall rupture). As a consequence, a curing process taking too long before the gasses are released may result in a hard and dense foam containing very few small cells [19].

The three main steps of a foaming process are nucleation, growth and stabilization of the cellular structure by physical or chemical means [20]. After CFA decomposition or a PBA has been injected, bubbles are produced, which is known as the nucleation step. The bubble nuclei develop into the final bubbles or cells during the growth step. The structure changes due to the development of numerous bubbles, while the cell wall becomes thinner. In order to stabilize the bubbles at a specific size, curing and/or cooling is performed to increase the matrix viscosity. Several physical characteristics, including solubility, viscosity, diffusivity, surface tension and glass transition temperature, have an effect on both processes. Controlling the nucleation and growth steps is very challenging because the foaming and curing processes are carried out simultaneously. To achieve a particular morphology and structure in rubber foams, it is essential to optimize the formulation (foaming agent, matrix type, nanofiller) and processing conditions (pressure, temperature, etc.). The effect of these processing variables is discussed in more detail in the following section.

3.1. Foaming Agents

For rubber foam manufacturing, two important families of FA (chemical and physical) are available, with each one having its advantages and disadvantages. Physical foaming agents (PFA) are generally low cost, do not react (inert) and leave no by-products that would influence the general foams' properties. However, most PFA (besides inert gases) are associated with safety hazards (explosion and flammability), environmental concerns (ozone depletion and greenhouse gases), require specific equipment (high pressure pumps and flow meters) and more complex processes due to their use. This is why, in some cases, it is more appealing to use chemical foaming agents (CFA), which are mostly powders. However, several CFA have health risks despite their widespread use and simple production. Additionally, their application has been constrained by unfavorable by-products and high cost. Nevertheless, CFA is the best, most economical and common choice in the foam industries.

PFA are inert and compressed gas or volatile liquids which are injected into the matrix at high pressure. After heating and depressurization, foaming occurs as a result of gas release or physical state change (liquid to gas). An excellent PFA needs to be thermally stable, nontoxic, odorless and nonflammable, having high solubility, diffusivity and vapor pressure. PFA molecules are divided into two types: inorganic and organic. Nitrogen (N_2), carbon dioxide (CO_2), water (H_2O) and air are the most common inorganic PFA. On the other hand, organic PFA are based on low molecular weight alkane (propane, butane, pentane, hexane), dichloromethane, dichloroethane and some hydrogenated chlorofluorocarbons (HCFC). Moreover, microspheres or micropearls were highly used more recently to prepare foams based on low molecular weight alkanes [21,22], and these materials can be included in the PFA category. Vlentini et al. prepared EPDM foams loaded with various foaming agents, including Expancel (E) and Hostatron (H) [23]. The foaming agent H (mixture of sodium, calcium and potassium bicarbonates, with a decomposition temperature of approximately 150 °C) was used at 1.3 phr, while the foaming agent E (isopentane microspheres, with an expanding temperature above 130 °C) was used at 14 phr to produce EPDM foams. The foaming agent E led to a homogeneous closed cell foam structure with cell sizes of approximately 20 μm, while the foaming agent H produced a mixture of closed/open cells with larger cell sizes (approximately 100 μm) with lower cell size uniformity.

Considering the challenges of using PFA and CFA, another method to produce rubber foams is particle leaching. In this case, water-soluble particles (e.g., potassium or sodium chloride) are added/dispersed in the rubber and then dissolved in a solvent (e.g., water), resulting in a porous cellular structure [24,25]. Although this method has a higher ability to control the cell size and cell density, it still faces several problems such as using a solvent (in compounding and also during leaching), always leads to an open-cell structure, needs high pressure and/or high temperature during compounding, and residual particles can still be present inside the matrix. Peng et al. used sodium chloride (NaCl) as a pore-forming agent inside a SR matrix to produce open cell foams for sound absorption applications [26].

In general, the foaming procedure consists of 4 steps: dissolution, bubble nucleation, bubble growth and curing. The $scCO_2$ has a number of advantages over CO_2 and N_2 due to its low critical temperature (T_c = 304.15 K) and pressure (P_c = 7.38 MPa), higher solubility and diffusion coefficient [27,28]. One example is ethylene vinyl acetate (EVA) foams produced by Jacobs's group using $scCO_2$ [29]. They reported that increasing the sorption pressure resulted in high density foams with small cell sizes. Similarly, Tessanan and his team showed that increasing the $scCO_2$ saturation time and pressure during NR foaming produced foams with smaller cell sizes and narrower cell size distributions [30].

There are two types of CFA: inorganic and organic. Ammonium, sodium and potassium carbonates, bicarbonates, nitrates and oleate are inorganic CFA. They mostly form open cell morphology in the presence of acids (accelerators) by releasing CO_2 and water above their thermal decomposition temperature (T_d). The majority of inorganic CFA have a slow endothermic decomposition. A typical dosing for CFA is in the range of 5–10 phr. Most organic CFA reactions are exothermic, which means that they produce energy (heat) above their decomposition temperature. CFA are derived from organic materials such as azodicarbonamide (ADC), 5-phenyl tetrazole (5-PT), *p*-toluenesulfonyl semicarbazide (TSSC), 4,4′-oxybis benzene sulfonyl hydrazide (OBSH) and *N,N*′-dinitroso pentamethylene tetramine (DPT), which produce a gas mixture of ammonia, carbon dioxide, nitrogen and water [31]. Suethao and coworkers produced open cell NR foams with various concentrations of potassium oleate (PO) as a foaming agent [32]. They claimed that decreasing the CFA content by 45% decreased the cell size by 50%, while the cell density was highly improved (800%), as reported in Figure 3.

When a CFA reaches its decomposing temperature, the molecules break down through a sequence of reactions, and the heat generated can activate nearby molecules. This autocatalytic reaction (cascade reactions) leads to a rapid expansion producing closed cells. The ability to control the decomposition temperature (approximately 210 °C for pure ADC) makes ADC an interesting and practical organic CFA. However, several options are available to decrease T_d. For example, decreasing the CFA particle size leads to more surface area for the thermal reaction to occur. Another possibility is to add some activators, such as zinc oxide (ZnO), zinc stearate ($ZnSt_2$), stearic acid, etc. However, the addition of ZnO to ADC not only decreases T_d, but also accelerates its decomposition rate [33]. ADC is known to be very efficient (231 cm^3 of gas/g) [34]. The mechanism of thermal decomposition is directly influenced by temperature and time. However, other factors, including the type, amount and particle size of the activators and CFA, as well as their level of dispersion, affect the thermal decomposition. CFA generate decomposition residues acting as nucleation agents for subsequent foaming, leading to a finer porous structure. Nevertheless, CFA can also have a detrimental impact on the material's properties or affect the color of the final product [35]. For example, the density of POE foams was reported to be mainly controlling the physical and mechanical properties [36]. As a result, increasing the ADC content from 1 to 13 phr decreased the density (820 to 57 kg/m^3), being less resilient (84–23%) with lower tensile strength (5.25 to 0.82 MPa) and elongation at break (602–127%). This trend was related to a higher gas volume resulting in lower mechanical properties.

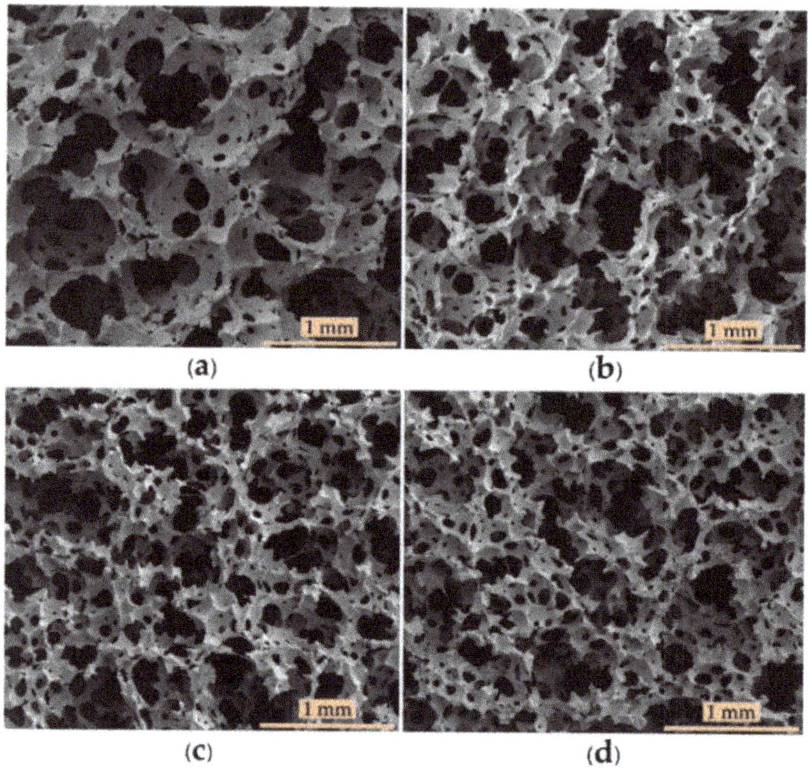

Figure 3. Micrographs of cellular NR based on different CFA contents: (**a**) original formulation, as well as different PO reduction: (**b**) 15%, (**c**) −30% and (**d**) 45% [32].

3.2. *Matrix*

Rubbers such as NR, EPDM, SBR, SR, etc., are the main matrices for rubber foams. Salmazo and coworkers investigated the effect of modified and unmodified NR matrix [37]. The modified compound was produced by epoxidized natural rubber (ENR). Both foams were prepared with ADC (10 phr) and cured under a range of electron beam irradiation doses, from 50 to 150 kGy. ENR foams were found to have higher nucleation rates and minimal cell degradation compared with foams made from NR. This was explained by the fact that the cured ENR foam included epoxide groups, which increased the curing level. By using a dip-coating technique, a low-cost and highly sensitive strain sensor of titanium carbide (MXene)/NR foam was prepared by assembling the $Ti_3C_2T_x$ nanosheets on the side and surface of NR foams (Figure 4a–e) [38]. According to the findings, the prepared smart materials showed extremely high sensitivity (GF = 14) to very low mechanical solicitation (low detection limit of 435 Pa), showing excellent adhesion between MXene and NR foam. In addition, the MXene/NR strain sensors exhibited a wide strain range from 0 to 80% and a low response time of 1 s (Figure 4f), which is of high importance to design wearable electronics, as the materials can detect a wide range of actions such as step monitoring and finger pressing (Figure 4g,h).

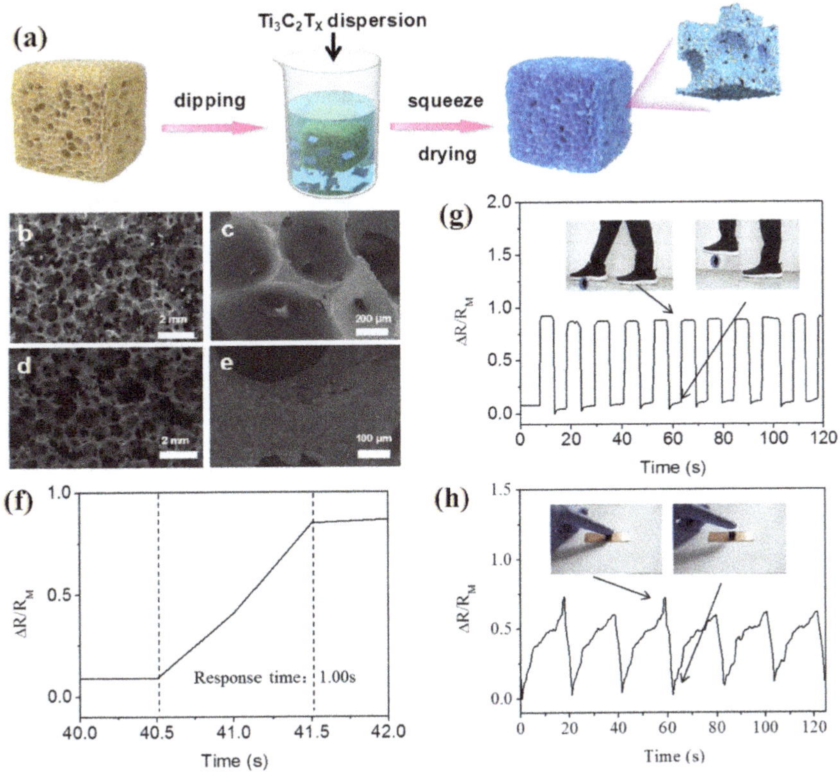

Figure 4. (**a**) The preparation steps for NR/MXene foams. (**b**,**c**) Micrographs of NR foams, (**d**,**e**) SEM images of NR/MXene foams, (**f**,**g**) the partial amplified signal (response time) and step monitoring from an adult and (**h**) a finger pressing application of the MXene/NR foam [38].

Lee and colleagues produced nanocomposite foams based on EPDM filled with halloysite nanotube (HNT) [39]. A batch reactor was used and the FA was $scCO_2$. They generated foams with high cell density (1.5×10^{10} cell/cm^3) and low cell size (7.8 μm). In this case, HNT were shown to be effective nucleating agents. Shao et al. reported a double crosslinking system by adding sulfur and dicumyl peroxide (DCP) at the same time for SBR (styrene–butadiene rubber) foaming [40]. The synergetic effect between both curing agents led to lower shrinkage (2.25%). The effect of single wall carbon nanotube (SWCNT) in EPDM/SBR foams was studied by Bahadar's group [41]. The incorporation of SWCNT (0.6% wt.) inside the EPDM/SBR blend increased the storage modulus (80%) while decreasing the loss modulus (27%).

In order to improve the characteristics (compression set, rebound resilience and tensile strength) of ethylene vinyl acetate (EVA) foams, ethylene-*1*-butene (EtBC) was added [42]. By combining NR with EVA, Kim et al. were able to produce foams having good tear strength and high rebound resilience at low density using the optimum temperature for curing [43]. The effect of the EVA:CPE ratio on the foaming and curing steps leading to improved mechanical characteristics was also investigated [44]. It was found that adding more EVA had no effect on the kinetics (curing time and scorch time). On the other hand, the ratio had a direct effect on hardness, rebound resilience and shrinkage. It was also observed that the cell density decreased for an EVA:CPE ratio from 0:100 to 50:50, but significantly increased from 30:70 to 10:00. Overall, the ER (void content) increased with increasing EVA content.

The viscoelastic properties of SR foams made with $scCO_2$ were examined by Liao et al. with respect to silica content, saturation temperature (T_s) and saturation pressure (P_s) [45]. They observed that because silica acted as a heterogeneous nucleation agent and viscosity enhancer, its concentration had a very important effect on both cell nucleation and cell growth. Moreover, by lowering the T_s (from 80 to 40 °C), cell nucleation was important because the cell density was increased (0.4–9.0 × 10^6 cells/cm^3) while the cell diameter was decreased (76 to 21 μm). Conversely, higher T_s (60 and 80 °C) induced cell rupture and coalescence associated with rapid cell growth (lower viscosity and elasticity). Finally, increasing P_s (from 10 to 14 MPa) increased the plasticization effect of $scCO_2$ and reduced the viscosity of the SR matrix, which again accelerated the cell growth rates. To investigate the relationship between cell sizes and mechanical properties, Luo et al. produced SR foams using spherical urea with various sizes (200–800 μm) as cell-forming agents [46]. They showed that increasing the cell size from 200–300 μm to 600–800 μm resulted in higher compressive strength from 0.257 MPa to 0.382 MPa. Because of the larger cell size (higher amount of gas volume), this improved the resistance of SR foams against compressive forces. On the contrary, the tensile results showed that increasing the cell size from 200–300 μm to 600–800 μm reduced the elongation at break (156–110%) and tensile strength (0.667 MPa to 0.433 MPa). These trends are expected since larger cell sizes generate higher stress concentration points, leading to lower tensile properties. In another work, POE foams were produced using different CFA contents (ADC: 2–5 phr) without any curing agent [47]. The cell size decreased by 29% and the cell density increased by 470% as the ADC content increased (2 to 4 phr), as reported in Figure 5. However, due to cell coalescence, 5 phr ADC produced bigger cells (148 μm) combined with less cell density (483 cells/mm^3). However, increasing the ADC content (2 to 5 phr) was found to decrease all the tensile properties: elongation at break (16%), strength at break (21%) and modulus (28%), as well as hardness (14%). In addition, a finite element method (FEM) was used to simulate the tensile behavior with good precision.

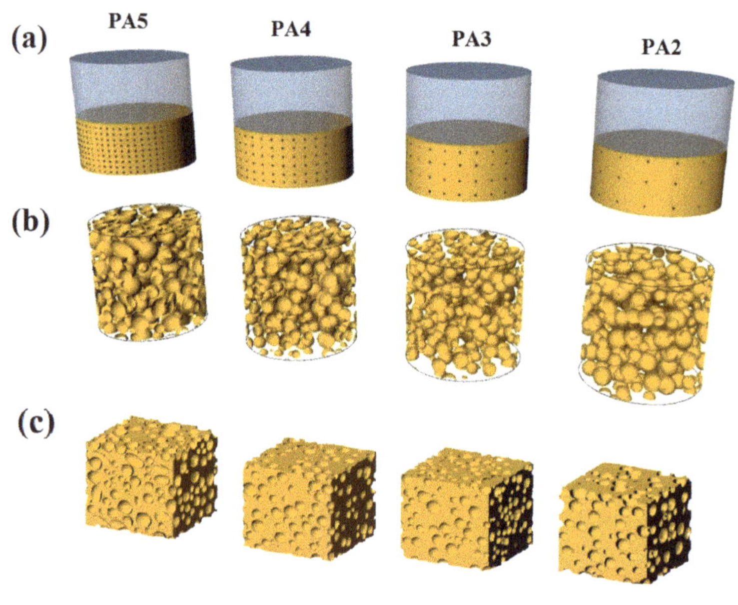

Figure 5. Schematic representation of the foaming process of POE with various ADC concentrations (2–5 phr, from right to left): (**a**) nucleation, (**b**) final structure in the mold and (**c**) final foam structure [47].

3.3. Nanofillers

Filler addition is always a starting point to improve the mechanical properties of any material. This is especially important for elastomers since the neat matrix usually has very low modulus and strength [48]. However, the addition of fillers modifies crosslinking parameters such as curing and scorch times, with both having a direct effect on the aging (stability) and mechanical properties. Rubber foams have been reinforced using a variety of fillers, but the most common ones are carbon black (CB), NG, CNT, nanofibers, NC and nanosilicates [3,49–51]. Several parameters must be determined to quantify the improvements, including the particle size and geometry, the dispersion level, aspect ratio and orientation. Other parameters are the chemical or physical bonding (type and number) with matrix molecules [52].

When CNT was replaced by NG (50:50 hybrid nanofiller), it was shown that the CNT/NR morphology changed from a discontinuous set of agglomerates into a continuous 3D network [53]. The thermal stability of EPDM foams was found to be enhanced with increasing CB concentration [54]. The viscoelasticity and foaming behavior of EPDM foams were reduced, while the curing and mechanical strength were enhanced. Vahidifar et al. developed a hybrid reinforcing system including NC and CB to study the effect of NC concentration (0–10 phr) on the crosslinking, morphological and mechanical performances of NR foams containing CB (10 phr) [10]. A rheological analysis confirmed that higher NC concentration reduced the scorch time and curing time (up to 50%). The tensile modulus and hardness of the compounds increased as the NC concentration increased, but the foam's resilience and elongation at break decreased. The production of foams with more uniform and smaller cells was easier by increasing the NC concentration from 0 to 5 phr. On the other hand, adding 7 phr NC resulted in a foam morphology made of two sections having various cell diameters and even producing a bimodal distribution. $scCO_2$ was also used to prepared SR/silica foams (saturated at 22 MPa and 50 °C for 1 h) to study the effect of silica concentration (40–70 phr) on the cellular morphology of microcellular closed cell SR foams [55]. According to the morphological results, the SR foam with 50 phr silica exhibited a maximum value of cell size (1.82 μm) and a minimum value of cell density (2.41×10^{10} cells/cm^3). However, higher silica content (70 phr) generated higher cell density (1.02×10^{11} cells/cm^3). These results indicate that the presence of silica is highly important to improve the heterogeneous nucleation of the cells, as well as acting as a reinforcing agent. To improve the mechanical and morphological characteristics of NR foams, Shojaei and colleagues used a physical hybrid system of NG and carbon nanotubes (CNT) [48]. As a result, increasing both CNT and GNS concentration from 0 to 2 phr decreased the cell diameter (800–80 μm in Figure 6) and improved the stress (50% deformation) from 150 to 440 kPa.

Pongmuksuwan and coworkers produced flexible electromagnetic absorber foams based on NR and CB/Fe_3O_4 as fillers [56]. Higher CB content (4–10 phr) led to lower average cell size (780–670 μm). A similar trend was observed for Fe_3O_4 addition. Compared to a foam containing Fe_3O_4, the foams produced with CB had higher tensile strength and modulus. This was because CB has some functional groups (hydroxyl or carboxyl groups) which improved its compatibility and interactions with NR chains. For example, the modulus of NR/CB and NR/Fe_3O_4 foams (10 phr of filler) gave 0.0015 MPa and 0.0007 MPa, respectively. Furthermore, the addition of CB (4–10 phr) increased the compression set by 25%. This showed that higher CB concentration decreases the recoverability of these foams. On the other hand, the incorporation of Fe_3O_4 did not have a significant effect on the compression set because of its limited interaction with NR chains. Hence, the compression set of NR/Fe_3O_4 increased from 18% to 26% by increasing Fe_3O_4 from 4 to 10 phr due to its rigidity which decreased the NR foams' elasticity. Phomark and coworkers used microcellulose (MC, 10–100 μm) and nanocellulose (NC, 10–3000 nm) fibers (5–20 phr) to prepare green and open cell NRLF (cell size: 10–500 μm) in the presence of PO as CFA (Figure 7) [57]. NC was found to have a better dispersion inside the NRLF matrix leading to better reinforcement compared to MC. By adding 5 phr MC or 15 phr NC, the highest tensile strengths were 0.43 MPa and 0.73 MPa, respectively. Compression tests also revealed that 15 phr NC increased the recovery to 34.9% (1.3-fold improvement), while no significant improvement was observed for MC-NRLF.

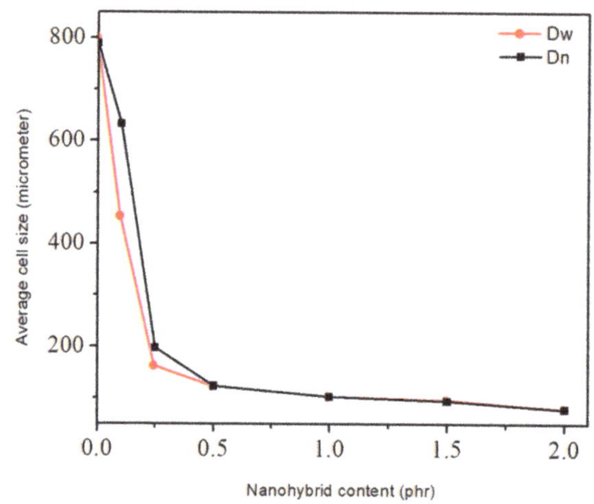

Figure 6. Average cell size of hybrid nanocomposite foams with respect to different CNT/GNS concentration [48].

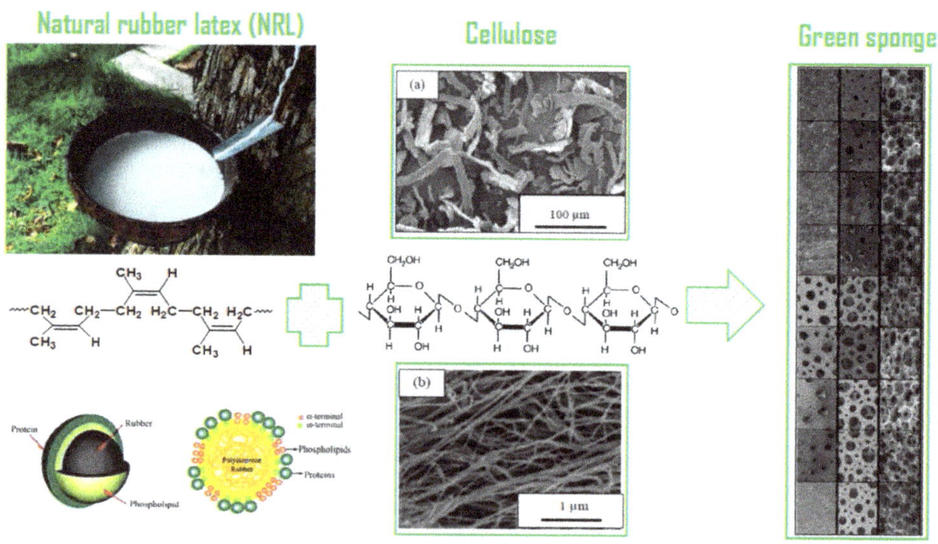

Figure 7. The structure and SEM images of (**a**) MC fibers and (**b**) NC fibers, as well as their green NRLF foams [57].

Different fillers, such as charcoal and silica (2–8 phr), were incorporated into NR to investigate the recovery properties of their foams [58]. Increasing the charcoal content led to larger cell size and lower cell density (Table 1), while the cell density increased and the cell size decreased with higher silica concentration. In contrast to charcoal, the recoverability of NR foam was improved (low compression set) as the silica content was increased (Figure 8). Hence, the ability for NR foams having charcoal concentrations above 2 phr to return to their original shape was reduced when compressed at 75% of their original thickness for

72 h. As the compression set decreased, elasticity increased. Thus, NR foams filled with silica showed higher elasticity than NR foams containing charcoal. This is as a result of silica's higher specific surface area having better filler–rubber interaction than charcoal.

Table 1. Morphological properties of NR foams with different types and concentrations of fillers (charcoal and silica) [58].

Sample Name	Average Cell Size (μm)	Cell Density (Cells/cm^3)
NRF (unfilled)	836	3041
NRF/2 Ch	860	2778
NRF/4 Ch	988	1807
NRF/6 Ch	1081	1379
NRF/8 Ch	1092	1333
NRF/2 Si	1079	1412
NRF/2 Si	909	2316
NRF/2 Si	876	2546
NRF/2 Si	673	5606

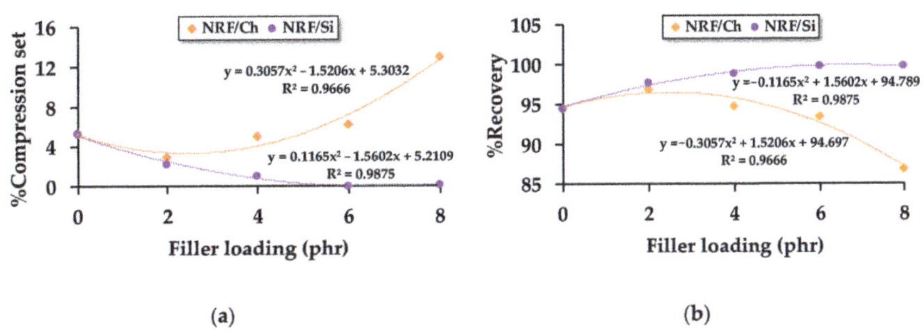

Figure 8. Compression properties of NR foams based on different types and concentrations of fillers (charcoal and silica): (**a**) compression set and (**b**) recovery [58].

In another work, Prasopdee and colleagues used cassava starch (Cs) as a low density filler (lower than charcoal and silica) to produce NR foams [59]. The foam density of NR/Cs was decreased compared to NRF filled with charcoal (14–16%) and silica (18–28%). The addition of Cs (4–12 phr) slightly decreased the cell size (0.54 to 0.52 mm) while improving the cell density (11,540 to 12,319 cells/cm^3). Although the cellular properties of the foams did not substantially change with increasing Cs content, the compression strength did increase (6.1 to 10.5 kPa). This behavior was related to higher crystallinity of Cs and chemical reactions between starch and vulcanization agents leading to higher stiffness.

3.4. Foaming and/or Saturation Temperature

Cell size and cell density (foam density) are directly impacted by the viscosity, which is mostly controlled by temperature [60]. While cell expansion requires high viscosity (low temperature) for easier nucleation, efficient foaming requires low viscosity (high temperature) for easier expansion. The formation of a cellular structure represents an equilibrium between two processes. First, the amount and kinetics of the gas generated from the FA decomposition acting as the blowing force leading to high bubble nucleation and growth. The second effect is an increase in the matrix viscosity related to the crosslinking reaction restricting further expansion and leading to stabilization. As a result, the interaction between the curing and foaming processes can help to identify the ideal foaming temperature. The relationship between NR foams' physical and mechanical characteristics and the foaming temperature (145–155 °C) was studied by Kim and coworkers [61]. As the foaming temperature was increased, the NR density decreased. Because the cell density within the rubber matrix increased with temperature, using lower temperatures (145 and 150 °C) decreased the densities compared to a higher temperature (155 °C). They also observed that for these NR foams, 155 °C was the optimum temperature for vulcanization and foaming. Because of the lower density, it was observed that the tensile strength decreased with increasing temperature, indicating higher foaming effectiveness. Additionally, the tensile modulus and tear strength both continuously decreased as the temperature increased.

Liu's group produced SR/silica (70 phr) foams at different T_s (50–80 °C) using $scCO_2$ (P_s of 22 MPa for 1 h) [55]. They observed that the cell size of foams saturated at 80 °C was 6.5 times compared to 50 °C. This was due to the lower sorption of CO_2 molecules caused by increasing T_s leading to lower nucleation spots and higher cell growth inside the matrix. Furthermore, a higher T_s decreased the elasticity/viscosity of the SR matrix leading to higher cell coalescence. According to Najib and colleagues, increasing the temperature generates higher gas pressure, stretching the cell walls even more [62]. This high cell wall expansion resulted in a substantially larger cell structure and a reduced foam density. Thus, there was less crosslinking and less solid phase (per unit volume). The cell size distribution (CSD) was improved, while the range of cell sizes was higher by increasing the foaming temperature. Compared to foams made at 150 °C and 160 °C, the foams made at 140 °C showed a more homogeneous CSD. Rostami et al. achieved functionally graded POE foams by changing the processing temperature (molding temperature), the average temperature (T_{avg}) and temperature difference (ΔT), as shown in Figure 9 [15]. By tailoring the cell size and cell density (across thickness), different thermal conductivities (0.125–0.180 W/m.K) were obtained. Increasing T_{avg} (from 207.5 to 215 °C) decreased the tensile properties, including the modulus, strength and elongation at break by 33%, 13% and 15%, respectively. On the contrary, increasing ΔT (from 10 to 40 °C) improved these properties by 14%, 26% and 10%, respectively. According to Pechurai and colleagues, raising the processing temperature reduced the scorch and cure times since the curing processes were faster [63]. Due to the direct effect of thermal breakdown on the heat produced (exotherm) or consumed (endotherm), which affects the crosslinking properties, the optimum curing temperature for a specific rubber matrix is not always the same, with or without CFA.

3.5. Foaming and Saturation Pressure

The amount of pressure applied when forming or molding a sample has a significant effect on how the rubber foam's morphology will develop. Kim and coworkers investigated the effect of foaming pressure on the structure and properties of NR foams [64]. Higher density (lower ER) and decreased foaming efficiency were produced when the foaming pressure was increased. With increasing foaming pressure, the hardness and tear strength steadily increased, but lower elongation at break was observed. $scCO_2$ is considered as the best PFA because it is an eco-friendly alternative and its P_s was shown to have a significant effect on the cellular microstructure. Moreover, using $scCO_2$, Tessanan and associates investigated different pressures (0–12.5 MPa) to determine the effect on NR foams' morphology [30]. According to their findings, increasing P_s decreased the cell diameter (below 10 μm) and narrowed the CSD. The main reason is that higher pressure decreases the activation energy for nucleation and more gas molecules (higher saturation/solubility) are available to produce the nuclei. A number of high temperature vulcanizates (HTV) SR microcellular foams were generated based on CO_2 [65]. Increasing P_s improved the CO_2 diffusivity in the HTV silicone rubber as well. Additionally, increasing the P_s made the matrix absorb more CO_2, leading to smaller cell diameter and higher cell density (higher gas concentration). Hence, Xiang and coworkers studied the effect of various P_s (16–22 MPa) of $scCO_2$ on the morphological properties of SR/silica (70 phr) foams prepared at 50 °C for 1 h [55]. Cell size decreased (2.11–0.708 μm), while the cell density increased ($0.296–1.02 \times 10^{11}$ cells/cm^3) if P_s increased (16–22 MPa). According to Xiang's group, it can be concluded that both T_s and P_s have a different effect on the morphological properties of rubber foams. Increasing P_s leads to higher cell density and lower cell size, while increasing T_s results in lower cell density and higher cell size in SR foams produced by $scCO_2$. Wang et al. investigated the solubility of different PFA (CO_2, N_2 and their mixture) and the morphology of SR foams [66]. Their results showed that for a single PFA, increasing the pressure increased their solubility in SR (Figure 10a). On the other hand, the CO_2 diffusivity and solubility were higher compared to N_2 under similar conditions (Figure 10b). They also observed that changing the CO_2:N_2 ratio can be used to modify the solubility and diffusion coefficient of the blowing agent system. In particular, a CO_2:N_2 ratio of 4:6 produced a highly uniform cellular structure with a foam density as low as 0.086 g/cm^3.

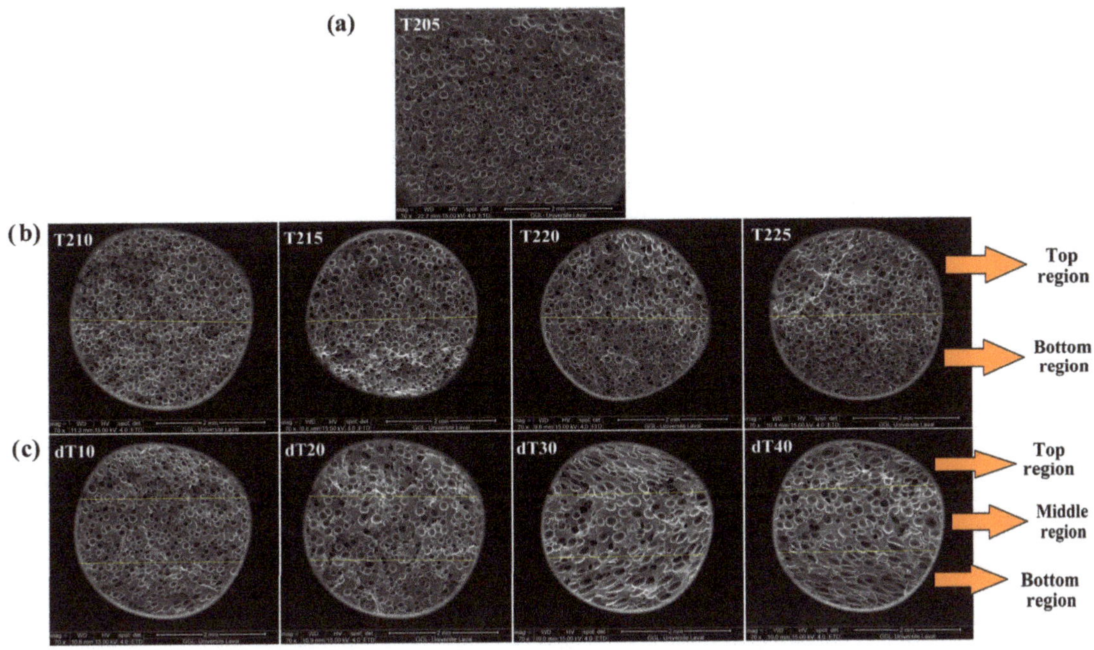

Figure 9. SEM images of: (**a**) uniform and graded POE foam at: (**b**) different T_{avg} of 207.5, 210, 212.5 and 215 °C (from left to right) and (**c**) different ΔT of 10, 20, 30 and 40 °C (from left to right) [15].

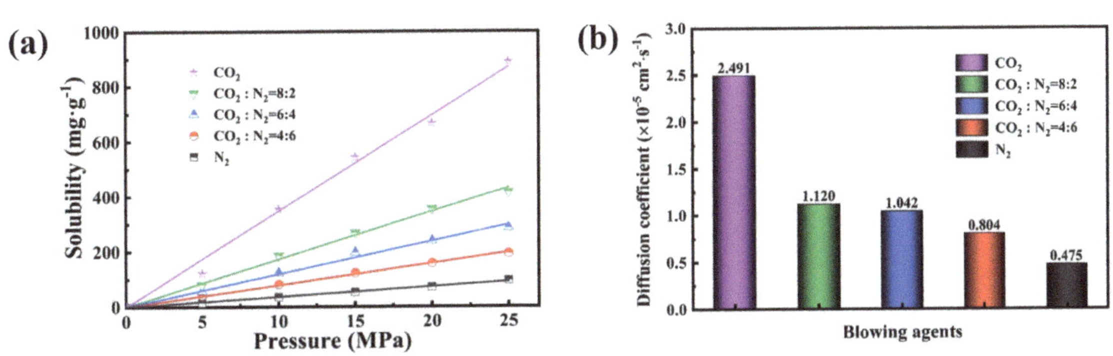

Figure 10. (**a**) Solubility as a function of gas pressure (50 °C) and (**b**) diffusion coefficient of the different PFA in SR foam at 50 °C and 10 MPa [66].

Zhu et al. proposed a novel $scCO_2$ aided post-vulcanization method to develop microcellular epoxidized natural rubber foam (f-ENR) with enhanced ductility, durability and cellular structure stability [67]. They observed that when P_s increased from 15 MPa to 25 MPa, the cell diameter decreased (1.70–1.30 μm) while the cell density increased (1.10×10^{11}–1.40×10^{11} cells/cm^3). However, their results showed that when the T_s was raised from 35 °C to 75 °C at 15 MPa, the cell density decreased (1.10×10^{11}–8.40×10^9 cells/cm^3) while the cell diameter increased (1.70–5.50 μm). These trends were attributed to the decreasing solubility of CO_2 in the ENR matrix when the T_s increases, thus limiting cell nucleation.

4. Conclusions

4.1. General Conclusions

Rubber foams are lightweight materials having a number of favorable characteristics, including low density, high elasticity and flexibility, high tear strength and great resistance to abrasion, as well as good thermal and acoustic insulation. These properties make rubber foams ideal for civil and industrial applications, with examples including automotive, aerospace, insulation, medical, construction, etc. Depending on the formulation, processing technique and foaming conditions used during their production, rubber foams can have different morphology (open or closed cell) with cell size ranging from millimeter to nanoscale. Table 2 presents a summary of the materials, foaming process parameters and physical and morphological properties of the studies included in this review paper.

Table 2. Summary of the materials, foaming process parameters and physical and morphological properties of rubber foams.

Matrix	Foaming Agent	Filler	Foaming Process	Density (g/cm^3)	Morphology	Reference
NR	ADC (4 phr)	NC (0–10 phr)	T (155 °C) P (50 bar) t (30 min) *	0.40	Closed cell CS (340–143 μm) CD (35–225 cells/mm^3) **	[10]
EPDM	OBSH (3 phr)	CNT (1–3 phr) NC (2–6 phr)	T (160 °C) P (50 bar) t (20 min)		Closed/open cell CD (5.21×10^{10}–6.28×10^{10} cells/cm^3)	[11]
EPDM	Micropearl (14 phr) Expancel (14 phr) Sodium chloride (305 phr)	CB (20 phr)	T (170 °C) P (2 bar) t (20 min)	0.23–0.42	Closed/open cell CS (60–80 μm)	[12]
POE	ADC (4 phr)		T_{avg} (205–215 °C) ΔT (0–40 °C) P (8.5 bar) T (12 min)	0.55–0.72	Closed cell CS (98–337 μm) CD (204–820 cells/mm^3)	[15]
SR	$scCO_2$	Silica (26 wt.%)	T (40–70 °C) P (10–15 MPa) t (1 h)	0.25–0.70	Closed cell CS (5–266 μm) CD (2.60×10^5–5.60×10^9 cells/cm^3)	[16]
NR	$NaHCO_3$ (3 to 12 phr)		T (150 °C) t (45 min)		Closed cell CS (420–620 μm)	[17]
SR		NG (1–4 wt.%)	T (60 °C) t (1 h)		Closed/open cell CS (191.00–79.40 μm) CD (2.13×10^5–7.15×10^5 cells/cm^3)	[18]
EPDM	Expancel (14 phr)		T (170 °C) P (2 bar) t (10 min)	0.48–0.59	Closed cell CS (20 μm)	[23]
EPDM	Hostatron (1.3 phr)		T (170 °C) P (2 bar) t (10 min)	0.57–0.73	Closed cell CS (100 μm)	[23]
SR	NaCl (0–100 phr)		T (150 °C) t (20 min)	0.44–0.63	Close/open cell CS (100–300 μm)	[26]
NR	PO (0–45%)		T (90 °C) t (2 h)	0.11–0.12	Open cell CS (273–548 μm) CD (10,241–82,450 cells/cm^3)	[32]
POE	ADC (1–13 phr)		T (170 °C) P (10 MPa) t (10 min)	0.06–0.82	Closed cell CD (3.79×10^9–6.29×10^{10} cells/cm^3)	[36]
NR	ADC (10 phr)		Electron beam irradiation (50–150 kGy) t (30–90 s)		Closed cell CS (22.60–74.72 μm) CD (4.92×10^5–249.60×10^5 cells/cm^3)	[37]
ENR	ADC (10 phr)		Electron beam irradiation (50–150 kGy) t (30–90 s)		Closed cell CS (6.83–74.57 μm) CD (4.19×10^5–1975.20×10^5 cells/cm^3)	[37]

Table 2. *Cont.*

Matrix	Foaming Agent	Filler	Foaming Process	Density (g/cm^3)	Morphology	Reference
EPDM	$scCO_2$	HNT (0–10 phr)	T (130 °C) P (30 MPa) t (24 h)	0.23–0.35	Closed cell CS (7.80–12.00 μm) CD (4.21×10^9–1.54×10^{10} cells/cm^3)	[39]
SBR	ADC (3 phr)		T (200 °C)	0.22–0.83	Closed/open cell CS (7.68–22.94 μm)	[40]
SR	$scCO_2$	Silica (30–50 phr)	T (40−80 °C) P (10−14 MPa) t (1 h)		Closed/open cell CS (4.24–77.10 μm) CD (0.40×10^6–1.40×10^9 cells/cm^3)	[45]
POE	ADC (2–5 phr)		T (205 °C) P (8.5 MPa) t (12 min)	0.61–0.75	Closed cell CS (109.40–153.10 μm) CD (103–591 cells/mm^3)	[47]
NR	ADC (6 phr)	CNT/NG (0–2 phr)	T (160 °C)		Closed cell CS (80–800 μm)	[48]
SR	$scCO_2$	Silica (40–70 phr)	T (50–80 °C) P (16–22 MPa) t (1 h)		Closed cell CS (0.71–4.60 μm) CD (0.40×10^{10}–1.02×10^{11} cells/cm^3)	[55]
NR	PO (1.5 phr)	Microcellulose/nanocellulose (5–20 phr)	T (100 °C) t (1 h)	0.15–0.31	Open cell (10–500 μm)	[57]
NR	PO (3.63%)	Cs (4–12 phr)	T (90 °C) t (2 h)	0.069–0.071	Open cell CS (0.54–0.52 μm) CD (11,540–12,319 cells/cm^3)	[59]
ENR	$scCO_2$		T (35 °C) P (15–25 MPa) t (1 h)		Closed cell CS(1.30–1.70 μm) CD (1.10×10^{11}–1.40×10^{11} cells/cm^3)	[67]
ENR	$scCO_2$		T (35–75 °C) P (15 MPa) t (1 h)		Closed cell CS (1.70–5.50 μm) CD (8.40×10^9–1.10×10^{11} cells/cm^3)	[67]

* T—temperature, P—pressure, t—time. ** CS—cell size, CD—cell density.

Three steps are important in the production of rubber foams: nucleation, growth and stabilization. Initially, gas nuclei are formed by a foaming agent (chemical or physical) inside the rubber matrix. These cells then grow in size as the curing proceeds to stabilize the morphology. Two important factors have an effect on the morphological, mechanical, physical and thermal properties. The first parameter is associated with formulation, while the second is the conditions used for processing. Type of rubbers, accelerators, foaming agents and fillers, along with their quantities, are associated with composition, while the steps of foaming (pressure, temperature, time, etc.) and crosslinking (pre-curing process, curing temperature and time) are associated with the production conditions. Most of the studies investigated how these factors affect the characteristics of rubber foams and they have been reviewed in this work. To fully understand all the relations between formulation, processing, structures and characteristics, further research is still required, as described next.

4.2. Opening for Future Work

The majority of the work performed in the early years of foam manufacture was devoted to developing different porous morphologies. Based on the development in nanotechnology, the focus switched to various nanoparticles and how they affect the characteristics of cellular polymeric materials. The study on nanocomposite rubber foams is still in its early stages and more work is needed to determine how the rubber foam's structure and properties are affected by the size, shape and surface chemistry of these nanoparticles. The theory of rubber foaming is still not completely understood, as the processes of foaming and curing are interrelated and very complex since they depend on several factors, such as pressure, temperature, concentration, time and pre/post-curing conditions, as well as the rubber type (single resin or blends). Finally, more work should be devoted to the modeling and simulation of all the steps involved in rubber foaming, including cell nucleation and growth in a rheologically complex multiphase material under unsteady and non-isothermal conditions.

Author Contributions: Conceptualization, D.R.; methodology, E.R.-T.-E.; software, E.R.-T.-E.; validation, D.R.; formal analysis, D.R.; investigation, E.R.-T.-E.; resources, D.R.; data curation, E.R.-T.-E.; writing—original draft preparation, E.R.-T.-E.; writing—review and editing, D.R.; visualization, E.R.-T.-E.; supervision, D.R.; project administration, D.R.; funding acquisition, D.R. All authors have read and agreed to the published version of the manuscript.

Funding: This research received no external funding.

Institutional Review Board Statement: Not applicable.

Informed Consent Statement: Not applicable.

Data Availability Statement: Not applicable.

Conflicts of Interest: The authors declare no conflict of interest.

References

1. Zhai, W.; Jiang, J.; Park, C.B. A review on physical foaming of thermoplastic and vulcanized elastomers. *Polym. Rev.* **2022**, *62*, 95–141. [CrossRef]
2. Jin, F.-L.; Zhao, M.; Park, M.; Park, S.-J. Recent trends of foaming in polymer processing: A review. *Polymers* **2019**, *11*, 953. [CrossRef] [PubMed]
3. Xie, Z.; Cai, Y.; Zhan, Y.; Meng, Y.; Li, Y.; Xie, Q.; Xia, H. Thermal insulating rubber foams embedded with segregated carbon nanotube networks for electromagnetic shielding applications. *Chem. Eng. J.* **2022**, *435*, 135118. [CrossRef]
4. Zhou, Y.; Tian, Y.; Peng, X. Applications and Challenges of Supercritical Foaming Technology. *Polymers* **2023**, *15*, 402. [CrossRef] [PubMed]
5. Suethao, S.; Ponloa, W.; Phongphanphanee, S.; Wong-Ekkabut, J.; Smitthipong, W. Current challenges in thermodynamic aspects of rubber foam. *Sci. Rep.* **2021**, *11*, 6097. [CrossRef] [PubMed]
6. Rostami-Tapeh-Esmaeil, E.; Vahidifar, A.; Esmizadeh, E.; Rodrigue, D. Chemistry, processing, properties, and applications of rubber foams. *Polymers* **2021**, *13*, 1565. [CrossRef] [PubMed]
7. Frisch, K.C. History of science and technology of polymeric foams. *J. Macromol. Sci. Chem.* **1981**, *15*, 1089–1112. [CrossRef]
8. Mosanenzadeh, S.G.; Naguib, H.E.; Park, C.B.; Atalla, N. Development, characterization, and modeling of environmentally friendly open-cell acoustic foams. *Polym. Eng. Sci.* **2013**, *53*, 1979–1989. [CrossRef]
9. Kong, W.-l.; Bao, J.-B.; Wang, J.; Hu, G.-H.; Xu, Y.; Zhao, L. Preparation of open-cell polymer foams by CO_2 assisted foaming of polymer blends. *Polymer* **2016**, *90*, 331–341. [CrossRef]
10. Vahidifar, A.; Esmizadeh, E.; Rostami, E.; Nouri Khorasani, S.; Rodrigue, D. Morphological, rheological, and mechanical properties of hybrid elastomeric foams based on natural rubber, nanoclay, and nanocarbon black. *Polym. Compos.* **2019**, *40*, 4289–4299. [CrossRef]
11. Shojaei-Dindarloo, A.; Karrabi, M.; Hamid, M.; Ghoreishy, R. Various nano-particles influences on structure, viscoelastic, Vulcanization and mechanical behaviour of EPDM nano-composite rubber foam. *Plast. Rubber Compos.* **2019**, *48*, 218–225. [CrossRef]
12. Zonta, E.; Valentini, F.; Dorigato, A.; Fambri, L.; Pegoretti, A. Evaluation of the salt leaching method for the production of ethylene propylene diene monomer rubber foams. *Polym. Eng. Sci.* **2021**, *61*, 136–153. [CrossRef]
13. Pang, Y.; Cao, Y.; Zheng, W.; Park, C.B. A comprehensive review of cell structure variation and general rules for polymer microcellular foams. *Chem. Eng. J.* **2022**, *430*, 132662. [CrossRef]
14. Azdast, T.; Hasanzadeh, R. Increasing cell density/decreasing cell size to produce microcellular and nanocellular thermoplastic foams: A review. *J. Cell. Plast.* **2021**, *57*, 769–797. [CrossRef]
15. Rostami-Tapeh-Esmaeil, E.; Shojaei, S.; Rodrigue, D. Mechanical and Thermal Properties of Functionally Graded Polyolefin Elastomer Foams. *Polymers* **2022**, *14*, 4124. [CrossRef] [PubMed]
16. Tang, W.; Liao, X.; Zhang, Y.; Li, J.; Wang, G.; Li, G. Mechanical–microstructure relationship and cellular failure mechanism of silicone rubber foam by the cell microstructure designed in supercritical CO_2. *J. Phys. Chem. C* **2019**, *123*, 26947–26956. [CrossRef]
17. Syahrin, S.; Zunaida, Z.; Hakimah, O.; Nuraqmar, S. Effect of blowing agent on compression and morphological properties of natural rubber latex foam. *AIP Conf. Proc.* **2020**, *2267*, 020037. [CrossRef]
18. Jia, F.; Liu, C.; Yang, B.; Lee, A.; Wu, L.; Ning, H. Preparation of Novel Graphene/Silicone Rubber Nanocomposite Dielectric Foams. *Polymers* **2022**, *14*, 3273. [CrossRef] [PubMed]
19. Vahidifar, A.; Esmizadeh, E.; Rodrigue, D. Effect of the simultaneous curing and foaming kinetics on the morphology development of polyisoprene closed cell foams. *Elastomery* **2018**, *22*, 3–18.
20. Standau, T.; Altstädt, V. Foams. In *Polypropylene Handbook*; Springer: Cham, Switzerland, 2019; pp. 579–641.
21. Marl, S.; Giesen, R.-U.; Heim, H.-P. Liquid silicone rubber foamed with thermoplastic expandable microspheres. *Materials* **2022**, *15*, 3779. [CrossRef] [PubMed]
22. Ryzhenkov, A.V.; Pogorelov, S.I.; Loginova, N.A.; Belyaeva, E.V.; Plestsheva, A.Y. Syntactic foams efficiency with the use of various microspheres for heat supply equipment and pipelines heat insulation. *Mod. Appl. Sci.* **2015**, *9*, 319. [CrossRef]

23. Valentini, F.; Dorigato, A.; Fambri, L.; Pegoretti, A. Novel EPDM/paraffin foams for thermal energy storage applications. *Rubber Chem. Technol.* **2021**, *94*, 432–448. [CrossRef]
24. Bianchi, M.; Valentini, F.; Fredi, G.; Dorigato, A.; Pegoretti, A. Thermo-mechanical behavior of novel EPDM foams containing a phase change material for thermal energy storage applications. *Polymers* **2022**, *14*, 4058. [CrossRef] [PubMed]
25. Jin, D.; Xiao, K.; Han, Y.; Xiang, G.; Zhou, Z.; Wang, J. A preparation method of porous surface nitrile butadiene rubber with low friction coefficient under water lubrication condition by salt leaching. *J. Appl. Polym. Sci.* **2021**, *138*, 50555. [CrossRef]
26. Peng, L.; Lei, L.; Liu, Y.; Du, L. Improved mechanical and sound absorption properties of open cell silicone rubber foam with NaCl as the pore-forming agent. *Materials* **2021**, *14*, 195. [CrossRef] [PubMed]
27. Yaqoob, A.A.; Ibrahim, M.N.M.; Umar, K.; Ahmad, A. Applications of Supercritical Carbon Dioxide in the Rubber Industry. In *Advanced Nanotechnology and Application of Supercritical Fluids*; Springer: Cham, Switzerland, 2020; pp. 199–218.
28. Zhang, W.; Deng, Z.; Yuan, H.; Luo, S.; Wen, H.; Liu, T. Preparation and properties of silicone rubber materials with foam/solid alternating multilayered structures. *Polym. J.* **2021**, *53*, 619–631. [CrossRef]
29. Jacobs, M.A.; Kemmere, M.F.; Keurentjes, J.T. Foam processing of poly (ethylene-*co*-vinyl acetate) rubber using supercritical carbon dioxide. *Polymer* **2004**, *45*, 7539–7547. [CrossRef]
30. Tessanan, W.; Phinyocheep, P.; Daniel, P.; Gibaud, A. Microcellular natural rubber using supercritical CO_2 technology. *J. Supercrit. Fluids* **2019**, *149*, 70–78. [CrossRef]
31. Coste, G.; Negrell, C.; Caillol, S. From gas release to foam synthesis, the second breath of blowing agents. *Eur. Polym. J.* **2020**, *140*, 110029. [CrossRef]
32. Suethao, S.; Phongphanphanee, S.; Wong-Ekkabut, J.; Smitthipong, W. The relationship between the morphology and elasticity of natural rubber foam based on the concentration of the chemical blowing agent. *Polymers* **2021**, *13*, 1091. [CrossRef]
33. Afrinaldi, B.; Vicarneltor, D.N.; Rudianto, R.P.; Hakim, A.R.; Muslim, O.F. Influence of Zinc Oxide Addition on Azodicarbonamide Thermal Decomposition in the Polyethylene/Ethylene Vinyl Acetate Foaming Release. *Mater. Sci. Forum. Trans. Tech. Pub.* **2021**, *1028*, 234–239. [CrossRef]
34. Heidari, A.; Fasihi, M. Cell structure-impact property relationship of polypropylene/thermoplastic elastomer blend foams. *Express Polym. Lett.* **2019**, *13*, 429–442. [CrossRef]
35. Kmetty, Á.; Litauszki, K.; Réti, D. Characterization of different chemical blowing agents and their applicability to produce poly (lactic acid) foams by extrusion. *Appl. Sci.* **2018**, *8*, 1960. [CrossRef]
36. Mao, Y.; Qi, R. Preparation of polyethylene–octene elastomer foams by compression molding. *J. Appl. Polym. Sci.* **2008**, *109*, 3249–3255. [CrossRef]
37. Salmazo, L.O.; Lopéz-Gil, A.; Ariff, Z.M.; Rodriguez-Mendez, M.L.; Job, A.E.; Rodriguez-Perez, M.A. Study of the foaming kinetics in epoxidized natural rubber foams crosslinked by electron beam irradiation. *Macromol. Chem. Phys.* **2018**, *219*, 1800295. [CrossRef]
38. Ding, H.; Luo, Z.; Kong, N.; Li, Z.; Zhao, P.; Zhang, J.; Tao, J. Constructing conductive titanium carbide nanosheet (MXene) network on natural rubber foam framework for flexible strain sensor. *J. Mater. Sci. Mater. Electron.* **2022**, *33*, 15563–15573. [CrossRef]
39. Lee, K.; Chang, Y.W.; Kim, S.W. Ethylene–propylene-diene terpolymer/halloysite nanocomposites: Thermal, mechanical properties, and foam processing. *J. Appl. Polym. Sci.* **2014**, *131*, 40307. [CrossRef]
40. Shao, L.; Ji, Z.-Y.; Ma, J.-Z.; Xue, C.-H.; Ma, Z.-L.; Zhang, J. The synergy of double cross-linking agents on the properties of styrene butadiene rubber foams. *Sci. Rep.* **2016**, *6*, 36931. [CrossRef] [PubMed]
41. Bahadar, A.; Zwawi, M. Development of SWCNTs-reinforced EPDM/SBR matrices for shock absorbing applications. *Mater. Res. Express* **2020**, *7*, 025310. [CrossRef]
42. Park, K.W.; Ray Chowdhury, S.; Park, C.C.; Kim, G.H. Effect of dispersion state of organoclay on cellular foam structure and mechanical properties of ethylene vinyl acetate copolymer/ethylene-1-butenecopolymer/organoclay nanocomposite foams. *J. Appl. Polym. Sci.* **2007**, *104*, 3879–3885. [CrossRef]
43. Kim, M.S.; Park, C.C.; Chowdhury, S.R.; Kim, G.H. Physical properties of ethylene vinyl acetate copolymer (EVA)/natural rubber (NR) blend based foam. *J. Appl. Polym. Sci.* **2004**, *94*, 2212–2216. [CrossRef]
44. Zhang, B.S.; Zhang, Z.X.; Lv, X.F.; Lu, B.X.; Xin, Z.X. Properties of chlorinated polyethylene rubber/ethylene vinyl acetate copolymer blend-based foam. *Polym. Eng. Sci.* **2012**, *52*, 218–224. [CrossRef]
45. Liao, X.; Xu, H.; Li, S.; Zhou, C.; Li, G.; Park, C.B. The effects of viscoelastic properties on the cellular morphology of silicone rubber foams generated by supercritical carbon dioxide. *RSC Adv.* **2015**, *5*, 106981–106988. [CrossRef]
46. Luo, S.K.; Ding, G.F.; Dai, B.; Yang, M.; Liu, F. Study on the cell structure and mechanical properties of methyl vinyl silicone rubber foam materials. *Adv. Mater. Res.* **2014**, *1004*, 297–306. [CrossRef]
47. Rostami-Tapeh-Esmaeil, E.; Heydari, A.; Vahidifar, A.; Esmizadeh, E.; Rodrigue, D. Experimental and Finite Element Simulation of Polyolefin Elastomer Foams Using Real 3D Structures: Effect of Foaming Agent Content. *Polymers* **2022**, *14*, 4692. [CrossRef]
48. Shojaie, S.; Vahidifar, A.; Naderi, G.; Shokri, E.; Mekonnen, T.H.; Esmizadeh, E. Physical Hybrid of Nanographene/Carbon Nanotubes as Reinforcing Agents of NR-Based Rubber Foam. *Polymers* **2021**, *13*, 2346. [CrossRef]
49. Afolabi, L.O.; Ariff, Z.M.; Hashim, S.F.S.; Alomayri, T.; Mahzan, S.; Kamarudin, K.-A.; Muhammad, I.D. Syntactic foams formulations, production techniques, and industry applications: A review. *J. Mater. Res. Technol.* **2020**, *9*, 10698–10718. [CrossRef]

50. Esmizadeh, E.; Vahidifar, A.; Rostami, E.; Nouri Khorasani, S.; Ghayoumi, M.; Khonakdar, H.A. Effect of carbon black on morphological and mechanical properties of rubber foams produced by a single-step method. *J. Appl. Res. Chem. Polym.Eng.* **2017**, *1*, 49–60.
51. Shao, Y.; Luo, C.; Deng, B.-w.; Yin, B.; Yang, M.-b. Flexible porous silicone rubber-nanofiber nanocomposites generated by supercritical carbon dioxide foaming for harvesting mechanical energy. *Nano Energy* **2020**, *67*, 104290. [CrossRef]
52. Leblanc, J.L. Rubber–filler interactions and rheological properties in filled compounds. *Prog. Polym. Sci.* **2002**, *27*, 627–687. [CrossRef]
53. Ponnamma, D.; Sadasivuni, K.K.; Strankowski, M.; Guo, Q.; Thomas, S. Synergistic effect of multi walled carbon nanotubes and reduced graphene oxides in natural rubber for sensing application. *Soft Matter* **2013**, *9*, 10343–10353. [CrossRef]
54. Bashir, M.A.; Iqbal, N.; Shahid, M.; Ahmed, R. Structural, viscoelastic, and vulcanization study of sponge ethylene–propylene–diene monomer composites with various carbon black loadings. *J. Appl. Polym. Sci.* **2014**, *131*, 39423. [CrossRef]
55. Xiang, B.; Deng, Z.; Zhang, F.; Wen, N.; Lei, Y.; Liu, T.; Luo, S. Microcellular silicone rubber foams: The influence of reinforcing agent on cellular morphology and nucleation. *Polym. Eng. Sci.* **2019**, *59*, 5–14. [CrossRef]
56. Pongmuksuwan, P.; Salayong, K.; Lertwiriyaprapa, T.; Kitisatorn, W. Electromagnetic absorption and mechanical properties of natural rubber composites based on conductive carbon black and Fe_3O_4. *Materials* **2022**, *15*, 6532. [CrossRef]
57. Phomrak, S.; Nimpaiboon, A.; Newby, B.-M.Z.; Phisalaphong, M. Natural rubber latex foam reinforced with micro-and nanofibrillated cellulose via Dunlop method. *Polymers* **2020**, *12*, 1959. [CrossRef]
58. Prasopdee, T.; Smitthipong, W. Effect of fillers on the recovery of rubber foam: From theory to applications. *Polymers* **2020**, *12*, 2745. [CrossRef] [PubMed]
59. Prasopdee, T.; Shah, D.U.; Smitthipong, W. Approaches toward high resilience rubber foams: Morphology–mechanics–thermodynamics relationships. *Macromol. Mater. Eng.* **2021**, *306*, 2100337. [CrossRef]
60. Sombatsompop, N. Analysis of cure characteristics on cross-link density and type, and viscoelastic properties of natural rubber. *Polym. Plast. Technol. Eng.* **1998**, *37*, 333–349. [CrossRef]
61. Kim, J.-H.; Koh, J.-S.; Choi, K.-C.; Yoon, J.-M.; Kim, S.-Y. Effects of foaming temperature and carbon black content on the cure characteristics and mechanical properties of natural rubber foams. *J. Indust. Eng. Chem.* **2007**, *13*, 198–205.
62. Najib, N.; Ariff, Z.; Bakar, A.; Sipaut, C. Correlation between the acoustic and dynamic mechanical properties of natural rubber foam: Effect of foaming temperature. *Mater. Des.* **2011**, *32*, 505–511. [CrossRef]
63. Pechurai, W.; Muansupan, T.; Seawlee, P. Effect of foaming temperature and blowing agent content on cure characteristics, mechanical and morphological properties of natural rubber foams. *Adv. Mater. Res.* **2014**, *844*, 454–457. [CrossRef]
64. Kim, J.-H.; Choi, K.-C.; Yoon, J.-M. The foaming characteristics and physical properties of natural rubber foams: Effects of carbon black content and foaming pressure. *J. Indust. Eng. Chem.* **2006**, *12*, 795–801.
65. Yang, Q.; Yu, H.; Song, L.; Lei, Y.; Zhang, F.; Lu, A.; Liu, T.; Luo, S. Solid-state microcellular high temperature vulcanized (HTV) silicone rubber foam with carbon dioxide. *J. Appl. Polym. Sci.* **2017**, *134*, 44807. [CrossRef]
66. Wang, B.; Liao, X.; Tang, W.; Li, S.; Jiang, Q.; Yang, J.; Li, J.; Li, R.; Tian, C.; Li, G. Cellular structure design by controlling the dissolution and diffusion behavior of gases in silicon rubber. *J. Supercrit. Fluids* **2022**, *186*, 105610. [CrossRef]
67. Zhu, J.; Li, X.; Weng, Y.; Tan, B.; Zhang, S. Fabrication of microcellular epoxidized natural rubber foam with superior ductility by designable chemical and physical crosslinking networks. *J. Supercrit. Fluids* **2022**, *181*, 105508. [CrossRef]

Disclaimer/Publisher's Note: The statements, opinions and data contained in all publications are solely those of the individual author(s) and contributor(s) and not of MDPI and/or the editor(s). MDPI and/or the editor(s) disclaim responsibility for any injury to people or property resulting from any ideas, methods, instructions or products referred to in the content.

materials

MDPI

Article

Theoretical and Practical Evaluation of the Feasibility of Zinc Evaporation from the Bottom Zinc Dross as a Valuable Secondary Material

Pauerová Katarína [1,*], Trpčevská Jarmila [1], Briančin Jaroslav [2] and Plešingerová Beatrice [3]

1 Faculty of Materials, Metallurgy and Recycling, Institute of Recycling Technologies, Technical University of Kosice, Letna 9, 042 00 Kosice, Slovakia
2 Institute of Geotechnics SAS, Slovak Academy of Sciences, Watsonova 45, 040 01 Kosice, Slovakia
3 Faculty of Materials, Metallurgy and Recycling, Institute of Metallurgy, Technical University of Kosice, Letna 9, 042 00 Kosice, Slovakia
* Correspondence: katarina.pauerova@tuke.sk

Abstract: This study presents a theoretical and practical evaluation of zinc evaporation from bottom zinc dross (hard zinc) as a secondary zinc source (zinc content approximately 94–97%), which originates in the batch hot-dip galvanizing process. The thermodynamics of the zinc evaporation process were studied under the normal pressure (100 kPa) in the inert atmosphere, using argon with flow rate 90 mL/min. Samples were subjected to the evaporation process for 5, 10 and 20 min under the temperature of 700 °C and 800 °C, respectively. For the theoretical thermodynamic study, HSC Chemistry 6.1 software was used and final products, as well as residuals after the evaporation process, were analyzed by SEM (Scanning Electron Microscopy) and EDX (Energy Dispersive X-ray). Calculated and experimental argon consumption in the process of zinc evaporation has been compared. A high purity zinc with efficiency over 99% was reached. Due to a dynamic regime, argon consumption at the temperature of 700 °C and 800 °C were 7 times and 3 times, respectively, less than calculated.

Keywords: bottom zinc dross; thermodynamic study; zinc evaporation; inert atmosphere

Citation: Katarína, P.; Jarmila, T.; Jaroslav, B.; Beatrice, P. Theoretical and Practical Evaluation of the Feasibility of Zinc Evaporation from the Bottom Zinc Dross as a Valuable Secondary Material. *Materials* **2022**, *15*, 8843. https://doi.org/10.3390/ma15248843

Academic Editor: Danuta Barnat-Hunek

Received: 16 November 2022
Accepted: 8 December 2022
Published: 11 December 2022

Publisher's Note: MDPI stays neutral with regard to jurisdictional claims in published maps and institutional affiliations.

Copyright: © 2022 by the authors. Licensee MDPI, Basel, Switzerland. This article is an open access article distributed under the terms and conditions of the Creative Commons Attribution (CC BY) license (https://creativecommons.org/licenses/by/4.0/).

1. Introduction

Zinc, as the fourth most produced metal worldwide, is a key element of industrial society. Zinc has a wide range of usage in crucial applications such as corrosion protection of steel in buildings, infrastructure, and vehicles [1,2]. Within these applications zinc is used in many forms such as high grade metal (coatings for steel), alloying elements (in brass), and chemical compounds, e.g., zinc oxide in tires [3,4]. In 2021, the world zinc reserves were estimated to be some 250 million tons. Australia owns the largest zinc reserves worldwide—an estimated 69 million tons [5]. Global zinc mine production reached nearly 13.8 million tons in the year 2021. China is the largest zinc miner and metal producer with 30.4% of the global amount [6]. In the same year, approximately 14.05 million tons of refined zinc was consumed worldwide [7].

Zinc, at the end of a product's life, can be recovered and recycled without the loss of its characteristics or value. Zinc containing products become a source of recycling feedstock at the end of their life, known as "old scrap". The old scrap is collected and processed based on scrap availability, metal composition and ease of processing. During the processing phase, zinc becomes available for recycling, as a "new scrap", due to potential losses during manufacturing and fabrication, e.g., drosses, residues, off-cuts. The current zinc end of life recycling rate reaches 34% [8–10].

Presently, approximately 70% of the zinc produced originates from mined ore and 30% from recycled or secondary zinc. More than 50% of zinc is used in galvanizing industry as zinc coatings. The hot-dip galvanizing process (HDG) is the most used method of zinc coating applications [10–12].

The hot-dip galvanizing process generates several wastes or secondary products, respectively. Solid wastes are more valuable due to its high zinc content. One kind of such waste is bottom zinc dross or so-called "hard zinc". Bottom zinc dross is created in the galvanizing process as the maximum iron solubility is exceeded, which is 0.03% in the molten zinc under the process temperature (approximately 450 °C). After exceeding this solubility, iron starts to precipitate into intermetallic compound in the form of $FeZn_{13}$. As intermetallic compound has a higher density than molten zinc, these compounds settle down on the bottom of the zinc kettle. Intermetallic compounds, together with molten zinc, are regularly withdrawn from the zinc kettle by a special mechanical dipper bucket. Bottom zinc dross is considered to be a valuable secondary zinc source due to its high zinc content (94–97%) [13,14].

Zinc recovery techniques are mainly based on the character of the zinc waste (metallic form, oxidic form, or complex compound). Pyrometallurgy, hydrometallurgy or its combination are used. Hydrometallurgy is preferred when zinc content is lower, due to economic advantages. Hydrometallurgical treatment, generally, involves the leaching—solvent, extraction—electrowinning route for the zinc recovery in the metallic form, as well as in the form of zinc compounds [15–19]. Several hydrometallurgical methods were devised to reclaim zinc values from zinc waste, generally. Hesham and Kamaleldin [20] focused on the extraction of zinc from blast-furnace dust using ammonium sulfate. Zinc oxide content in the studied sample was 49.6%. Zinc was presented in the form of hydrozincite, hemimorphite, smithsonite and sphalerite. In the first place, the dust sample was roasted at 850 °C to obtain a more stable zinc form (ZnO) that is more susceptible towards ammonium sulfate leaching at low temperatures. Under the set optimal conditions for the formation of soluble zinc compounds, with molar ratio 1:8 of roasted zinc dust (ZnO) and $(NH_4)_2SO_4$ under the temperature of 350 °C, up to 95% of zinc as hydroxide was leached with 0.5 M sulfuric acid. Wang et al. [21] leached basic oxygen steelmaking filter cake in organic acids (oxalic, citric, acetic, propionic, butyric and valeric) to gain zinc from sample, selectively. The content of zinc in the filter cake was on average 6.5% in the form of zinc oxide and zinc ferrite. Butyric acid exhibited excellent selectivity among tested acids, with up to 49.7% of the zinc being leached from the filter cake. Mehmet et al. [22] studied optimization possibilities of selective zinc leaching from electric arc furnace steelmaking dust using response surface methodology. Zinc content in the sample was 26.95% in the form of ZnO and complex oxidic compounds bonded with iron, manganese, magnesium, or titan. The most influential zinc leaching factors were determined as acid concentration and quadratic factors of acid concentration by using Anova. The proposed criteria in which zinc recovery is greater than 70% and iron recovery is lower than 10% was: acid concentration between around 1.6–3.1 $mol \cdot L^{-1}$, leaching duration of 56.42 min and L/S ratio of 5. Xie et al. [23] studied zinc extraction from industrial waste residue by conventional acid leaching. Zinc content in the waste was 24.27%. The effect of reaction time, sulfuric acid concentration, leaching temperature, stirring speed, and liquid solid ratio on zinc leaching rate were studied. Authors reached 86.34% of zinc recovery under the condition of sulfuric acid concentration 0.61 M, reaction temperature 25 °C, liquid-solid ratio 4:1 (mL/g), stirring speed 400 rpm, and leaching time 30 min. Radzyminska-Lenarcik et al. [24] recovered zinc from metallurgic waste sludges. Zinc bearing sludge contained 11.0–13.0% of zinc. Hydrochloric, sulfuric, and lactic acids, as well as ammonia and NaOH solutions were studied for the highest amount of zinc recovery. Extraction methods such as electrolysis and solvent extraction were applied. The most effective leaching solutions were the concentrated ammonia, 30% NaOH, and 80% lactic acid. By electrolysis 92–99% of metal zinc was recovered, and using solvent extraction recovered 96–99% of zinc depending on the solution pH.

Lorenzo et al. [25] recovered electrolytic zinc or zinc sulfate from galvanizing zinc dross by means of acid leaching, followed by using solvent extraction. The author declared patent for this technique, and its research is extensively described in detail. Galvanizing dross with zinc content of 63%, of which 3% is metallic zinc, was put under the leaching process in organic acid reagent of D2EHPA under the temperature of 40 °C. The zinc reaching yield was within 50 min and pH 2.7 was 97.7%. Besar et al. [26] processed zinc dross originating in the hot-dip galvanizing by hydrometallurgy to obtain zinc oxide. Zinc content of the dross was 96.5% in its metallic form. Variables as process time (30, 60 and 90 min), temperature (150 °C, 170 °C, 190 °C), concentration of glacial acetic acid (20%, 40%, 60%) were studied. The best process conditions were at 90 min under the temperature of 150 °C, and a glacial acetic acid concentration of 60%. Under set conditions, a yield of 62.45% was obtained and zinc content in zinc oxide was 70.23%. However, the pyrometallurgy seems to be more reasonable when zinc concentration is higher [16–19]. The best available pyrometallurgical technology for zinc-bearing residue treatment is the Waelz process using a rotary kiln. The Waelz process is characterized by the volatilization of zinc from an oxidized solid mixture (zinc bearing waste) through reduction by coke or coal. The Waelz kiln operates at a rotational speed of approximately 1 rpm and a slope of 2–3%. The air enters the kiln at the slag discharge end. Solid charge is dried and heated up until a reaction starts. At about 800 °C zinc oxides start to reduce and thus zinc volatilizes into a gas phase. Subsequently, zinc vapors are re-oxidized into ZnO and transported as solid dust with the counter current process air flow into a dust chamber [27]. The Waelz kiln process was also used for the carbothermic reduction of zinc containing waste by author Zhang et al. [28]. In the research zinc oxides from electric arc furnace dust were investigated. Zinc content in the dust was 21.5% in the form of franklinite and zincite. Dust was reduced by carbon at temperatures between 800–1300 °C. The ZnO reduction and zinc evaporation occurred in the temperature range of 1000–1100 °C. At a temperature of 1100 °C, 99.11% of zinc was evaporated. Xue Denga et al. [29] applied the evaporation method on the blast furnace dust, followed by condensation and separation with vacuum carbothermal reduction into the metallic zinc. The content of ZnO in the dust material was in the amount of 7.9%, of which metallic zinc was 6.34%. Authors conducted the first group of vacuum carbothermal reduction at the temperature of 800 °C, within the time of 90 min and carbon additions of 8%, 10%, and 12%. The second group of experiments was conducted with a reduction temperature of 900 °C, within the time of 90 min and carbon additions of 6%, 8%, and 8% with 2% B_2O_3 additive. When the range of the carbon addition was 10–14% and the temperature reached 900 °C, the volatilization rate of metallic zinc exceeded 99.6%. Jintao et al. [30] processed copper smelter dust containing 73.89% of ZnO by evaporation and subsequent condensation of zinc vapors from dust. According to their results, 99.94% of zinc oxide powder was transformed into zinc vapors through carbothermal reduction at the temperature of 1373–1573 K for 30–60 min. Zinc recovery from a manganese battery was also investigated by Zhan et al. [31]. Manganese batteries with zinc content of 14.32% were processed by vacuum evaporation under the temperature of 1123 K and oxygen-control oxidation with 12.5% oxygen content and 21 L/min nitrogen flow rate to prepare a nano-zinc oxide with high added value. Zinc dross from the hot-dip galvanizing process was studied by author Wang et al. [32] through the method of super-gravity separation. Zinc dross contained 65.4% of zinc, mostly presented as pure metallic zinc, and the rest zinc was bonded in the intermetallic compounds form of $Fe_2Al_5Zn_{0.75}$. Studied factors were gravity coefficient (15, 50, 100, 300, 500, 800, 1000), separating time (15, 60, 120, 180, and 350 s), and separating temperature (430, 460, 510, 560, and 610 °C). Over 79% of zinc was recovered with a high purity of about 99% at gravity coefficient 500, within a time of 180 s and temperature of 510 °C.

Bottom zinc dross from the hot-dip galvanizing process was put under investigation in several studies. Bottom zinc dross was investigated using electrochemical and pyrometallurgical methods by Ghayad et al. [33]. Dross containing 96.18% of zinc was dissolved in concentrated sulfuric acid to create zinc sulfate solution (100 g Zn/L). Dissolution was performed under the temperature of 55 °C for 1.5 h. Solution was adjusted by hydrogen peroxide to remove iron. The obtained filtrated zinc sulfate solution (with content of Zn^{+2}

in amount of 32.1% besides Al^{+3} and Fe^{+2} in amount of 0.99% and 0.53%, respectively) was then used for electrowinning and electrorefining of zinc. Authors studied variable factors such as current density (40–90 mA/cm^2 rising by 10 mA/cm^2) and time (120, 270, 360, 480, and 600 min). The highest zinc efficiency of 90% was obtained at the current density of 40 mA/cm^2, within time of 30 min and under the temperature of 25 °C. Zinc deposit was in high purity of 99.99%. A second research technique of zinc recovery (same author's research [34]) using pyrometallurgical technique was conducted in the heating furnace. The dross sample was placed in the heating crucible, covered with thermal cement, and heated to the temperature of 900 °C. Under the given temperature formed zinc vapors passed through the silicon carbide tube and subjected to cooling into zinc melt. The studied parameter was a temperature and time of evaporation process. The highest zinc recovery (82%) was obtained under the temperature up to 1000 °C within 1.5 h. Further increases in time were not effective in higher zinc recovery. Obtained zinc was of high purity (99.95%). Prasad [34] prepared electrolytic zinc powder from bottom zinc dross in the sodium zincate solution. Content of zinc in the studied sample was 94.5%, presented in the form of metallic zinc and the intermetallic compound of Fe-Zn. To obtain best results, the author studied several parameters influencing electrowinning process, such as effect of zinc concentration in the electrolyte (16, 30, 45, 60, and 75 g/L of zinc), effect of NaOH concentration (180–220 g/L of NaOH), effect of anode to cathode distance (20–60 mm), temperature within time (30–50 °C within 0–125 min), effect of current density on the electro deposition of zinc powder (2–10 amp/dm^2), and the specific power consumption. A high purity of zinc powder (99.5%) was obtained at room temperature in efficiency of 86% at 5 amp/dm^2 current density, 1.2 V impressed voltage, concentration of electrolyte 16 g/L Zn and 220 g/L NaOH. Sinha et al. [35] established a leaching-precipitation-crystallization route to produce high-grade zinc sulfate and phosphate salts, along with by-products. Zinc content in the studied sample was 97.5%. Firstly, zinc dross was leached in a sulfuric acid at a specified concentration (9–12% v/v was varied) at ambient temperature for 24 h. After this process, the leach liquor was drained out from the leaching tank and filtered. Then, liquor was put under the precipitation process to remove iron and other impurities. The purified liquor was further used for crystallization of zinc sulfate salts (temperature set at 70 °C and 30 °C for stability range of crystallization of $ZnSO_4 \cdot H_2O$ and $ZnSO_4 \cdot 7H_2O$, respectively) and precipitation of zinc phosphate salts (considering pH ~3.5 and ~4.5, flower shape, sheet-like, and spherical aggregate lumps like morphologies were observed).

In practice, the production of ZnO is the most common way of zinc recovery from the bottom zinc dross. There is no established process for bottom zinc dross treatment to zinc recovery in its metallic form to keep the zinc in the galvanizing loop. The present study focuses on the development of the most effective and economical way for high purity zinc recovery from bottom zinc dross by zinc evaporation and its condensation. This can be achieved by detailed research of the zinc evaporation process in the dynamic system using a lower temperature than of the zinc boiling point. Before the overall process can be evaluated in terms of economic viability, several tasks and partial investigation must be conducted. The first main task is a thermodynamic study of the zinc evaporation process from bottom zinc dross. For this purpose, thermodynamics were used as a preliminary study for an effective and economical way of successful zinc recovery from the bottom zinc dross. Theoretical and experimental results were evaluated.

2. Experimental Section

2.1. Materials

Samples of bottom zinc dross (Figure 1a) were drilled, and zinc chips were obtained for experimental usage (Figure 1b).

(**a**) (**b**)

Figure 1. (**a**) Bottom zinc dross sample with dimension of approximately 20 × 20 × 7 cm and (**b**) drilled zinc chips as dose for experiments.

The sample was analyzed to determine zinc and the content of other elements (Table 1) through AAS (atomic absorption spectrometry) analysis using model ContrAA 700. Phase analysis was realized by XRD (X-ray diffraction) using model XPERT PRO RV-11/2010, PANalytical, as shown in Figure 2 and Table 2. Microstructure of bottom zinc dross using light microscopy (Olympus BX53M) after etching can be seen in Figure 3.

Table 1. Chemical composition of the bottom zinc dross sample by AAS analysis.

Element	Zn	Fe	Pb	Al	Ni
Amount [%]	>94–97	2.2–3.4	0.7–1.5	0.8–1	0.3–0.7

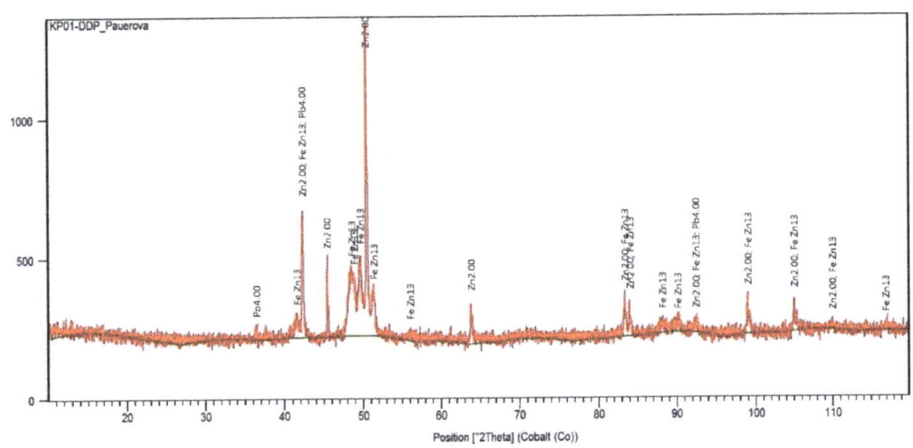

Figure 2. XRD pattern of the bottom zinc dross sample [36].

Table 2. Results of the XRD analysis—phase composition of the sample.

Ref. Code	Compound Name	Chemical Formula
96-901-1600	Zinc	Zn
03-065-1238	Iron Zinc	$FeZn_{13}$
96-900-8478	Lead	Pb

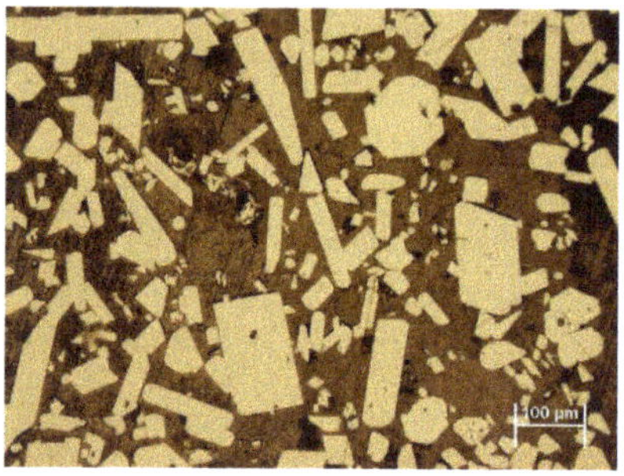

Figure 3. Microstructure of bottom zinc dross after etching (dark color represents a zinc matrix and light color objects represent intermetallic compounds of $FeZn_{13}$).

The light objects are intermetallic compounds of $FeZn_{13}$. The dark color represents a zinc matrix that was formed by the solidification of excess zinc melt, which was captured when removing intermetallic compounds ($FeZn_{13}$) from the bottom of the galvanizing kettle.

2.2. Analysis of Zinc Evaporating Conditions

A thermodynamic study available in the literature and by HSC Chemistry software 6.1 was carried out to set experimental conditions. Parameters such as zinc evaporation temperature, inert gas addition, time of evaporation, and zinc condensation were studied.

The theoretical amount of inert gas (e.g., Ar) needed for evaporation of one gram of Zn (0.0153 mol Zn) is calculated from the thermodynamic data (Table 3; HSC 6.1) according to the Equation:

$$n_{Ar} = \frac{n_{Zn} \cdot p_{Ar}}{p_{Zn}}, \ [\text{mol}] \tag{1}$$

and recalculated to Ar volume at 25 °C ($V_{Ar} = n_{Ar} \cdot V_m(25\ °C)$). n_{Ar}—amount of substance [mol], n_{Zn}—amount of substance of Zn(g) vapors [mol], p_{Ar}—equilibrium vapor pressure of Ar [Pa], V_{Ar}—volume of inert Ar gas [dm^3], V_m(25 °C)—24.5 dm^3 by one mol of ideal gas. Calculated parameters data within selected temperatures are listed in Table 3.

Table 3. Inert gas amount (nAr, V_{Ar}) at equilibrium state in system (Zn+Ar) after the evaporation of 1 g Zn (i.e., 0.0153 mol Zn) at selected temperatures.

Temperature [°C]	Equilibrium Constant of Phase Transformation Zn(l) = Zn(g) [-]	Equilibrium Partial Pressure [Pa]		Amounts of Inert Ar Gas [mol]	Volume of Ar at 25 °C [dm^3]
t	K	$p_{(Zn)}$	$p_{(Ar)}$	n_{Ar}	V_{Ar} (25 °C)
500	0.00183	183	99,817	8.4	204
600	0.0153	1530	98,470	0.98	24
700	0.0814	8140	91,860	0.17	4.2
800	0.314	31,350	68,650	0.033	0.82
900	0.949	94,940	5060	0.0008	0.02

The melting temperature of Zn is 419.5 °C. With increasing temperature, the partial zinc vapor pressure over the zinc melt increase. Reaching temperatures 750–800 °C, zinc starts to evaporate in greater amounts. Reaching zinc boiling point (907 °C), the zinc vapor curve rises rapidly. Inert Ar gas prevents oxidation of zinc vapors during the evaporation process.

The equilibrium state of Zn in system (0.0153 mol Zn: 0.001 mol Ar) and (0.0153 mol Zn: 1 mol Ar) in the temperature range of 500–950 °C is compared in Figure 4a,b (HSC 6.1 diagrams). The calculation of p_{Zn} dependence on temperature was conducted regarding how condensed Fe impurities affected the activity of Zn(l) ($a \neq 1$).

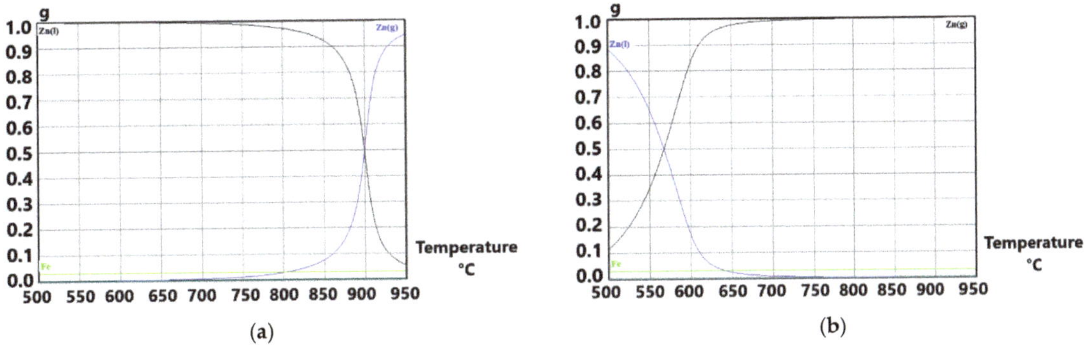

Figure 4. Change of p(Zn) with temperature in the system of (1 g Zn: 0.03 g Fe: Ar): (**a**) $1 \cdot 10^{-3}$ mol Ar and (**b**) 1 mol Ar (1 g Zn = 0.0153 mol).

Decreasing the overall pressure in the system (below 100 kPa) moves the zinc boiling point to lower values. Similarly, it also works with the inert gas flow over the molten zinc surface. In the reactor, the passing inert gas takes zinc vapors out of the evaporation zone, and thus constantly disrupts the effort to achieve equilibrium partial pressure in the system. It means that the addition of a greater amount of argon into the system allows evaporation of a greater amount of zinc. The larger the difference between the equilibrium and non-equilibrium state, the faster the Zn(g) evaporates.

2.3. Experimental Procedure

For the experimental process of zinc evaporation, the pipe resistance laboratory furnace was used. The apparatus for evaporation and subsequent condensation of zinc vapors is schematized in Figure 5.

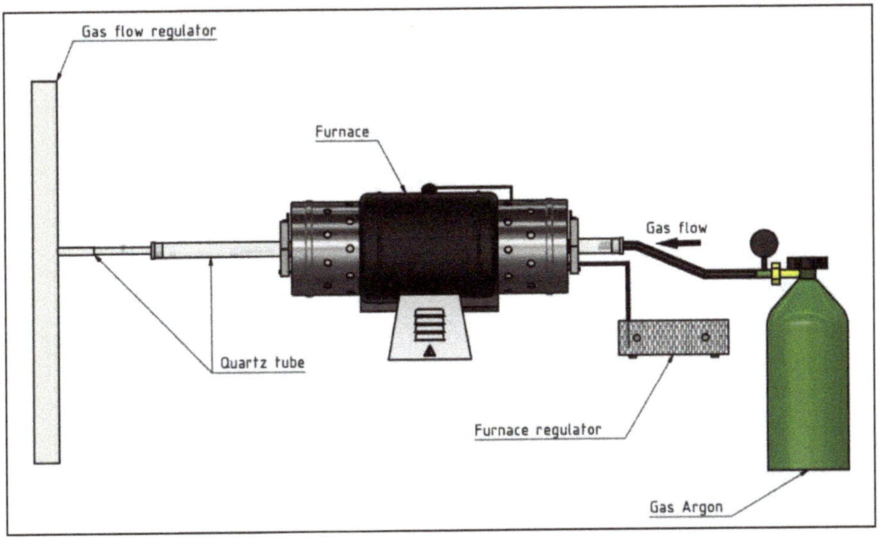

Figure 5. The scheme of apparatus built for experiments: (from left to right) gas flow regulator, quartz tube of diameter 20 mm, quartz tube of diameter 12 mm (inserted in the 20 mm quartz tube), the pipe resistance furnace, furnace regulator, and gas bomb.

A bottom zinc dross sample in the amount of approximately 3 g was put into quartz tube with a diameter of 12 mm. This quartz tube (12 mm) was then inserted into a larger quartz tube with a diameter of 20 mm (according to Figure 5). Tubes were partly inserted into the furnace. The sample was inserted into the furnace when required temperature was reached (700 °C and 800 °C, respectively). During heating, all apparatuses were blown using argon to ensure inert atmosphere. The argon flow was set up on a flow rate of 90–100 mL/min (mostly 91 mL/min—dependent on the process changes). Evaporation process temperatures were 700 °C and 800 °C, respectively. The time of the evaporation process was observed for periods of 5, 10 and 20 min. At first, the initial experiments under the temperature of 700 °C were conducted to observe sample behavior in the furnace during the evaporation process. The thermodynamics of zinc evaporation showed only feasibility of the evaporation, not the overall time of the zinc evaporation process. After the process, samples were cooled down by removing quartz tube from the furnace at a speed of 2 cm/min, keeping the argon atmosphere. Finally, the theoretical and experimental efficiency of zinc evaporation was compared. The final product (zinc), as well as residue, were evaluated by SEM and EDX.

3. Results and Discussion

3.1. Thermodynamic Study

Based on the thermodynamic calculation of zinc evaporation and the argon consumption within temperatures (Table 3), two temperatures were chosen for experiments (700 °C and 800 °C). In terms of a detailed simulation of the process, the comparison of the dependence of zinc vapors equilibrium pressure on the argon amount in the system was simulated by HSC software for the temperature of 700 °C and 800 °C, respectively.

Figure 6a,b shows the simulation of zinc evaporation in the argon flow atmosphere at temperature 700 °C (Figure 6a) and 800 °C (Figure 6b), respectively. The calculation in HSC software presumes that condensed phases are impure (a ≠ 1). Inert atmosphere ensures there is no zinc oxidation during the process.

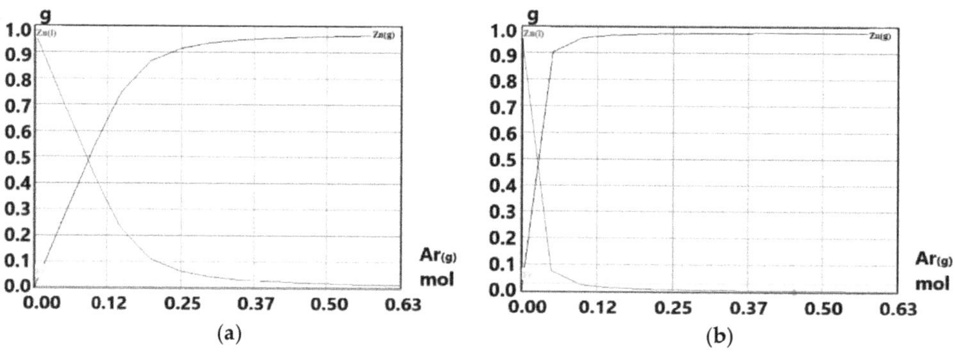

Figure 6. Comparison of dependence of Zn(g) equilibrium partial pressure on the Ar amount in system (1 g Zn: 0.03 g Fe: Ar): (**a**) under the temperature of 700 °C and (**b**) under the temperature of 800 °C. Condensed phases are impure (a ≠ 1).

From the graph in Figure 6a, when considering impure zinc in the system, it is necessary to supply ≈0.5 mol (≈12.4 L at 25 °C) of argon to evaporate almost one gram of zinc at the temperature of 700 °C. Increase of temperature to 800 °C (Figure 6b) decreases spent Ar amount by four times (≈0.12 mol, which is ≈2.97 L at 25 °C). Comparing the evaporation process by HSC software (Figure 6) with the calculation from Table 3, there is a difference in the spent Ar argon. Calculation presented in Table 3 is based on a pure zinc system, whereas the calculation made by HSC software considers the system to be impure (iron content). Experiments were conducted with a sample of 3 g, so theoretically there should be 18.6 L and 1.2 L of argon consumption within the temperature of 700 °C and

800 °C, respectively (according to the evaporation process simulation as shown in Figure 6). Real argon consumption is listed in Table 4.

Table 4. Zinc efficiency within temperature and time of evaporation.

Temperature [°C]	Time [min]	Initial Dose of Sample [g]	Zn Content in Dose [g]	Impurities [g]	Residue after Evaporation [g]	Efficiency [%]	$Ar_{(g)}$ Consumption [L]
700	20	3.03	2.94	0.09	0.09	**100**	**1.8**
	10	3.05	2.96	0.09	0.33	**91.9**	**0.9**
800	20	3.09	2.99	0.1	0.13	**99**	**1.8**
	10	3.08	2.98	0.1	0.13	**99**	**0.9**
	5	3.01	2.92	0.09	0.14	**98.3**	**0.45**

Generally, the size of chips determines the rate of zinc evaporation. The smaller the particles are, the bigger the reaction surface is, and thus the higher the rate of evaporation. Impurities in the chips may also influence the evaporation process. The main impurity is iron bonded with zinc into an intermetallic compound. According to Marder ([37] p. 10–12) during the heating, the zeta (ζ) phase, $FeZn_{13}$, decomposes within the temperature of 530 °C creating delta (δ) phase, $FeZn_{10}$. Then delta phase decomposes into gamma (γ) phase, Fe_3Zn_{10}, within the temperature of 665 °C. Gamma phase is stable within the temperature of 782 °C. Exceeding this temperature, phase decomposes into liquid Zn and αFe. Based on the knowledge of individual phase decomposition, it can be deduced that evaporation process over the temperature of 700 °C is partially influenced by the presence of gamma phase. Zinc bonded into an intermetallic compound evaporates by smaller velocity than pure zinc does. Presence of impurity (iron) slows down the overall zinc evaporation process. The evaporation process under the temperature of 800 °C should not be influenced by the intermetallic compound's presence, as the zinc is released from the intermetallic compound with iron. However, generally, during the zinc evaporation, iron concentrates presented in the residue. As this iron concentrates in the residue and at the same time the zinc reduces (in the input dose), the final phase of zinc evaporation slows down with time.

3.2. Experimental Part

The condensation area of zinc product as well as residue after evaporation (in the quartz tube) are shown in Figure 7. During the process, two forms of zinc product were formed (Figure 8). Zinc foil as well as zinc drops were formed when zinc vapors condensed. The reason for different zinc products in the single evaporation process is probably due to unstable argon flow rate during the evaporation process. When argon flow rate increased, zinc foil formed.

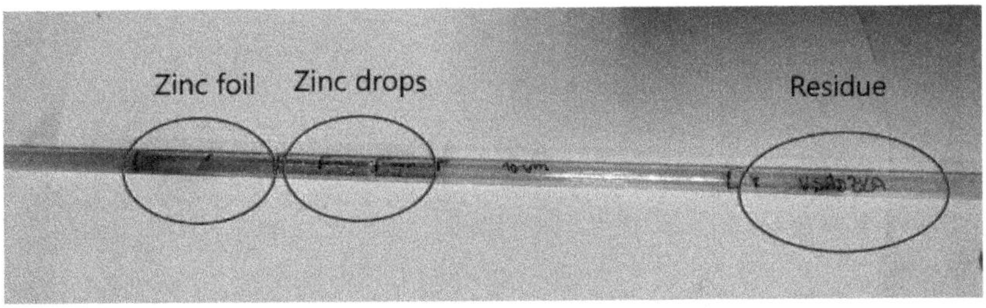

Figure 7. Quartz tube with a dose after evaporation process: zinc foil, zinc drops, and the residue.

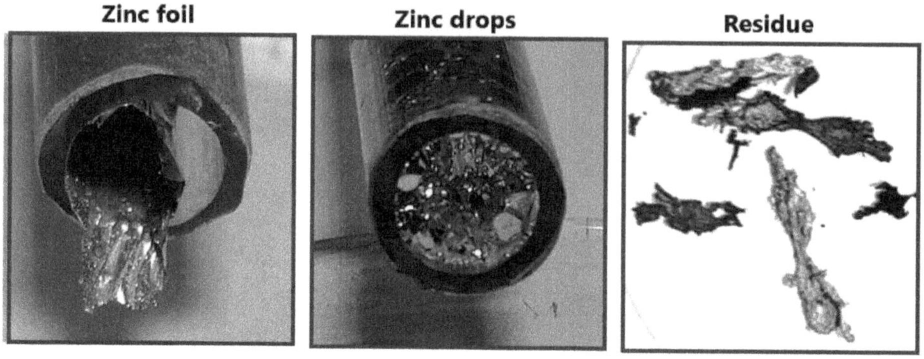

Figure 8. Zinc products (zinc foil, zinc drops) and residue.

Morphology of analyzed samples can be seen in Figure 9a,b (zinc foil), Figure 10a,b (zinc drops), and Figure 11a,b (residue after evaporation). The description of morphology is below.

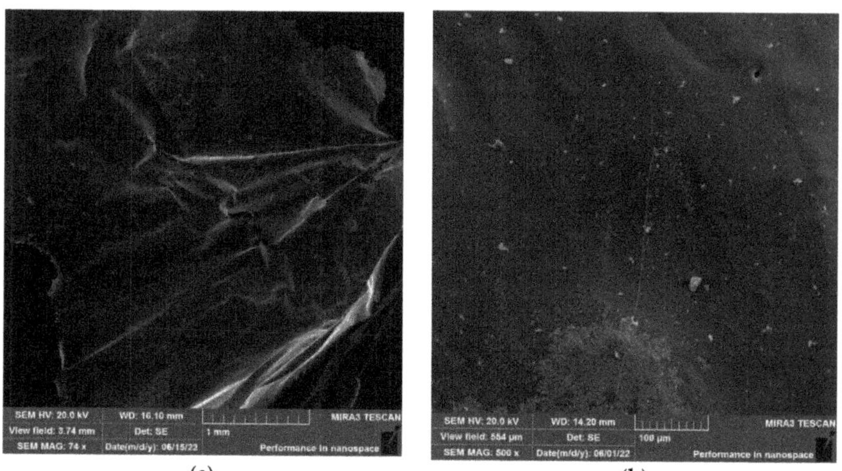

(**a**) (**b**)

Figure 9. Morphology of the obtained zinc product (evaporation process under the temperature of 700 °C) in form of "foil" at (**a**) 1 mm magnification and (**b**) 100 μm magnification.

Results of SEM analysis of the product in the form of "foil" is shown in Figure 9a,b. The morphology of this product is smooth and consistent.

Results of SEM analysis of the product in the form of "drop" is shown in Figure 10a,b. Spherical products of various sizes have a smooth surface.

Results of SEM analysis of the residue is shown in Figure 11a,b. In Figure 11a, the "needle-like" morphology of residue after evaporation can be observed.

EDX analysis of the zinc foil, zinc drop, as well as the residue after evaporation process under the temperature of 700 °C are shown in Figure 12 ((a) zinc foil, (b) zinc drop, (c) residue).

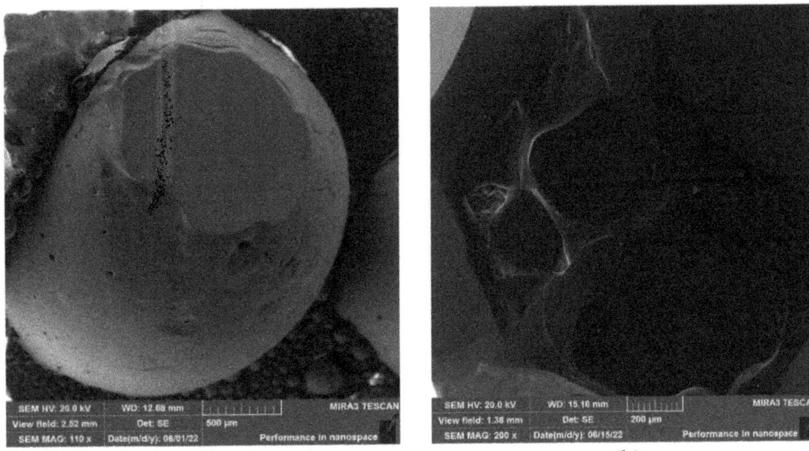

(a) (b)

Figure 10. Morphology of the obtained zinc product (evaporation process under the temperature of 700 °C) in the form of "drop" at (**a**) 500 μm magnification and (**b**) 200 μm magnification.

(a) (b)

Figure 11. Morphology of the residue after the evaporation process under the temperature of 700 °C at (**a**) 500 μm magnification and (**b**) 5 μm magnification.

Zinc products (Figure 12a,b) are high grade purity without any oxidation. No other elements were evaporated during the process. All iron and other elements remained in the residue (Figure 12c). Zinc content in the residue was negligible.

The morphology of the zinc product sample, as well as the residue after the evaporation process at the temperature of 800 °C, can be seen in the Figure 13 ((a) zinc drop, (b) residue).

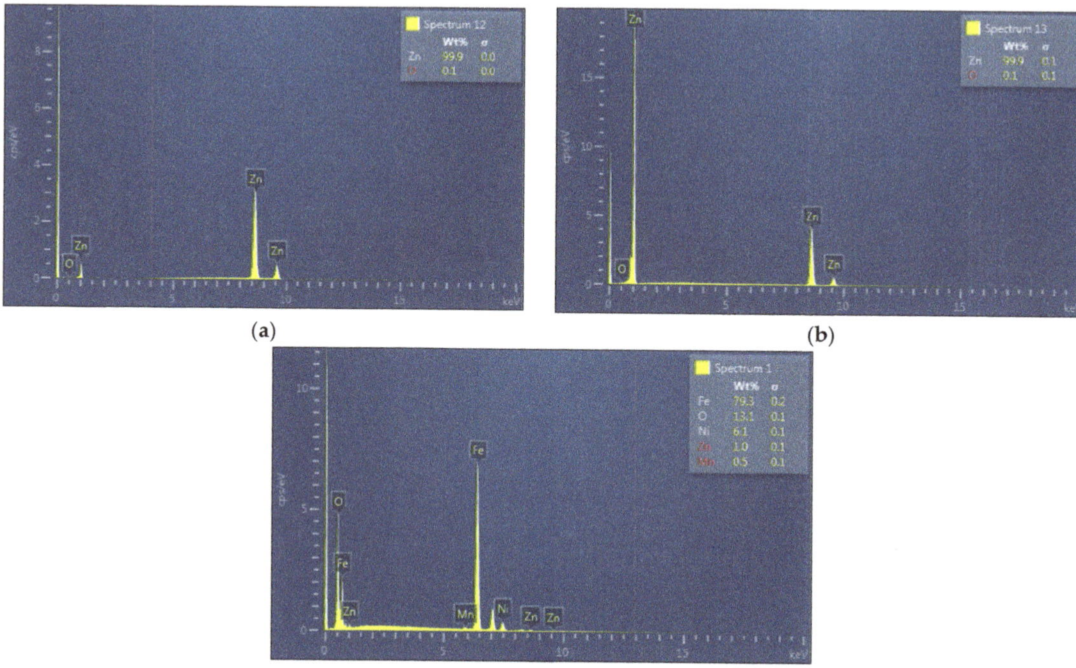

Figure 12. EDX of (**a**) zinc foil and (**b**) zinc drop at 700 °C. (**c**) EDX of the residue after evaporation at 700 °C.

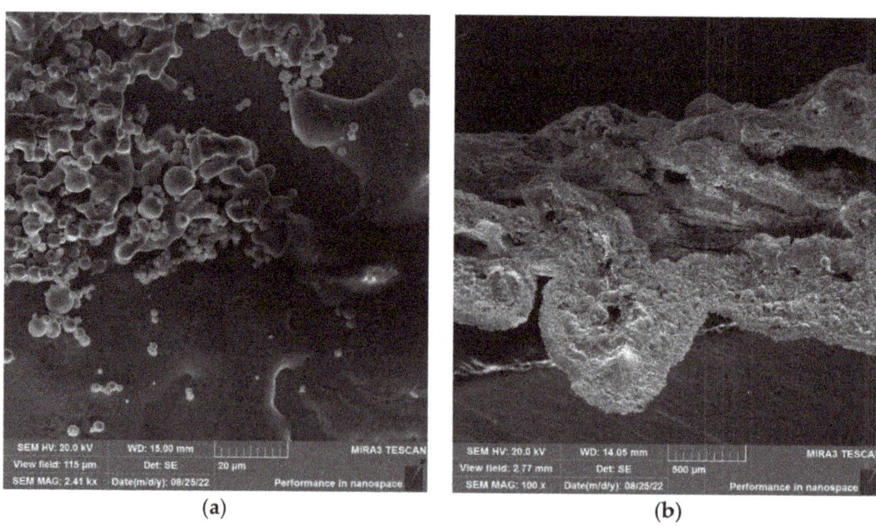

Figure 13. Morphology of (**a**) zinc drop product and (**b**) the residue after evaporation by 800 °C.

The result of SEM analysis of the zinc product is shown in Figure 13a, where the individual and connected spherical particles can be observed. The result of EDX analysis of these samples is shown in Figure 14 ((a) zinc drop, (b) residue).

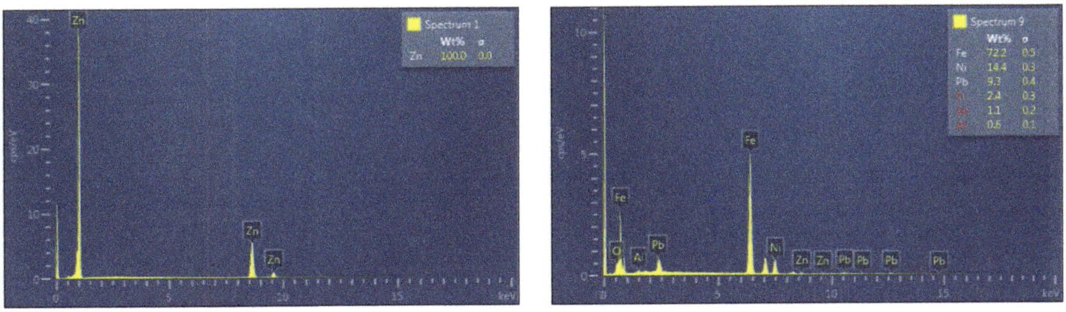

(**a**) (**b**)

Figure 14. EDX of (**a**) zinc drop product and (**b**) the residue after evaporation by 800 °C.

As it is shown in Figure 14a, a high purity zinc product was obtained. No impurities passed to the final product. A negligible amount of zinc remained in the residue (Figure 14b).

Experimental results within selected temperatures and time of evaporation can be seen in Table 4.

From Table 4, a high zinc efficiency was obtained under the set evaporation times. Within the temperature of 700 °C, the time of 10 min was sufficient for evaporation of 92% zinc. This level of efficiency is about to be considered to energy consumption in half the time of evaporation process. During the process of zinc evaporation, iron concentrates in the residue and thus the zinc evaporation may slow down. Comparing zinc efficiency within the temperature of 700 °C and 800 °C in Table 4, it cannot be clearly stated that evaporation at the temperature of 700 °C is more efficient (100% of Zn) than the process under 800 °C (99% of Zn) within the same evaporation time (20 min). Differences in the zinc efficiency are very small, almost negligible, considering the input dose amount, and possible heterogeneity of the sample in the realized experiments. Within the temperature of 800 °C, all zinc was evaporated within the time of 20 and 10 min. Argon consumption for each experiment is also stated in Table 4. The real argon consumption (1.8 L of $Ar_{(g)}$ by 700 °C and 0.9 L of $Ar_{(g)}$ by 800 °C) within the dynamic experimental configuration and the argon gas flow of 90 mL/min trough 12 mm quartz tube is significantly lower than theoretical argon consumption (12.4 L of $Ar_{(g)}$ by 700 °C and 2.97 L of $Ar_{(g)}$ by 800 °C) calculated for equilibrium state in the static test configuration. This significantly lower carrier argon consumption is a result of dynamic regime of the experiment, and thus relates to the velocity of zinc vapors removal from the melt zinc surface in the evaporation zone. The gas flow permanently disrupts the equilibrium state over the molten zinc surface and accelerates the rate of its evaporation significantly. The volume and the velocity of the argon gas is an important parameter for the evaporation process itself. The velocity of the gas flow depends on the flow of the gas, as well as the quartz tube diameter.

The results of pilot experiments confirmed that high pure zinc from bottom zinc dross can be obtained by evaporation and its condensation, quite simply. In the flow of argon gas within a temperature of 700–800 °C the process runs quickly. The zinc evaporation rate depends on both the temperature and the velocity of inert gas flow above the melt. With the increasing velocity of the gas flow, the condensation area of vapors will also shift further along from the evaporation zone. The molten zinc surface is another important parameter influencing the process efficiency.

As expected, during experiments significantly lower amounts of inert gas was consumed, due to the dynamic mode of the process. In this case, it was proved that it is possible to apply decreasing temperature, thus decreasing argon flow rate for process optimalization.

Within this given apparatus configuration (Figure 5), either the flow rate of argon or temperature can be decreased and thus the argon consumption can be optimized. The conducted experiments, selected conditions and test arrangement are not sufficient for a kinetic study of zinc evaporation and the complex evaluation of the overall process efficiency. However, obtained results indicate experimental proceedings for further kinetic study. The conducted research is crucial for kinetics of zinc evaporation as a significant factor for the overall process of optimalization. Kinetic research can address several points, such as argon flow rate within selected quartz tube diameter, defined input surface of the sample, temperature, and time of evaporation process.

In the case of a successful result of the bottom zinc dross refinement and its economic feasibility, this process could be implemented on a semi-operational scale in cooperation with a bottom zinc dross producer. Producers are currently interested in this ongoing research. This research is applied research for a specific type of waste refinement.

4. Conclusions

Based on a thermodynamic and experimental study of zinc evaporation from the bottom dross, the following conclusions were derived:

1. The thermodynamic study pointed to the choice of zinc evaporation temperature due to argon consumption (700 °C and 800 °C, respectively);
2. Based on theoretical calculation, the argon consumption at a temperature of 800 °C is four times lower (2.97 L of Ar(g)) than that at a temperature of 700 °C (12.4 L of Ar(g));
3. A high purity zinc product (100% of zinc) can condense in the form of "foil" or "drop" based on the argon flow rate;
4. Argon consumption during the experimental procedure was significantly lower than theoretically calculated due to the dynamic regime. Real argon consumption at a temperature of 800 °C was 0.9 L and at a temperature of 700 °C was 1.8 L, respectively.

Nowadays, the processing of bottom zinc dross in a recycling loop plays an important role for sustainability of the secondary materials. As pure zinc is to be obtained under the economic viable condition, its reuse in the HDG process is ensured.

Author Contributions: Writing—original draft preparation, P.K.; conceptualization, P.K.; methodology, B.J. and P.B.; investigation, P.K. and B.J.; resources, P.K.; validation, T.J. and P.B.; writing—review and editing, P.K., T.J. and P.B.; supervision, T.J. and P.B. All authors have read and agreed to the published version of the manuscript.

Funding: This research was funded by the project VEGA 1/0641/20.

Institutional Review Board Statement: Not applicable.

Informed Consent Statement: Not applicable.

Data Availability Statement: Not applicable.

Conflicts of Interest: The authors declare no conflict of interest.

References

1. Trpčevská, J. *Zinok, Jeho Aplikácia, Výroba a Recyklácia*; Fakulta Materiálov, Metalurgie a Recyklácie, Technická Univerzita v Košiciach: Košice, Slovakia, 2018; ISBN 978-80-553-2997-0.
2. Zinc Alloys: Properties, Production, Processing and Applications. Available online: https://matmatch.com/learn/material/zinc-alloy (accessed on 4 December 2022).
3. Rostek, L.; Espinoya, L.T.; Goldman, D.; Loibl, A. A dynamic material flow analysis of the global anthropogenic zinc cycle: Providing a quantitative basis for circularity discussions. *Resour. Conserv. Recycl.* **2022**, *180*, 106154. [CrossRef]
4. Grégoire, M.; Barbara, R.K. The anthropogenic cycle of zinc: Status quo and perspectives. *Resour. Conserv. Recycl.* **2017**, *123*, 1–10.
5. Garside, M. Global Zinc Reserves by Country 2021. Statista, April 2022. Zinc Reserves Worldwide by Country 2021. Statista. Available online: https://www.statista.com/statistics/273639/global-zinc-reserves-by-country/ (accessed on 15 October 2022).
6. Zinc Metal Production Volume Worldwide 2005–2021. Zinc: Global Metal Production. 2021. Available online: https://www.statista.com/statistics/264878/world-production-of-zinc-metal/ (accessed on 3 December 2022).
7. Garside, M. Global Zinc Consumption 2005–2021. Statista, July 2022. Zinc Metal Consumption 2021. Statista. Available online: https://www.statista.com/statistics/264884/world-zinc-usage/ (accessed on 15 October 2022).

8. International Zinc Association. Zinc Recycling, Closing the Loop. Available online: https://sustainability.zinc.org/wp-content/uploads/sites/17/2022/03/Closing-the-Loop_VF_3_22.pdf (accessed on 18 October 2022).
9. International Zinc Association. Recycling, Zinc Sustainability. Recycling—ZINC Sustainability. Available online: https://sustainability.zinc.org/recycling/ (accessed on 25 October 2022).
10. Encore Recycling. Recycling Zinc. Available online: https://encorerecyclers.com/2017/10/26/recycling-zinc/ (accessed on 26 October 2022).
11. American Galvanizers Association. Zinc Recycling. Available online: galvanizeit.org (accessed on 26 October 2022).
12. American Galvanizers Association. Zinc Uses. Available online: https://galvanizeit.org/hot-dip-galvanizing/what-is-zinc/zinc-uses (accessed on 23 October 2022).
13. Katarína, B.; Jarmila, T.; Jana, P.; Martina, L. Zinc waste treatment originated during hot-dip galvanizing. *World Metall. Erzmetall* **2017**, *70*, 223–226.
14. Blašková, K.; Trpčevská, J.; Dorková, M. Characterization of the hard zinc originated from the hot-dip galvanizing. *Holist. Approach Environ.* **2018**, *8*, 107–113.
15. Recylex. Zinc Recycling and Production. Available online: https://recylex.eu/en/zinc (accessed on 22 October 2022).
16. Anderson, C.G. *Pyrometallurgy in Book Reference Module in Materials Science and Materials Engineering*; Elsevier: Amsterdam, The Netherlands, 2016. [CrossRef]
17. Fathi, H. *Textbook of Pyrometallurgy*; Laval University: Quebec City, QC, Canada, 2002; 592p, ISBN 2-922686-05-1.
18. Fathi, H. *Volume 2: Hydrometallurgy*; Gordon & Breach: Philadelphia, PA, USA, 1970; 468p, ISBN 0-677-01-7804.
19. Havlik, T. *Hydrometallurgy: Principles and Applications*; Elsevier: Amsterdam, The Netherlands, 2014; 552p, ISBN 978-1845694074.
20. Hesham, S.; Kamaledin, M.H. Extraction of zinc from blast-furnace dust using ammonium sulfate. *J. Chem. Technol. Biotechnol.* **2004**. [CrossRef]
21. Wang, J.; Wang, Z.; Zhang, Z.; Zhang, G. Zinc Removal from Basic Oxygen Steelmaking Filter Cake by Leaching with Organic Acids. *Met. Mater. Trans. B* **2019**, *50*, 480–490. [CrossRef]
22. KUL, M.; OSKAY, K.O.; ŞİMŞİR, M.; SÜBÜTAY, H.; KIRGEZEN, H. Optimization of selective leaching of Zn from electric arc furnace steelmaking dust using response surface methodology. *Trans. Nonferrous Met. Soc. China* **2015**, *25*, 2753–2762. [CrossRef]
23. Xie, T.; Sun, C.; Li, G.; Luo, Y.; Zheng, X.; Ma, A. Zinc Extraction from Industrial Waste Residue by Conventional Acid Leaching. In *Characterization of Minerals, Metals and Materials*; Springer: Berlin/Heidelberg, Germany, 2021; pp. 121–129.
24. Radzyminska-Lenarcik, E.; Sulewski, M.K. Recovery of Zinc from Metallurgic Waste Sludges. *Pol. J. Environ. Stud.* **2015**, *24*, 1277–1282. [CrossRef]
25. Lorenzo, D.M.; Vega, J.M.R.; Nogueira, E.D. Process for the Production of Electrolytic Zinc or High Purity Zinc Salts from Secondary Zinc Raw-Materials. U.S. Patent US4401531A, 1983.
26. Besar, B.; Perindustrian, K. Proses ekstraksi seng oksida dari seng dross (extraction process of zinc oxide from zinc dross using siti Augustina). Proses Ekstraksi Seng Oksida Siti Augustina. *J. Kim. Dan Kemasan* **2016**, *38*, 103–108.
27. Matthes, J.; Waibel, P.; Keller, H.B. A new infrared camera-based technology for the optimization of the Waelz process for zinc recycling. *Miner. Eng.* **2011**, *24*, 944–949. [CrossRef]
28. Zhang, H.-N.; Li, J.-L.; Xu, A.-J.; Yang, Q.-X.; He, D.-F.; Tian, N.-Y. Carbothermic Reduction of Zinc and Iron Oxides in Electric Arc Furnace Dust. *J. Iron Steel Res. Int.* **2014**, *21*, 427–432. [CrossRef]
29. Deng, X.; Huang, R.; Xiaodong, L.; Yang, J.; Yang, J. Separation and recovery of metallic zinc and iron concentrate from blast furnace dust by vacuum carbothermal reduction. *Process. Saf. Environ. Prot.* **2022**, *162*, 746–751. [CrossRef]
30. Gao, J.; Huang, Z.; Wang, Z.; Guo, Z. Recovery of crown zinc and metallic copper from copper smelter dust by evaporation, condensation and super-gravity separation. *Sep. Purif. Technol.* **2020**, *231*, 115925. [CrossRef]
31. Zhan, L.; Ouyang, L.; Zhenming, X. Preparing nano-zinc oxide with high-added-value from waste zinc manganese battery by vacuum evaporation and oxygen-control oxidation. *J. Clean. Prod.* **2020**, *251*, 119691. [CrossRef]
32. Wang, Z.; Gao, J.; Shi, A.; Meng, L.; Guo, Z. Recovery of zinc from galvanizing dross by a method of super-gravity separation. *J. Alloy. Compoungs* **2018**, *735*, 1997–2006. [CrossRef]
33. Ghayad, I.; Al Ansary, A.; Aziz, Z.A.; Al Akhar, A. Recovery of zinc from zinc dross using pyrometallurgical and electrochemical methods. *Egypt. J. Chem.* **2019**, *62*, 373–384. [CrossRef]
34. Sanjay, P. Zinc powder preparation from zinc dross. *Mater. Today Proc.* **2022**, *67*, 336–341. [CrossRef]
35. Shivedra, S.; Choudhari, R.; Mishra, D.; Shekhar, S.; Agrawal, A.; Sahu, K.K. Valorisation of waste galvanizing dross: Emphasis on recovery of zinc with zero effluent strategy. *J. Environ. Manag.* **2020**, *256*, 109985. [CrossRef]
36. Powder Diffraction file PDF. 2. *XRD Software Analysis*, 2022.
37. Marder, A.R. The metallurgy of zinc-coated steel. *Prog. Mater. Sci.* **2000**, *45*, 191–271. [CrossRef]

materials

MDPI

Article

Determining the Mechanical Properties of Solid Plates Obtained from the Recycling of Cable Waste

Maciej Wędrychowicz [1,*], Władysław Papacz [2], Janusz Walkowiak [2], Adam Bydałek [1], Andrzej Piotrowicz [1], Tomasz Skrzekut [3], Jagoda Kurowiak [1], Piotr Noga [3] and Mirosław Kostrzewa [4]

1 Institute of Materials and Biomedical Engineering, Faculty of Mechanical Engineering, University of Zielona Gora, Prof. Z. Szafrana Street 4, 65-516 Zielona Góra, Poland
2 Institute of Mechanical Engineering, Faculty of Mechanical Engineering, University of Zielona Gora, Prof. Z. Szafrana Street 4, 65-516 Zielona Góra, Poland
3 Faculty of Non-Ferrous Metals, AGH University of Science and Technology, A. Mickiewicza Av. 30, 30-059 Cracow, Poland
4 Eko Harpoon Recycling sp. z o. o., Cząstków Mazowiecki 128, 05-152 Czosnów, Poland
* Correspondence: m.wedrychowicz@iimb.uz.zgora.pl

Abstract: In this article, the possibility of obtaining a solid plate from waste cable sheaths, by mechanical recycling, i.e., grinding, plasticising and pressing, is discussed—waste cable sheaths being pure PVC with a slight admixture of silicone. Press moulding was carried out under the following conditions: temperature 135 °C, heating duration 1 h and applied pressure 10 MPa. The yield point of the obtained solid plate obtained was 15.0 + −0.6 MPa, flexural strength 0.94 MPa, yield point 0.47 MPa and Charpy's impact strength 5.1 kJ/m^2. The resulting solid plate does not differ significantly from the input material, in terms of mechanical strength, so, from the point of view of strength, that is, from a technical point of view, such promising processing of waste cables can be carried out successfully in industrial practice.

Keywords: plastic materials; recycling/recovery; mechanical processing; mechanical properties

Citation: Wędrychowicz, M.; Papacz, W.; Walkowiak, J.; Bydałek, A.; Piotrowicz, A.; Skrzekut, T.; Kurowiak, J.; Noga, P.; Kostrzewa, M. Determining the Mechanical Properties of Solid Plates Obtained from the Recycling of Cable Waste. *Materials* **2022**, *15*, 9019. https://doi.org/10.3390/ma15249019

Academic Editors: Pavol Liptai, Jaroslav Briančin and Maroš Halama

Received: 10 November 2022
Accepted: 12 December 2022
Published: 16 December 2022

Publisher's Note: MDPI stays neutral with regard to jurisdictional claims in published maps and institutional affiliations.

Copyright: © 2022 by the authors. Licensee MDPI, Basel, Switzerland. This article is an open access article distributed under the terms and conditions of the Creative Commons Attribution (CC BY) license (https://creativecommons.org/licenses/by/4.0/).

1. Introduction

Due to the resistant degradation process and thus negative impact on nature and also the huge post-consumption volume, plastic material wastes are a severe environmental problem. At present, the emission of plastics into the environment is intense and continuous. We can distinguish the "anthroposphere", i.e., a separate synthetic-origin geological stratum rich in components that do not occur naturally in nature, which includes the so-called microplastics; it is detected not only in the anthroposphere but also in the biosphere and penetrates into water reservoirs, the soil and the air [1,2].

Plastic waste consists of polymers such as ethylene terephthalate (PET), high-density polyethylene (HDPE), low-density polyethylene (LDPE), polyvinyl chloride (PVC), polyamides (PA) and polystyrene (PS) [3,4], etc. Polymers are large-molecule synthetic materials made up of long chains of repeating units called monomers, consisting of carbon, hydrogen and heteroatoms such as chlorine, nitrogen and others. Due to the possibility of large-scale production, polymers are produced primarily from petroleum-based substances but less often from secondary materials. The continuous demand for plastics has made the accumulation of plastic waste in landfills take up a lot of usable space; in addition, the growing demand for plastics has led to the depletion of crude oil as part of a non-renewable fossil fuel. The simplified structure of polymers, especially in electronic devices, and the low cost of plastic-including products results in a short period of use (product life cycle), after which they become waste. The production of solid plastic waste is increasing day by day and is now around 150 million tonnes per year [5,6]. The recycling of plastics plays an important role in the management of polymer waste, thus allowing the recovery of secondary raw materials in the form of regranulates or chemicals. A crucial step in achieving a circular economy is recycling with high added value [7,8]. One of the most critical steps in waste management, in terms of cost and as material recovery efficiency, is the segregation and separation of waste.

1.1. Comparative Analysis of Mechanical Properties of Recycled Plastics and Economic Justification for Their Use

An analysis [9] of the profitability, the profit and loss of the investment in a technological line for the mechanical recycling of PVC from cable waste, was carried out over a 3-year period from the start of the investment. The assumptions of this analysis were as follows: annual PVC processing in a range of 803–972 tons, the amount of waste generated 20–24 tons, the cost of investment 25 thousand USD and operating expenses 5 thousand USD. Already after the second year, the investment is becoming profitable with a profit of 24 thousand USD, which almost completely compensates for the capital expenditure of the first year. Several other factors, such as the annual increase in the price of PVC, the costs associated with the disposal of PVC and the investment leverage, make the investment even more profitable with the profit in the third year being estimated at over 26 thousand USD. Of course, such an analysis should be conducted individually for each recycling plant, as this takes into account such aspects as local laws and taxes which differ depending on the region. Nevertheless, it shows clearly that the idea of mechanically recycling PVC from cable waste can be economically justified.

Although recycled materials may have physical properties similar to primary plastics, the resulting financial savings are negligible because the mechanical properties of most plastics deteriorate significantly after many processing cycles [10,11], which may contradict the above paragraph somewhat and constitute a counterpoint to it. One characteristic of the so-called mechanical recycling of plastics is that the material is degraded along with the next processing cycle [12]. Moreover, the permanent cross-linking of polymer structures is not suitable for their multiple processing, but these can be used, for example, as an additive to modified asphalt materials, contributing to the overall improvement in the material [13–16]. In order to obtain higher mechanical properties of the material, appropriate modification should be applied, e.g., physical modification [17–21]. The studies published so far have shown that the application of an appropriate compatibility method can be successfully used to modify a regenerated rubber mixture, containing various thermoplastics [22]. On the other hand, utility items, made partially or entirely from recycled materials can be used wherever excessive and specific physico-chemical properties are not required or where a short lifetime is expected for such items. In such a case, the use of a secondary material for the production of products can be much more commercially and economically justified; however, the increasing reduction in the value of the material, in such a production cycle scenario, should be taken into consideration. In the research published so far, it is necessary to mention the references that have shown the high efficiency of combining waste polyethylene with a copolymer and an elastomer in manufacturing a material that can be effectively used in industry [23,24]. Marossy [25] and Záraga [26] analysed the effect of chlorinated polyethylene in PVC/PE blends and found that polyethylene improves its rigidity if the concentration of chlorinated polyethylene is not too high. Kollar and co-authors [27] demonstrated that a blend of polyvinyl chloride with polyethylene can be obtained by melt blending using high- and low-density polyethylene, without the use of a stabilizer. Xu [28] studied the separation of the plasticiser of polyvinyl chloride from copper wires obtained from thin electrical cables and investigated the effect of the plasticiser before and after extraction, finding similar properties.

1.2. Recycling of Cable Waste and Its Use

A large part of the research on copper cable recycling is focussed on developing better technological and organisational solutions so as to be able to recover as many material fractions as possible. The problem is that the recovered waste is not suitable for industrial use. Therefore, numerous studies are carried out on the development of a method to improve the properties of polymer recyclates, in order to find suitable industrial applications for them. Currently, plastic recyclates are most often used in the process of injection and extrusion, which do not yet have their equivalent in industrial usage [29,30]. Recycling cables from a scrap of electric and electronic equipment is a complex problem because there is a wide range of plastics from standard, pure thermoplastics to cross-linked polymers with their numerous mixtures such as glass and carbon fibre, talc or calcium carbonate. In order to reduce the costs related to segregation, it can be proposed to eliminate the costly separation in favour of processing mixed waste into full-value products using hot pressing technology [31,32]. Several works [33–35] deal strictly with research on the mechanical recycling process and the strength properties of the utility materials obtained from PVC with a varied history of use and forms, namely, raw and recycled, as well as tubes and bottles. The results of this work show an increase in viscosity with an increase in the content of recycled PVC. Optimal tensile strength and impact strengths were determined using SEM fracture photo-micrography: hardness increases with the increasing density of the composite. In his article, Tatsuno [36] presented the results of plastics reinforced by carbon fibre. This study explained the behaviour of fibre deformation and press load during formation and the relationship between material temperature, press load and mechanical strength. The use of constant pressure on the press has been beneficial for obtaining good mechanical strength. Diaz and Ortega [37] presented a concept using rotary moulding, in order to reprocess polymers derived from cable waste. Hou, Ye and Mai [38] presented the results of the production of advanced composite elements for civil aircraft by the compression moulding method, using a polyetherimide composite reinforced with carbon fibre. As a result of pressed moulding, aileron ribs and hinges for civil aircraft with high load resistance were obtained. The use of both physical and chemical modifications improves the mixing effect of plastics and thus obtains the homogeneous properties of composite products [39]. Similar conclusions have been made in other works [40,41]. Another example of polymer plate production is the article by Vidales-Barriguete et al. [42]. The aim of their work was to analyse the physical and mechanical properties of mixed gypsum boards with waste plastic cable sheaths. The results of this work indicate a significant increase in the elasticity of plates with plastic waste when the reduction in cracking is taken into account, along with compliance with the minimum value of bending strength and a slight improvement in the heat conduction coefficient, resulting from lower energy demand. As stated in the study by Vidales-Barriguete, gypsum boards with the addition of plastic could be a suitable alternative to the gypsum boards on the market, contributing to sustainable construction, not only in new constructions but also in renovated buildings. Yet another article [42] presents an alternative material for acoustic barriers, anti-sound barriers, produced partly from plastic waste, rather than entirely from wood or steel. Materials made of recycled plastic do not differ significantly in utility, nor as regards conservation and aesthetic properties, but are also cheaper to manufacture and use (up to 194 USD/m^2 vs. up to 269 USD/m^2). A similar issue concerning the production of plates from secondary materials was discussed in the paper [43]. The boards made of wood and plywood were compared with the material obtained from recycled material from the paper industry, which also includes plastics. Each plate was obtained in the same way: the input material was placed in a mould, heated and pressed on a hydraulic press. The waste boards had similar mechanical properties to the agglomerated wood boards. However, they were distinguished by their higher flexibility and a significant difference in swelling. Water absorption tests showed that the waste board performed better than the agglomerated wood and plywood boards.

To the best of our knowledge, no research has focused on the development of secondary plastic plate manufacturing technologies that could be transferred into industrial manufacturing. In general, the above-mentioned research focussed on finding ways to improve certain properties of polymers by using fillers or modifiers in the injection and extrusion process as well as to apply sustainability and circular economy criteria to newly developed "eco-products".

The aim of this work is to develop a solution for the recycling of cable waste by designing a procedure for converting it into solid sheets and examining its mechanical properties. The following tasks arise from the above purpose:

- Estimation of the fractions of individual plastics per batch of waste delivered;
- Manufacture of a plate-type product enabling the acquisition of research samples;
- Performing selected mechanical tests to assess the quality of the plate material obtained from recyclates along with an analysis and evaluation of the test results;
- A proposal for the use of recyclates from the dominant plastic fractions and fibrous composites, resulting mainly from cable waste.

2. Materials and Methods

Waste plastics, also known as PW, were supplied by the recycling company Eco Harpoon Sp. z. o.o with registered offices in Cząstków Mazowiecki (Poland). The share of waste plastics in the material delivered was determined on the basis of the organoleptic identification undertaken. The waste was in the form of cut and torn outer and inner wire covering. Figure 1 shows an example of the electric cable bundles provided by the company.

Figure 1. Examples of scrap electronic equipment delivered for testing.

As mentioned above, the identification of plastics was carried out using a simplified method, the so-called organoleptic method which consists in observing the effects accompanying the combustion of a sample of the material in the burner flame. The combustion method was used, generally, to segregate the types of polymers, namely, thermoplastics, which, under the influence of temperature, soften, deform and melt, and duroplasts, which do not exhibit such characteristics. During the combustion tests, the following was established:

(a) The flammability of plastic material;
(b) The flame colour and colour scheme, i.e., the type of flame;
(c) The behaviour of the material in the flame;
(d) The smell of fumes after the sample is extinguished.

Due to the specific behaviour of PVC in the flame burner, the organoleptic method was used to quickly distinguish non-comminuted cable sheaths. The test resulted in separated (in Figure 1) PVC covers (right side) and only a few polysiloxane covers (left side). The organoleptic method is not an ideal method for identifying plastics, if only because of poor individual interpretation of tester; that is, the statements "irritating smell" and "sweet smell" can be quite vague when referring to test results and ambiguous for several independent testers. The organoleptic method was used only for the initial grouping of the material that may be present in the tested material. In order to check the content of the above-mentioned types of polymers, Fourier transform infrared (FTIR) spectroscopy was additionally performed using the Thermo scientific nicolet iS50 spectroscope, which is discussed in more detail in Section 3.1.

Subsequently, electric cables were subjected to pre-treatment processes, including shredding and separation processes, resulting in a fraction of the waste plastics (PW), namely, the so-called "speck", being devoid of some metals, i.e., aluminium. Shredding stage was performed with the Wire Shredder Stokermill K750 (STOKKERMILL Recycling, Udine, Italy). Subsequently, speck was characterised in terms of its size and quantity using sieves with different mesh sizes ranging from 0.1 mm to 3 mm. Percentage distribution of individual fractions (together with the comminuted plastics) is shown in Figure 2b. Initial weight of speck was 463.101 g.

(a)

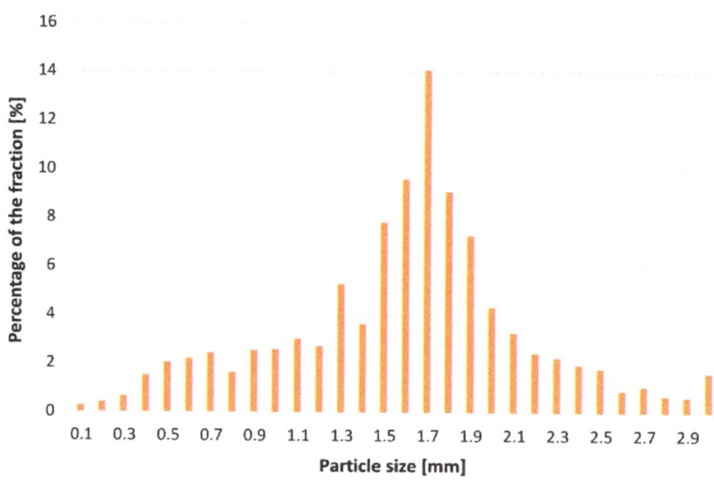

(b)

Figure 2. (**a**) Input material ("speck") to the moulding process, obtained in the shredding process. (**b**) Percentage distribution size of the fractions by particle size.

2.1. Sample Preparation

As already mentioned in the previous section, so-called "speck" was used in the press moulding for the production of plates. If the "speck" comprises shredded polysiloxane insulations, it will remain in the same aggregate during plasticising and pressing, thus forming inclusions. Thick-walled plates were made using the press moulding technology. The following treatments can be distinguished in the technological process:

- Preparation of the mould for the process, in particular securing all surfaces of the mould in contact with the processed material by using an anti-adhesive agent to protect it from sticking;
- Preparation of the plastic for processing, i.e., heating in the heating chamber at a temperature of 60 °C for 2 h to evaporate moisture from the surface of the material;
- Filling the mould with an appropriate amount of material in the form of the "regrind" obtained for testing and heating it to the appropriate temperature, as well as closing the mould with a punch plate with a preload;
- Press moulding, such as the punch load with appropriate force for a specified time vis-à-vis the plasticised material;
- Cooling the mould with the product, removing the punch plate, disassembling one side plate of the mould and then removing the product.

The technological station for the production of plates is shown in Figure 3. The test station consists of a metal mould with a section of resistance heaters embedded in it and an electronic heating system with temperature control, mounted on the table of a hydraulic press with a trigger acting on the punch plate. The task of the heating and temperature control system was to heat the mould to the temperature previously adopted by appropriate programming of the control system. The value of the mould temperature, as set and the current temperature of the processed material, measured in its geometric centre, was read on the front panel of the programmer. The total power of the heaters mounted in the form was 5.5 kW.

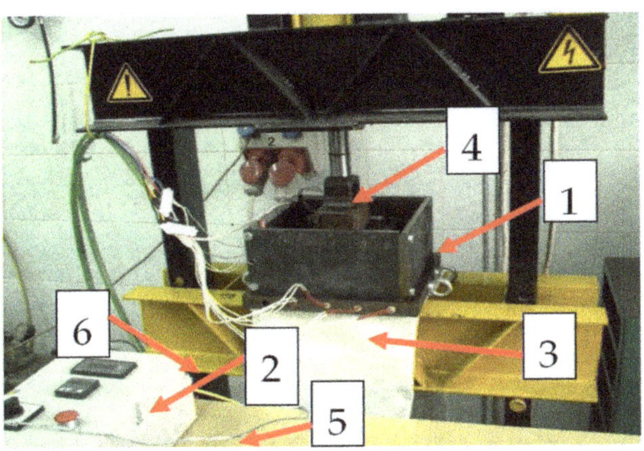

Figure 3. Test station for the production of plates obtained from the recycling of cable waste: 1—metal mould, 2—heating system with temperature control, 3—resistance heaters, 4—pressure pin of the hydraulic cylinder, 5—control thermo-couple cable, 6—base of the hydraulic press.

The parameters of the PVC sheet press moulding process were established, taking into account the following facts:

- Glass transition temperature T_g is about 85 °C, for material with a high plasticiser content up to 75 °C;
- In the temperature range of 140–170 °C, degradation of the plastic begins, and hydrogen is released.

The following parameters of process were adopted:

- Material temperature T_W = 135 °C;
- Heating time t = 1 h;
- Pressure p = 10 MPa.

The temperature of the material before forming was measured in the geometric centre of the volume of backfilled-material-buried "speck", while in the heating phase, the material was preloaded with the pressure coming from the weight of the punch, i.e., approximately 0.02 MPa. As a result of press moulding, a plate with a thickness of 1.4 cm, 17.5 cm wide and 20 cm long, as shown in Figure 4, was obtained.

Figure 4. Plate obtained in the press moulding process.

As a result of the pressing process, the plasticised material flowed with partial mixing of the particles of the plasticised material (Figure 4). However, one cannot talk about the homogenisation of the structure as it is in the plasticizing system of the extruder. During the heating, the fragmented particles of the cable sheaths were only adhesively connected with a small initial pressure. In the flow direction (Figure 5) of the material, the strength properties may be higher than in the transverse direction. In addition, the material is not homogeneous throughout the mass, so there may be local differences in the values of some properties. Waste covers probably come from different manufacturers and can be significantly different in terms of the amount and type of additives, including plasticizers.

2.2. *Strength Tests of Samples from Waste Plastic*

2.2.1. Test of Static Compressive Strength

The static compression test was performed in accordance with the standard ISO 15527:2022 [44]. Samples with a square cross-section of 12 × 12 mm and a height of 5 mm were prepared for the compression tests.

The yield point R_e was calculated according to the following relationship:

$$R_e = \frac{F}{A} \tag{1}$$

where *F*—strength at yield point, and *A*—initial cross-sectional area of the sample.

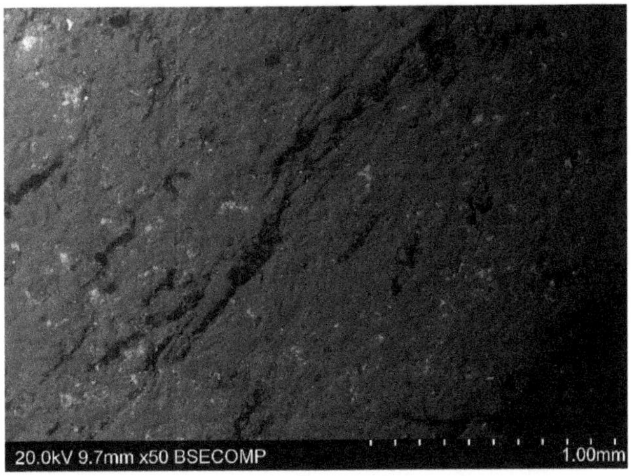

Figure 5. Scanning electron microscope (SEM) micrograph of the plate obtained in the press moulding process.

2.2.2. Test of Static Bend Strength

The static bend strength test was performed in accordance with the standard ISO 178:2019 [45]. For three-point bending tests, samples with a length of 120 mm and cross-sectional dimensions of 15 × 12 mm were prepared.

Bending strength R_g was calculated according to the following formula:

$$R_g = \frac{M}{W} = 3 \cdot F \cdot l / 2 \cdot b \cdot h^2 \quad (2)$$

where *M*—bending moment, *W*—strength index, *F*—maximum bending force, *l*—distance between supports, and *b*, *h*—width and height of the sample.

2.2.3. Charpy Impact Test

The material impact tests were carried out on a Charpy apparatus. The Charpy impact test was performed in accordance with the standard ISO 179-1:2010 [46]. Samples of material with dimensions $l \times b \times h$ = 120 × 15 × 10 mm were cut in the same way as the samples for bending tests. Data and results are given in instrument scaling units and then converted to the SI system.

The impact strength was calculated according to the following relationship:

$$a_n = \frac{A_n}{b \cdot h} \quad (3)$$

where *b*, *h*—width and thickness of the sample, and A_n—work needed to break the sample.

3. Results and Discussion

3.1. Identification of the Plastic

The materials were identified and separated into fractions, and their masses were weighed (Table 1). In the batch delivered, plasticised polyvinyl chloride (PVC-P) predominated and constituted approximately 99.3%. The remaining part is polysiloxane (silicone) (SI) in the form of three bonded sheaths that insulate the electric wiring.

Table 1. Results of identification and mass measurements of the waste delivered for testing.

Item	Effects—Burning in the Burner Flame	Identification	Share		
			Quantity pcs	**Weight, g**	**Percent, %**
1	The density is higher than that of water; self-extinguishing; does not drip; yellow centre of the flame with a green border at the bottom; the smell of hydrochloric acid; when removed from the flame—white smoke	Polyvinyl chloride (PVC)	126	847.6	99.3
2	The density is higher than that of water; burns with a bright yellow flame; no smell; after extinguishing—white ash	Polysiloxane (silicone) (SI)	3	6.0	0.7
In total			129	853.6	100

In order to accurately conduct the analysis of the tested waste cables, the FTIR analysis was performed. The absorption spectra ranged from 400 to 4000 cm^{-1} using an ATR detector with a resolution of 16 scans per spectrum and an optical resolution of 4 cm^{-1}. In order to maintain the reliability and repeatability of the results, the ATR crystal was cleaned with alcohol-soaked tissue after each measurement. Before each measurement, the background spectrum was measured and collected. The tests were carried out in ambient conditions (room temperature 23 °C, air humidity 40%). The FTIR−ATR analysis was performed for six samples, and then the result was averaged. The measurement results are shown in Figure 6.

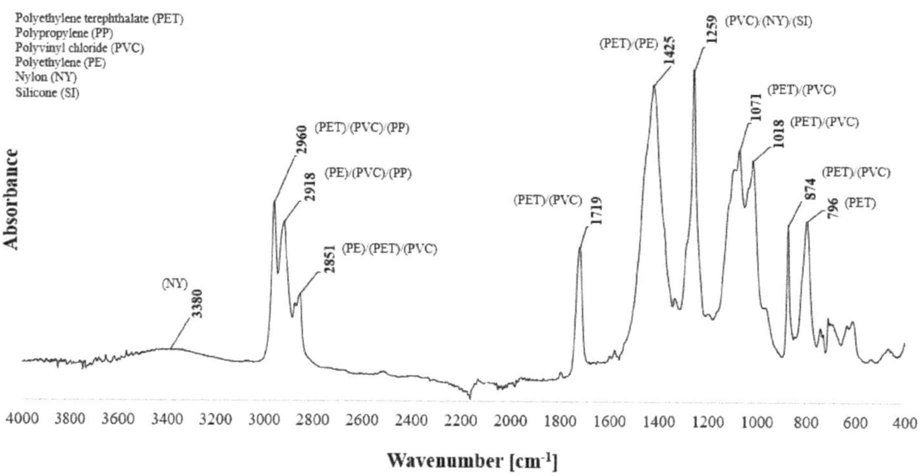

Figure 6. FTIR−ATR spectrum for recycled cable waste.

Describing from the left of the spectrum, the characteristic broadly distributed peak at 3380 cm^{-1} is characteristic of the stretching vibrations of the -OH group. The peaks at 2918 cm^{-1} and 1425 cm^{-1} are characteristic of the occurrence of the asymmetric stretching vibrations and deformation vibrations of the -CH_3 group. The slightly lower peaks of 2851 cm^{-1} and 1259 cm^{-1} are characteristic of the occurrence of the -CH group, responsible for deformation and bending vibrations. The peaks at 2960 cm^{-1}, 1071 cm^{-1} and in the range from 874 cm^{-1} to 796 cm^{-1} are responsible for the vibration of the -CH_2 group. The present peak at 1719 cm^{-1} and 1018 cm^{-1} is characteristic of the stretching -CO group. The collected spectrum and its characterisation are consistent and confirm literature reports [47–51]. In summary, the FTIR−ATR analysis identified six polymer components and one calcium carbonate filler compound.

3.2. Compression Test Results

On the basis of compression tests, yield point R_e was calculated. The results of the measurements and calculations are presented in Table 2. The waveforms of compressive stresses are presented in Figure 7.

Table 2. The results of the calculation of the plate yield point.

Item	Compression			
	Sample No.	**F_{pl} (kN)**	**ΔL_{pl} (mm)**	**R_e (MPa)**
1.	S1	2.96	6.48	15.0
2.	S2	2.46	6.52	14.1
3.	S3	2.38	6.42	15.1
4.	S4	2.41	6.47	15.6
5.	S5	2.24	6.58	15.4
6.	S6	2.22	6.55	14.9

Final result: R_e = 15.0 ± 0.6 MPa.

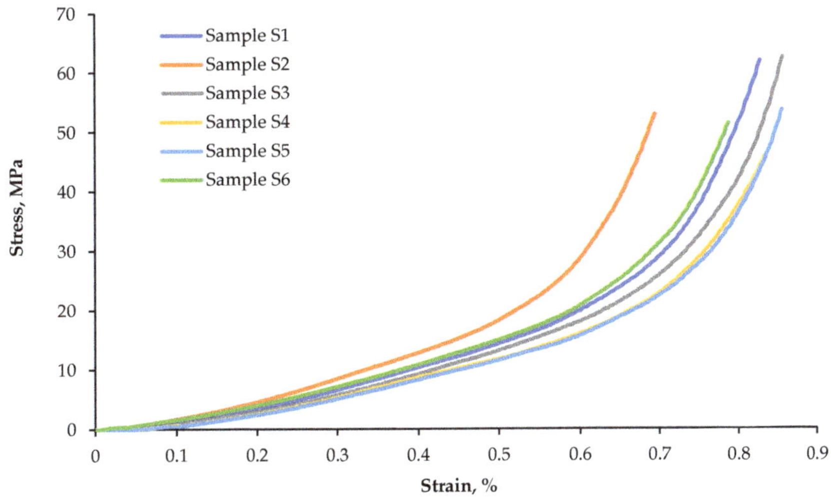

Figure 7. Waveforms of compressive stresses as a function of the deformation of specimens S1–S6.

The standard deviation obtained from the test is S_x = 0.5196, the critical value of the T-Student distribution, with k = n–1 degrees of freedom, is t_α = 2.5706, and the significance level is α_{ist} = 0.05.

The stress at yield point R_e, in the compression test, on the samples taken from the plates, varies in the range from 14.1 to 15.6 MPa. The average value is R_e = 15.0 ± 0.6 MPa. The measurement error is relatively small, less than 5%. Changes in the value of compressive stresses, as a function of deformation in a sample and used as an example, are shown in Figure 7.

3.3. Bending Test Results

On the basis of bending tests, strength R_m and yield point R_e were calculated. The results of measurements and calculations are presented in Table 3.

Table 3. Results of calculations of the bending strength and the yield point of plates.

Item	Three-Point Bending Test						
	Sample No.	F_{pl} (kN)	ΔL_{pl} (mm)	F_{max} (kN)	ΔL_{max} (mm)	R_m (MPa)	R_e (MPa)
1.	G1	6.43	4.01	13.67	17.01	0.91	0.40
2.	G2	7.43	3.72	15.08	16.02	0.96	0.41
3.	G3	6.33	5.01	12.52	16.23	0.93	0.46
4.	G4	7.79	5.12	15.80	16.84	0.90	0.52
5.	G5	8.47	4.89	16.73	16.88	1.16	0.56
6.	G6	7.89	4.82	13.65	17.02	0.80	0.48

Final result: $R_m = 0.94 \pm 0.14$ MPa for: $S_x = 0.1367$, $t_\alpha = 2.5706$. Final result: $R_e = 0.47 \pm 0.07$ MPa for: $S_x = 0.0621$, $t_\alpha = 2.5706$.

The bending strength of the plate material varies from 0.8 to 1.16 MPa, while the value of the yield strength ranges from 0.40 to 0.56 MPa. The average values of the bending strength and yield point are as follows: $R_m = 0.94 \pm 0.14$ MPa, $R_e = 0.47 \pm 0.07$ MPa.

In the case of the values of R_m as well as R_e, measurement errors are relatively large, in the range of a dozen percent. The results of the load—displacement curves in three-point bending test are shown in Figure 8.

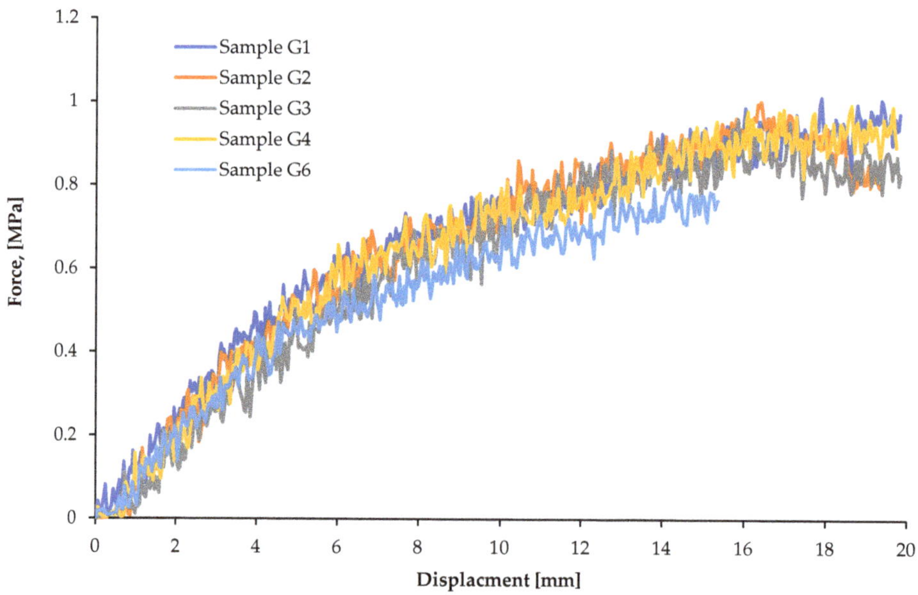

Figure 8. Load—displacement curves in three-point bending test (samples G1—G6).

3.4. Impact Test Results

On the basis of tests carried out on the Charpy apparatus, the impact strength was calculated a_n. The results of the measurements and calculations are shown in Table 4.

Table 4. Results of measurements and calculations of impact strength according to Charpy.

Item	Charpy's Impact Test				
	Sample No.	A_n (J)	$b_x h$ (cm^2)	a_n (J/cm^2)	a_n (kJ/m^2)
1.	U1	0.78	1.62	0.481	4.8
2.	U2	0.75	1.53	0.490	4.9
3.	U3	0.90	1.60	0.562	5.6
	U4	0.75	1.53	0.490	4.9
5.	U5	0.80	1.63	0.491	4.9
6.	U6	0.85	1.53	0.556	5.6

Final result: $a_n = 5.1 \pm 0.4$ kJ/m^2 for: $S_x = 0.3768$, $t_\alpha = 2.5706$.

3.5. Static Tensile Test and Hardness

Results of the uniaxial stretching test are shown in Figure 9. Three stretching tests were performed on standard, paddle-shaped samples, according to the standard ISO 37:2005 [52]. Based on literature data [53], pure polyvinyl chloride reaches a value in the range of 0.785 to even 156 MPa, while Shore hardness ranges from 35 to 110MPa. It depends on the addition of plasticizers to the material. For example, Tygon polymer B-44-4X contains a small amount of phthalate plasticizers (orthophthalate), which reduce the intermolecular interaction and increase the mobility of polymer chains. Thus, they reduce hardness and increase its flexibility. Comparing the results of the static tensile test with other polymers, such as Silopren E VPI 4036 G, Tygon polymer R-40, Siloprene E3078, etc., the properties of the obtained material are similar to the commercial Tygon Polymer R-1000 [53]. Its tensile strength is about 8 MPa, and the hardness is 40 Shore. The presented graphs show that the tensile strains are elastic to the failure point with deformations exceeding 100% (Table 5). Tensile strength is typical for this group of materials.

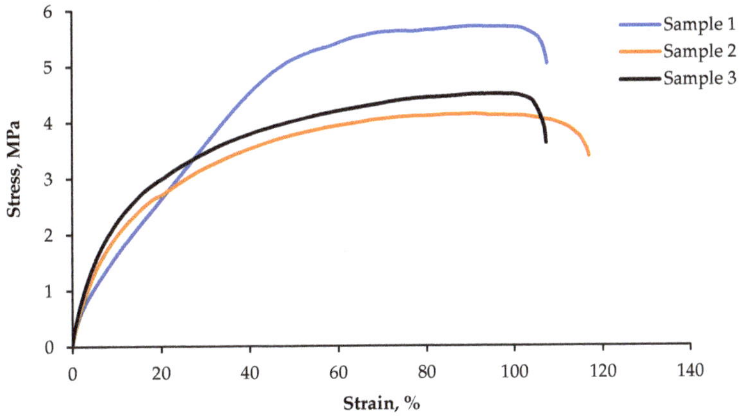

Figure 9. Tensile curves of samples made from the test plate.

Table 5. Mechanical properties, established on the basis of the tensile test.

Item	R_m (MPa)	A (%)
Sample 1	5.70	101
Sample 2	4.14	110
Sample 3	3.47	102
Average	4.44	105

The measurement of the hardness of elastomeric materials such as various types of rubbers and plastics can be performed using the Shore method, in accordance with ISO 7619-1:210 [54]. The Sauter Hardness Test HDD 100-1 was used for hardness testing. The Shore hardness (D scale) was 36.

4. Conclusions

On the basis of the test results obtained for recycled PVC, it was found that the products manufactured from them after the shredding process should have the structure of solid or perforated plates and thus will be resistant to compressive stresses. Yield strength R_e for PVC varies in a very wide range from 0.3 Pa to approximately 70 MPa depending on the content of plasticisers which can be compared to the results obtained in [55,56]. The strength value of 15.0 MPa, obtained in the tests, can be considered quite good, suitable for materials of various grades and probably of various manufacturers, which were additionally used for several years or more. During this time, PVC insulations were exposed to overheating and environmental influences, such as UV radiation.

A significant influence of plasticisers on mechanical properties was observed. For example, the shear modulus of elasticity (Young) of PVC at T = 20 °C was more than 1200 MPa for 0% DOP content (di-octyl phthalate) and approximately 45 MPA for the 30% DOP content, whereas for a 50% DOP content, only 15–20 MPa, similar to [57]. There are also PVC grades on the market with E = 3000 MPa.

Similar to the results of the compression tests, the result *R*, obtained in the bending test S_m = 0.94 MPa, is typical for plasticised products, as described in [58]. For non-plasticised products, R_m can be about 90 MPa [59]. The Charpy impact strength of the product (plate) of a_n = 5.1 kJ/m^2 is comparable to materials made of granulate as described in [60–62].

The results obtained from the analysis performed by FTIR−ATRinfrared spectroscopy and analysed with the available library database confirmed the presence of polymers such as polyvinyl chloride (PVC), silicone, poly(ethylene terephthalate) (PET), polypropylene (PP), polyethylene (PE), nylon and trace amounts of calcium carbonate, which is a fairly common inorganic filler in cables. The remaining unidentified noise collected in the absorption spectrum is most likely due to impurities or the presence of other polymers appearing in trace amounts.

The test results confirm the possibility of producing simple structures such as a solid plate, despite the fact that the material is neither perfectly plasticised nor homogeneous. Under manufacturing conditions, it is possible to plasticise the material in the extrusion process, and, at a later stage, the material stream with a slightly lower temperature can be widened with sections of rollers or pressed without additional heating. It was also stated that in the case of press moulding under laboratory conditions, PVC cannot be heated to a sufficiently high temperature for a long time due to the possibility of it degrading. The obtained results encourage research aimed at developing a technology for making prototype products. Despite the worse strength and processing parameters of recyclates, through skilful dosing of reinforcing materials, composites with the required physical and chemical properties can be obtained.

Due to the fact that the mechanical properties are not very high, compared to other materials, the use of the obtained plates is, of course, limited, and they should only be used in the case of low bearing. However, taking into account other physico-chemical properties, such as low electrical conductivity or low wettability, as well as the final shape (plates or blocks), the authors suggest that such objects could be used as electrical insulators (which, in fact, this material originally was) or as water shields such as windscreen wipers, roofing, etc.).

Based on the research results presented, the following conclusions can be drawn:

- Using the method of plastic consolidation, it is possible to obtain solid plates or with recesses from waste materials, meeting the mechanical requirements of some types of products, e.g., platforms;

- Materials after plastic consolidation show increased plasticity compared to the material of plates made of the original granulate;
- The microstructural tests carried out showed no defects inside the analysed plates, such as cracks or delamination, which confirms the correct selection of pressing parameters.

It should be noted that the plates used in water and sewage infrastructure, road infrastructure, etc., under real conditions, are mainly loaded with compressive and bending forces. The results obtained from the research are promising for the applications mentioned.

Author Contributions: Conceptualisation, A.P., M.W., T.S. and P.N.; methodology, J.W. and W.P.; validation, A.P. and A.B.; formal analysis, A.P., M.W. and W.P.; investigation, J.W., A.B., J.K. and W.P.; resources, M.W. and M.K.; data curation, J.K., T.S. and P.N.; writing—original draft preparation, M.W. and A.P.; writing—review and editing, A.P., J.W. and A.B.; visualisation, M.W., P.N. and T.S.; supervision, A.B.; funding acquisition M.K., A.B. and M.W. All authors have read and agreed to the published version of the manuscript.

Funding: This research was funded by the Polish National Center for Research and Development, grant number POIR.01.01.01-00-0362/19.

Institutional Review Board Statement: Not applicable.

Informed Consent Statement: Not applicable.

Data Availability Statement: Data sharing is not applicable to this article.

Acknowledgments: The authors would like to thank ECO Harpoon Recycling sp. z o. o. for supplying plastic waste material.

Conflicts of Interest: The authors declare no conflict of interest.

References

1. Thrift, E.; Porter, A.; Galloway, T.S.; Coomber, F.G.; Mathews, F. Ingestion of plastic by terrestrial small mammals. *Sci. Total environ.* **2022**, *842*, 156679. [CrossRef] [PubMed]
2. Nocoń, W.; Moraczewska-Majkut, K.; Wiśniowska, E. Microplastics in water level of contamination and threats connected with presence of these micropollutants. *Technol. Water* **2018**, *4*, 24–29.
3. Chen, X.; Yan, N.A. Brief overview of renewable plastics. *Mater. Today Sustain.* **2020**, *7–8*, 100031. [CrossRef]
4. Geyer, R.; Jambeck, J.; Law, K. Production, use, and fate of all plastics ever made. *Sci. Adv.* **2017**, *3*, e1700782. [CrossRef]
5. Ojogbo, E.; Ogunsona, E.O.; Mekonnen, T.H. Chemical and physical modifications of starch for renewable polymeric materials. *Mater. Today Sustain.* **2020**, *7–8*, 100028. [CrossRef]
6. Rahimi, A.; Garcia, J. Chemical recycling of waste plastics for new materials production. *Nat. Rev. Chem.* **2017**, *1*, 0046. [CrossRef]
7. Sharuddin, S.D.A.; Abnisa, F.; Daud, W.M.A.; Aroua, M.K. A Review on pyrolysis of plastic wastes. *Energy Convers. Manag.* **2016**, *155*, 308–326. [CrossRef]
8. Czarnecka-Komorowska, D.; Wiszumirska, K. Sustainability design of plastic packaging for the Circular Economy. *J. Chem. Technol. Polym. Proces.* **2020**, *65*, 8–17. [CrossRef]
9. Janajreh, I.; Alshrah, M.; Zamzam, S. Mechanical recycling of PVC plastic waste streams from cable industry: A case study. *Sustain. Cities Soc.* **2015**, *18*, 13–20. [CrossRef]
10. Corona, B.; Shen, L.; Reike, D.; Carreon, J.R.; Worrel, E. Towards sustainable development through the circular economy—A review and critical assessment on current circularity metrics. *Resour. Conserv. Recycl.* **2019**, *151*, 104498. [CrossRef]
11. Vollmer, I.; Jenks, M.J.F.; Roelands, M.C.P.; White, R.; Harmelen, T.; Wild, P.; Laan, G.; Meirer, F.; Keurentjes, J.T.F.; Weckhuysen, B. Beyond Mechanical Recycling: Giving New Life to Plastic Waste. *Angew. Chem. Int. Ed. Engl.* **2020**, *59*, 15402–15423. [CrossRef] [PubMed]
12. Zhang, C.; Hu, M.; Maio, F.; Sprecher, B.; Yang, X.; Tukker, A. An overview of the waste hierarchy framework for analyzing the circularity in construction and demolition waste management in Europe. *Sci. Total Environ.* **2022**, *803*, 149892. [CrossRef] [PubMed]
13. Czarnecka-Komorowska, D.; Nowak-Grzebyta, J.; Gawdzińska, K.; Mysiukiewicz, O.; Tomasik, M. Polyethylene/Polyamide Blends Made of Waste with Compatibilizer: Processing, Morphology, Rheological and Thermo-Mechanical Behavior. *Polymers* **2021**, *13*, 2385. [CrossRef] [PubMed]
14. Fazli, A.; Rodrigue, D. Waste Rubber Recycling: A Review on the Evolution and Properties of Thermoplastic Elastomers. *Materials* **2020**, *13*, 782. [CrossRef]
15. Pedreño-Rojas, M.A.; Rodríguez-Liñán, C.; Flores-Colen, I.; De Brito, J. Use of Polycarbonate Waste as Aggregate in Recycled, Gypsum Plasters. *Materials* **2020**, *13*, 3042. [CrossRef]

16. Gibreil, H.A.A.; Feng, C.P. Effects of high-density polyethylene and crumb rubber powder as modifiers on properties of hot mix asphalt. *Int. Res. J. Eng. Technol.* **2017**, *142*, 101–108. [CrossRef]
17. Behl, A.; Sharma, G.; Kumar, G. A sustainable approach: Utilization of waste PVC in asphalting of roads. *Constr. Build. Mater.* **2014**, *54*, 113–117. [CrossRef]
18. Ziari, H.; Nasiri, E.; Amini, A.; Ferdosian, O. The effect of EAF dust and waste PVC on moisture sensitivity, ruttingresistance, and fatigue performance of asphalt binders and mixtures. *Constr. Build. Mater.* **2019**, *203*, 188–200. [CrossRef]
19. Penava, N.V.; Rek, V.; Houra, I.F. Effect of EPDM as a compatibilizer on mechanical properties and morphology of PP/LDPE blends. *J. Elastomers Plast.* **2012**, *47*, 391–403. [CrossRef]
20. Xavier, S.F. Properties and Performance of Polymer Blends. In *Polymer Blends Handbook—Part II*; Springer: New York, NY, USA, 2014; Volume 2, pp. 123–194.
21. Taguet, A.; Cassagnau, P.; Lopez- Cuesta, J.-M. Structuration, selective dispersion and compatibilizing effect of (nano)fillers in polymer blends. *Prog. Polym. Sci.* **2014**, *39*, 1526–1563. [CrossRef]
22. Yu, F.; Huang, H.-X. Simultaneously toughening and reinforcing poly(lactic acid)/thermoplastic polyurethane blend via enhancing interfacial adhesion by hydrophobic silica nanoparticles. *Polym. Test.* **2015**, *45*, 107–113. [CrossRef]
23. Ouyang, Y.; Mauri, M.; Pourrahimi, A.M.; Ostergren, I.; Lund, A.; Gkourmpis, T.; Prieto, O.; Xu, X.; Hagstrand, P.-O.; Muller, C. Recyclable Polyethylene Insulation via Reactive Compounding with a Maleic Anhydride-Grafted Polypropylene. *Appl. Polym. Mater.* **2020**, *2*, 2389–2396. [CrossRef]
24. Sonnier, R.; Leroy, E.; Clerc, L.; Bergeret, A.; Lopez-Cuesta, J.-M.; Bretelle, A.-S.; Ienny, P. Compatibilizing thermoplastic/ground tyre rubber powder blends: Efficiency and limits. *Polym Test.* **2008**, *27*, 901–907. [CrossRef]
25. Adhikari, B.; De, D.; Maiti, S. Reclamation and recycling of waste rubber. *Prog. Polym. Sci.* **2000**, *25*, 909–948. [CrossRef]
26. Marossy, K.; Barczy, P. Improvement of the Properties of Polyethylene with Chlorinated Polyethylene (CPE). *Polym. Polym. Compos.* **2003**, *11*, 115–122. [CrossRef]
27. Zarraga, A.; Munoz, M.E.; Pena, J.J.; Santamaria, A. Rheological Effects of the Incorporation of Chlorinated Polyethylene Compatibilizers in a HDPE/PVC Blend. *Polym. Eng. Sci.* **2001**, *41*, 1893–1902. [CrossRef]
28. Kollar, M.; Zsoldos, G. Investigating poly-(vinyl-chloride)-polyethylene blends by thermal methods. *J. Therm. Anal. Calorim.* **2012**, *107*, 645–650. [CrossRef]
29. Xu, J.; Tazawa, N.; Kumagai, S.; Kameda, T.; Saito, Y.; Yoshioka, T. Simultaneous recovery of high-purity copper and polyvinyl chloride from thin electric cables by plasticizer extraction and ball milling. *RSC Adv.* **2018**, *8*, 6893–6903. [CrossRef]
30. Smol, M.; Duda, J.; Czaplicka-Kotas, A.; Szołdrowska, D. Transformation towards Circular Economy (CE) in Municipal Waste Management System: Model Solutions for Poland. *Sustainability* **2020**, *12*, 4561. [CrossRef]
31. Janajreh, I.; Alshrah, M. Remolding of Cross-Linked Polyethylen Cable Waste: Thermal and Mechanical Property Assessment. *IJTEE* **2013**, *5*, 191–198.
32. Qudaih, R.; Janajreh, I.; Vukusic, S.E. Recycling of cross-linked polyethylene cable waste via particulate infusion. *Adv. Sustain. Manuf.* **2011**, *4*, 141–148.
33. Sombatsompop, N.; Thongsang, S. Rheology, morphology and mechanical and thermal properties of recycled PVC pipes. *J. Appl. Polym. Sci.* **2001**, *82*, 2478–2486. [CrossRef]
34. Pikoń, K.; Poranek, N.; Czajkowski, A.; Łaźniewska-Piekarczyk, B. Poland's Proposal for a Safe Solution of Waste Treatment during the COVID-19 Pandemic and Circular Economy Connection. *Appl. Sci.* **2021**, *11*, 3939. [CrossRef]
35. Vidales-Barriguete, A.; Santa-Cruz-Astorqui, J.; Pina-Ramirez, C.; Kosior-Kazberuk, M.; Kalinowska-Wichrowska, K.; Atanes-Sanchez, E. Study of the Mechanical and Physical Behavior of Gypsum Boards with Plastic Cable Waste Aggregates and Their Apllication to Construction Panels. *Materials* **2021**, *14*, 2255. [CrossRef] [PubMed]
36. Hag-Elsafi, O.; Elwell, D.J.; Hiris, M.; Glath, G. Noise Barriers Using Recycled-Plastic Lumber. *Transp. Res. Rec.* **1999**, *1670*, 49–58. [CrossRef]
37. Pelegrini, M.; Pinheiro, G.; Valle, J.A.B. Plates made with solid waste from the recycled paper industry. *J. Waste Manag.* **2010**, *30*, 268–273. [CrossRef]
38. Tatsuno, D.; Yoneyama, T.; Kawamoto, K.; Okamoto, M. Hot press forming of thermoplastic CFRP sheets. *Procedia Manuf.* **2018**, *15*, 1730–1737. [CrossRef]
39. Noga, P.; Piotrowicz, A.; Skrzekut, T.; Zwoliński, A.; Strzępek, P. Effect of Various Forms of Aluminium 6082 on the Mechanical Properties, Microstructure and Surface Modification of the Profile ater Extrusion Process. *Materials* **2021**, *14*, 5066. [CrossRef]
40. Tatsuno, D.; Yoneyama, T.; Kawamoto, K.; Okamoto, M. Effect of sie die pressure and adaptive die temperature control in press forming of U-beam using carbon fiber-reinforced PA6 sheets. *J. Compos. Mater.* **2017**, *50*, 4273–4286. [CrossRef]
41. Diaz, S.; Ortega, Z.; McCourt, M.; Kearns, M.P.; Benitez, A.N. Recycling of polymeric fraction of cable waste by rotational moulding. *Waste Manag.* **2018**, *76*, 199–206. [CrossRef]
42. Hou, M.; Ye, L.; Mai, T.W. Manufacturing Process and Mechanical Properties of Thermoplastic Composite Components. *J. Mater. Process. Technol.* **1997**, *63*, 334–338. [CrossRef]
43. Formela, K.; Korol, J.; Saeb, M.R. Interfacially modified LDPE/GTR composites with non-polar elastomers: From microstructure to macro-behavior. *Polym. Test.* **2015**, *42*, 89–98. [CrossRef]

44. *ISO 15527:2022*; Plastics—Compression-Moulded Sheets of Polyethylene (PE-UHMW, PE-HD)—Requirements and Test Methods. International Organization for Standardization: Geneva, Switzerland, 2022. Available online: https://www.iso.org/standard/80208.html (accessed on 19 June 2022).
45. *ISO 178:2019*; Plastics—Determination of Flexural Properties. International Organization for Standardization: Geneva, Switzerland, 2019. Available online: https://www.iso.org/standard/70513.html (accessed on 30 June 2022).
46. *ISO 179-1:2010*; Plastics—Determination of Charpy Impact Properties—Part 1: Non-instrumented impact test. International Organization for Standardization: Geneva, Switzerland, 2010. Available online: https://www.iso.org/standard/44852.html (accessed on 30 July 2022).
47. Fan, C.; Huang, Y.Z.; Lin, J.N.; Li, J. Microplastic constituent identification from admixtures by Fourier-transform infrared (FTIR) spectroscopy: The use of polyethylene terephthalate (PET), polyethylene (PE), polypropylene (PP), polyvinyl chloride (PVC) and nylon (NY) as the model constituents. *Environ. Technol. Innov.* **2021**, *23*, 101798. [CrossRef]
48. Toczek, K.; Lipińska, M.; Pietrasik, J. Smart TPE Materials Based on Recycled Rubber Shred. *Materials* **2021**, *14*, 6237. [CrossRef] [PubMed]
49. Dodi, G.; Popescu, D.; Cojocaru, F.D.; Aradoaei, M.; Ciobanu, R.C.; Mihai, C.T. Use of Fourier-Transform Infrared Spectroscopy for DNA Identification on Recycled PET Composite Substrate. *Appl. Sci.* **2022**, *12*, 4371. [CrossRef]
50. Bellasi, A.; Binda, G.; Pozzi, A.; Galafassi, S.; Volta, P.; Bettinetti, R. Microplastic Contamination in Freshwater Environments: A Review, Focusing on Interactions with Sediments and Benthic Organisms. *Environments* **2020**, *7*, 30. [CrossRef]
51. Anwar, U.H.; Soufi, K.Y.; Al-Hadhrami, L.M.; Shemsi, A.M. Failure investigation of an underground low voltage XLPE insulated cable. Anti-Corros. *Methods Mater.* **2015**, *62*, 281–287. [CrossRef]
52. *ISO 37:2005*; Rubber, Vulcanized or Thermoplastic—Determination of Tensile Stress-Strain Properties. International Organization for Standardization: Geneva, Switzerland, 2005. Available online: https://www.iso.org/standard/40200.html (accessed on 30 July 2022).
53. Matweb Material Property Data. Available online: https://www.matweb.com/index.aspx (accessed on 30 July 2022).
54. *ISO 7619-1:2010*; Rubber, Vulcanized or Thermoplastic—Determination of Indentation Hardness—Part 1: Durometer Method (Shore Hardness). International Organization for Standardization: Geneva, Switzerland, 2010. Available online: https://www.iso.org/standard/50756.html (accessed on 30 July 2022).
55. Hahladakis, J.N.; Velis, C.A.; Weber, R.; Iacovidou, E.; Purnell, P. An overview of chemical additices present in plastics: Migration, release, fate and environmental impact during their use, disposal and recycling. *J. Hazard. Mater.* **2018**, *344*, 179–199. [CrossRef]
56. Gnatowski, A.; Kijo-Kleczkowska, A.; Suchecki, Ł.; Palutkiewicz, P.; Krzywański, J. Analysis of Thermomechanical Properties of Polyethylene with Cement Addition. *Materials* **2022**, *15*, 1587. [CrossRef]
57. John, M.J.; Francis, B.; Varughese, K.T.; Thomas, S. Effect of chemical modification on properties of hybrid fiber biocomposites. *Compos. A Appl. Sci. Manuf.* **2008**, *39*, 352–363. [CrossRef]
58. Margolis, J.M. Chapter 3 Elastomeric Materials and Processes. In *Modern Plastics Handbook*; Harper, C.A., Ed.; McGraw-Hill: New York, NY, USA, 1997; pp. 190–241.
59. Saechtling, H. *International Plastic Handbook*; Scientific and Technical Publishing House: Warsaw, Poland, 1998.
60. Krawczuk, D. Polymers for thermoforming and their properties. *Polymer Process.* **2012**, *18*, 103–112.
61. Ragaert, K.; Delva, L.; Van Geem, K. Mechanical and chemical recycling of solid plastic waste. *Waste Manag.* **2017**, *69*, 24–58. [CrossRef] [PubMed]
62. Singh, N.; Hui, D.; Singh, R.; Ahuja, I.P.S.; Feo, L.; Fraternali, F. Recycling of plastic solid waste: A state of art review and future applications. *Compos. B Eng.* **2017**, *115*, 409–422. [CrossRef]

materials

MDPI

Article

One-Step Electrochemical Synthesis and Surface Reconstruction of NiCoP as an Electrocatalyst for Bifunctional Water Splitting

Minhao Sheng, Yawei Yang *, Xiaoqing Bin and Wenxiu Que *

Electronic Materials Research Laboratory, International Center for Dielectric Research, Shaanxi Engineering Research Center of Advanced Energy Materials and Devices, School of Electronic Science and Engineering, Xi'an Jiaotong University, Xi'an 710049, China

* Correspondence: ywyang@xjtu.edu.cn (Y.Y.); wxque@xjtu.edu.cn (W.Q.)

Abstract: We adopted a simple one-step electrochemical deposition to acquire an efficient nickel cobalt phosphorus (NiCoP) catalyst, which avoided the high temperature phosphatization engineering involved in the traditional synthesis method. The effects of electrolyte composition and deposition time on electrocatalytic performance were studied systematically. The as-prepared NiCoP achieved the lowest overpotential (η_{10} = 111 mV in the acidic condition and η_{10} = 120 mV in the alkaline condition) for the hydrogen evolution reaction (HER). Under 1 M KOH conditions, optimal oxygen evolution reaction (OER) activity (η_{10} = 276 mV) was also observed. Furthermore, the bifunctional NiCoP catalyst enabled a high-efficiency overall water-splitting by applying an external potential of 1.69 V. The surface valence and structural evolution of NiCoP samples with slowly decaying stability under alkaline conditions are revealed by XPS. The NiCoP is reconstructed into the $Ni(Co)(OH)_2$ (for HER) and Ni(Co)OOH (for OER) on the surface with P element loss, acting as real "active sites".

Keywords: electrocatalyst; electrodeposition; NiCoP; water electrolysis; surface reconstruction

Citation: Sheng, M.; Yang, Y.; Bin, X.; Que, W. One-Step Electrochemical Synthesis and Surface Reconstruction of NiCoP as an Electrocatalyst for Bifunctional Water Splitting. *Materials* **2023**, *16*, 1529. https://doi.org/10.3390/ma16041529

Academic Editor: Rosalinda Inguanta

Received: 15 January 2023
Revised: 7 February 2023
Accepted: 9 February 2023
Published: 11 February 2023

Copyright: © 2023 by the authors. Licensee MDPI, Basel, Switzerland. This article is an open access article distributed under the terms and conditions of the Creative Commons Attribution (CC BY) license (https://creativecommons.org/licenses/by/4.0/).

1. Introduction

Since the policy of "carbon peak and carbon neutrality" was advocated, energy conservation and emission reduction have become the consensus of all mankind. The development of green energy technology is extremely critical to reduce carbon dioxide emissions [1]. The preparation of green hydrogen occupies an important strategic position in the future. Electrolysis of water to hydrogen has the potential to transfer a large amount of renewable energy electricity to the decarbonized green hydrogen industrial sector [2,3].

Conventional water electrolysis reactions involve two electrochemical semi-reactions at the anode and cathode, in which electrochemical kinetics are controlled by HER and OER, respectively. The theoretical decomposition voltage of electrolytic water is 1.23 V. Nevertheless, in practical electrolytic cells, higher energy supply is inevitably required owing to the slow kinetic limiting step of HER and OER [4]. Therefore, the development of various highly active electrolytic water catalysts to reduce the reaction activation energy can significantly reduce the cost of hydrogen production.

In terms of catalytic activity, the commercial market tends to prefer mature noble metal catalysts (Pt, Ir, and Ru), but their high cost, limited reserves, and easy aggregation characteristics urge the development of non-noble metal catalysts [5–8]. Transition-metal-based electrocatalysts have shown great potential for overall water splitting over the past few years. Various oxides (MoO_x) [9], hydroxides ($Ni(OH)_2$) [10,11], sulfides (MoS_2) [12], selenides ($MoSe_2$) [13], nitrides (VN) [14], and carbides (MXene) [15,16] materials have been extensively reported as water splitting electrocatalysts. Among the aforementioned catalysts, transition metal phosphides (TMPs) possess a hydrogenase-like structure, and are regarded as promising nonprecious electrocatalysts for overall water splitting [17–22]. Single metal phosphides have been widely reported in the field of water electrolysis [23]. Liu et al. [24] employed NiS_2 single-crystal octahedrons as a precursor and designed a 6 nm thin-wall hollow metallic Ni_2P and metalloid NiP_2 polymorph through phosphorization. Hierarchically porous W-doped CoP nanoflake arrays on carbon cloth (W-CoP NAs/CC) were synthesized for pH-universal HER [25]. In addition, bimetallic phosphides can improve the electronic states of single metal and highly extend catalytic performance as a result of a synergetic effect. Lin et al. [19] proposed a novel in situ doping-induced lattice strain strategy to synthesize NiCoP/S nanocrystals (NCs), which significantly improved HER performance in a wide pH range. Ye et al. [20] constructed porous NiCoP heterostructures on Ni foam using the hydrothermal and phosphating method. However, the present phosphide preparation process almost always involves high temperature phosphating aftertreatment of transition metal species. The complexity of the multi-step process and the dangerous volatile PH_3 gas hinder the reproducibility and green friendliness of large-scale catalyst preparation [26,27].

Based on the above considerations, we synthesized highly active NiCoP microspheres on a self-supported conductive carbon cloth by employing a simple one-step electrochemical deposition method. Electrochemical deposition is an effective strategy for catalyst synthesis because of its advantages of being green, safe, controllable, and efficient. Compared with the post-treatment high-temperature phosphating process, we can directly obtain NiCoP active substances by skipping this tedious step. We systematically evaluated the differences between the catalytic properties of single metal (NiP and CoP) and double transition metal phosphide (NiCoP), and optimized the electrochemical deposition time of NiCoP synthesis. Through compositional optimization, NiCoP achieved the lowest overpotential (η_{10} = 111 mV in the acidic condition and η_{10} = 120 mV in the alkaline condition) for HER. Under alkaline conditions, excellent OER activity (η_{10} = 276 mV) was also observed. The overall water splitting device composed of NiCoP/NiCoP was assembled with an initial cell voltage of 1.69 V to achieve a current density of 10 mA cm^{-2} in 1.0 M KOH. Ex situ XPS characterization captured the dynamic structure evolution and surface remodeling during water electrolysis catalysis, indicating that the dissolution of P from the surface of NiCoP to form hydroxide (HER) and oxyhydroxide (OER) remodelers slowly inhibited the electrolysis reaction. Based on these investigations, we believe that rapid electrochemical synthesis of low-cost NiCoP is expected to show potential in the field of water splitting. In addition, our insight into the structural evolution of phosphide after prolonged electrolysis helps to reveal the true active site.

2. Methodology

The carbon cloth (CC) was pretreated with acetone, ethyl alcohol, and ultrapure water; subjected to ultrasonic treatment for 20 min; and dried in an oven at 60 °C for 24 h. As shown in Figure 1, the NiCoP catalyst was obtained by one-step electrochemical deposition at room temperature without any heat-treatment steps. Specifically, the configuration scheme of the precursor solution was as follows: 2 mM $CoCl_2{\cdot}6H_2O$, 2 mM $NiCl_2{\cdot}6H_2O$, and 10 mM $NaH_2PO_2{\cdot}H_2O$ were dissolved in 20 mL of water. The evenly mixed solution was transferred to an electrolytic cell. A three-electrode system was used for electrochemical deposition. The working electrode was carbon cloth (1 cm × 1 cm size), with a carbon rod as the counter electrode and a saturated calomel electrode ($Hg/HgCl_2$, SCE) as the reference electrode. We applied a constant voltage (−1 V vs. SCE) and different times (30, 60, and 90 min) for electrochemical deposition to determine the optimal catalyst performance. As a comparison, the corresponding precursor salts can be removed for the synthesis process of NiP and CoP. The electrodeposition time was 60 min. For electrochemical deposition associated with NiCoP catalysts, the specific quality load is provided in Table 1. Specific reagent information, physical characterization, and electrochemical tests are provided in the Supporting Information.

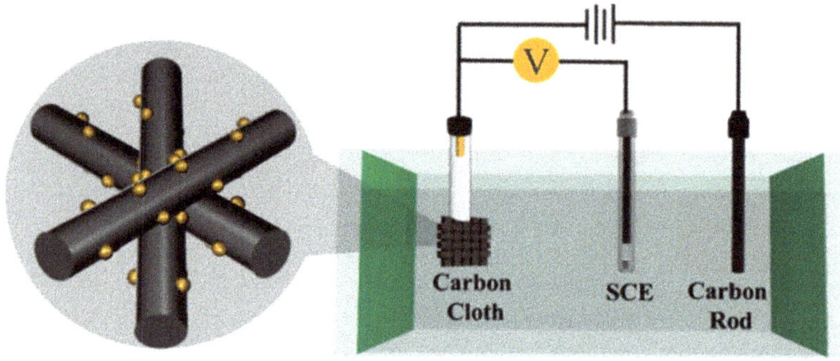

Figure 1. Electrodeposition synthesis procedure of NiCoP.

Table 1. The mass load of the relevant catalyst.

Catalysts	Mass Loading (mg cm^{-2})
NiP	1.8 ± 0.2
CoP	1.5 ± 0.2
NiCoP 0.5 h	0.7 ± 0.2
NiCoP 1 h	2.0 ± 0.2
NiCoP 1.5 h	4.2 ± 0.2

3. Results and Discussion

3.1. Physicochemical Characterization

First, XRD characterization (Figure 2a) was used to determine the composition of the phase. Except for the two broad peaks (~25° and 44°) attributed to the conductive substrate carbon cloth, there was no obvious NiCoP diffraction peak in the NiCoP catalyst after electrochemical deposition 1 h. For CoP and NiP samples, the XRD peak shape and peak position were roughly the same as those of NiCoP. The XRD results showed that the phosphating compounds obtained by electrochemistry are amorphous, which was quite different from the samples obtained by phosphides at a high temperature. The morphology of the catalyst is shown in Figure 2b–d and Figures S1–S3, which demonstrated the successful electrochemical growth of relevant phosphides on carbon cloth substrates. Under the same electrodeposition conditions of 1 h, NiP, CoP, and NiCoP particles have a slightly different shape. For NiP, micron-sized spherical particles were bound together and evenly attached to the surface of the CC, and the average size of NiP particles was close to 3 μm. This demonstrated the close contact between the electrochemically deposited NiP and the carbon cloth substrate, implying mechanical stability of the catalyst but a limited contact area with the electrolyte. As for CoP, relatively dispersed nanosheets and stacked flower-like particles were observed in CoP. In addition, large patches of bare and smooth carbon cloth indicated that the catalytically active site of growing active nanocrystals is limited on the surface of CC under the same conditions of electrochemical deposition. Small uniform NiCoP nanoparticles were closely arranged on the CC. As shown in Figures S1–S3, the EDS results displayed that the relevant elements were uniformly distributed in NiP, CoP, and NiCoP, and the atomic ratio (Table S1) of Ni/P was 4:1, that of Co/P was 5.93:1, and that of Ni/Co/P was 2.64:10.64:1.

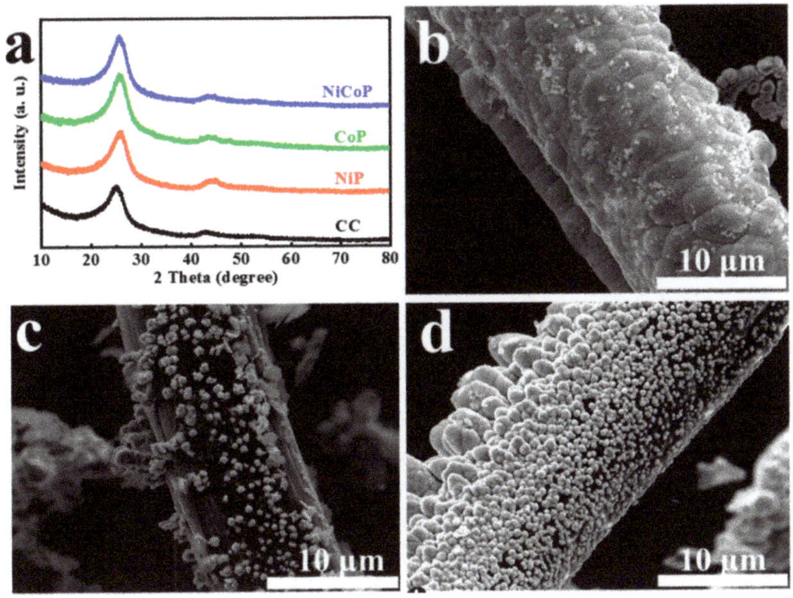

Figure 2. (**a**) XRD of CC, NiP, CoP, and NiCoP. SEM images of (**b**) NiP, (**c**) CoP, and (**d**) NiCoP.

X-ray photoelectron spectroscopy (XPS) characterization was also carried out to further illustrate the species information and surface valence of each element in electrochemically deposited phosphating compounds. The survey spectra (Figure S4) of NiCoP showed the existence of Ni, Co, P, and O peaks. For the Ni 2p spectrum (Figure 3a), it can be fitted into six peaks. The peaks of 873.1 eV and 855.3 eV are attributed to Ni^{2+} $2p_{1/2}$ and Ni^{2+} $2p_{3/2}$, respectively. Another pair of peak binding energies of 874.4 eV and 856.4 eV belong to Ni^{3+} $2p_{1/2}$ and Ni^{3+} $2p_{3/2}$, respectively [28,29]. The final pair of diffraction peaks of 879.6 eV and 861.5 eV are assigned as satellite characteristic peaks of Ni $2p_{1/2}$ and Ni $2p_{3/2}$ components, respectively [29]. For the Co 2p spectrum (Figure 3b), there are four decoupled peaks. The binding energies of 780.9 eV and 796.5 eV are certified as Co $2p_{3/2}$ and Co $2p_{1/2}$, respectively. Likewise, two shake-up satellite peaks are centered at 803.3 and 786.0 eV [23,30]. The existence of Ni^{2+} and Co^{2+} can accelerate electrolytic water reaction by weakening the O-H bond. Furthermore, the P 2p region (Figure 3c) consists of P-O (132.9 eV) and M-P (129.9 eV) bonds and the O 1s peak (Figure 3d) is centered at 531.2 eV, which suggests the coexistence of the metal phosphides and the oxide species on the NiCoP surface [25,31]. Therefore, the above results just confirm that one-step electrochemical deposition can obtain NiCoP, avoiding the tedious operation steps brought by high-temperature phosphating. A large number of works in the literature have suggested that positively charged $Ni^{\delta+}$ /$Co^{\delta+}$ and negatively charged $P^{\delta-}$ active sites are formed in phosphating compounds, accelerating charge transfer and reducing internal resistance, and thus are favorable to HER and OER activity [28]. Regarding the NiP XPS spectrum (Figure S5), the Ni 2p spectrum can be fitted by the doublet characteristic of Ni^{2+} (855.3 eV and 873.1 eV) and Ni^{3+} (856.5 eV and 874.5 eV) as well as two shake-up satellites (861.6 eV and 879.7 eV). For the CoP sample, there are two shake-up satellite peaks (785.2 eV and 803.3 eV), Co $2p_{1/2}$ (796.2 eV), and Co $2p_{3/2}$ (780.2 eV) in the XPS spectrum of Co. In addition, the presence of P elements in both NiP and CoP XPS confirms the formation of mono-metal phosphates.

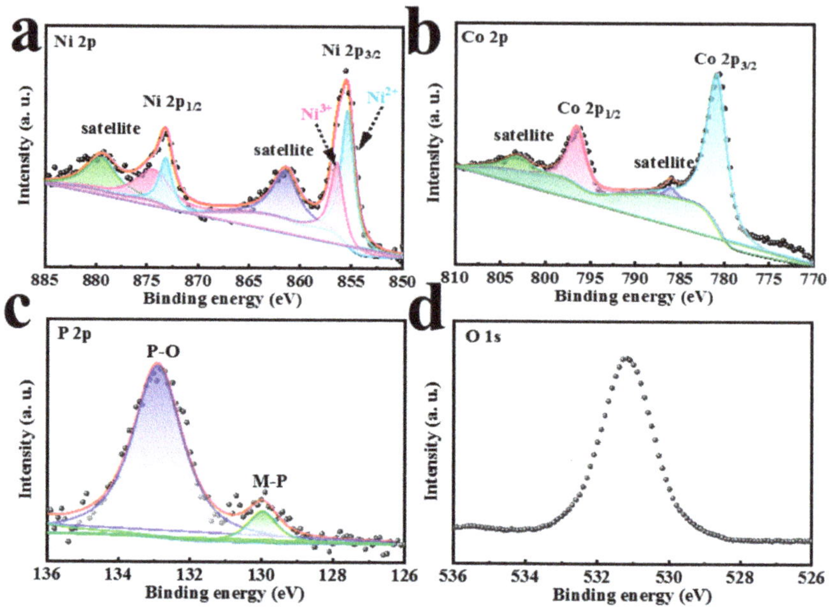

Figure 3. XPS spectra of (**a**) Ni 2p, (**b**) Co 2p, (**c**) P 2p, and (**d**) O 1 s of NiCoP.

3.2. Electrochemical Measurements

Subsequently, HER activities of mono-metal NiP and CoP and bimetallic NiCoP during 1 h of electrochemical deposition were evaluated in an acidic three-electrode system (0.5 M H_2SO_4). As shown in Figure 4a, on the LSV curve, except for PtC, NiCoP for 1 h was the optimal HER catalytic activity (as low as 110 mV vs. RHE overpotential at 10 mA cm^{-2}), superior to NiP (154 mV vs. RHE) and CoP (189 mV vs. RHE). The electrochemical deposition time also had a significant effect on the performance of the NiCoP catalyst. In order to obtain 10 mA cm^{-2} current density, an overpotential of 140 mV versus RHE needs to be applied for NiCoP for 0.5 h. For 1 h and 1.5 h, there is little difference in their catalytic performance in NiCoP. A sharp increase in current (Figure S6) during the 1 h electrodeposition showed that the precursor transforms into NiCoP. After 1 h, the current density was generally stable because the increase in the thickness of the catalyst layer affected its conductivity instead. Therefore, from the perspective of catalytic activity and time efficiency, 1 h of electrochemical deposition of NiCoP was determined for the follow-up study. Notably, in order to reveal the kinetics of the electrochemical reaction, the Tafel slope of related catalysts was determined and is presented in Figure 4b. The Tafel slopes of NiP, CoP, and NiCoP 1 h were 74.73, 80.96, and 69.49 mV dec^{-1}, respectively, which verified the lowest kinetic energy barrier as well as a fast reaction speed in NiCoP, and meant that HER complied with the Volmer–Heyrovsky mechanism. Further, the Tafel slope of NiCoP decreased with the increase in deposition time from 0.5 to 1.5 h. The NiCoP 1.5 h (51.06 mV dec^{-1}) sample even showed reaction kinetics similar to that of PtC (56.01 mV dec^{-1}). The electrochemically effective surface area (ECSA) represented the number of active sites of electrocatalysts, estimated via the cyclic voltammetry (CV) method (Figure S7). As shown in Figure 4c, the double-layer capacitance (C_{dl}) value of NiP, CoP, NiCoP 0.5 h, NiCoP 1 h, and NiCoP 1.5 h was 9.98, 1.06, 19.67, 37.10, and 33.27 mF cm^{-2}, respectively, in which the value of the NiCoP 1 h was obviously the largest and manifested more active sites. The electrochemical impedance spectra (EIS) of the catalysts (Figure 4d) in the region of high and low frequencies were then investigated to elucidate charge transfer resistance (R_{ct}). It can be seen that the value of fitting R_{ct} of NiCoP 1 h was 8.14 Ω (Table S2) based on an equivalent circuit model (Figure S8), which indicates a faster electron transfer rate. It was much lower than that of NiP (30.19 Ω), CoP (371.00 Ω), NiCoP 0.5 h (326.00 Ω), and NiCoP 1.5 h (20.58 Ω). The above results all confirmed that the bitransition metal phosphide promoted the electrochemical HER catalytic reaction compared with NiP or CoP alone. The electron transfer rate and reactive site of NiCoP can be further regulated by an appropriate electrochemical deposition time.

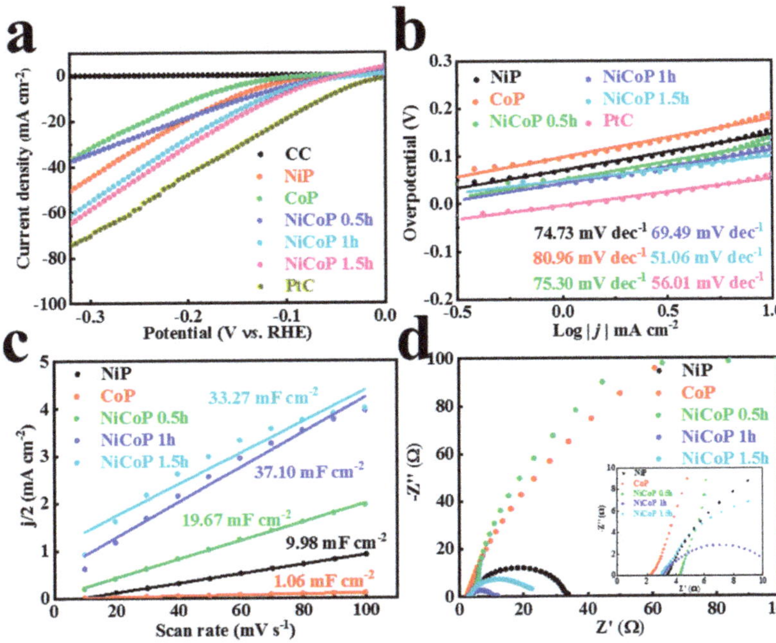

Figure 4. (**a**) LSV curves and (**b**) Tafel plots of CC, NiP, CoP, NiCoP 0.5–1.5 h, and PtC. (**c**) C_{dl} and (**d**) EIS Nyquist plot of NiP, CoP, and NiCoP 0.5–1.5 h in 0.5 M H_2SO_4 for HER; the inset presents the plot enlarged over the high frequency range.

Similarly, in order to show the universality of the catalyst in different electrolytes, we also evaluated the HER performance of NiCoP under alkaline conditions of 1 M KOH. Three phosphides, including NiP, CoP, and NiCoP, under the same 1 h deposition time were tested for HER activity (Figure 5a). Similar to acidic conditions, NiCoP showed the best performance of 120 mV overpotential, even better than commercial PtC (173 mV). The Tafel slopes (Figure 5b) for NiP, CoP, and NiCoP are 110.39, 148.12, and 118.42 mV dec^{-1}, respectively. These values suggest that the HER of these catalyst proceeds according to the Volmer–Heyrovsky reaction under alkaline conditions. Unsurprisingly, the fitting curve is roughly linear; the highest C_{dl} of 32.90 mF cm^{-2} is still displayed by NiCoP, followed by NiP (13.92 mF cm^{-2}) and CoP (0.28 mF cm^{-2}) in that order (Figures 5c and S9). According to the R_{ct} value (Figures 5d and S3) of the EIS fitting circuit, NiCoP exhibited the lowest R_{ct} of 44.55 Ω among NiP (578.50 Ω) and CoP (2419.00 Ω) monometallic phosphides. The remarkably different low frequency behavior of NiCoP may be attributable to the electrochemical double layer at the surface, whereas the physical origin may be both a single or double Schottky barrier in the substrate–catalyst heterojunction or a hydroxide interface (NiCoP surface reconstruction in an alkaline environment) material created at the boundary [32].

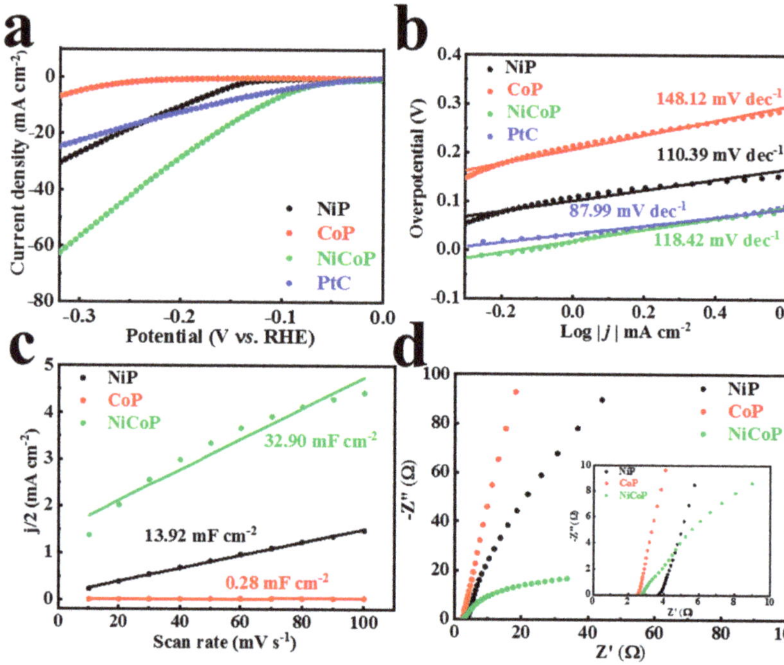

Figure 5. (**a**) LSV curves and (**b**) Tafel plots of NiP, CoP, NiCoP 1 h, and PtC. (**c**) C_{dl} and (**d**) EIS Nyquist plot of NiP, CoP, and NiCoP 1 h in 1 M KOH for HER; the inset presents the plot enlarged over the high frequency range.

In addition to excellent HER characteristics, the OER characteristics of materials also have an important impact on the final water electrolysis efficiency. Compared with the simple HER reaction mechanism, it is often more difficult to break through the reaction barrier of OER to develop an active catalyst. To illustrate the prospective of one-step electrochemical deposition of NiCoP in water electrolysis, the catalytic activity of OER at 1 M KOH was studied. NiCoP presented an outstanding OER activity. It only required a low overpotential of 276 mV to achieve a current density 10 mA cm^{-2} (Figure 6a), which was superior to those of NiP (342 mV) and CoP (341 mV). The Tafel slopes based on polarization curves are presented in Figure 6b. The Tafel slope for NiP, CoP, and NiCoP was 140.46, 132.87, and 111.41 mV dec^{-1}, respectively, which surpassed commercial IrO_2 (178.20 mV dec^{-1}) toward OER. In an alkaline solution, the fitting curve (Figures 6c and S10) is roughly linear. The C_{dl} value of NiCoP is up to 22.70 mF cm^{-2}, higher than that of NiP (13.04 mF cm^{-2}) and CoP (1.24 mF cm^{-2}). Similarly, NiCoP has the fastest electron transfer capability, with R_{ct} as low as 1.43 Ω (Figure 6d and Table S4). All electrochemical data are provided in Tables S5 and S6.

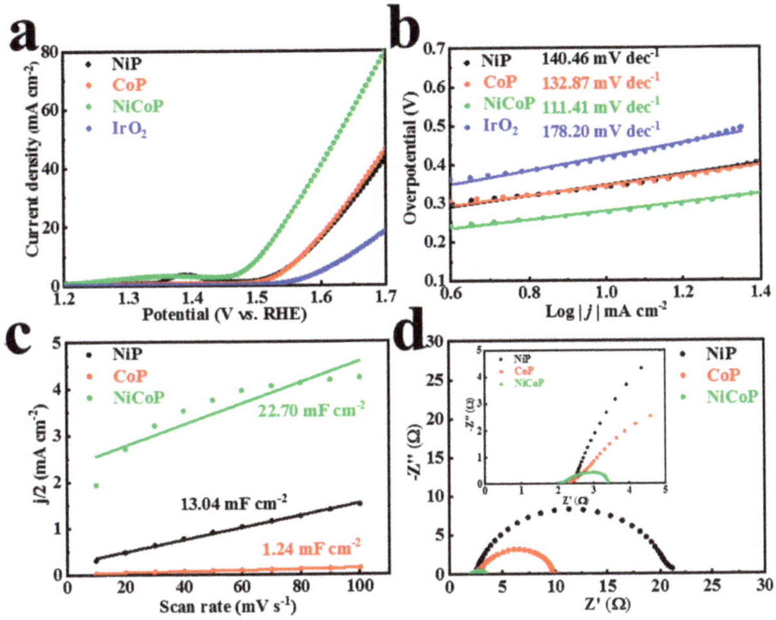

Figure 6. (**a**) LSV curves and (**b**) Tafel plots of NiP, CoP, NiCoP 1 h, and IrO_2. (**c**) The double−layer capacitance (C_{dl}) and (**d**) EIS Nyquist plot of NiP, CoP, and NiCoP 1 h in 1 M KOH for OER; the inset presents the plot enlarged over the high frequency range.

In view of the excellent HER and OER performance of NiCoP, we assembled an overall water splitting device, in which NiCoP served as both the anode and cathode in 1 M KOH. Impressively, NiCoP/NiCoP afforded current densities of 10 mA cm^{-2} when the cell voltage was 1.69 V (Figure 7), with combined overpotentials of 460 mV. Impressively, symmetrical NiCoP/NiCoP catalysts were superior to the commercial PtC/IrO_2 couple. It is comparable to the conventional transition metal phosphide reported recently (Table S7). The stability evaluation of catalysts is of great significance for the application of water splitting. As is illustrated in Figure 7a, NiCoP/NiCoP showed superior durability in the overall water splitting reaction with slight overpotential fluctuation at 36 h of continuous electrolysis. In the absence of continuous N_2 injection to eliminate the electrode bubble operation, the overpotential at 10 mA cm^{-2} current density increased by about 70 mV after a long period of electrolysis. The chronopotentiometry curve clearly indicated that commercial PtC/IrO_2 catalysts deactivated significantly after 12 h of the stability test as a result of the formation of large bubbles. In order to confirm this speculation, the commercial catalyst can also remain stable during continuous electrolysis when N_2 gas is introduced continuously to eliminate bubbles. For NiCoP, the phenomenon is less pronounced. This is mainly because of the structure of the catalyst surface and the difference in hydrophilic and hydrophobic properties. Thus, NiCoP still shows an advantage over commercial catalysts, especially in the actual water electrolysis environment. The structural evolution of the tested samples was also revealed by physical characterization in order to understand the transformation of catalyst activity. The XRD (Figure S11a) of NiCoP as cathode after HER durability still revealed no additional obvious phase formation, and the anodic NiCoP also performed almost identically after the OER cycle. According to SEM images in Figure S11b, the NiCoP of the cathode was transformed into an agglomerated nanoflowers morphology. For the anode (Figure S11c), the particles were still spherical, but larger than the original particle size of NiCoP. These results indicate that structural reconstruction of NiCoP may occur, but the performance retention rate has not completely failed during the long-term operation of water splitting. According to the XPS test (Figure 7c–f and Figure S12), we can obtain the change in valence state and structural reconstruction of NiCoP chemicals in the cathode and anode of overall water splitting during a long time period of electrolysis. After HER, there are no obvious peak position changes in the Ni and Co 2p peak of NiCoP cathode (Figure 7c,d). Three groups of Ni 2p (Ni^{2+}: 873.1 eV and 855.4 eV, Ni^{3+}: 874.4 eV and 856.4 eV, satellite: 879.7 eV and 861.2 eV) and Co 2p (Co $2p_{1/2}$: 796.5 eV, Co $2p_{3/2}$: 780.9 eV, and satellite: 803.1 eV and 788.4 eV) fitting peaks show that the transition metal element valence state of NiCoP has no change in the process of HER. It is worth noting that the XPS peak of P element (Figure S11b) after HER is almost invisible and the characteristic peak (M-OH and M-O) of O 1s (Figure S11c) located at binding energies of 532.4 eV and 531.0 eV appears. The above results indicate that the surface of NiCoP has been reconstructed to $Ni(OH)_2$ or $Co(OH)_2$ with P loss of the pristine electrode after the HER test [33]. For the anode of NiCoP, the valence state of Ni and Co has elevated compared with the original sample according to the binding energy (Figure 7e,f) of Ni(Co) 2p. Specifically, the Ni 2p distinct peak at 874.7/857.0 eV (Ni $2p_{1/2}$/Ni $2p_{3/2}$) and 881.7/862.9 eV (satellite), which corresponds to NiOOH, underwent a complete positively shift after OER [34,35]. The XPS location of Co 2p also appears to display an obvious positive shift, representing the existing of CoOOH [36,37]. Similarly, the P peak also disappears and the O 1s peaks shift positively to 532.8 eV (M-OH) and 532.0 eV (M-O) (Figure S11d,e). The above results demonstrate that NiCoP can be reconstructed into high-valence NiCo oxyhydroxides by the electrochemical OER process [26]. Thus, the evolution of surface hydroxides or oxyhydroxides can continue to undergo electrocatalytic reactions in spite of the attenuation of water splitting activity because of the absence of P element.

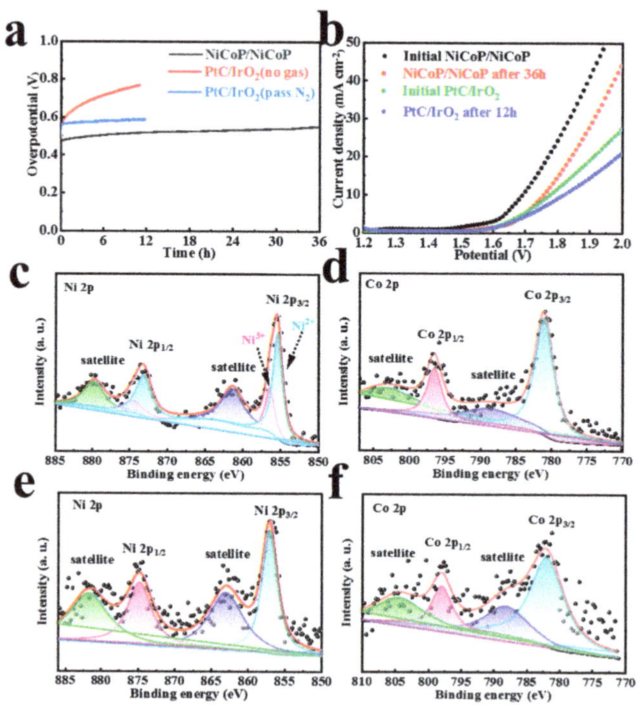

Figure 7. (**a**) Water splitting stability test of NiCoP/NiCoP and PtC/IrO_2. (**b**) LSV curves of NiCoP/NiCoP and PtC/IrO_2 before and after overall water splitting. High-resolution XPS spectra of (**c**) Ni 2p and (**d**) Co 2p after HER in 1.0 M KOH. High-resolution XPS spectra of (**e**) Ni 2p and (**f**) Co 2p after OER in 1.0 M KOH.

4. Conclusions

Highly active phosphides (NiP, CoP, and NiCoP) were obtained through a simple one-step electrodeposition method. Compared with the late high temperature phosphating treatment, a new scheme was added for the preparation of phosphide materials. The effects of electrolyte composition and deposition time on electrocatalytic performance were studied systematically. Through compositional optimization, NiCoP achieved the high catalytic activity for HER in different pH solutions, only requiring an overpotential of 111 mV in 0.5 M H_2SO_4 and 120 mV in 1 M KOH at 10 mA cm^{-2}. Under alkaline conditions, excellent OER activity was also observed, with an overpotential of 276 mV at 10 mA cm^{-2} without IR correction. The overall water splitting device composed of NiCoP/NiCoP was assembled with an initial cell voltage of 1.69 V to achieve 10 mA cm^{-2} current density in 1.0 M KOH. Subsequently, the surface valence and structural evolution of NiCoP samples with slowly decaying stability under alkaline conditions are revealed. NiCoP is reconstructed into $Ni(Co)(OH)_2$ hydroxides (for HER) and Ni(Co)OOH oxyhydroxides (for OER) on the surface with P element loss, acting as real "active sites".

Supplementary Materials: The following supporting information can be downloaded at: https://www.mdpi.com/article/10.3390/ma16041529/s1, Figure S1: SEM and EDS elemental mapping of NiP, Figure S2: SEM and EDS elemental mapping of nanoparticle CoP, Figure S3: SEM and EDS elemental mapping of nanoparticle NiCoP, Figure S4: XPS survey scan spectra of NiCoP, Figure S5: High resolution NiP XPS spectra of (a) Ni 2p and (c) P 2p. High resolution CoP XPS spectra of (b) Co 2p and (d) P 2p, Figure S6: The relationship between current and time during the NiCoP 1.5 h electrodepostion, Figure S7: CV curves of (a) NiP, (b) CoP, (c) NiCoP 0.5 h, (d) NiCoP 1 h, and (e) NiCoP 1.5 h at various scan rates in 0.5 M H_2SO_4, Figure S8: An equivalent circuit model based on electrochemical impedance spectrums, Figure S9: CV curves of (a) NiP, (b) CoP, and (c) NiCoP 1 h at various scan rates in 1 M KOH for HER, Figure S10: CV curves of (a) NiP, (b) CoP, and (c) NiCoP 1 h at various scan rates in 1 M KOH for OER, Figure S11: (a) XRD patterns of NiCoP before and after electrocatalytic reaction (HER and OER) in 1 M KOH. SEM images of NiCoP after (b) HER and (c) OER in 1 M KOH, Figure S12: (a) XPS survey spectrum of the NiCoP after water splitting. High resolution XPS spectra of (b) P 2p and (c) O 1s after HER in 1.0 M KOH. High resolution XPS spectra of (d) P 2p and (e) O 1s after OER in 1.0 M KOH, Table S1: Summary of the atomic percentages of NiP, CoP, NiCoP obtained from the EDS spectrum, Table S2: The fitted parameters of the EIS data of NiP, CoP, NiCoP 0.5 h, NiCoP 1.0 h, and NiCoP 1.5 h catalysts for HER in 0.5 M H_2SO_4, Table S3: The fitted parameters of the EIS data of NiP, CoP, NiCoP 1 h catalysts for HER in 1.0 M KOH, Table S4: The fitted parameters of the EIS data of NiP, CoP, NiCoP 1 h catalysts for OER in 1.0 M KOH, Table S5: Summary of the HER performance of all electrocatalysts, Table S6: Summary of the OER performance of all electrocatalysts, Table S7: Comparison of the catalytic activity of NiCoP with recently reported non-precious metal electrocatalysts toward the water-splitting in the 1 M KOH. References [38–48] are cited in the Supplementary Material.

Author Contributions: Methodology, M.S. and X.B.; Formal analysis, M.S.; Investigation, W.Q.; Writing—original draft, M.S.; Writing—review and editing, Y.Y. All authors have read and agreed to the published version of the manuscript.

Funding: This work was supported by the Key Science and Technology Developing Project of Shaanxi Province (2020KWZ-004 to W.Q.), the Natural Science Foundation of Shaanxi Province (No. 2021JQ-059 to Y.Y.), and Fundamental Research Funds for the Central Universities (No. xjh012020041 to Y.Y.).

Institutional Review Board Statement: Not applicable.

Informed Consent Statement: Not applicable.

Data Availability Statement: The data presented in this study are available on request from the corresponding author.

Conflicts of Interest: There are no conflict to declare.

References

1. Díaz-Sainz, G.; Fernández-Caso, K.; Lagarteira, T.; Delgado, S.; Alvarez-Guerra, M.; Mendes, A.; Irabien, A. Coupling continuous CO_2 electroreduction to formate with efficient Ni-based anodes. *J. Environ. Chem. Eng.* **2023**, *11*, 109171. [CrossRef]
2. Liu, G.; Wang, K.; Gao, X.; He, D.; Li, J. Fabrication of mesoporous $NiFe_2O_4$ nanorods as efficient oxygen evolution catalyst for water splitting. *Electrochim. Acta* **2016**, *211*, 871–878. [CrossRef]
3. Wu, J.; Ge, X.; Li, Z.; Cao, D.; Xiao, J. Highly dispersed NiCoP nanoparticles on carbon nanotubes modified nickel foam for efficient electrocatalytic hydrogen production. *Electrochim. Acta* **2017**, *252*, 101–108. [CrossRef]
4. Tang, Y.; Yang, C.; Xu, X.; Kang, Y.; Henzie, J.; Que, W.; Yamauchi, Y. MXene Nanoarchitectonics: Defect-Engineered 2D MXenes towards Enhanced Electrochemical Water Splitting. *Adv. Energy Mater.* **2022**, *12*, 2103867. [CrossRef]
5. Liu, L.; Wang, Y.; Zhao, Y.; Wang, Y.; Zhang, Z.; Wu, T.; Qin, W.; Liu, S.; Jia, B.; Wu, H.; et al. Ultrahigh Pt-Mass-Activity Hydrogen Evolution Catalyst Electrodeposited from Bulk Pt. *Adv. Funct. Mater.* **2022**, *32*, 2112207. [CrossRef]
6. Li, Y.; Gu, Q.; Johannessen, B.; Zheng, Z.; Li, C.; Luo, Y.; Zhang, Z.; Zhang, Q.; Fan, H.; Luo, W.; et al. Synergistic Pt doping and phase conversion engineering in two-dimensional MoS_2 for efficient hydrogen evolution. *Nano Energy* **2021**, *84*, 105898. [CrossRef]
7. Jian, X.; Zhang, M.-M.; Li, R.; Liu, J.-X.; Fu, F.; Liang, Z.-H. Atomically dispersed ultralow-platinum loading on $Ti_3C_2T_x$ MXene as efficient catalyst for hydrogen evolution reaction. *Electrochim. Acta* **2022**, *411*, 140091. [CrossRef]
8. Li, S.; Xie, H.; Dong, Q.; Jing, S.; Li, T.; Xu, L.; Hu, L. Synthesizing Carbon-Supported, High-Loading, Ultra-Small Pt_3Ni Nanoparticles via Tuning the Surface Electrostatic Effect. *Small Struct.* **2022**, *4*, 2200176. [CrossRef]
9. Zhang, J.Y.; Liang, J.; Mei, B.; Lan, K.; Zu, L.; Zhao, T.; Ma, Y.; Chen, Y.; Lv, Z.; Yang, Y.; et al. Synthesis of Ni/NiO@MoO_{3-x} Composite Nanoarrays for High Current Density Hydrogen Evolution Reaction. *Adv. Energy Mater.* **2022**, *12*, 2200001. [CrossRef]
10. Yu, M.; Zheng, J.; Guo, M. La-doped NiFe-LDH coupled with hierarchical vertically aligned MXene frameworks for efficient overall water splitting. *J. Energy Chem.* **2022**, *70*, 472–479. [CrossRef]
11. Gonçalves, J.M.; Matias, T.A.; Saravia, L.P.H.; Nakamura, M.; Bernardes, J.S.; Bertotti, M.; Araki, K. Synergic effects enhance the catalytic properties of alpha-$Ni(OH)_2$-FeOCPc@rGO composite for oxygen evolution reaction. *Electrochim. Acta* **2018**, *267*, 161–169. [CrossRef]
12. Sheng, M.; Bin, X.; Yang, Y.; Tang, Y.; Que, W. Defect engineering-driven phase structure design of 2H@1T MoS_2 for electrochemical hydrogen evolution reaction. *Mater. Lett.* **2021**, *311*, 131624. [CrossRef]
13. Zong, H.; Yu, K.; Zhu, Z. Heterostructure nanohybrids of Ni-doped $MoSe_2$ coupled with Ti_2NT_x toward efficient overall water splitting. *Electrochim. Acta* **2020**, *353*, 136598. [CrossRef]
14. Meng, K.; Wen, S.; Liu, L.; Jia, Z.; Wang, Y.; Shao, Z.; Qi, T. Vertically Grown MoS_2 Nanoplates on VN with an Enlarged Surface Area as an Efficient and Stable Electrocatalyst for HER. *ACS Appl. Energy Mater.* **2019**, *2*, 2854–2861. [CrossRef]
15. Sheng, M.; Bin, X.; Yang, Y.; Tang, Y.; Que, W. In Situ Electrosynthesis of MAX-Derived Electrocatalysts for Superior Hydrogen Evolution Reaction. *Small* **2022**, *18*, 2203471. [CrossRef]
16. Tang, Y.; Yang, C.; Sheng, M.; Yin, X.; Que, W. Synergistically Coupling Phosphorus-Doped Molybdenum Carbide with MXene as a Highly Efficient and Stable Electrocatalyst for Hydrogen Evolution Reaction. *ACS Sustainable Chem. Eng.* **2020**, *8*, 12990–12998. [CrossRef]
17. Li, L.; Zou, W.; Ye, Q.; Li, Q.; Feng, Q.; Wei, J.; Xu, X.; Wang, F. Quasi-parallel nickel cobalt phosphide nanosheet arrays as highly efficient electrocatalyst for hydrogen evolution and overall water splitting at large current densities. *J. Power Sources* **2021**, *516*, 230657. [CrossRef]
18. Lu, S.-S.; Zhang, L.-M.; Dong, Y.-W.; Zhang, J.-Q.; Yan, X.-T.; Sun, D.-F.; Shang, X.; Chi, J.-Q.; Chai, Y.-M.; Dong, B. Tungsten-doped Ni–Co phosphides with multiple catalytic sites as efficient electrocatalysts for overall water splitting. *J. Mater. Chem. A* **2019**, *7*, 16859–16866. [CrossRef]
19. Lin, Y.; Chen, X.; Tuo, Y.; Pan, Y.; Zhang, J. In-situ doping-induced lattice strain of NiCoP/S nanocrystals for robust wide pH hydrogen evolution electrocatalysis and supercapacitor. *J. Energy Chem.* **2022**, *70*, 27–35. [CrossRef]
20. Ye, F.; Yang, Y.; Liu, P.; Feng, Y.; Cao, Y.; Cao, D.; Ta, L.; Ma, X.; Xu, C. In-situ porous flake heterostructured NiCoP/Ni foam as electrocatalyst for hydrogen evolution reaction. *Electrochim. Acta* **2022**, *423*, 140578. [CrossRef]
21. Sayed, D.M.; El-Nagar, G.A.; Sayed, S.Y.; El-Anadouli, B.E.; El-Deab, M.S. Activation/deactivation behavior of nano-NiO_x based anodes towards the OER: Influence of temperature. *Electrochim. Acta* **2018**, *276*, 176–183. [CrossRef]
22. Zhao, J.; Zhang, J.J.; Li, Z.Y.; Bu, X.H. Recent Progress on NiFe-Based Electrocatalysts for the Oxygen Evolution Reaction. *Small* **2020**, *16*, e2003916. [CrossRef]
23. Sun, H.; Min, Y.; Yang, W.; Lian, Y.; Lin, L.; Feng, K.; Deng, Z.; Chen, M.; Zhong, J.; Xu, L.; et al. Morphological and Electronic Tuning of Ni_2P Through Iron Doping towards Highly Efficient Water Splitting. *ACS Catal.* **2019**, *9*, 8882–8892. [CrossRef]
24. Liu, T.; Li, A.; Wang, C.; Zhou, W.; Liu, S.; Guo, L. Interfacial Electron Transfer of Ni_2P–NiP_2 Polymorphs Inducing Enhanced Electrochemical Properties. *Adv. Mater.* **2018**, *30*, 1803590. [CrossRef] [PubMed]
25. Wang, X.; Chen, Y.; Yu, B.; Wang, Z.; Wang, H.; Sun, B.; Li, W.; Yang, D.; Zhang, W. Hierarchically Porous W-Doped CoP Nanoflake Arrays as Highly Efficient and Stable Electrocatalyst for pH-Universal Hydrogen Evolution. *Small* **2019**, *15*, 1902613. [CrossRef]
26. Wang, T.; Liu, X.; Yan, Z.; Teng, Y.; Li, R.; Zhang, J.; Peng, T. Facile Preparation Process of NiCoP–$NiCoSe_2$ Nano-Bilayer Films for Oxygen Evolution Reaction with High Efficiency and Long Duration. *ACS Sustain. Chem. Eng.* **2019**, *8*, 1240–1251. [CrossRef]

27. Du, C.; Yang, L.; Yang, F.; Cheng, G.; Luo, W. Nest-like NiCoP for Highly Efficient Overall Water Splitting. *ACS Catal.* **2017**, *7*, 4131–4137. [CrossRef]
28. Cao, Y.; Chen, Z.; Ye, F.; Yang, Y.; Wang, K.; Wang, Z.; Yin, L.; Xu, C. One-step synthesis of amorphous NiCoP nanoparticles by electrodeposition as highly efficient electrocatalyst for hydrogen evolution reaction in alkaline solution. *J. Alloys Compd.* **2022**, *896*, 163103. [CrossRef]
29. Jeung, Y.; Jung, H.; Kim, D.; Roh, H.; Lim, C.; Han, J.W.; Yong, K. 2D-structured V-doped Ni(Co,Fe) phosphides with enhanced charge transfer and reactive sites for highly efficient overall water splitting electrocatalysts. *J. Mater. Chem. A* **2021**, *9*, 12203–12213. [CrossRef]
30. He, P.; Yu, X.-Y.; Lou, X.W. Carbon-Incorporated Nickel–Cobalt Mixed Metal Phosphide Nanoboxes with Enhanced Electrocatalytic Activity for Oxygen Evolution. *Angew. Chem. Int. Ed.* **2017**, *56*, 3897–3900. [CrossRef]
31. Li, J.; Yan, M.; Zhou, X.; Huang, Z.-Q.; Xia, Z.; Chang, C.-R.; Ma, Y.; Qu, Y. Mechanistic Insights on Ternary $Ni_{2-x}Co_xP$ for Hydrogen Evolution and Their Hybrids with Graphene as Highly Efficient and Robust Catalysts for Overall Water Splitting. *Adv. Funct. Mater.* **2016**, *26*, 6785–6796. [CrossRef]
32. Pehlivan, İ.B.; Arvizu, M.A.; Qiu, Z.; Niklasson, G.A.; Edvinsson, T. Impedance Spectroscopy Modeling of Nickel–Molybdenum Alloys on Porous and Flat Substrates for Applications in Water Splitting. *J. Phys. Chem. C* **2019**, *123*, 23890–23897. [CrossRef]
33. Rajeshkhanna, G.; Singh, T.I.; Kim, N.H.; Lee, J.H. Remarkable Bifunctional Oxygen and Hydrogen Evolution Electrocatalytic Activities with Trace-Level Fe Doping in Ni- and Co-Layered Double Hydroxides for Overall Water-Splitting. *ACS Appl. Mater. Inter.* **2018**, *10*, 42453–42468. [CrossRef] [PubMed]
34. Li, J.; Li, J.; Zhou, X.; Xia, Z.; Gao, W.; Ma, Y.; Qu, Y. Highly Efficient and Robust Nickel Phosphides as Bifunctional Electrocatalysts for Overall Water-Splitting. *ACS Appl. Mater. Inter.* **2016**, *8*, 10826–10834. [CrossRef]
35. Jin, Y.; Huang, S.; Yue, X.; Du, H.; Shen, P.K. Mo- and Fe-Modified $Ni(OH)_2$/NiOOH Nanosheets as Highly Active and Stable Electrocatalysts for Oxygen Evolution Reaction. *ACS Catal.* **2018**, *8*, 2359–2363. [CrossRef]
36. Meng, C.; Lin, M.; Sun, X.; Chen, X.; Chen, X.; Du, X.; Zhou, Y. Laser synthesis of oxygen vacancy-modified CoOOH for highly efficient oxygen evolution. *Chem. Commun.* **2019**, *55*, 2904–2907. [CrossRef]
37. Guo, C.; Liu, J.; Quan, F.; Zhang, S.; Yu, T.; Mo, Y.; Guo, R.; Liu, X.; Liu, Y.; Mu, W.; et al. Ce-doped self-assembled ultrathin CoOOH nanosheets as efficient oxygen evolution reaction electrocatalyst. *J. Alloys Compd.* **2022**, *920*, 165898. [CrossRef]
38. Xu, K.; Cheng, H.; Lv, H.; Wang, J.; Liu, L.; Liu, S.; Wu, X.; Chu, W.; Wu, C.; Xie, Y. Controllable Surface Reorganization Engineering on Cobalt Phosphide Nanowire Arrays for Efficient Alkaline Hydrogen Evolution Reaction. *Adv. Mater.* **2018**, *30*, 1703322. [CrossRef]
39. Li, J.; Wei, G.; Zhu, Y.; Xi, Y.; Pan, X.; Ji, Y.; Zatovsky, I.V.; Han, W. Hierarchical NiCoP nanocone arrays supported on Ni foam as an efficient and stable bifunctional electrocatalyst for overall water splitting. *J. Mater. Chem. A* **2017**, *5*, 14828–14837. [CrossRef]
40. Li, Y.; Zhang, H.; Jiang, M.; Zhang, Q.; He, P.; Sun, X. 3D Self-Supported Fe-Doped Ni_2P Nanosheet Arrays as Bifunctional Catalysts for Overall Water Splitting. *Adv. Funct. Mater.* **2017**, *27*, 1702513. [CrossRef]
41. Ma, D.; Li, R.; Zheng, Z.; Jia, Z.; Meng, K.; Wang, Y.; Zhu, G.; Zhang, H.; Qi, T. NiCoP/CoP Nanoparticles Supported on Ti4O7 as the Electrocatalyst Possessing an Excellent Catalytic Performance toward the Hydrogen Evolution Reaction. *ACS Sustain. Chem. Eng.* **2018**, *6*, 14275–14282. [CrossRef]
42. Zhang, H.; Li, X.; Hähnel, A.; Naumann, V.; Lin, C.; Azimi, S.; Schweizer, S.L.; Maijenburg, A.W.; Wehrspohn, R.B. Bifunctional Heterostructure Assembly of NiFe LDH Nanosheets on NiCoP Nanowires for Highly Efficient and Stable Overall Water Splitting. *Adv. Funct. Mater.* **2018**, *28*, 1706847. [CrossRef]
43. Wang, J.G.; Hua, W.; Li, M.; Liu, H.; Shao, M.; Wei, B. Structurally Engineered Hyperbranched NiCoP Arrays with Superior Electrocatalytic Activities toward Highly Efficient Overall Water Splitting. *ACS Appl. Mater Interfaces* **2018**, *10*, 41237–41245. [CrossRef] [PubMed]
44. Li, W.; Zhang, S.; Fan, Q.; Zhang, F.; Xu, S. Hierarchically scaffolded CoP/CoP2 nanoparticles: Controllable synthesis and their application as a well-matched bifunctional electrocatalyst for overall water splitting. *Nanoscale* **2017**, *9*, 5677–5685. [CrossRef] [PubMed]
45. Tang, S.; Wang, X.; Zhang, Y.; Courte, M.; Fan, H.J.; Fichou, D. Combining Co_3S_4 and Ni:Co_3S_4 nanowires as efficient catalysts for overall water splitting: An experimental and theoretical study. *Nanoscale* **2019**, *11*, 2202–2210. [CrossRef]
46. Darband, G.B.; Aliofkhazraei, M.; Hyun, S.; Rouhaghdam, A.S.; Shanmugam, S. Electrodeposited Ni Co P hierarchical nanostructure as a cost-effective and durable electrocatalyst with superior activity for bifunctional water splitting. *J. Power Sources* **2019**, *429*, 156–167. [CrossRef]
47. Li, H.; Li, Q.; Wen, P.; Williams, T.B.; Adhikari, S.; Dun, C.; Lu, C.; Itanze, D.; Jiang, L.; Carroll, D.L.; et al. Colloidal Cobalt Phosphide Nanocrystals as Trifunctional Electrocatalysts for Overall Water Splitting Powered by a Zinc-Air Battery. *Adv. Mater.* **2018**, *30*, 1705796. [CrossRef]
48. Zhang, X.; Zhang, S.; Li, J.; Wang, E. One-step synthesis of well-structured NiS–$Ni_2P_2S_6$ nanosheets on nickel foam for efficient overall water splitting. *J. Mater. Chem. A* **2017**, *5*, 22131–22136. [CrossRef]

Disclaimer/Publisher's Note: The statements, opinions and data contained in all publications are solely those of the individual author(s) and contributor(s) and not of MDPI and/or the editor(s). MDPI and/or the editor(s) disclaim responsibility for any injury to people or property resulting from any ideas, methods, instructions or products referred to in the content.

materials

MDPI

Article

Development of High Temperature Water Sorbents Based on Zeolites, Dolomite, Lanthanum Oxide and Coke

Esther Acha *, Ion Agirre and V. Laura Barrio

Department of Chemical and Environmental Engineering, Faculty of Engineering of Bilbao, University of the Basque Country (UPV/EHU), Plaza Ingeniero Torres Quevedo 1, 48013 Bilbao, Spain
* Correspondence: esther.acha@ehu.eus; Tel.: +34-94-601-4050; Fax: +34-94-601-4179

Abstract: Methanation is gaining attention as it produces green methane from CO_2 and H_2, through Power-to-Gas technology. This process could be improved by in situ water sorption. The main difficulty for this process intensification is to find effective water sorbents at useful reaction temperatures (275–400 °C). The present work comprises the study of the water sorption capacity of different materials at 25–400 °C. The sorption capacity of the most studied solid sorbents (zeolites 3A & 4A) was compared to other materials such as dolomite, La_2O_3 and cokes. In trying to improve their stability and sorption capacity at high temperatures, all these materials were modified with alkaline-earth metals (Ba, Ca & Mg). Lanthana-Ba and dolomite sorbents were the most promising materials, reaching water sorption values of 120 and 102 $mg_{H2O}/g_{sorbent}$, respectively, even at 300 °C, i.e., values 10-times higher than the achieved ones with zeolites 3A or 4A under the same operating conditions. At these high temperatures, around 300 °C, the water sorption process was concluded to be closer to chemisorption than to physisorption.

Keywords: water sorption; high temperature sorbents; zeolites; lanthanum oxide; dolomite; cokes

Citation: Acha, E.; Agirre, I.; Barrio, V.L. Development of High Temperature Water Sorbents Based on Zeolites, Dolomite, Lanthanum Oxide and Coke. *Materials* **2023**, *16*, 2933. https://doi.org/10.3390/ma16072933

Academic Editors: Pavol Liptai, Jaroslav Briančin and Maroš Halama

Received: 3 March 2023
Revised: 4 April 2023
Accepted: 4 April 2023
Published: 6 April 2023

Copyright: © 2023 by the authors. Licensee MDPI, Basel, Switzerland. This article is an open access article distributed under the terms and conditions of the Creative Commons Attribution (CC BY) license (https://creativecommons.org/licenses/by/4.0/).

1. Introduction

Currently, the world's energy market is relying on oil. The need for energy is only rising, and forecasts show it will be increasing due to the growth of the human population and the development of countries [1,2]. Natural oil is a depleting resource and will not be able to sustain the way in which humans consume it forever. It also produces harsh by-products, like carbon dioxide emissions into the environment, which can be considered responsible for changes to the earth's climate [3]. Therefore, a transition to renewable energies is necessary to reduce the use of fossil fuels in power generation, and therefore, reduce emissions of polluting gases to combat climate change. The Paris Agreement on climate change states that, by 2030, 27% of total energy consumption must come from renewable sources and greenhouse gas emission must be reduced 40%. The European Union ratified this agreement and presented a set of measures (Clean Energy for all Europeans) in November 2016.

Renewable energy sources, such as solar and wind plants, often cannot meet the demand for electricity; conversely, however, it may happen that during the night, when the electricity demand decreases, there can be an overproduction or surplus of green electricity. Storage of electricity is a solution for this problem [4]. Electrochemical cells or batteries can offer a private or temporary solution, but the use of batteries to store large mass amounts of electricity is practically impossible with current technologies and is extremely expensive. A promising alternative storage method is the Power-to-Gas (PtG) concept. This concept relies on the use of overproduced energy to electrolytically split water into hydrogen and oxygen. The hydrogen gas can react with carbon dioxide to form the highly energetic methane or biomethane gas. This biomethane can be stored in methane tanks or even be injected to the natural gas grid due to its similar properties to the natural gas [5,6]. Some authors studied the feasibility of different configurations (i.e., combination of fermentation or gasification with methanation, combination of flue gas separation with ammonia synthesis and methanation) concluding that the price of biomethane could be competitive with the development of methanation technology [7,8]. However, the existing natural gas grids have strict limitations on the amount of hydrogen gas allowed into them [9]. This methanation process, also known as Sabatier's reaction, is a highly exothermic reaction and water molecules are produced as byproduct [10].

$$CO_2 + 4H_2 \leftrightarrow CH_4 + 2H_2O \quad \Delta H^o = -165 \text{ kJ/mol} \tag{1}$$

This overall reaction is the sum of the "reverse water shift reaction" and the methanation of CO, which implies the production of CO as an intermediate product [11,12].

$$CO_2 + H_2 \leftrightarrow CO + H_2O \quad \Delta H^o = 41 \text{ kJ/mol} \tag{2}$$

$$CO + 3H_2 \leftrightarrow CH_4 + H_2O \quad \Delta H^o = -206 \text{ kJ/mol} \tag{3}$$

Sabatier reaction involves kinetic limitations, and therefore, the use of a catalyst is necessary. Many metals (Ni, Co, Fe, Ru, Rh, Ir, Pd, Pt, Zr, Ce) and supports (TiO_2, SiO_2, Al_2O_3, etc.) have been tested as catalysts [6,13–16], concluding that Ni, Ru, Rh are the best active metals to catalyze the Sabatier's reaction [14]. Among these metals, Ni is the most used one due to its low price. High reaction rates and high conversions—around 90% and close to chemical equilibrium—can also be achieved with commercial catalyst [17], but in most of cases, under temperatures above 250–275 °C.

Borgschulte et al. [17] and Walspurger et al. [18] were one of the first to report water sorption enhanced methanation results in order overcome the thermodynamic limitations of this reaction. More concretely, these authors used an adsorbent to adsorb the byproduct water inside the catalytic bed to shift the reaction to the products side, and therefore, increase the methane yield. Borgschulte et al. [17] used zeolite 5A as adsorbent while Walspurger et al. [18] used zeolite 4A. Both works showed how the equilibrium conversions were overcome at the studied temperature due to sorption (100% of conversion in case of Walspurger et al. [18]). However, in both studies, the adsorbent was saturated in the very first minutes of the reaction, which is an indication of low water sorption capacities of used adsorbents in the reaction operation conditions. Faria et al. [19] also demonstrated the thermodynamic effect of the in situ water removal. More recently, a study has been published on the optimization and understanding of the operating conditions of a sorption enhanced methanation process, with special emphasis on the thermal profile inside the reactor [20]; Zeolite 4A, CaO and La_2O_3 were the studied sorbents. As well as in the aforementioned studies, the water sorption capacity was very limited at methanation conditions (300 °C, approximately), except pure CaO; howeverm this material presented plugging problems inside the reactor. Moreover, it was observed a CO_2 sorption or carbonation phenomena in the very first minutes of the process. More recently, some other authors have published similar works combining different catalysts and adsorbents. Gómez et al. [21] studied the use of zeolite 5A as adsorbent, Coppola et al. [22] employed zeolite 3A and granular CaO, which reacts with water to form $Ca(OH)_2$, and Pieterse et al. [23] tested the performance of zeolites 4A and 13X. In all the cases, the saturation was achieved after just a few minutes (5–8 min).

It is clear that further study is required in order to find suitable materials with a higher water sorption capacity in order to adsorb water in such conditions—but over a longer time. According to the different studies published in the open literature, most of them are based on non-modified reference materials. The present paper compares and studies the water sorption capacity of a selection of materials, based on the previously described literature or other well-known sorption industrial applications, as non-modified raw materials as well as the modification of these ones. For this purpose, non-modified zeolite 4A, zeolite 3A, zeolite ZSM5, lanthanum oxide, dolomite and carbonaceous materials were employed [24,25];

afterwards, these sorbents were modified by different alkaline-earth metals (Ba, Ca, and Mg), which are known to affect the sorption capacity of water [26–29]. This study has been carried out in a systematic way with the objective of finding the most suitable sorbent with the largest sorption capacity as possible, to be employed in the future in methanation reaction at high temperatures. All these materials have been deeply characterized. The CO_2 competitive sorption or material carbonatation phenomena has been left for future studies.

2. Material and Methods

2.1. Preparation of the Adsorbents

Different sorbents were tested for water sorption: commercial zeolites (3A -Fluka-, 4A -Sigma Aldrich- and ZSM5 -Zeolyst CBV 5524G-), lanthanum oxide (Scharlau > 98%), dolomite (Franefoss Miljøkalk A/S, Norway, 98.5 wt% $CaMg(CO_3)_2$) and carbonaceous materials (Norit Activated carbon from Acros, and self-produced charcoal by slow pyrolysis of pine-wood at 900 °C). Some of these materials were modified with 10 wt% of Ba, Ca and Mg and their sorption capacity was also tested. For this purpose, the metal precursors employed were $Ba(NO_3)_2$ (Scharlau, >99%), $CaCl_2$ (Panreac, 95%), $Ca(NO_3)_2{\cdot}4H_2O$ (Scharlau, >99%) and $Mg(NO_3)_2{\cdot}6H_2O$ (Scharlau, >99%). These modified sorbents were prepared by wet impregnation in a rotary evaporator (Heidolph Laborota 4000). An adequate amount of material and metal precursor were mixed with 10 mL of distilled water per gram of material, which allowed for the obtaining of desired composition of the modified adsorbent. The suspension was kept in stirring during the night and the excess water was evaporated at 45 °C under vacuum. The obtained solid was dried in an oven at 110 °C for 12 h. Finally, it was calcined in air atmosphere at 700 °C in a muffle furnace, using 3 °C/min heating rate, and maintained at this temperature for 4 h. With this calcination step the metal oxide, which is the hygroscopic phase, was obtained [30]. Before sorption analysis and characterization, the samples were pressed and sieved to obtain a particle size between 0.42 and 0.50 mm. This particle size was selected because it is the one to be used in future work in the fixed-bed reactor when testing methanation reaction with the most promising water sorbents, as different particle diameters may offer different internal diffusion limitations.

2.2. Measurement of Sorption Capacity of the Sorbents

The water sorption capacity of the prepared sorbents was analyzed by thermogravimetric analysis (TGA) (see scheme in Figure 1). Around 100 milligrams of sample were placed in the TGA crucible. Samples were firstly heated up to 400 °C under 50 NmL/min of nitrogen to desorb any moisture, and then temperature was decreased to 25 °C. After this initial pre-treatment, the nitrogen flow was redirected, before entering to the furnace, through three gas bubblers filled with water, generating a moisture-saturated N_2 flow at 25 °C (1.56 NmL/min). This flow was passed over the sample until it was saturated with water. Then, the temperature was gradually increased to 100, 200, 300 and 400 °C to determine the water sorption capacity at each temperature. This analysis makes it possible to compare the water sorption capacity of all the materials under the same conditions. Each temperature was kept constant during a 2 h period. The weight gain with respect to the initial dry sample is employed to calculate the sorption capacity of the tested materials.

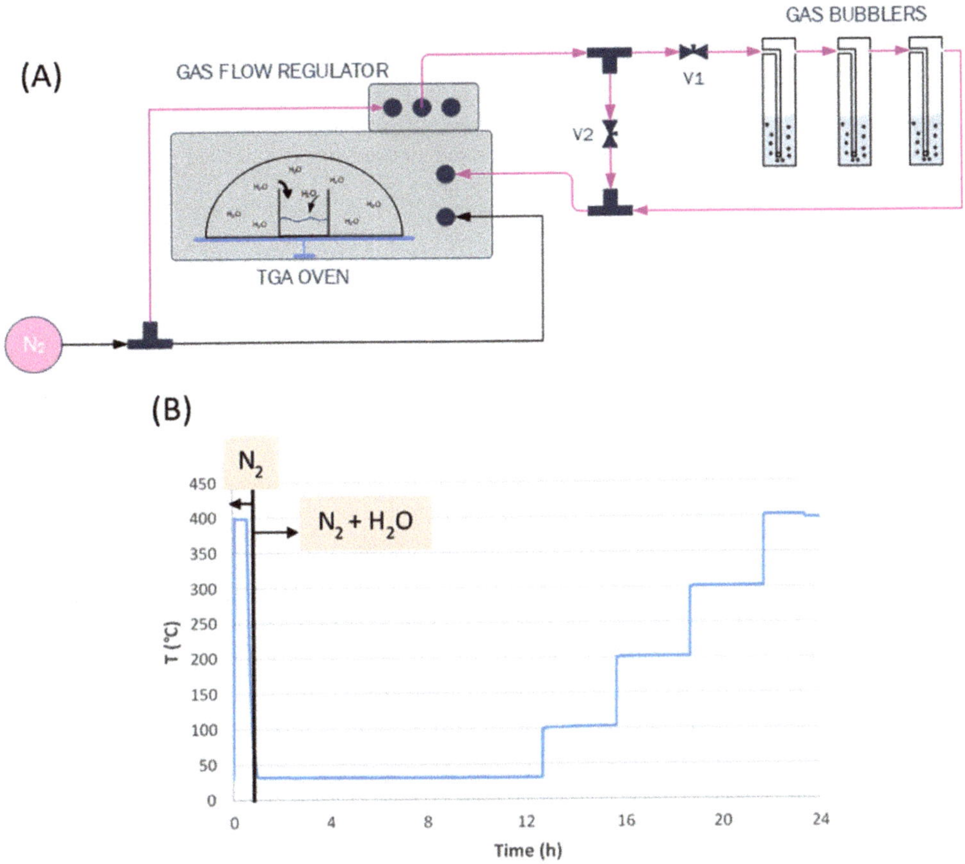

Figure 1. (**A**) Scheme of the experimental design employed for sorption measurements. (**B**) Temperature program employed during the water sorption measurements.

2.3. Characterization of the Sorbents

The physicochemical properties of the prepared sorbents were determined by various characterization techniques, which are listed and described below. The characterization of the samples was performed after being calcined at 700 °C during a 4 h period.

N_2 adsorption-desorption isotherms were performed at 77 K in Autosorb 1C-TCD equipment that allows the determination of textural properties (pore volume, BET surface area and average pore diameter). Prior to sample analysis, the samples were degassed at 300 °C for 12 h. The Brunauer-Emmett-Teller (BET) method was used to calculate the surface area. Pore size distribution was calculated using the Barrett-Joyner-Halenda (BJH) method.

Temperature-programmed desorption (TPD) of NH_3 was used to determine the acidic characteristics of the samples. The equipment used was a Micromeritics® AutoChem II instrument (Micromeritics Instrument Corporation, Norcross, USA). Prior to analysis the samples were heated to 700 °C in a 5% H_2–Ar mixture, flushed with He for 30 min and then cooled to 100 °C. Afterwards, NH_3 loading was kept for 30 min. The physically adsorbed NH_3 was then removed by flushing He at 150 °C until no further desorption was recorded. Finally, the release of chemisorbed NH_3 was recorded increasing the temperature from 150 °C to 895 °C at 10 °C/min. The total moles of adsorbed ammonia were calculated with the area below the curve and a straight baseline over the temperature range 150–895 °C.

CO chemisorption was performed to analyze the reactivity or the sorbents. It was determined by CO-pulse chemisorption in a Micromeritics® AutoChem II instrument. Samples were pre-treated in 5% H_2–Ar mixture at 700 °C (to simulate the pre-treatment that will be done in future works with the solids when operating in sorption enhanced Sabatier reaction) and CO chemisorption capacity was measured at 35 °C in a 5% CO–He mixture until the observed peaks were identical.

Temperature-programmed reduction (TPR) of H_2 was employed for analyzing the reactivity of the sorbents with hydrogen. It was measured in the Micromeritics® Autochem II. A continuous flow of 5% H_2–Ar was passed over around 0.1 g of sieved sample and the temperature was increased from 25 °C to 900 °C.

X-ray diffraction of the prepared sorbents was performed to identify the crystalline species using a Seifert XRD 3000P diffractometer, equipped with a PWBragge-Brentano q/2q 2200 goniometer, bent graphite monochromator and automatic slit with Cu Kα radiation.

3. Results and Discussion

3.1. TGA

Initially, the water sorption capacity of the pure calcined materials at different temperatures was tested. Figures 2 and 3 show the results of all the non-modified materials. The results are given as milligrams of adsorbed water per gram of dry sorbent. The time in which the temperature was increased (from 25 to 100–200–300–400 °C) is indicated in each figure with vertical lines. It can be observed that each material takes a different time to be saturated in the beginning of the analysis, at 25 °C (to reach weight stabilization). Therefore, the total time elapsed in these tests varies depending on the material tested. As expected, the maximum water sorption capacity was observed at 25 °C, at the lowest temperature. Afterwards, following each temperature increase, the sample weight kept stable or decreased due to desorption of part of the previously adsorbed water.

When comparing the pure sorbents, different sorption capacities and kinetics can be observed. Zeolites are highly hydrophilic sorbents due to their electrostatic charged framework and the abundance of extra-framework cations [31]. Among the tested zeolites, the ZSM5 is the one showing the lowest sorption capacity at 25 °C (Figure 2), around 80 $mg_{H2O}/g_{sorbent}$, but the sorption kinetics are similar to that of zeolite 4A, getting the maximum sorption capacity in around some minutes. Zeolite 3A and zeolite 4A reached similar water sorption capacity at 25 °C of around 220–240 $mg_{H2O}/g_{sorbent}$. Zeolite 3A showed lower sorption rate, probably due to the smaller pore size, which could lead to limitations in mass transport. When temperature was increased to 100 °C the sorption capacity strongly decreased in all the zeolites and it continued decreasing at 200 °C, getting almost negligible values at the desired temperatures of around 300 °C. The Si/Al ratio of zeolites plays an important role in their sorption capacity, as the aluminum-rich zeolites are the ones showing higher sorption capacities [31]. Hence, zeolite 4A and 3A show a Si/Al ratio equal to 1 [32], while the used ZSM5 zeolite presents a Si/Al ratio of 25. This is in accordance with the experimental sorption capacity observed.

Figure 3 shows two different types of behavior when comparing the water sorption capacity of biocoke, commercial activated carbon Norit, lanthana and dolomite. Activated carbon Norit showed higher water sorption capacity at 25 °C than the biocoke, but still very low. In both cases, the sorption capacity almost disappeared when temperature increased to 100 °C. On the contrary, both lanthana and dolomite showed an intermediate sorption capacity at 25 °C (145 $mg_{H2O}/g_{sorbent}$ vs. 220–240 $mg_{H2O}/g_{sorbent}$ obtained with zeolites 4A & 3A). It must be mentioned that although lanthana and dolomite were the ones showing the slowest sorption rates at 25 °C, they showed greater binding energy with water, since it remained adsorbed at higher temperatures. In the case of lanthana, its sorption capacity was maintained at 145 $mg_{H2O}/g_{sorbent}$ even at 200 °C, although as the temperature continued to increase the sorption capacity dropped to 45 $mg_{H2O}/g_{sorbent}$. Dolomite sorption capacity at 25, 100 and 200 °C remained below the values showed by the lanthana, but at 300 °C it was much higher (125 $mg_{H2O}/g_{sorbent}$). The Ca and Mg composition of dolomite proved very adequate for water sorption at high temperatures, being the only pure material, among those analyzed, that had a remarkable sorption capacity at this temperature. Dolomite structure may be affected by temperature and atmosphere changing from magnesian calcite to periclase (MgO) and lime (CaO), as observed in the literature [33,34] and the XRD results explained in Section 3.2.5. This variable structure may allow the creation of very stable hydrated salts at certain temperature ranges, with elevated dehydration temperature.

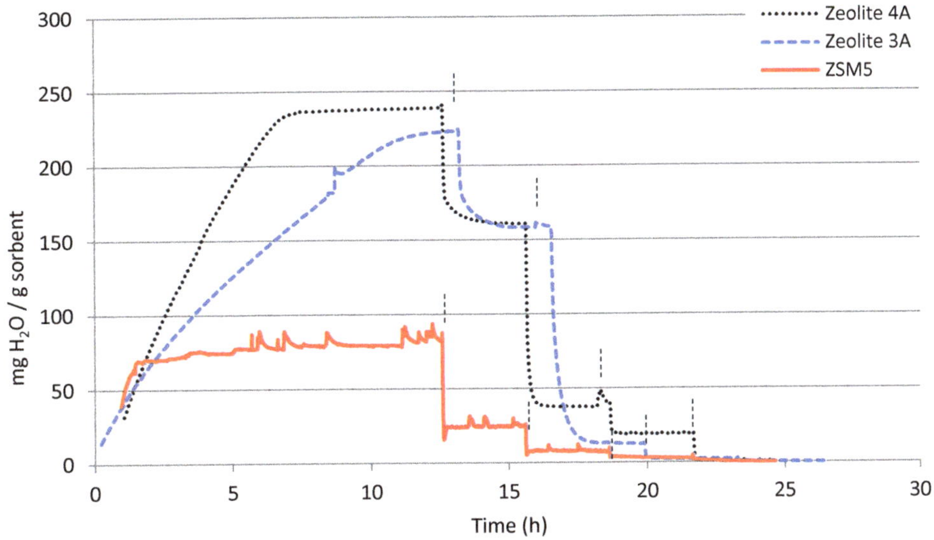

Figure 2. Water sorption capacity of calcined zeolite 4A, zeolite 3A and ZSM5. The vertical lines in each graph indicate the temperature change: 25 °C, 100 °C, 200 °C, 300 °C and 400 °C.

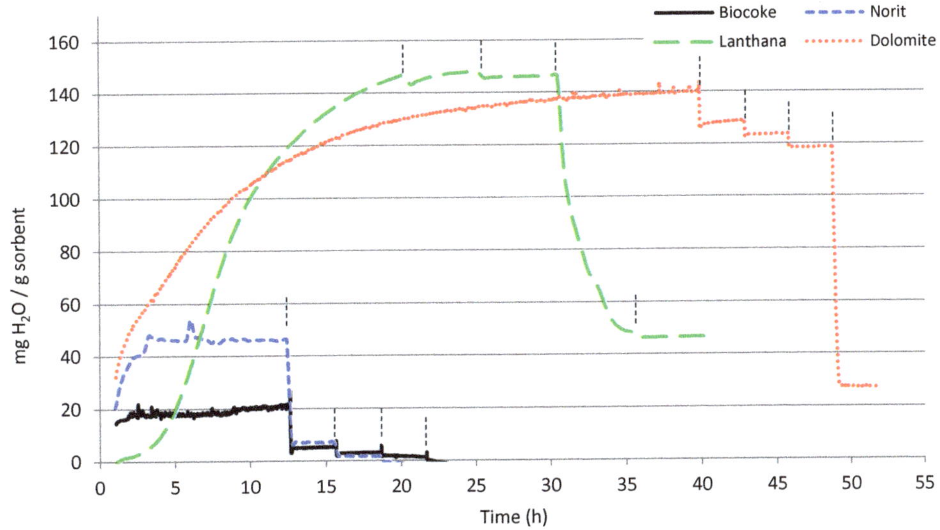

Figure 3. Water sorption capacity of the calcined lanthana, dolimite, biocoke and commercial activated carbon Norit. The vertical lines in each graph indicate the temperature change: 25 °C, 100 °C, 200 °C, 300 °C and 400 °C.

In order to improve the sorption capacity at elevated temperatures, some of the materials were modified with different alkaline earth metals. These modifications could have an effect on the sorption kinetics as well as on the capability to adsorb water at different temperatures. As explained above, the materials analyzed in this work have the ultimate purpose of using them as adsorbents in a continuous fixed bed reactor, where the methanation reaction takes place at around 300 °C. For this reason, not only the sorption capacity at such high temperatures is a critical parameter, but also the sorption rate. It must be taken into account that the contact time among the water molecules and the adsorbent will be very short when the reaction and sorption will take place in the reactor.

In Figure 4, the modifications done to the cokes are shown and in Figure 5 the modifications done to lanthana and zeolite 4A. Table 1: a summary of the maximum water capacity obtained for those pure and modified sorbents is given, with the purpose of favoring the comparison and discussion of the results. Modification with Ba and Mg seemed to even decrease the water sorption capacity of the commercial activated carbon Norit (Figure 4), which may be an indication of active surface coverage. However, the modification with Ca did enormously improve the sorption capacity of the biocoke at 25 °C, increasing from 19 to 476 $mg_{H2O}/g_{sorbent}$. This sorption capacity increase was still slightly noticeable at 100 °C, but at higher temperatures the capacity turned again negligible.

Zeolite 4A was modified with Ca using two different promoters: $CaCl_2$ and $Ca(NO_3)_2$. The modification with $CaCl_2$ showed much greater sorption capacity at 25 °C than when it was modified with $Ca(NO_3)_2$, although this difference was smaller at 100 °C and almost insignificant at higher temperatures (Figure 5). In any case, as shown in Table 1, the sorption capacity of the modified zeolite 4A was in all cases less than the zeolite 4A. It seems that the deposited metals partially covered the active surface of zeolite. By modifying the lanthanum with Mg, the sorption kinetics accelerated at 25 °C, while it slowed down when modified with Ba (Figure 5). The comparison of the sorption capacities of these two materials showed that the addition of Ba allowed obtaining higher sorption capacities even at 300 °C. In the case of lanthana (Table 1), the sorption capacities were greater at 25, 100 and 200 °C, but decreased rapidly at 300 °C. However, after modification with Ba, the sorption capacity at temperatures up to 300 °C was fairly stable.

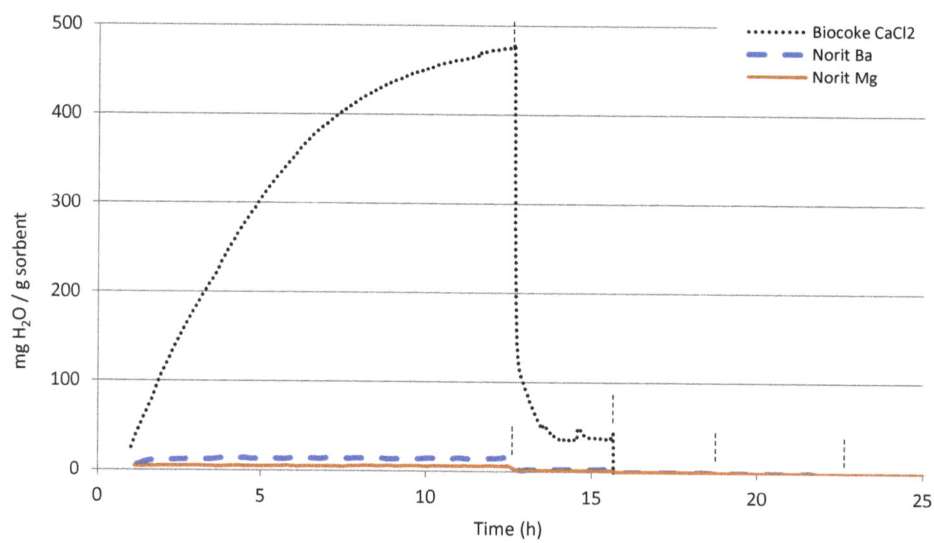

Figure 4. Water sorption capacity of the calcined and modified biocoke and commercial activated carbon Norit. The vertical lines in each graph indicate the temperature change: 25 °C, 100 °C, 200 °C, 300 °C and 400 °C.

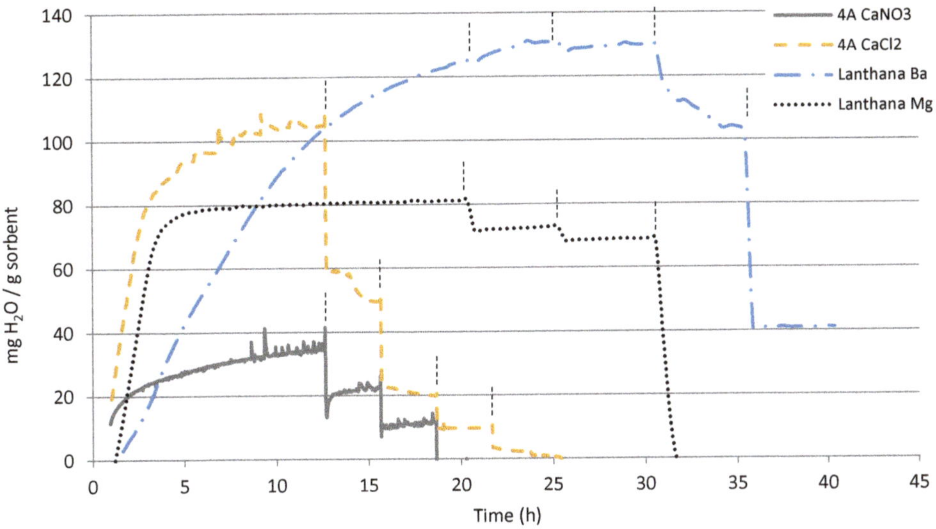

Figure 5. Water sorption capacity of the calcined and modified zeolite 4A and lanthana. The vertical lines in each graph indicate the temperature change: 25 °C, 100 °C, 200 °C, 300 °C and 400 °C.

Table 1. Maximum water sorption capacity (mg of adsorbed water per sorbent g) reached at different temperatures for all the tested calcined sorbents.

mg H_2O/g Sorbent	25 °C	100 °C	200 °C	300 °C	400 °C
Biocoke	19	6	4	<1	<1
Biocoke Ca ($CaCl_2$)	473	37	<1	<1	<1
Norit	46	7	8	<1	<1
Norit Mg	6	2	<1	<1	<1
Norit Ba	15	3	<1	<1	<1
ZSM5	80	24	8	3	<1
Zeolite 3A	222	160	12	3	<1
Zeolite 4A	163	83	26	11	<1
Zeolite 4A Ca ($CaCl_2$)	105	44	20	10	2
Zeolite 4A Ca ($Ca(NO_3)_2$)	46	33	22	7	<1
Lanthana	147	148	146	48	47
Lanthana Mg	81	73	69	<1	<1
Lanthana Ba	125	131	130	102	41
Dolomite	140	128	124	118	27

3.2. *Characterization*

3.2.1. Textural Properties

Taking into account that sorption processes are surface processes, the specific surface area and morphology of the samples are critical parameters of the sorbents. In Table 2 the multipoint BET surface area, BJH pore volume and BJH average pore size diameter are given for most of the tested calcined materials calculated with N_2 adsorption–desorption isotherms. They are all mesoporous materials, with particle size in the range of 2–50 nm. Comparing the cokes, it is especially remarkable the difference in surface area. The surface morphology of biocokes obtained by pyrolysis of biomass is very dependent of the temperature and procedure employed for its generation. The biocoke tested as sorbent was generated from wood biomass at 900 °C, yielding very high surface areas [35]. When impregnating Mg over Norit the surface area increased slightly, pore volume almost doubled and pore diameter also increased. As explained in the experimental section, after the calcination in air at 700 °C Mg atoms are expected to be as MgO. The MgO could be deposited generating porous aggregates in the surface. The surface area of MgO depends on the compound employed for its production and on the preparation method, but BET surface areas of even 400 m^2/g are reported in the literature [36]. The increase of the average pore size of Norit Mg can be due to the small pores being covered by Mg deposition, in addition to the pores of the MgO itself, which may be bigger. These bigger pores may be the reason for the bigger increase of volume than area after impregnation. In order to discuss in detail the surface change due to the Mg deposition, in Figure 6 the surface area distribution versus pore diameter of biocoke, Norit and Norit Mg is depicted. In this graph, it is easily observable that the pure cokes had very different total areas, but with very similar pore diameter range and very homogeneous pores. However, after Mg deposition, the smallest pores disappear in Norit and the window of available pore diameters was wider.

Regarding zeolites, it can be observed in Table 2 that the textural properties of ZSM5 zeolite and zeolites 4A and 3A were quite different. On the one hand, the ZSM5 had a greater surface area and pore volume compared to zeolites 3A and 4A. This fact could imply greater water sorption, but on the other hand, the smaller average pore size diameter that it showed (see Figure 7) could lead to greater difficulties in water accessing the smallest pores [31]. Additionally, the less favorable above-mentioned Al/Si ratio resulted in a much lower water sorption capacity as measured by TGA (see Figure 2). The high surface area may had improved the mass transport of water, which was noticeable in the quick saturation of the sorbent at 25 °C (see Figure 2). Comparing the surface area distribution of these materials, given in Figure 7, it can be observed a homogeneous pore size distribution in the case of pure zeolites, while the Ca modified zeolite presented a wider range of pore diameter. It seems that the Ca deposition modified the zeolite surface partially blocking the pores and generating new porosity starting in values of about the size of pure CaO [37,38].

Table 2. Surface characterization of pure and modified calcined sorbents by N_2 adsorption–desorption isotherms: BET surface area, BJH cumulative desorption pore volume and BJH desorption pore diameter.

	BET Surface Area (m^2/g)	**BJH Pore Volume (cm^3/g)**	**BJH Pore Size Diameter (Å)**
Biocoke	624	0.22	38
Norit	21	0.02	38
Norit Mg	22	0.05	43
ZSM5	251	0.08	19
Zeolite 3A	3	0.01	43
Zeolite 4A	7	0.03	31
Zeolite 4A Ca ($CaCl_2$)	7	0.02	43
Lanthana	14	0.05	39
Lanthana Mg	15	0.04	34
Lanthana Ba	15	0.04	38
Dolomite	39	0.14	35

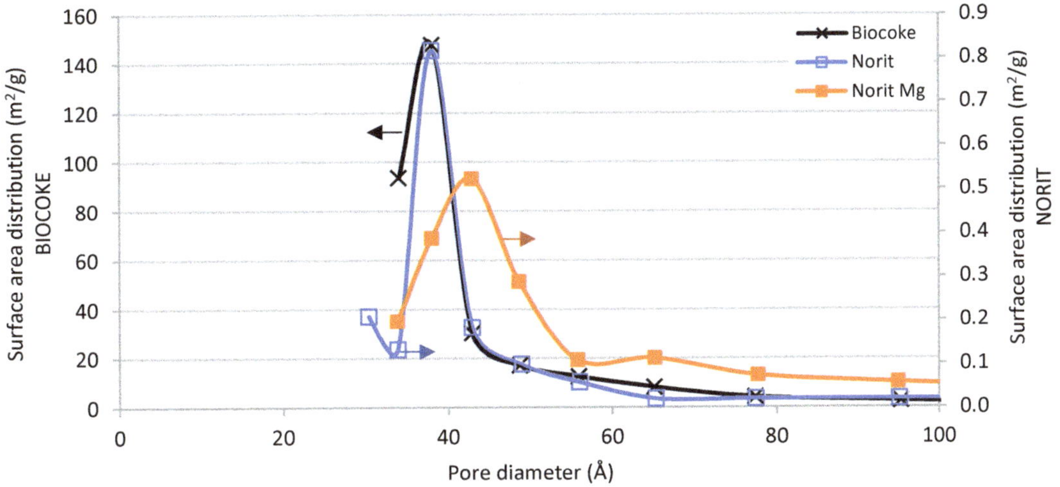

Figure 6. Comparison of the surface area distribution for biocoke (left axis), commercial activated carbon Norit (right axis) and Mg modified Norit (right axis). Obtained from the N_2 adsorption–desorption isotherm data. (BJH method desorption surface area).

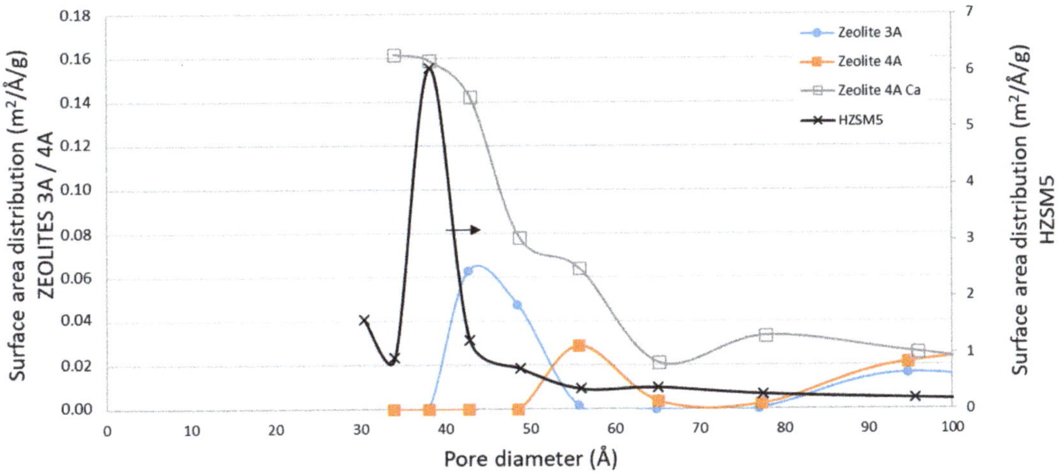

Figure 7. Comparison of the surface area distribution for ZSM5 (right axis), zeolite 3A (left axis), zeolite 4A (left axis) and zeolite 4A Ca ($CaCl_2$) (left axis). Obtained from the N_2 adsorption–desorption isotherm data. (BJH method desorption surface area).

When comparing the surface morphology of lanthana and dolomite, higher surface area in the case of dolomite can be observed (in Table 2). Taking into account the TGA results from Figure 2. it can be concluded that, even though the surface area of dolomite was higher than that of lanthana, the sorption kinetics were slower at 25 °C and it took longer to get a saturated surface, probably due to smaller pore diameters. The calculated BET surface area of the lanthana, lanthana Mg and lanthana Ba sorbents was very similar (Table 2), but when comparing the area calculated with BJH method differences were observed. The BJH surface area was 29.5 m^2/g for lanthana, 11.6 m^2/g for lanthana Mg and 11.8 m^2/g for lanthana Ba. In Supplementary Material, the cumulative BJH surface area versus pore size distribution of these three materials is given (Figures S1–S3).

It is important to remark that gas adsorption is a tool for characterization and comparison of the texture of porous solids, and BET and BJH areas are different methods for pores' analysis. BJH method can be employed for comparison of mesoporous materials with medium/big mesopores [39]. The comparison of BJH areas among the lanthana-modified sorbents plotted in Figure 8 shows that the addition of Mg and Ba partially blocked the pores in the range of 30–70 Å in a similar way, also decreasing the total BJH area and volume. When comparing the metal addition on lanthana with the observation in Norit (Figure 6), it is remarkable that in this case the BaO and MgO did not seem to create porous structures when impregnated over lanthana, which may be due to better dispersion of smaller deposited particles. The water sorption capacity of the modified lanthana was smaller at 25, 100 and 200 °C, maybe affected by the aforementioned decrease of the surface area associated to the smallest pores. Nevertheless, at 300 °C, the water sorption capacity of the Ba modified lanthana was still high (102 $g_{H2O}/g_{sorbent}$) and reaching a similar value than that of the dolomite (119 $g_{H2O}/g_{sorbent}$), indicating that the Ba addition to lanthana had stabilized the adsorbed water molecules.

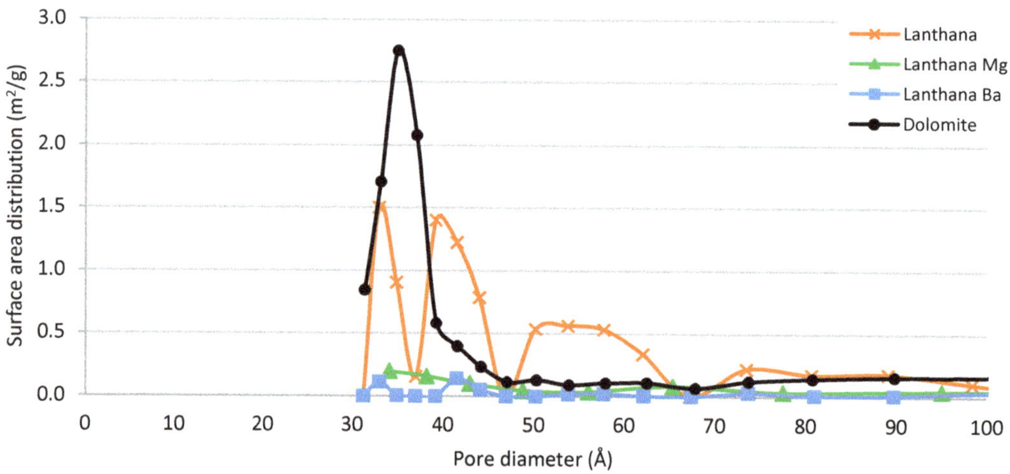

Figure 8. Comparison of the surface area distribution for dolomite, lanthana, lanthana Mg and lanthana Ba. Obtained from the N_2 adsorption–desorption isotherm data. (BJH method desorption surface area).

3.2.2. Temperature-Programmed Desorption of Ammonia, NH_3-TPD

The acidity of the calcined materials was determined to analyze its possible effect on their water sorption capacity. Ammonia molecule, due to its basic character, can easily be adsorbed at the acid sites of different strengths on the surface of the material. The total acidity can be correlated with the total moles of ammonia adsorbed, given in Table 3. The complete NH_3-TPD profile of the characterized sorbents is depicted in Supplementary Material (Figures S4–S6).

Acidity is expected to have a positive influence in water sorption capacity of the sorbents. Some authors state that Lewis acid sites are responsible for water sorption at low partial pressure/vapor pressure ratios (P/P_0), while the strong Brönsted sites are the responsible ones at high P/P_0 ratios [31]. Comparing the total acidity values, it can be observed that the most acidic unmodified materials are the zeolites. The total acidity of the unmodified materials was not directly correlated with their sorption capacity at either 25 °C or higher temperatures. The inherent Lewis acidity of cations is increased with the charge and decreased with the size of the cation. The charge of Ca, Ba and Mg is the same, 2+; however, the size is $r_{Mg} < r_{Ca} < r_{Ba}$. Therefore, the Lewis acidity may be expected to be the highest for the Mg modified sorbents. By NH_3–TPD measurement, it is only possible to calculate the total acidity of the sample and to distinguish the acid site strength, but not the type of acidic site.

Table 3. Acidity of the calcined sorbents, measured by NH_3–TPD.

	Total Acidity mmol $NH_3/g_{sorbent}$
Biocoke	0.214
Biocoke Ca ($CaCl_2$)	0.059
Norit	0.162
Norit Mg	0.191
Norit Ba	0.198
ZSM5	0.595
Zeolite 3A	0.494
Zeolite 4A	0.644
Zeolite 4A Ca ($CaCl_2$)	0.197
Lanthana	0.232
Lanthana Mg	0.346
Lanthana Ba	0.713
Dolomite	0.336

Cokes did not show any significant acidity peak, even though when they were modified, the calculated total acidity was distributed all along the temperature range tested. In the case of zeolites, the unmodified samples presented weak acidity in the range of 150–400 °C, and only in the case of zeolite 3A was a small peak detected at high temperature, also indicating strong acid sites. When modifying the zeolite 4A with Ca, the acidity peak disappeared and the total acidity decreased, probably due to surface coverage. This was also concluded from the measurement of textural properties (decrease of the surface area and the pore volume, Table 2). The addition of Ca also implied a strong acidity decrease in the biocoke. However, when comparing the water sorption capacity in Ca-modified adsorbents two very different trends were observed. On the one hand, the water sorption capacity decreased in the case of zeolite 4A Ca, and on the other hand, it greatly increased in case of the biocoke Ca sample (Table 1).

Pure lanthana presented almost no acidity, as it is expected from its basicity nature [40]. Dopants can affect the acidic/basic properties, which was observed by acidity increase after addition of Mg and Ba. When analyzing the water sorption capacity, the increase showed by lanthana at high temperatures—due to the addition of Ba—is remarkable. In the case of modified lanthana sorbents, the acidity increase with Ba may indicate an increase of Brönsted acid sites, as the Lewis acidity is expected to be the lowest as explained previously. In Supplementary Material (Figure S6) it can be observed that after the Ba addition a broad peak in the range of 250–550 °C appeared, which could be considered medium strength acidity. In the case of dolomite, it showed higher total acidity than pure lanthana, and more homogeneously distributed along all the temperatures.

3.2.3. CO Chemisorption

CO can be employed as adsorbate specie to try to compare the surface reactivity of different materials. The prepared calcined materials showed low adsorption capacity of CO, as shown in Table 4. The chemisorbed CO moles decreased in the modified biocoke, Norit and zeolite 4A, while it increased in the modified lanthana samples. This difference in behavior is probably because lanthana was very inert to CO, so when the metal was added its reactivity increased. Several authors have previously studied the reactivity of alkaline earth metals. Bajdich et al. [41] studied the CO adsorption energy on alkaline-earth metal oxides and they determined that the CO adsorption energy was weaker when moving from MgO to BaO (MgO > CaO > SrO > BaO). The decrease of CO chemisorption in the order Mg > Ca > Ba was also identified in Pt modified zeolites [42]. This effect was attributed to a larger electron charge transfer to platinum clusters, as the metal dispersion measured by hydrogen chemisorption in the final material did not change.

Table 4. CO cumulative quantity per gram of calcined sorbents, measured by CO chemisorption.

	CO Cumulative Quantity µmol/g
Biocoke	1.01
Biocoke Ca ($CaCl_2$)	0.24
Norit	1.01
Norit Mg	0.67
Norit Ba	0.83
Zeolite 3A	0.42
Zeolite 4A	1.03
Zeolite 4A Ca ($CaCl_2$)	0.90
ZSM5	1.93
Lanthana	0.04
Lanthana Ba	2.05
Lanthana Mg	0.18
Dolomite	3.20

In Table 4, it can be observed that, according to our results, when solids were modified with Ba they presented higher chemisorption of CO than when the modification was done with Mg. A significant increase can be observed in the lanthana Ba sorbent. The higher reactivity of modified lanthana sorbents could be related with the better dispersion as deduced from the N_2 sorption–desorption results' discussion. Comparing for each material, and at each temperature, the chemisorbed CO with the measured water sorption capacity, a direct correlation is not observed. In Supplementary Material, the correlations at 25, 100, 200 and 300 °C are given (Figures S7–S10). At 300 °C, the materials that showed higher water sorption capacity were lanthana Ba and dolomite, which also presented high CO chemisorbed quantity. On the contrary, the zeolite ZSM5 showed low water sorption capacity at 300 °C, even though the CO quantity was similar to that of Lanthana Ba. This implies that the water adsorption on the sorbents is not employing the same active sites as those for CO chemisorption. Therefore, the reactivity measured by CO chemisorption is not representative of water sorption capacity.

3.2.4. Temperature-Programmed Reduction of H_2, TPR-H_2

As explained along the manuscript, the prepared sorbents will be tested in the near future in a lab-scale pilot plant to carry out a sorption enhanced catalytic methanation process. The catalysts employed in that reaction usually require a prior reduction step at 700 °C, in order to generate the active phase of the metal, and this catalyst is physically mixed with the adsorbent. Moreover, during the methanation, a hydrogen-rich reaction mixture (reducing atmosphere) is continuously fed to the reactor. In the case of the sorbents, usually the active phase for water sorption is the oxide form. Therefore, TPR–H_2 test is important to understand if the sorbent will be affected by the reduction pre-treatment performed to the mixture of catalyst adsorbent in the pilot plant prior to methanation reaction. The complete TPR–H_2 profiles of the sorbents are given in Supplementary Material (Figures S11–S13).

In coke-based sorbents, big differences were observed when being treated with the H_2-containing mixture (see Figure S11). Biocoke was very stable; it showed no peaks. Norit activated carbon showed quite high hydrogen consume in temperatures around 500–600 °C, which was also detected in the work by Mouat et al. [43]. This reactivity with hydrogen was strongly affected by the addition of Mg, disappearing the original peak and appearing the two reduction peaks. MgO is not expected to be reduced with H_2 [44], and therefore, those new peaks may be due to the interaction or covering of Mg. The interaction of H_2 with the zeolite-based sorbents was very low, which can be concluded by the low values of the signals shown in Figure S12, according to the expected behavior [45]. The zeolite modified with Ca did not show any significant reducibility change, compared to the pure zeolite 4A. The lanthana-based sorbents did not show any interaction with H_2, as depicted in Figure S13. When modifying lanthana with different metals, the effect of H_2 may increase and shift to higher temperatures [40,46]; however, in our case, when modifying lanthana with Mg and Ba, insignificant change was observed. Therefore, the TPR results indicate that the CaO, BaO and MgO phases will remain as oxides in the sorbent even after the reduction step of the catalyst when applying them for methanation enhanced reaction.

3.2.5. X-ray Diffraction

The XRD diffractograms of the calcined zeolite 3A, zeolite 4A, zeolite 4A Ca ($CaCl_2$) and zeolite 4A Ca ($Ca(NO_3)_2$) are given in Figure 9. In Supplementary Material, the standard PDF of the identified compounds in each analyzed sample are given (Figures S14–S17). In the zeolite 3A, the main peaks were identified as Zeolite A, syn (Ref. Code 01-076-1509), while in the Zeolite 4A they were identified as Sodium Aluminum Phosphate Silicate Hydrate (Ref. Code 00-038-0323). In the modified zeolite 4A Ca sample, the zeolite structure seemed to be affected by the incorporation of the metal, as the main compound was identified as Zeolite A, syn (Ref. Code 01-076-2422), with the highest intensity peak at 2Theta value of 31.8. In this modified sample, Ca was identified as metallic Calcium (Ref. Code 00-001-0735), with main peaks at 2Theta values of 27.3, 45.5, 56.5 and 75.3. In the modified zeolite with $Ca(NO_3)_2$ salt, the identified compounds were: Zeolite A., syn (Ref. Code 01-082-1257), and $CaCO_3$ (Ref. code 01-072-1651) 29.4, 35.9. In this modified zeolite, the presence of metallic Ca was not detected, only the carbonate specie.

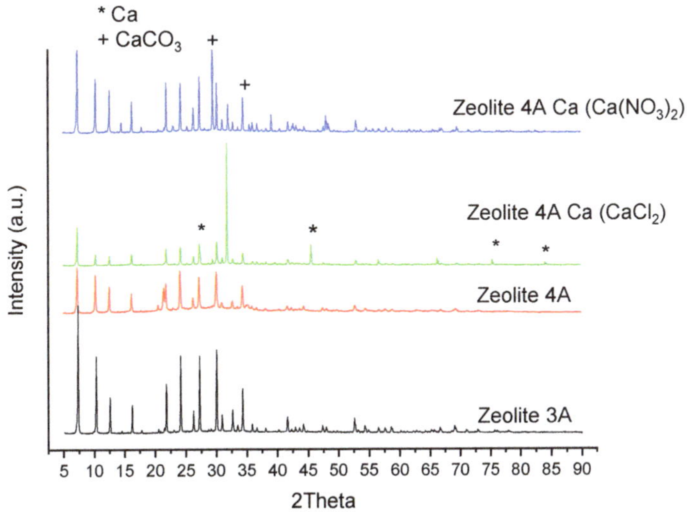

Figure 9. XRD patterns of calcined zeolite 3A, zeolite 4A and zeolite 4A Ca ($CaCl_2$).

In Figures 10 and 11, the XRD patterns of calcined lanthana, modified lanthana Mg and dolomite are given. Their standard PDF of the identified compounds are given in Supplementary Information (Figures S18–S20). In the unmodified lanthana sample, the compound identified was Lanthanum Hydroxide (Ref. Code 00-036-1481), indicating the high hydration tendency of this material. Regarding the Mg-modified lanthana sample, the addition of Mg did not affect the structure of the material, as Lanthanum Hydroxide was still identified, together with Magnesium Oxide (MgO) (Ref. Code 00-001-1235) with two main peaks at 2Theta values of 43.1 and 62.4. Regarding dolomite, the identified compounds Calcium hydroxide (Ref. Code 01-084-1275), lime CaO (Ref. Code 01-074-1226) and periclase MgO (Ref. code 00-045-0945) were detected, indicating again the hydration tendency of this material which was also observed in the sorption tests (Table 1).

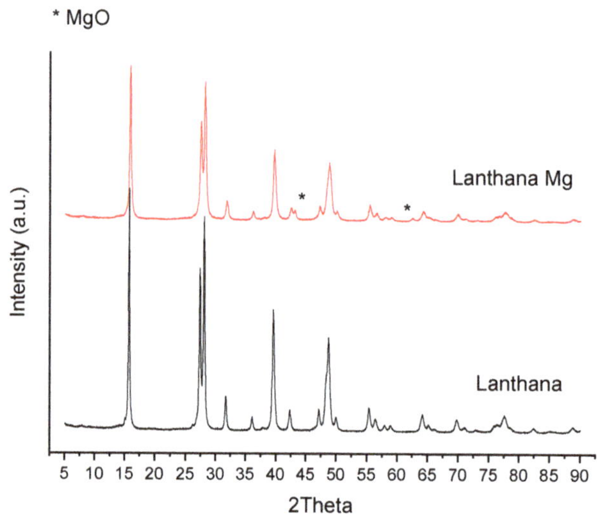

Figure 10. XRD patterns of calcined lanthana and lanthana Mg.

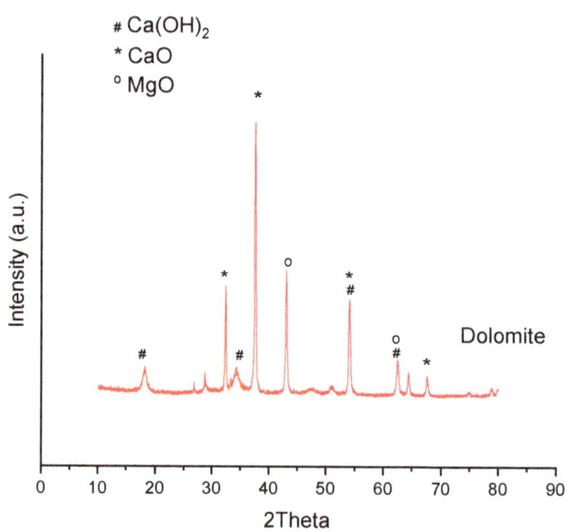

Figure 11. XRD patterns of calcined dolomite.

4. Conclusions

The water sorption capacity of the tested materials showed differences in kinetics and sorption capacity at 25 °C. Cokes presented very low sorption capacity, even if the Biocoke Ca was the one with the highest sorption capacity at 25 °C (473 $mg_{H2O}/g_{sorbent}$), it quickly decreased to 37 $mg_{H2O}/g_{sorbent}$ when increasing temperature. Among the tested zeolites (zeolite 3A, zeolite 4A and ZSM5) the 3A and 4A showed quite high sorption capacity at 25 °C (around 220 $mg_{H2O}/g_{sorbent}$), 160 $mg_{H2O}/g_{sorbent}$ at 100 °C and decreased to very low values at 200 °C. The modification of the zeolite 4A with two different Ca promoters did not result in better behavior. Lanthana and dolomite had intermedium sorption capacity at 25 °C (140–150 $mg_{H2O}/g_{sorbent}$); however, lanthana kept that value until 200 °C and dolomite even until 300 °C. Modifying lanthana with Mg did not show better performance, but Ba addition made the adsorbed water much more stable at 300 °C. Thus, lanthana-Ba and dolomite sorbents reached water sorption values of 120 and 102 $mg_{H2O}/g_{sorbent}$, respectively, even at 300 °C, i.e., values 10 times higher than the achieved ones with zeolites 3A or 4A under the same operating conditions, which would enhance the in situ sorption process in methanation reaction getting longer times to reach the breakthrough point.

Surface area was not the only critical parameter for water sorption at the tested conditions, even though it is a superficial process. This indicates that the water sorption at high temperatures is closer to chemisorption than to physisorption. Therefore, not only surface is required, but also adequate active centers, where the water adsorbed is stable. All the tested materials were mesoporous, with quite high pore diameter (30–60 Å). The surface area and pore diameter could enhance the mass transport and benefit the kinetics of the adsorption process. Kinetics are important because the future application of these sorbents is in a continuous fixed bed reactor, with limited contact time between water and the sorbent.

The acidity of the materials was quite low in general, and just zeolite 4A and lanthana Ba had a clear acidity peak at low–medium temperatures (200–550 °C), indicating weak/medium strength acidity. These two materials were promising for water sorption, because they showed high water sorption capacity. CO chemisorption technique was not found to give correlations between water reactivity in the sorbent and adsorbed CO.

Further work will be done in the future in order to modify the dolomite and analyze if it is possible to increase the adsorbed water quantity or keep its adsorption capacity at higher temperatures. The possible concentration polarization phenomena that may occur in the real system when adsorbing water from a mixture containing other gases such as H_2, CO, CO_2 and CH_4 will also be taken into account.

Supplementary Materials: The following supporting information can be downloaded at: https://www.mdpi.com/article/10.3390/ma16072933/s1, Details on N_2 sorption-desorption analyses (Figures S1–S3); NH_3-TPD graphs (Figures S4–S6); Correlation of the CO chemisorption and water sorption capacity at different temperatures (Figures S7–S10); H_2-TPR profiles (Figures S11–S13); XRD standard patterns (Figures S14–S20).

Author Contributions: Conceptualization, E.A.; Methodology, I.A. and V.L.B.; Investigation, E.A. and I.A.; Writing—original draft, E.A. and I.A.; Writing—review & editing, E.A. and I.A.; Project administration, V.L.B.; Funding acquisition, V.L.B. All authors have read and agreed to the published version of the manuscript.

Funding: This research was funded by University of the Basque Country (UPV/EHU), Spanish Ministry of Economy and Innovation and European Union through the European Regional Development Fund (FEDER) (Projects: ENE2017-82250-R and PID2020-112889RB-I00 ENE), and Basque Country Government (Project: IT1554-22).

Institutional Review Board Statement: Not applicable.

Informed Consent Statement: Not applicable.

Data Availability Statement: Not applicable.

Acknowledgments: The authors also acknowledge the contribution of J. Cambra (retired).

Conflicts of Interest: The authors declare no conflict of interest.

Nomenclature

BET	Brunauer-Emmett-Teller
BJH	Barrett-Joyner-Halenda
PtG	Power to Gas
TGA	thermo-gravimetric analysis
TPD	temperature-programmed desorption
TPR	temperature-programmed reduction
XRD	X-ray diffraction

References

1. Aydin, G. Modeling of Energy Consumption Based on Economic and Demographic Factors: The Case of Turkey with Projections. *Renew. Sustain. Energy Rev.* **2014**, *35*, 382–389. [CrossRef]
2. Aydin, G. Regression Models for Forecasting Global Oil Production. *Pet. Sci. Technol.* **2015**, *33*, 1822–1828. [CrossRef]
3. Xia, A.; Zhu, X.; Liao, Q. Hydrogen Production from Biological Sources. In *Encyclopedia of Sustainability Science and Technology*; Meyers, R.A., Ed.; Springer: New York, NY, USA, 2017; pp. 833–863, ISBN 978-1-4939-2493-6.
4. Safari, F.; Dincer, I. Assessment and Optimization of an Integrated Wind Power System for Hydrogen and Methane Production. *Energy Convers. Manag.* **2018**, *177*, 693–703. [CrossRef]
5. *European Biogas Association (EBA) EBA's Biomethane Fact Sheet*; European Biogas Association: Brussels, Belgium, 2014.
6. García-García, I.; Izquierdo, U.; Barrio, V.L.; Arias, P.L.; Cambra, J.F. Power-to-Gas: Storing Surplus Electrical Energy. Study of Al_2O_3 Support Modification. *Int. J. Hydrog. Energy* **2016**, *41*, 19587–19594. [CrossRef]
7. Skorek-Osikowska, A.; Martín-Gamboa, M.; Dufour, J. Thermodynamic, Economic and Environmental Assessment of Renewable Natural Gas Production Systems. *Energy Convers. Manag. X* **2020**, *7*, 100046. [CrossRef]
8. Castellani, B.; Rinaldi, S.; Morini, E.; Nastasi, B.; Rossi, F. Flue Gas Treatment by Power-to-Gas Integration for Methane and Ammonia Synthesis—Energy and Environmental Analysis. *Energy Convers. Manag.* **2018**, *171*, 626–634. [CrossRef]
9. Sarić, M.; Dijkstra, J.W.; Walspurger, S.; Haije, W.G. *The Potential of "Power to Gas" Technology Integrated with Biomethane Production*; ECN: Petten, The Netherlands, 2014.
10. Walker, S.B.; Van Lanen, D.; Fowler, M.; Mukherjee, U. Economic Analysis with Respect to Power-to-Gas Energy Storage with Consideration of Various Market Mechanisms. *Int. J. Hydrog. Energy* **2016**, *41*, 7754–7765. [CrossRef]
11. Baraj, E.; Vagaský, S.; Hlinčik, T.; Ciahotný, K.; Tekáč, V. Reaction Mechanisms of Carbon Dioxide Methanation. *Chem. Pap.* **2016**, *70*, 395–403. [CrossRef]
12. Lehner, M.; Tichler, R.; Steinmüller, H.; Koppe, M. Methanation. In *Power-to-Gas: Technology and Business Models "Methanation"*; Springer: Berlin/Heidelberg, Germany, 2014; pp. 41–61, ISBN 978-3-319-03994-7.
13. Aziz, M.A.A.; Jalil, A.A.; Triwahyono, S.; Sidik, S.M. Methanation of Carbon Dioxide on Metal-Promoted Mesostructured Silica Nanoparticles. *Appl. Catal. A Gen.* **2014**, *486*, 115–122. [CrossRef]
14. Beierlein, D.; Häussermann, D.; Pfeifer, M.; Schwarz, T.; Stöwe, K.; Traa, Y.; Klemm, E. Is the CO_2 Methanation on Highly Loaded Ni-Al 2 O 3 Catalysts Really Structure-Sensitive? *Appl. Catal. B Environ.* **2019**, *247*, 200–219. [CrossRef]
15. De Leitenburg, C.; Trovarelli, A.; Kašpar, J. A Temperature-Programmed and Transient Kinetic Study of CO_2 Activation and Methanation over CeO_2 Supported Noble Metals. *J. Catal.* **1997**, *166*, 98–107. [CrossRef]
16. Panagiotopoulou, P.; Kondarides, D.I.; Verykios, X.E. Selective Methanation of CO over Supported Noble Metal Catalysts: Effects of the Nature of the Metallic Phase on Catalytic Performance. *Appl. Catal. A Gen.* **2008**, *344*, 45–54. [CrossRef]
17. Borgschulte, A.; Gallandat, N.; Probst, B.; Suter, R.; Callini, E.; Ferri, D.; Arroyo, Y.; Erni, R.; Geerlings, H.; Züttel, A. Sorption Enhanced CO2 Methanation. *Phys. Chem. Chem. Phys.* **2013**, *15*, 9620–9625. [CrossRef]
18. Walspurger, S.; Elzinga, G.D.; Dijkstra, J.W.; Sarić, M.; Haije, W.G. Sorption Enhanced Methanation for Substitute Natural Gas Production: Experimental Results and Thermodynamic Considerations. *Chem. Eng. J.* **2014**, *242*, 379–386. [CrossRef]
19. Catarina Faria, A.; Miguel, C.V.; Madeira, L.M. Thermodynamic Analysis of the CO_2 Methanation Reaction with in Situ Water Removal for Biogas Upgrading. *J. CO2 Util.* **2018**, *26*, 271–280. [CrossRef]
20. Agirre, I.; Acha, E.; Cambra, J.F.; Barrio, V.L. Water Sorption Enhanced CO_2 Methanation Process: Optimization of Reaction Conditions and Study of Various Sorbents. *Chem. Eng. Sci.* **2021**, *237*, 116546–116556. [CrossRef]
21. Gómez, L.; Martínez, I.; Navarro, M.V.; García, T.; Murillo, R. Sorption-Enhanced CO and CO_2 Methanation (SEM) for the Production of High Purity Methane. *Chem. Eng. J.* **2022**, *440*, 135842. [CrossRef]
22. Coppola, A.; Massa, F.; Salatino, P.; Scala, F. Evaluation of Two Sorbents for the Sorption-Enhanced Methanation in a Dual Fluidized Bed System. *Biomass Convers. Biorefin.* **2021**, *11*, 111–119. [CrossRef]
23. Pieterse, J.A.Z.; Elzinga, G.D.; Booneveld, S.; van Kampen, J.; Boon, J. Reactive Water Sorbents for the Sorption-Enhanced Reverse Water–Gas Shift. *Catal. Lett.* **2022**, *152*, 460–466. [CrossRef]

24. Lin, H.; Liu, Y.; Deng, J.; Jing, L.; Dai, H. Methane Combustion over the Porous Oxides and Supported Noble Metal Catalysts. *Catalysts* **2023**, *13*, 427. [CrossRef]
25. Mikšík, F.; Miyazaki, T.; Thu, K.; Miyawaki, J.; Nakabayashi, K.; Wijayanta, A.T.; Rahmawati, F. Enhancing water adsorption capacity of acorn nutshell based activated carbon for adsorption thermal energy storage application. *Energy Rep.* **2020**, *6*, 255–263.
26. Tang, Y.; Kourtellaris, A.; Tasiopoulos, A.J.; Teat, S.J.; Dubbeldam, D.; Rothenberg, G.; Tanase, S. Selective CO_2 adsorption in water-stable alkaline-earth based metal–organic frameworks. *Inorg. Chem. Front.* **2018**, *5*, 541–549. [CrossRef]
27. Da Silva, A.L.; Wu, L.; Caliman, L.B.; Castro, R.H.; Navrotsky, A.; Gouvêa, D. Energetics of CO_2 and H_2O adsorption on alkaline earth metal doped TiO_2. *Phys. Chem. Chem. Phys.* **2020**, *22*, 15600–15607. [CrossRef] [PubMed]
28. Aristov, Y.I. *Selective Water Sorbents, a New Family of Materials for Adsorption Cooling/Heating: State of the Art*; V Minsk International Seminar "Hear Pipes, Heat Pumps, Refrigerators": Minsk, Belarus, 2003.
29. Glasser, L. Thermodynamics of Inorganic Hydration and of Humidity Control, with an Extensive Database of Salt Hydrate Pairs. *J. Chem. Eng. Data* **2014**, *59*, 526–530. [CrossRef]
30. Ropp, R.C.C. Chapter 3—Group 16 (O, S, Se, Te) Alkaline Earth Compounds. In *Encyclopedia of the Alkaline Earth Compounds*; Elsevier: Amsterdam, The Netherlands, 2013; pp. 105–197, ISBN 978-0-444-59550-8.
31. Ng, E.-P.; Mintova, S. Nanoporous Materials with Enhanced Hydrophilicity and High Water Sorption Capacity. *Microporous Mesoporous Mater.* **2008**, *114*, 1–26. [CrossRef]
32. Munthali, M.W.; Elsheikh, M.A.; Johan, E.; Matsue, N. Proton Adsorption Selectivity of Zeolites in Aqueous Media: Effect of Si/Al Ratio of Zeolites. *Molecules* **2014**, *19*, 20468–20481. [CrossRef]
33. Olszak-Humienik, M.; Jablonski, M. Thermal behavior of natural dolomite. *J. Therm. Anal. Calorim.* **2015**, *119*, 2239–2248. [CrossRef]
34. Acha, E.; Chen, D.; Cambra, J.F. Comparison of novel olivine supported catalysts for high purity hydrogen production by CO_2 sorption enhanced steam reforming. *J. CO2 Util.* **2020**, *42*, 101295. [CrossRef]
35. Adrados, A.; De Marco, I.; Lopez-Urionabarrenechea, A.; Solar, J.; Caballero, B.M.; Gastelu, N. Biomass Pyrolysis Solids as Reducing Agents: Comparison with Commercial Reducing Agents. *Materials* **2016**, *9*, 3. [CrossRef]
36. Rouquerol, J.; Rouquerol, F.; Llewellyn, P.; Maurin, G.; Sing, K.S.W. (Eds.) 11—Adsorption by Metal Oxides. In *Adsorption by Powders and Porous Solids: Principle, Methodology and Applications*, 2nd ed.; Academic Press: Oxford, UK, 2014; pp. 393–465, ISBN 978-0-08-097035-6.
37. Mansir, N.; Hwa Teo, S.; Lokman Ibrahim, M.; Yun Hin, T.-Y. Synthesis and Application of Waste Egg Shell Derived CaO Supported W-Mo Mixed Oxide Catalysts for FAME Production from Waste Cooking Oil: Effect of Stoichiometry. *Energy Convers. Manag.* **2017**, *151*, 216–226. [CrossRef]
38. Antzara, A.; Heracleous, E.; Lemonidou, A.A. Development of CaO-Based Mixed Oxides as Stable Sorbents for Post-Combustion CO2 Capture Via Carbonate Looping. *Energy Procedia* **2014**, *63*, 2160–2169. [CrossRef]
39. Thommes, M.; Kaneko, K.; Neimark, A.V.; Olivier, J.P.; Rodriguez-Reinoso, F.; Rouquerol, J.; Sing, K.S.W. Physisorption of Gases, with Special Reference to the Evaluation of Surface Area and Pore Size Distribution (IUPAC Technical Report). *Pure Appl. Chem.* **2015**, *87*, 1051–1069. [CrossRef]
40. Pandey, A.; Jain, G.; Vyas, D.; Irusta, S.; Sharma, S. Nonreducible, Basic La_2O_3 to Reducible, Acidic La2–XSbxO_3 with Significant Oxygen Storage Capacity, Lower Band Gap, and Effect on the Catalytic Activity. *J. Phys. Chem. C* **2017**, *121*, 481–489. [CrossRef]
41. Bajdich, M.; Nørskov, J.K.; Vojvodic, A. Surface Energetics of Alkaline-Earth Metal Oxides: Trends in Stability and Adsorption of Small Molecules. *Phys. Rev. B* **2015**, *91*, 155401. [CrossRef]
42. Karge, H.G.; Weitkamp, J. (Eds.) *Post-Synthesis Modification I:3 (Molecular Sieves)*; Springer: Berlin/Heidelberg, Germany, 2002.
43. Mouat, A.R.; Lohr, T.L.; Wegener, E.C.; Miller, J.T.; Delferro, M.; Stair, P.C.; Marks, T.J. Reactivity of a Carbon-Supported Single-Site Molybdenum Dioxo Catalyst for Biodiesel Synthesis. *ACS Catal.* **2016**, *6*, 6762–6769. [CrossRef]
44. Carabineiro, S.A.C.; Bogdanchikova, N.; Pestryakov, A.; Tavares, P.B.; Fernandes, L.S.G.; Figueiredo, J.L. Gold Nanoparticles Supported on Magnesium Oxide for CO Oxidation. *Nanoscale Res. Lett.* **2011**, *6*, 435. [CrossRef]
45. Lee, B.J.; Hur, Y.G.; Kim, D.H.; Lee, S.H.; Lee, K.-Y. Non-Oxidative Aromatization and Ethylene Formation over Ga/HZSM-5 Catalysts Using a Mixed Feed of Methane and Ethane. *Fuel* **2019**, *253*, 449–459. [CrossRef]
46. Paul, B.; Khatun, R.; Sharma, S.K.; Adak, S.; Singh, G.; Das, D.; Siddiqui, N.; Bhandari, S.; Joshi, V.; Sasaki, T.; et al. Fabrication of Au Nanoparticles Supported on One-Dimensional La_2O Nanorods for Selective Esterification of Methacrolein to Methyl Methacrylate with Molecular Oxygen. *ACS Sustain. Chem. Eng.* **2019**, *7*, 3982–3994. [CrossRef]

Disclaimer/Publisher's Note: The statements, opinions and data contained in all publications are solely those of the individual author(s) and contributor(s) and not of MDPI and/or the editor(s). MDPI and/or the editor(s) disclaim responsibility for any injury to people or property resulting from any ideas, methods, instructions or products referred to in the content.

materials

MDPI

Article

A Comprehensive View of the Optimization of Chromium (VI) Processing through the Application of Electrocoagulation Using a Pair of Steel Electrodes

Ľubomír Pikna [1,*], Mária Heželová [1], Dagmar Remeteiová [1], Silvia Ružičková [1], Róbert Findorák [1] and Jaroslav Briančin [2]

1 Faculty of Materials, Metallurgy and Recycling, Technical University of Kosice, Letná 9, 042 00 Kosice, Slovakia
2 Institute of Geotechnics, Slovak Academy of Sciences, Watsonova 45, 040 01 Kosice, Slovakia
* Correspondence: lubomir.pikna@tuke.sk

Abstract: In the presented article, an electrocoagulation method using a steel cathode and a steel anode was used to obtain chromium from laboratory-prepared model solutions with known compositions. The study aimed to analyze the effect of solution conductivity, pH, and 100% efficiency of chromium removal from the solution, as well as the highest possible Cr/Fe ratio in the final solid product throughout the process of electrocoagulation. Different concentrations of chromium (VI) (100, 1000, and 2500 mg/L) and different pH values (4.5, 6, and 8) were investigated. Various solution conductivities were provided by the addition of 1000, 2000, and 3000 mg/L of NaCl to the studied solutions. Chromium removal efficiency equal to 100% was achieved for all studied model solutions for different experiment times, depending on the selected current intensity. The final solid product contained up to 15% chromium in the form of mixed FeCr hydroxides obtained under optimal experimental conditions: pH = 6, I = 0.1 A, and c (NaCl) = 3000 mg/L. The experiment indicated the advisability of using a pulsed change of electrode polarity, which led to a reduction in the time of the electrocoagulation process. The results may help in the rapid adjustment of the conditions for further electrocoagulation experiments, and they can be used as the optimization experimental matrix.

Keywords: chromium; electrocoagulation; metal recovery; secondary raw materials; optimization

Citation: Pikna, Ľ.; Heželová, M.; Remeteiová, D.; Ružičková, S.; Findorák, R.; Briančin, J. A Comprehensive View of the Optimization of Chromium (VI) Processing through the Application of Electrocoagulation Using a Pair of Steel Electrodes. *Materials* **2023**, *16*, 3027. https://doi.org/10.3390/ma16083027

Academic Editor: Frank Czerwinski

Received: 15 March 2023
Revised: 3 April 2023
Accepted: 8 April 2023
Published: 11 April 2023

Copyright: © 2023 by the authors. Licensee MDPI, Basel, Switzerland. This article is an open access article distributed under the terms and conditions of the Creative Commons Attribution (CC BY) license (https://creativecommons.org/licenses/by/4.0/).

1. Introduction

The steel industry not only produces various types of steel but also produces a large amount of slag and other waste. According to the type of steel produced, slag can contain up to 5% chromium. Therefore, slag can serve as a source of chromium and other metals, once the recovery methods are ecological and of a reasonable cost. Hydrometallurgically obtained slag leachates can be processed by the electrocoagulation method (EC). EC essentially represents in situ generation of a coagulation agent using appropriate electrodes and an electric current. The principle of the method is generally available in numerous areas of the available electrochemical literature and its implementation depends on the type of compound to be removed and the pair of electrodes used. EC is suitable for recovering a variety of metals and compounds from leachate or wastewater.

Copper, nickel [1], fluoride [2], chromium [1,3], various organic pollutants [4–6], phosphorus [6], boron [7], nickel, lead, mercury, zinc, arsenic, cadmium, and chromium [8] are some of the components that can be treated or removed from dyehouse or distillery wastewaters, leachate of environmental samples, leachate of industrial waste, or underground water using EC with a combination of different working electrodes.

The effect of anode and cathode materials on metal ions has been confirmed by many papers [9–11]. The degree of efficiency of the EC process is different for each metal when changing the pair of electrodes. Studies on alternative foam types of electrodes or electrodes made from metal chips have also been published, and in one study, the efficiency of the EC method on greywater was 93% and 87% compared to the 84% efficiency of a traditional aluminum plate electrode [12].

In an effort to reduce the amount of chemicals used in the processing of various raw materials or in chemical analysis, the question arises as to whether it is not more appropriate to use electrocoagulation instead of the direct application of various chemical coagulation agents for water decontamination. Ackbal et al. focused, in their article, on a comparison of the EC method and "conventional" methods in the removal of heavy metals (Cu, Cr, and Ni) [1]. As a result, EC proved to be an equally effective and ultimately cheaper method with lower chemical consumption. Hybrid types of electrocoagulation such as photo-, peroxi-, and photo-peroxi-electrocoagulation with the use of a Fe electrode proved to be a possible modification of the process with promising results in the removal of organic pollutants from distillery industrial effluent [9].

For the remediation of tannery wastewater, the combination of EC with other pre-treatment methods was discussed by Verma et al. [13]. Among them, ultrasonication, adsorption, biological treatment, and membrane processes were used to improve EC performance. The use of an additional method is associated not only with advantages but may be associated with an excessive amount of energy used, the formation of by-products, and the required UV energy for a long period.

Many authors have focused on its application for the removal of trivalent or hexavalent chromium from wastewater solutions of any type. Al-Qodah et al., in their review [8], provide a detailed overview of works focused on EC for heavy metals removal (Cr(III), Cr(VI), Cu, Cd, Zn, Ni, As(III), Pb, and Hg) under different experimental conditions with Fe, Al, Cu, stainless steel, or Zn electrodes and its combinations. The Cr(III) or Cr(VI) concentrations in the reviewed studies were from 10 mg/L to 2235 mg/L (with most of them being under 500 mg/L) with removal efficiencies from 42% to 100% or 7.4% for the highest-mentioned concentration.

The authors of studies usually monitor the effectiveness of the decontamination process in their chosen time frame, often with high current densities, focusing on the removal of the contaminant, regardless of the composition of the solid product or the intensity of the electric current used, including side effects, such as heating the solution at higher current intensities.

The approach of the present study is different. The work focuses on the optimization of the process of obtaining chromium from the model solutions in the form of a further processable product and, at the same time, obtaining the decontaminated water.

Numerous variations of the experimental conditions lead to the formulation of the best conditions for combinations of current intensity, the concentration of chromium(VI) solution, conductivity, and pH value. The overall optimization matrix was compiled from the results. TGA and SEM-EDX analyses were carried out for verification of the obtained EC product. Overall, the present study is expected to help the rapid adjustment of EC experimental conditions for future laboratory and real sample analyses.

2. Materials and Methods

Model solutions for electrocoagulation experiments with Cr(VI) concentrations of 100, 1000, and 2500 mg/L were prepared from potassium dichromate.

A laboratory power supply EA-PS 3032-10B (ELEKTRO-AUTOMATIK), with a max output 10 A and 32 V–DC power source, a shaft stirrer WISESTIR HT50DX with rpm regulation, and a pH meter ORION 3 STAR (THERMO) were used.

Steel plates ST37-2 with active surface dimensions of 50 × 40 × 2 mm were directly dipped in the solution. The overall dimensions of the electrodes were 150 × 40 × 2 mm. The distance between the electrodes was 5 cm.

The influence of NaCl was studied for concentrations of 1000, 2000, and 3000 mg/L of NaCl.

The influence of pH was verified for the values 4.5, 6, and 8. The initial pH was adjusted at the beginning of the experiment and regulated throughout the duration of the experiment. For this purpose, 1:1 hydrochloric acid and/or 1 mol/dm^3 NaOH solutions were used.

In the EC process, current intensities of 0.1, 0.5, 1, and 2 A (mean current densities: 4.76, 23.8, 47.6, and 95.2 mA/cm^2, respectively) were tested.

The volume of solutions equal to 350 mL and the rotations of the shaft stirrer of 250 rpm were constant for each experiment. All of the reagents used were of analytical purity grade.

The scheme of the experimental arrangement is shown in Figure 1.

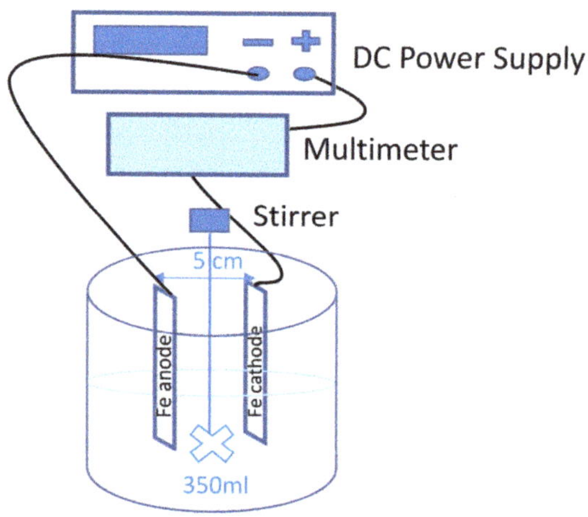

Figure 1. Scheme of the experimental arrangement.

The EC experiments lasted for the time period needed for the absolute removal of chromium from the solution. As a result of using the same volume of each sample and the condition of total removal of Cr from the solution, different sampling time intervals were chosen depending on the chromium concentration in the solution and the applied current intensity.

Chromium in the filtrates, as well as Cr and Fe in the solid products, was determined by atomic absorption spectrometry (VARIAN 20+); in all tables, graphs, and calculations, a concentration of 0 mg/L means that it was below the LOQ of AAS.

Solid products were characterized by TGA, SEM, and EDX analyses. The derivatograph C, MOM, was used for TGA measurements of the change in product weights of the direct EC process under heating conditions. The thermal gradient of sample heating was 10 °C/min.

A scanning electron microscope (SEM), TESCAN MIRA 3 FE, with microanalytical EDX equipment, OXFORD INSTRUMENTS, was used for EC product characterization.

The overall combination of experimented model solutions is given by rows 1–22 in Table 1. It contains the current, conductivity, concentration of NaCl and Cr(VI), and voltage (which changes during the measurement depending on the conductivity of the solution; the table shows the average voltage values).

Table 1. Investigated model solutions experimental set.

	c (Cr VI) (mg/L)	I (A)	U (V)	c (NaCl) (mg/L)	G (mS)	pH
1.	1000	0.1	3	3000	7.2	6
2.	1000	0.5	8.5	3000	7.2	6
3.	1000	1	16	3000	7.2	6
4.	1000	2	25	3000	7.2	6
5.	1000	0.1	2.2	3000	7.2	8
6.	1000	0.5	6.7	3000	7.2	8
7.	1000	1	12.7	3000	7.2	8
8.	1000	2	19	3000	7.2	8
9.	1000	0.1	2.5	3000	7.2	4.5
10.	1000	0.5	8.5	3000	7.2	4.5
11.	1000	1	13	3000	7.2	4.5
12.	1000	2	25	3000	7.2	4.5
13.	100	0.5	22	1000	2.2	6
14.	100	0.5	15.5	2000	3.5	6
15.	100	0.5	12	3000	4.7	6
16.	1000	1	29	1000	4	6
17.	1000	1	23	2000	6	6
18.	1000	1	16	3000	7.2	6
19.	2500	1	12.5	1000	10.4	6
20.	2500	1	12.5	2000	11.2	6
21.	2500	1	10.3	3000	12.2	6
22. *	1000	0.1	3	3000	7.2	6

* The experimental conditions are the same as in row 1, but the electrode polarity was changed at 5 min intervals.

3. Results and Discussion

In the process of optimization, the influence of pH, current intensity, and NaCl concentration on the overall time of total Cr removal from the solution, the energy consumption, as well as the Cr/Fe ratio in solid products, was observed. In the process of electrocoagulation by using Fe-Fe electrodes, hexavalent chromium dissolved in water was reduced in the solution by divalent iron ions produced by the dissolution of the Fe anode. The high efficiency of Cr removal by electrocoagulation using the Fe-Fe electrode pair has also been confirmed by the works of other authors [10,14,15].

Another monitored parameter was the temperature of the solution, which increased with the intensity of the current. In solutions with 1000 mg/L of chromium and 3000 mg/L of NaCl, the mean increases were 1, 5, 15, and 30 °C for I = 0.1, 0.5, 1, and 2 A, respectively.

3.1. Time Dependence of the EC Process

The duration of the EC process was customized to the aim of total, 100% efficiency, Cr removal. Therefore, different sampling times were also chosen.

Figure 2 shows the curves of decreasing Cr concentration during the EC process for the pH values 4.5, 6, and 8 and for the current intensities of 0.1, 0.5, 1, and 2 A. The calculated concentration of Cr was 1000 mg/L for the preparation of the model solution.

Experiments were performed with a NaCl concentration equal to 3000 mg/L. Table 2 summarizes the time (in minutes) when the Cr concentration reached zero as a function of pH and amperage.

Table 2. Minutes until the chromium concentration was zero in the solution.

c (Cr VI) (mg/L)	pH	0.1 A	0.5 A	1 A	2 A
100	6	60	15	8.5	n/a
1000	4.5	480	180	100	50
	6	450	120	80	40
	8	660	200	120	90
2500	6	1320	420	240	135

The EC process was completed earlier when the solution was kept at pH = 6 for all current intensities. A higher current intensity caused a shorter time of Cr removal, measured in tens of minutes for I = 1 A or 2 A compared to hundreds of minutes for I = 0.1 A. On the other hand, a higher current intensity was associated with an increase in solution temperature and a decrease in the Cr/Fe ratio—which is crucial for this type of Cr-containing product.

The shorter duration of the electrocoagulation process at pH = 6 is also reflected in the total consumption of electric energy (Figure 3). After converting to 1 g of solid product, the average energy consumption was 1.1, 3.58, 8.75, and 11.07 Wh for currents of 0.1, 0.5, 1, and 2 A and pH = 6. There is a significant difference in energy consumption for the lowest and highest current intensity used for all of the studied pHs. Taking into account the results obtained, it can be concluded that for the Fe-Fe electrode combination, pH = 6 is optimal, as has already been observed by other authors [16]. A chromium removal efficiency equal to 100% was also achieved at pH = 8, similar to the results described by Handman et al. [17], as well as the results reported by the authors Keshmirizadeh et al. and Dermentzis et al. [18,19] at pH = 4.5. However, the results presented here show that environments with different pHs of 4.5 to 8 are usable. The total time may be unimportant if the goal is to completely remove Cr from the solution using a small amount of energy and, at the same time, to obtain a solid product with the highest possible Cr/Fe ratio, which can be further processed. In addition, Dermentzis et al., Aguilar-Ascón et al., and Rodriguez et al. [19–21] describe faster removal of chromium when using a higher current, even at the expense of higher energy consumption or heating of the solution. In these works, the priority was set as 100% effectiveness of Cr removal from the solution as well as, the achievement of the highest possible Cr/Fe ratio in the solid product. The prepared solid products, with a high portion of Cr, can be used as a source of chromium for metallurgical applications, whereby the chromium, originally found in the metallurgical waste (slag), can be returned to the production process.

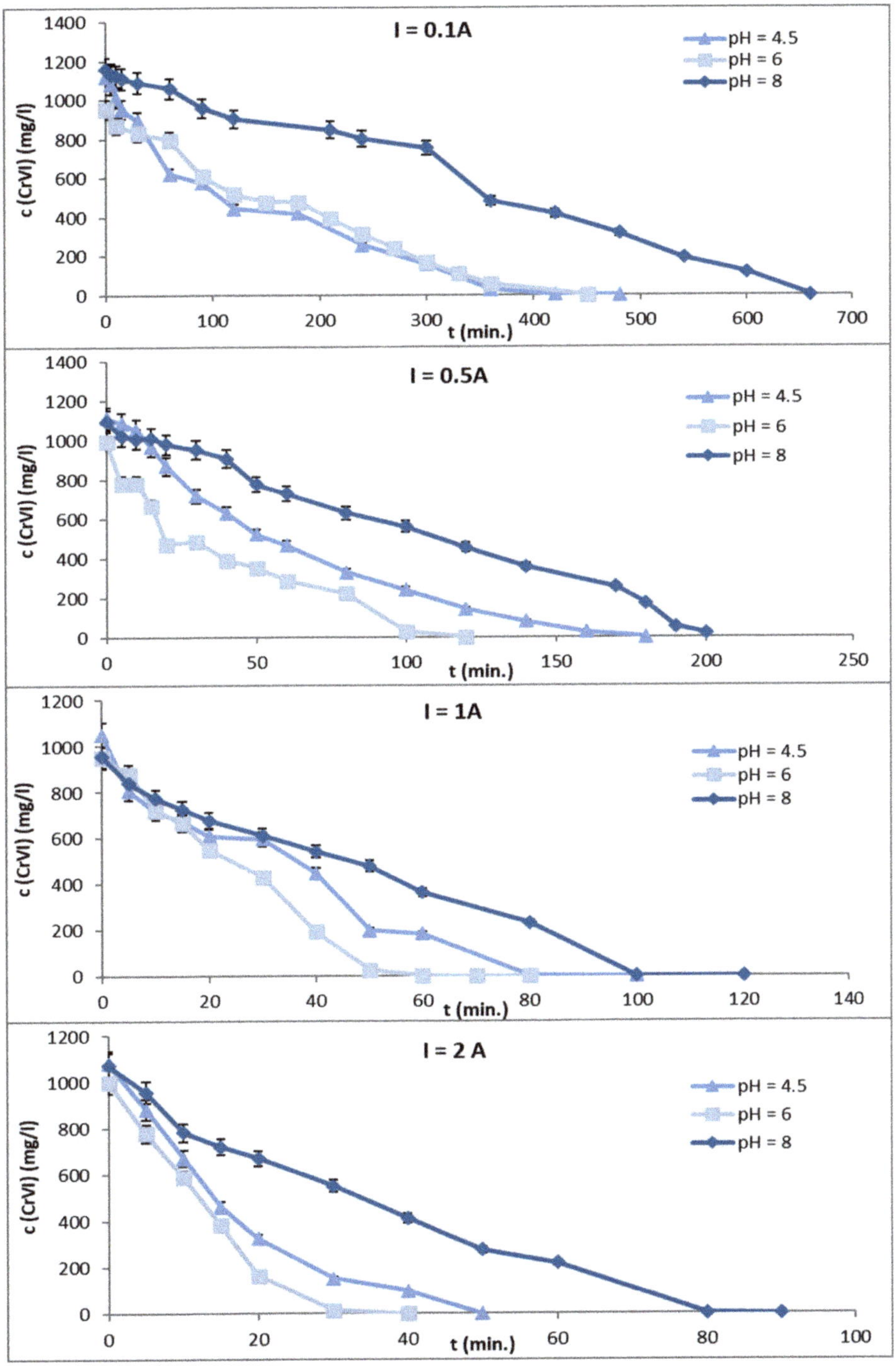

Figure 2. Decrease in Cr concentration in model solutions at pH 4.5, 6, and 8 for current intensities of 0.1, 0.5, 1, and 2 A.

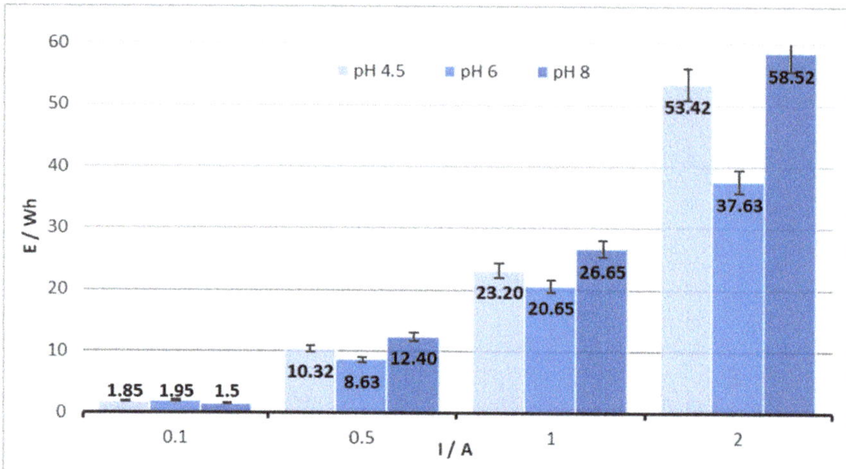

Figure 3. Energy consumption for the solution c(Cr(VI)) = 1000 mg/L and c(NaCl) = 3000 mg/L.

Prevention of electrode passivation, anode corrosion, and the liberation of Fe(II) ions into the solution, as well as improving the conductivity of the solution are the functions of NaCl's presence in the solutions leading to successful Cr removal from the solution. In the experiments, concentrations of 1000, 2000, or 3000 mg/L of NaCl were used combined with 100, 1000, or 2500 mg/L of Cr(VI). No significant effect of NaCl concentration on the duration of the EC process to reach zero Cr concentration in the solution was observed. On the basis of the observation of the experiments, it was found that a certain concentration of NaCl (at least 1000 mg/L), especially at a concentration of 100 mg/L Cr, is necessary for sufficient conductivity of the solution and maintaining the current intensity at a certain level due to the design limitations of the voltage source used. Sufficient conductivity is also necessary due to energy consumption. Higher conductivity of the solution requires a lower voltage to attain the required current intensity and, as a result, lower electric energy consumption. While maintaining parameters such as pH and chromium concentration with a difference only in NaCl concentration, a three-fold decrease in NaCl concentration can cause a reduction in energy consumption by 20 to 50%, depending on the intensity of the current applied to the given solution.

3.2. The Cr/Fe Ratio Observation

Table 3 summarizes the composition of the solid electrocoagulation product connected to the monitored parameters for the experimental conditions in rows 1–22 according to Table 1.

Table 3. Information about the composition of the solid products obtained under different (Table 1) experimental conditions.

	Solid Product Composition		
	c (Cr) (g/kg)	**c (Fe) (g/kg)**	**Cr/Fe Ratio**
1.	138.9	384.3	0.3614
2.	107.5	367.9	0.2922
3.	95.8	388.3	0.2467
4.	79.6	437.8	0.1818
5.	84.9	349	0.2434

Table 3. *Cont.*

	Solid Product Composition		
	c (Cr) (g/kg)	c (Fe) (g/kg)	Cr/Fe Ratio
6.	80.5	373.3	0.2163
7.	74.3	366	0.2028
8.	77.3	381.4	0.2026
9.	91.7	371.3	0.2469
10.	89.1	374.9	0.2377
11.	75.1	357.9	0.2098
12.	75.8	378.6	0.2002
13.	52.9	506.6	0.1044
14.	72.2	537.9	0.1342
15.	83.2	406.5	0.2047
16.	102	367.6	0.2775
17.	107.7	374.9	0.2873
18.	95.8	388.3	0.2467
19.	87.1	360.6	0.2417
20.	106.6	424.3	0.2513
21.	105.4	441.1	0.2390
22. *	142.4	356.9	0.3989

* The experimental conditions are the same as in row 1, but the electrode polarity was changed at 5 min intervals.

Figure 4 shows the dependence of the Cr/Fe ratio in the solid product on the concentration of NaCl in the solution. There is no evident trend in the relationship between NaCl and the Cr/Fe ratio for all the Cr concentrations studied, and the values are within the deviation interval for each concentration.

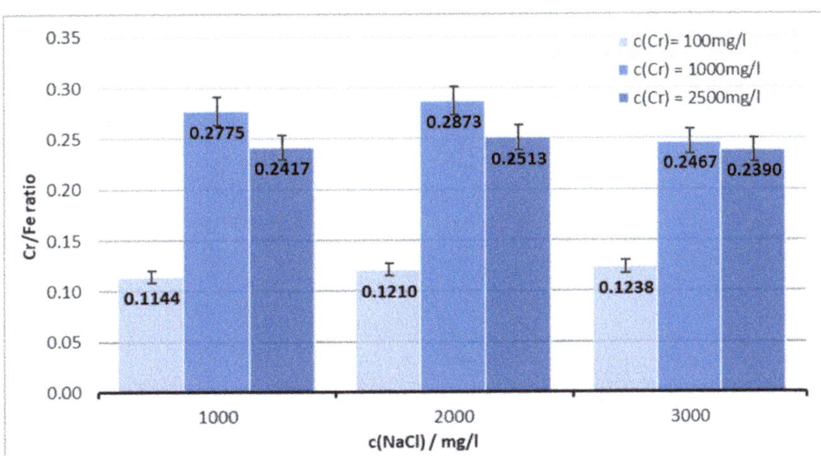

Figure 4. The Cr/Fe ratio's dependence on NaCl concentration for pH = 6 and I = 1 A.

Another observation is the dependence of the Cr/Fe ratio on the pH of the solution from which the product was created. As can be seen in Figure 5, for a concentration of Cr 1000 mg/and a concentration of NaCl 3000 mg/L, the pH of the solution significantly affects the composition of the final solid product and the best Cr/Fe ratio in the product is for pH = 6. At this pH, an approximately 30% percent higher Cr/Fe ratio was achieved compared to pH 4.5 or 8 at a current of 0.1. The observation was similar for currents 0.5 A and 1 A, although with a percentage difference of approximately 20% and 16%. However, when high current intensities (2 A) were used, this positive effect was lost and, as indicated by the data in Figure 5, pH did not influence the Cr/Fe ratio.

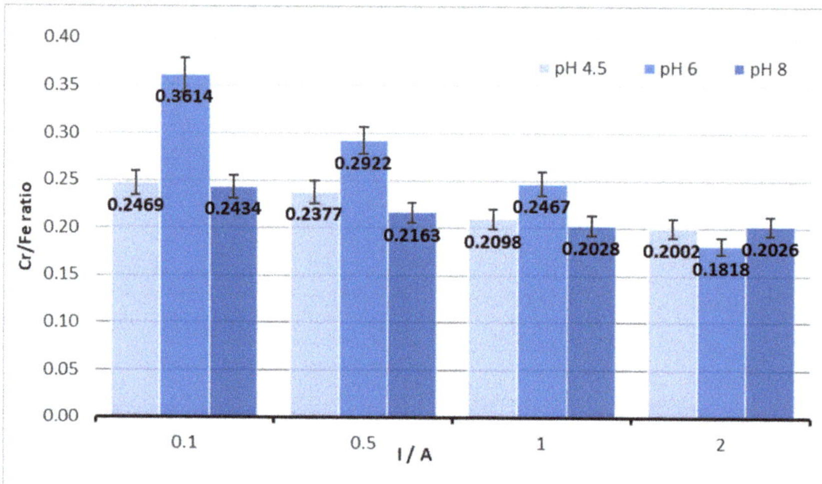

Figure 5. The Cr/Fe ratio's dependence on pH for c(Cr(VI)) = 1000 mg/L and c(NaCl) = 3000 mg/L for different amperages.

In an effort to further improve the Cr/Fe ratio in the solid product in favor of chromium, an experiment in which the polarity of the electrodes was changed at 5 min intervals was carried out. An experiment was carried out for pH = 6 and I = 0.1 A. Although in the case of constant connection, the Cr/Fe ratio was 0.3614, when the polarity of the electrodes was repeatedly changed, a value of 0.3989 was achieved. This represents an increase of 10.3%. Therefore, it is possible to suppose that low currents and a higher frequency of electrode polarity change could contribute to an increase in the proportion of Cr in the solid product.

The dependence of the Cr/Fe ratio on amperage is depicted in Figure 6. A significant decrease in the Cr/Fe ratio for a given pH, mainly for the solution with a chromium concentration of 100 mg/L, can be observed. For a solution with a relatively low concentration of chromium (100 mg/L) and under the chosen experimental conditions described above (mainly a NaCl concentration maximum of 3000 mg/L), as well as with the hardware limitations of the DC supply source used, it was not possible to carry out the experiment at a current of 2 A mainly due to the relatively low conductivity of the solution. Application of a high current on the solution with a low chromium concentration led to the removal of all chromium present and could produce a large amount of Fe ions inside the solution within a short time. As a consequence, the Cr/Fe ratio could be unsatisfactory. For the solutions with the lowest concentration of chromium (100 mg/L), the biggest differences in the Cr/Fe ratio were observed depending on the current. On the other hand, for the highest studied concentration of 2500 mg/L, minimal differences in intensity depending on the applied current were observed.

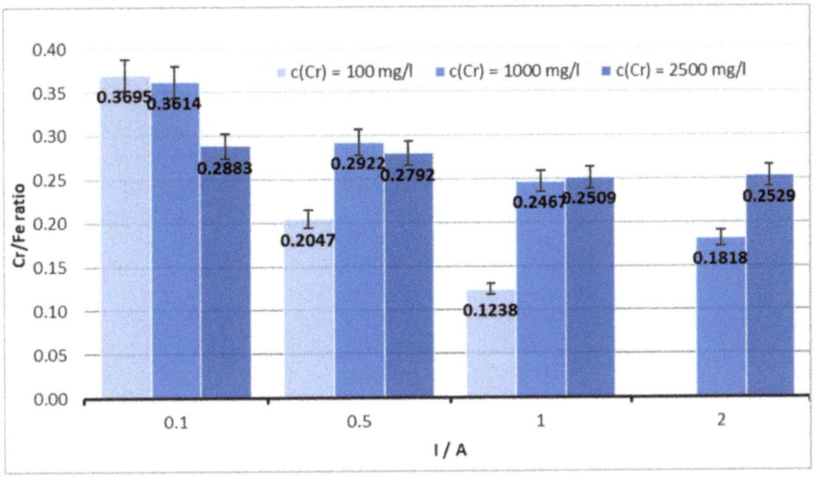

Figure 6. The Cr/Fe ratio's dependence on amperage for different concentrations of Cr(VI), pH = 6, c (NaCl) = 3000 mg/L.

3.3. Overall Experimental Matrix of the Obtained Results

On the basis of the extensive variation in the experimental conditions, an overview of the results in Table 4 was created to help with the rapid determination of the advantages and disadvantages that could influence the purpose of electrocoagulation. The + sign means suitable conditions for electrocoagulation; on the other hand, the − sign describes unsuitable conditions for achieving the desired effect.

Table 4. Scheme of suitable (+) and unsuitable (−) experimental conditions for the EC process of low, medium, and high concentrations of chromium solutions.

Monitored Parameter	c(Cr(VI))/mg/L	pH	G/mS	I/A			
				0.1	0.5	1	2
Duration of EC/min	100	6.0	4.7	−	+	+	n/a
	1000	4.5	7.2	−	−	+	+
		6.0		−	+	+	+
		8.0		−	−	+	+
	2500	6.0	12.2	−	−	−	+
Energy consumption/Wh	100	6.0	4.7	+	+	+	n/a
	1000	4.5	7.2	+	+	−	−
		6.0		+	+	+	−
		8.0		+	+	−	−
	2500	6.0	12.2	+	+	−	−
The Cr/Fe ratio	100	6.0	4.7	+	−	−	n/a
	1000	4.5	7.2	+	−	−	−
		6.0		+	+	+	−
		8.0		+	−	−	−
	2500	6.0	12.2	+	+	+	+

3.4. Solid Product Characterization

On the basis of the XRD measurement, the solid product was evaluated as amorphous, probably in the form of iron and chromium hydroxides. Therefore, thermogravimetric analysis (TGA) was performed to obtain more detailed information on the composition of the product. The aim was to find out if the product only has the expected composition in the form of hydroxides and oxides-hydroxides or if it also contains some impurities originating from the chemicals used in electrocoagulation. Establishing its behavior at elevated temperatures and establishing the temperature for further processing of the solid product to crystallin oxidic form were the goals of the study as well. The sample shows thermal instability up to 800 °C (Figure 7) with a total mass loss of approximately 22% (confirmed by repeated measurements). It is a continuous function of weight loss with a change in its rate, while in the first step, the maximum mass gradient is at a temperature of 192 °C, and the temperature interval of this change starts approximately from 100 °C to 300 °C when the peaks of the second and third steps of decomposition occur with two maxima around 482 °C and 576 °C, respectively. In the mentioned interval of 300–680 °C, there can be observed a slight decrease in mass loss. In the entire monitored interval of thermal heating, an endothermic effect is recorded, which is probably related to the release of bound hydrated water to Fe and Cr oxides, which confirmed our assumption about the composition of the solid product.

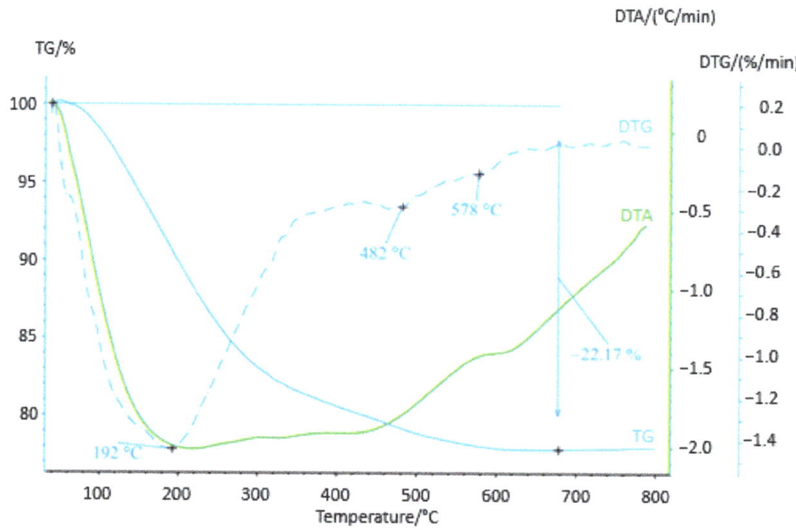

Figure 7. TGA analysis of the solid product of the EC process obtained from the model solution c(Cr(VI) = 2500 mg/L), c(NaCl) = 3000 mg/L, pH = 6, I = 0.1 A.

SEM-EDX analysis (Figure 8) was performed for the composition determination in the micro area for solid products prepared under the conditions described in Table 1, rows 21 and 22. According to the EDX spectra measured in the SEM images, marked areas of elements Fe, Cr, O, K, and Cl and under 0.5% Si, Al, or Mn were each identified. From spectrum 1 in Figure 8A (the entire area shown in the first SEM image), chromium, iron, and oxygen cover 96% of all identified elements, and at spectrum 9 in Figure 8B (small area in the second SEM image), it was 96,7%. The amounts of chromium determined by the EDX analysis were 14.6% or 15.4% for the mentioned areas, and the Cr/Fe ratios calculated from the EDX results were 0.2640 and 0.3298, respectively. However, the composition and amounts of Fe and Cr were in good agreement with those in the AAS analysis (compared with the calculated Cr/Fe ratio of AAS 0.2390 and 0.3989) mentioned in Table 3, rows

21 and 22, respectively, taking into account local microanalysis of the solid product and bulk analysis of the dissolved solid product.

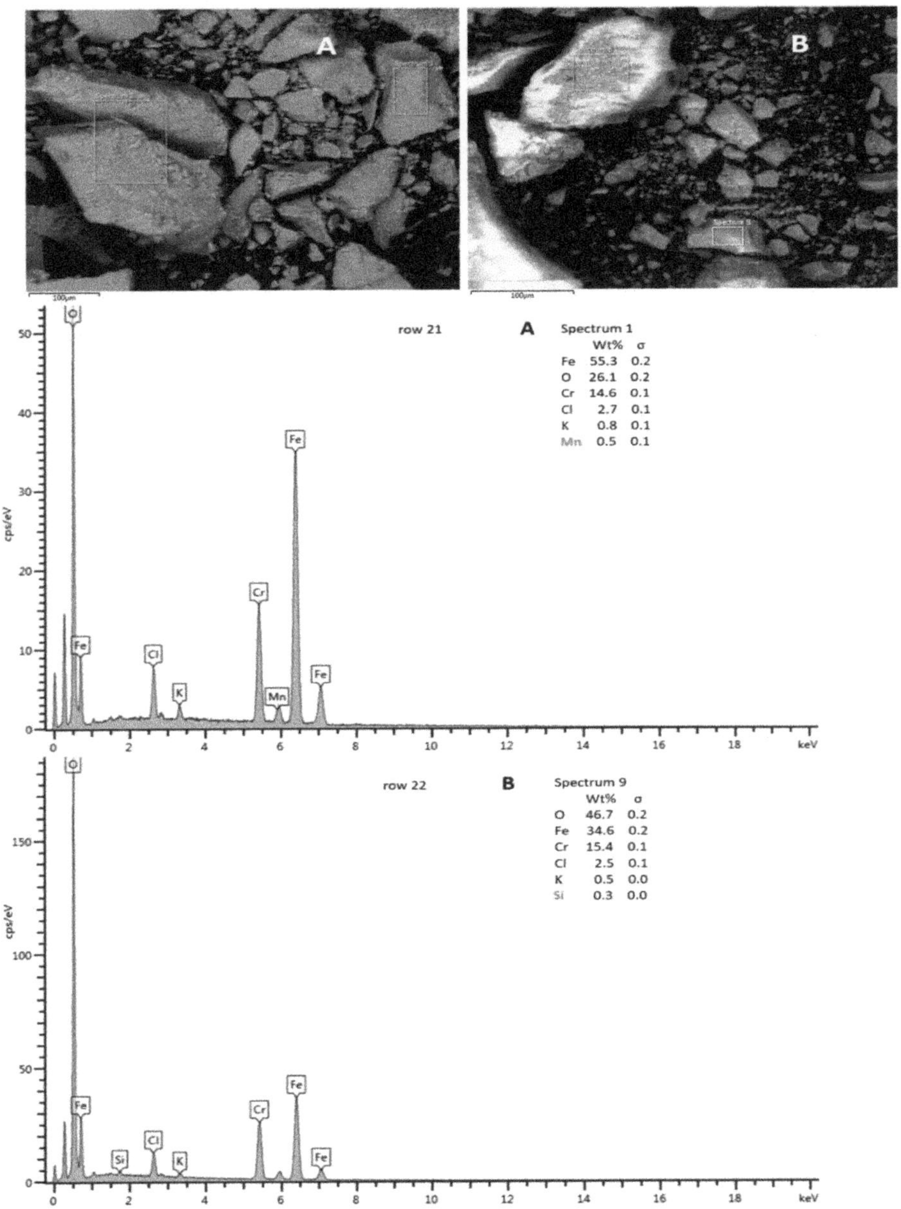

Figure 8. SEM scan and EDX analysis of solid product prepared under the conditions described in Table 1, row 21 (**A**) and SEM scan and EDX analysis of solid product prepared under the conditions described in Table 1, row 22 (**B**).

4. Conclusions

1. Electrocoagulation with a steel cathode and a steel anode was applied to the treatment of model solutions containing chromium (VI) of 100, 1000, and 2500 mg/L. The production of an EC product with a high proportion of chromium was the important investigated parameter.
2. Based on the experimental results, it was observed that Fe-Fe electrodes pair with an applied current intensity of 0.1 A, an inter-electrode distance of 5 cm, solution pH = 6, and a NaCl concentration of 3000 mg/L are the ideal operating conditions to obtain a solid product containing up to 15% Cr. The optimal concentration of NaCl is important due to the sufficient conductivity of the solution and the associated energy consumption.
3. The duration of EC experiments increased in the order of the applied current intensities: 2 A, 1 A, 0.5, and 0.1 A. In reverse order, the ratio of Cr in the EC products decreased. The use of low currents appears suitable because of a better ratio of Cr/Fe, as well as lower energy consumption and almost no overheating of the solution during relatively long experiments with high concentrations of chromium in the solution. However, if the intention is to quickly remove the contaminant, regardless of the composition of the product, it is possible to use high currents. This is primarily the case for many environmental studies that deal with the removal of contaminants.
4. The influence of NaCl concentration on the proportion of chromium in the solid product has not been confirmed.

The experimental result matrix in Table 4 could serve as a tool for the rapid set-up of EC experiments. It is presumed to be useful for the facilitation of other analyses leading to the removal of chromium, mainly within environmental analyses.

Overall, the experimental results indicate that the EC process is able to be an effective tool for obtaining Cr-rich products. For this purpose, the serious limitations of the study are having a low or very high pH, the application of high current intensities connected with the solution overheating, and insufficient conductivity leading to high energy consumption.

The benefit of the work is the observation of an increase in the Cr/Fe ratio when the polarity of the electrodes was repeatedly changed. An increase of 10.3% was observed in this parameter. This observation opens the way to the use of pulsed electrocoagulation, the results of which can be compared with this study in the future stage of research.

From an environmental point of view, 100% efficiency in removing Cr is a useful tool for the remediation of Cr-contaminated wastewater. This is the goal of our future EC research.

Author Contributions: Methodology, Ľ.P. and D.R.; Validation, D.R.; Formal analysis, M.H. and S.R.; Investigation, R.F. and J.B.; Resources, Ľ.P. and M.H.; Writing—original draft, Ľ.P.; Writing—review & editing, M.H. All authors have read and agreed to the published version of the manuscript.

Funding: This research was funded by project VEGA of Ministry of Education, Science, Research and Sport of the Slovak Republic, grant number 1/0008/21.

Institutional Review Board Statement: Not applicable.

Informed Consent Statement: Not applicable.

Data Availability Statement: Not applicable.

Conflicts of Interest: The authors declare no conflict of interest.

References

1. Ackbal, F.; Camci, S. Comparison of Electrocoagulation and Chemical Coagulation for Heavy Metal Removal. *Chem. Eng. Technol.* **2010**, *33*, 1655–1664. [CrossRef]
2. Betancor-Abreu, A.; Mena, V.F.; González, S.; Delgado, S.; Souto, R.M.; Santana, J.J. Design and optimization of an electrocoagulation reactor for fluoride remediation in underground water sources for human consumption. *J. Water Process Eng.* **2019**, *31*, 100865. [CrossRef]

3. Bagrii, V.A.; Demchenko, V.Y.; Remez, S.V.; Bashtan, S.Y. Electrocoagulation purification of water of Cr^{6+} compounds. *J. Water Chem. Technol.* **2012**, *35*, 243–245. [CrossRef]
4. Al-Qodah, Z.; Al-Shannag, M.; Bani-Melhem, K.; Assirey, E.; Yahya, M.A.; Al-Shawabkeh, A. Free radical-assisted electrocoagulation processes for wastewater treatment. *Environ. Chem. Lett.* **2018**, *16*, 695–714. [CrossRef]
5. Kobya, M.; Gengec, E.; Demirbas, E. Operating parameters and costs assessments of a real dyehouse wastewater effluent treated by a continuous electrocoagulation process. *Chem. Eng. Process.* **2016**, *101*, 87–100. [CrossRef]
6. Tchamango, S.R.; Ngayo, K.W.; Belibi, P.D.B.; Nkouam, F.; Ngassoum, M.B. Treatment of a dairy effluent by classical electrocoagulation and indirect electrocoagulation with aluminum electrodes. *Sep. Sci. Technol.* **2020**, *56*, 1128–1139. [CrossRef]
7. Kartikaningsih, D.; Shih, Y.J.; Huang, Y.H. Boron removal from boric acid wastewater by electrocoagulation using aluminum as sacrificial anode. *Sustain. Environ. Res.* **2016**, *26*, 150–155. [CrossRef]
8. Al-Qodah, Z.; Al-Shannag, M. Heavy metal ions removal from wastewater using electrocoagulation processes: A comprehensive review. *Sep. Sci. Technol.* **2017**, *52*, 2649–2676. [CrossRef]
9. Asaithambi, P.; Sajjadi, B.; Aziz, A.R.A.; Daud, W.M.A.W. Performance evaluation of hybrid electrocoagulation process parameters for the treatment of distillery industrial effluent. *Process Saf. Environ. Prot.* **2016**, *104*, 406–412. [CrossRef]
10. Das, D.; Nandi, B.K. Treatment of iron ore beneficiation plant process water by electrocoagulation. *Arab. J. Chem.* **2021**, *14*, 102902. [CrossRef]
11. Hu, Q.; He, L.; Lan, R.; Feng, C.; Pei, X. Recent advances in phosphate removal from municipal wasterwater by electrocoagulation process: A review. *Sep. Purif. Technol.* **2023**, *308*, 122944. [CrossRef]
12. Sharifi, N.S.; Karimi-Jashni, A. Development and application of novel high throughput metal waste chips and foam electrodes for electrocoagulation treatment of greywater. *Environ. Technol.* **2023**, *44*, 528–539. [CrossRef]
13. Verma, R.K.; Gautam, K.; Agrahari, S.; Kumar, S. Potential of electrocoagulation technology for the treatment of tannery industrial effluents: A brief review. *Chem. Process Eng.* **2022**, *43*, 217–222.
14. Maiti, S.K.; Hazra, T.; Debsarkar, A. Removal of total chromium from landfill leachate by electrocoagulation process using iron electrodes: Effect of some major operational and chemical parameters. *J. Indian Chem. Soc.* **2020**, *97*, 2453–2460.
15. Peng, H.; Guo, J. Removal of chromium from wastewater by membrane filtration, chemical precipitation, ion exchange, adsorption electrocoagulation, electrochemical reduction, electrodialysis, electrodeionization, photocatalysis and nanotechnology: A review. *Environ. Chem. Lett.* **2020**, *18*, 2055–2068. [CrossRef]
16. Thirugnanasambandham, K.; Shine, K. Investigation on the removal of chromium from wastewater using electrocoagulation. *Int. J. Chem. React. Eng.* **2018**, *16*, 20170155. [CrossRef]
17. Handman, S.S.; El-Naas, M.H. Characterization of the removal of Chromium(VI) from groundwater by electrocoagulation. *J. Ind. Eng. Chem.* **2014**, *20*, 2775–2781.
18. Keshmirizadeh, E.; Yousefi, S.; Rofouei, M.K. An investigation on the new operational parameter effective in Cr (VI) removal efficiency: A study on electrocoagulation by alternating pulse current. *J. Hazard. Mater.* **2011**, *190*, 119–124. [CrossRef]
19. Dermentzis, K.; Christoforidis, A.; Valsamidou, E.; Lazaridou, A.; Kokkinos, N. Removal of hexavalent chromium from electroplating wastewater by electrocoagulation with iron electrodes. *Glob. Nest J.* **2011**, *13*, 412–418.
20. Aguilar-Ascón, E.; Marrufo-Saldaña, L.; Neyra-Ascón, W. Reduction of Total Chromium Levels from Raw Tannery Wastewater via Electrocoagulation using Response Surface Methodology. *J. Ecol. Eng.* **2019**, *20*, 217–224. [CrossRef]
21. Rodriguez, M.G.; Mendoza, V.; Puebla, H.; Martinez, S.A. Removal of Cr(VI) from wastewaters at semi-industrial electrochemical reactors with rotating ring electrodes. *J. Hazard. Mater.* **2009**, *163*, 1221–1229. [CrossRef] [PubMed]

Disclaimer/Publisher's Note: The statements, opinions and data contained in all publications are solely those of the individual author(s) and contributor(s) and not of MDPI and/or the editor(s). MDPI and/or the editor(s) disclaim responsibility for any injury to people or property resulting from any ideas, methods, instructions or products referred to in the content.

materials

MDPI

Article

The Influence of Manganese Addition on the Properties of Biodegradable Zinc-Manganese-Calcium Alloys

Wanda Mamrilla [1,2], Zuzana Molčanová [2], Beáta Ballóková [2], Miroslav Džupon [2], Róbert Džunda [2], Dávid Csík [2,3], Štefan Michalik [4], Maksym Lisnichuk [5] and Karel Saksl [2,3,*]

1 Department of Biomedical Engineering and Measurement, Faculty of Mechanical Engineering, Technical University of Košice, Letná 9, 040 01 Košice, Slovakia; wanda.mamrilla@tuke.sk
2 Slovak Academic of Science, Institute of Materials Research, Watsonova 47, 040 01 Košice, Slovakia; molcanova@saske.sk (Z.M.); bballokova@saske.sk (B.B.); mdzupon@saske.sk (M.D.); rdzunda@saske.sk (R.D.); dcsik@saske.sk (D.C.)
3 Faculty of Materials, Metallurgy and Recycling, Institute of Materials and Quality, Technical University of Košice, Letná 9, 042 00 Košice, Slovakia
4 Diamond Light Source Ltd., Harwell Science and Innovation Campus, Didcot OX11 0DE, UK; stefan.michalik@diamond.ac.uk
5 Faculty of Science, Institute of Physics, Pavol Jozef Šafárik University in Košice, Park Angelinum 9, 040 01 Košice, Slovakia; maksym.lisnichuk@upjs.sk
* Correspondence: karel.saksl@tuke.sk; Tel.: +421-55-602-2023

Abstract: This study focuses on the preparation and characterization of zinc-based alloys containing magnesium, calcium, and manganese. The alloys were prepared by the melting of pure elements, casting them into graphite molds, and thermo-mechanically treating them via hot extrusion. The phase compositions of the samples were analyzed using X-ray diffraction technique and SEM/EDX analysis. The analysis confirmed that in addition to the Zn matrix, the materials are reinforced by the $CaZn_{13}$, $MgZn_2$, and Mn-based precipitates. The mechanical properties of the alloys were ascertained by tensile, compressive, and bending tests, measurement of the samples microhardness and elastic modulus. The results indicate that an increase in Mn content leads to an increase in the maximum stress experienced under both tension and compression. However, the plastic deformation of the alloys decreases with increasing Mn content. This study provides valuable insights into the microstructural changes and mechanical behavior of zinc-based alloys containing magnesium, calcium, and manganese, which can be used to design alloys for specific biomedical applications.

Keywords: zinc-based alloys; microstructure; mechanical properties; hot extrusion; synchrotron data; EDX analysis

Citation: Mamrilla, W.; Molčanová, Z.; Ballóková, B.; Džupon, M.; Džunda, R.; Csík, D.; Michalik, Š.; Lisnichuk, M.; Saksl, K. The Influence of Manganese Addition on the Properties of Biodegradable Zinc-Manganese-Calcium Alloys. *Materials* **2023**, *16*, 4655. https://doi.org/10.3390/ma16134655

Academic Editor: Lili Tan

Received: 14 June 2023
Revised: 23 June 2023
Accepted: 26 June 2023
Published: 28 June 2023

Copyright: © 2023 by the authors. Licensee MDPI, Basel, Switzerland. This article is an open access article distributed under the terms and conditions of the Creative Commons Attribution (CC BY) license (https://creativecommons.org/licenses/by/4.0/).

1. Introduction

Biodegradable metals and alloys are materials that have the ability to degrade and dissolve over time in a biological environment. These materials offer several advantages for various applications, particularly in the field of biomedical engineering. They can be used for temporary implants, providing support, and/or treatment to the affected area while gradually degrading and being replaced by natural tissue. This eliminates the need for additional surgeries to remove permanent implants, reducing complications and promoting faster healing [1,2]. The alloys currently under investigation are predominantly comprised of iron (Fe), magnesium (Mg), and zinc (Zn) [3]. Although Fe alloys have good mechanical properties, their large elastic moduli can result in a stress-shielding effect. Additionally, their relatively slow degradation rates may result in the accumulation of corrosion products within the body, which could potentially trigger inflammation [4]. Mg alloys possess low densities and high specific strengths. However, their rapid degradation rates and release of hydrogen during the degradation process impose limitations on their potential applications [5,6]. Zinc alloys, on the other hand, exhibit a moderate corrosion potential, falling between those of iron and magnesium (Mg) alloys, which makes their degradation rates comparatively more suitable for certain applications [7]. Hence, Zn alloys are recognized as an advanced biodegradable material for the upcoming intracorporeal implants. Nevertheless, the mechanical properties of pure Zn are unsatisfactory and their enhancement can be achieved through effective alloying and deformation processing [8,9]. Previous studies have indicated that magnesium is an exceptionally effective element for alloying with Zn alloys, greatly improving their mechanical performance [10]. Calcium similarly enhances the mechanical properties of the alloys, making them stronger and more durable. Calcium also promotes bone regeneration and healing, facilitating better integration with the surrounding bone tissue and improving the overall performance of implants. Additionally, calcium plays a crucial role in the regulation of cellular processes, including cell adhesion and proliferation, which are important for tissue growth and healing. Manganese refines the grain size enhancing the strength of the biodegradable zinc alloys. Furthermore, manganese plays a role in influencing the corrosion rate of these alloys. It helps mitigate the corrosion process and extend the lifespan of the implants. This corrosion control is essential for ensuring the long-term stability and reliability of the zinc biodegradable alloys within the human body metabolism of amino acids and carbohydrates [11].

Multiple criteria have been established for materials utilized for bone implants, including the requirement for an ultimate tensile strength (UTS) between 200 and 300 MPa, an elongation at fracture of 15 to 20%, and a elastic modulus (E) falling within the range of 10 to 30 GPa [12]. Venezuela et al. proposed target values for orthopedic internal fixation devices, which include a yield strength (YS) greater than 230 MPa, an ultimate tensile strength (UTS) greater than 300 MPa, a elastic modulus (E) ranging from 10 to 20 GPa, and an elongation at fracture (ε) of 15 to 18% [13].

In a study conducted by Yang et al., Zn-Mg-Ca ternary alloys were investigated. The addition of Mg had the most significant positive impact on the mechanical properties of the alloys. The precipitation of the intermetallic phase, Mg_2Zn_{11}, resulted in a desirable hardening effect through dispersion strengthening. Among the studied alloys, $ZnMg_{1.5}Ca_{0.1}$ demonstrated the highest ultimate tensile strength of 442 MPa after undergoing hot extrusion. On the other hand, the $ZnMg_1Ca_{0.1}$ alloy exhibited the highest elongation of $\varepsilon = 5.44\%$ [14].

Liu et al. investigated Zn-1Mg-0.1Mn alloys and observed exceptional mechanical performance after hot rolling [15]. These alloys exhibited notable improvements, such as yield strength of 195 MPa, an ultimate tensile strength of 299 MPa, an elongation of 26%, and a hardness of 107.82 Hv.

System Zn3Mg0.7Y was studied by Panaghie et al. preparing by induction casting. Materials performed increased Young modulus 15.07, materials stiffness 1.65 and micorhardness 1.10 compared to pure Zn which achieved values of 2.9, 0.82 and 0.17, respectively [16].

Zhang-Zhi et al. aimed to improve mechanical properties by adding 0.4 wt% of Ag, Cu or Ca to biodegradable Zn-0.8Mn alloy. The ductility and the tensile strength were improved after extrusion. The YS, UTS and ε for Zn-0.8Mn-0.4Ag were 156 MPa, 251 MPa and 62%, respectively. For Zn-0.8Mn-0.4Cu alloy it was 191 MPa, 308 MPa and 39%, respectively. The highest values of YS 253 MPa and UTS 343 MPa are obtained for Zn-0.8Mn-0.4Ca alloy with ε of 8% [17].

This paper focuses on examining the impact of Mn on the microstructure and mechanical properties of the Zn-Mg-Ca ternary alloys. The objective is to establish a correlation between the chemical composition, microstructure, and mechanical properties of these alloys, with the aim of providing valuable data to optimize the composition design and processing of Zn-based biodegradable alloys containing Mn. It is a so far unstudied new system consisting of four biocompatible elements. The mechanical properties are improved not only by addition of Mn but laso by the processing technique.

2. Materials and Methods

2.1. Preparation of Alloys

The melt-stirring technique was used to fabricate experimental material. The raw metal materials were melted from high-purity zinc (Heneken 99.995%), magnesium (Alfa Aesar 99.98%), calcium (Alfa Aesar 99.5%), and manganese (Alfa Aesar 99.3%). Melting was performed in electric furnace at 650 °C under protective atmosphere of argon gas. The alloys were cast into a preheated graphite cylinder-shaped mold using the gravitational casting method. The as-cast ingots in the next step underwent a hot extrusion at temperature of 250 °C. The extrusion was carried out with an extrusion ratio of 21:1 and a speed of 3 mm/s. The resulting extruded rods had a circular cross-section with a diameter of 4 mm, and they were naturally cooled to room temperature in calm air. Each prepared alloy composition had a constant weight percentage of 0.4% for both Mg and Ca, while the weight percentage of Mn varied from 0 to 1.1 wt.%. The chemical compositions of prepared alloys were determined by EDX analyses (Table 1). As from the table can be seen the composition of all the prepared alloys exhibited a difference of less than 0.07 wt.% compared to their nominal composition.

Table 1. Chemical composition of the prepared alloys.

Nominal Composition [wt%]	Actual Composition [wt%]			
	Zn	Mg	Ca	Mn
Zn-0.4Mg-0.4Ca	99.17	0.44	0.39	-
Zn-0.4Mg-0.4Ca-0.2Mn	99.05	0.41	0.39	0.15
Zn-0.4Mg-0.4Ca-0.4Mn	98.8	0.47	0.38	0.35
Zn-0.4Mg-0.4Ca-0.6Mn	98.58	0.44	0.43	0.55
Zn-0.4Mg-0.4Ca-0.8Mn	98.45	0.38	0.42	0.75
Zn-0.4Mg-0.4Ca-1.1Mn	98.04	0.48	0.38	1.1

2.2. Microstructure and Phase Characterization

Metallographic analyses were performed in order to evaluate uniformity of the microstructure and to ascertain the average size of matrix grains. The specimens were first embedded in an electrically conductive material (PolyFast, Struers), followed by grinding with SiC (P-800) abrasive papers. In the final step, was polishing using a diamond abrasive paste containing particles of size up to 0.25 μm. The microstructure of the alloys was revealed by etching of the samples polished surface in the Nital etchant (Ethanol 96% and HNO_3 4%). The microstructure of both the as-cast and extruded specimens was examined using two different techniques. The first technique involved the use of the Olympus GX 71 optical microscope, which provides a macroscopic view of the sample to study its overall grain structure and phase distribution. For more detailed analysis, a scanning electron microscope (TESCAN VEGA-3 LMU, TESCAN Brno, s.r.o., Brno, Czech Republic) equipped by EDX spectrometer (Bruker Nano Xflash 410M, Billerica, MA, USA) was employed. The phase composition of the hot extruded samples was determined by analyzing X-ray diffraction (XRD) patterns. These XRD patterns were obtained from synchrotron measurements conducted at the I12-JEEP beamline [18] located at the Diamond Light Source in the Harwell Science and Innovation Campus in Oxfordshire, UK. The X-ray diffraction (XRD) measurements were conducted by irradiating the samples with an X-ray beam of cross-section 0.5 mm × 0.5 mm and photon energy of 108.69 keV. Diffraction images were collected in transmission geometry employing a 2M CdTe Pilatus detector situated in an asymmetric position in respect of an incoming X-ray beam. 2D diffraction images were azimuthally integrated into 1D intensity curves as a function of a scattering angle 2θ using the DAWN software, 2.26.0 [19].

Thin foils were prepared for detailed observation of the microstructure in a TEM. The experimental samples from hot extruded alloys were prepared by grinding/polishing and ion milling employing Ion Slicer EM-09100IS (JEOL, Tokyo, Japan). TEM images were acquired by means of High-resolution Scanning-transmission Electron Microscope (S)TEM JEOL 2100F (JEOL, Japan), equipped with a Schottky field emission gun, and operated at an acceleration voltage of 200 kV. For phase identification, the Selected Area Electron Diffraction (SAED) technique was used.

2.3. Mechanical Testing

The tensile strength and deformation properties of the alloys after extrusion were tested by the uniaxial tensile strength test. For the experiment, specimens of diameter 3 mm and a length of 15 mm were prepared according to the ASTM-E8-E8M standard [20]. The compressive strength was evaluated on cylindrical specimens with a diameter of $\phi = 4$ mm and a length of 8 mm, and for the bending test, $\phi = 4$ mm × 55 mm according to the ASTM E9-89 standard [21]. All mechanical properties were evaluated at room temperature by using a universal material testing machine (TIRATEST 2300, TIRA, Schalkau, Germany). The crosshead speed used for the tensile test was 0.1 mm/min, 0.2 mm/min for the compressive test, and 1 mm/min for the bending test. Each composition underwent a minimum of five tests to ensure the reproducibility of the results. After, the tests fracture surfaces of the specimens were examined using the Jeol JSM 7000F scanning electron microscopy. Microhardness measurements were performed on the hot extruded samples using a Wilson-Wolper Tukon 1102 microhardness tester with a Vickers indenter. The Vickers hardness was determined according to ISO 6507 [22]. The applied load was 0.1 kg with a dwelling time of 10 s.

3. Results

3.1. Microstructure of the Zn-Mg-Ca and Zn-Mg-Ca-Mn Alloys

The microstructure of the as-cast state experimental materials was examined, revealing polyhedral grains with an approximate size of 40 µm. After hot extrusion, the grains reduce to approximately 10 µm. The hot extrusion process allows grain refinement due to ongoing dynamic recrystallization, as well as fragmentation of the eutectic structure. Figure 1a,b shows the microstructures of the Zn-0.4Mg-0.4Ca alloy in the as-cast state and after extrusion, respectively. Figure 1c,d display the microstructures of the Zn-0.4Mg-0.4Ca-0.8Mn samples. Notably, rectangular precipitates are visible in the Mn-containing samples after deformation.

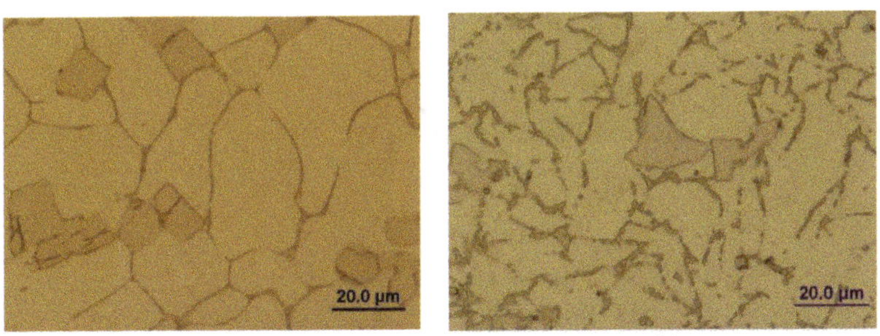

(a) (b)

Figure 1. *Cont.*

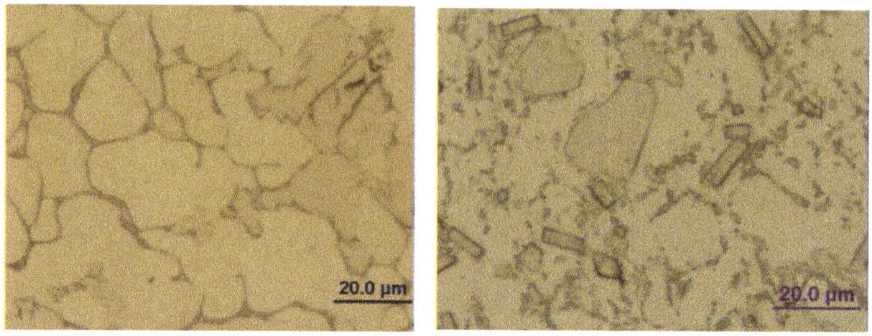

(c) (d)

Figure 1. Optical micrographs of (**a**) Zn-0.4Mg-0.4Ca as-cast alloy, (**b**) Zn-0.4Mg-0.4Ca extruded alloy, (**c**) Zn-0.4Mg-0.4Ca-0.8Mn as-cast alloy, (**d**) Zn-0.4Mg-0.4Ca-0.8Mn extruded alloy.

The EDX analysis conducted on the microstructure revealed the presence of various phases in the alloys, including Zn, $CaZn_{13}$, $MgZn_2$, and Mn. Figure 2a depicts the microstructure of the as-cast alloy, which consists of primary grains and eutectic structures at their boundaries. Primary grains form during the solidification process, while eutectic structures arise due to the eutectic reaction between solid phases during the cooling process. Irregular precipitates observed at the boundaries of the primary grains in Figure 2b,c were confirmed. The EDX analysis suggests presence of the $MgZn_2$, $CaZn_{13}$, and Mn-containing precipitates.

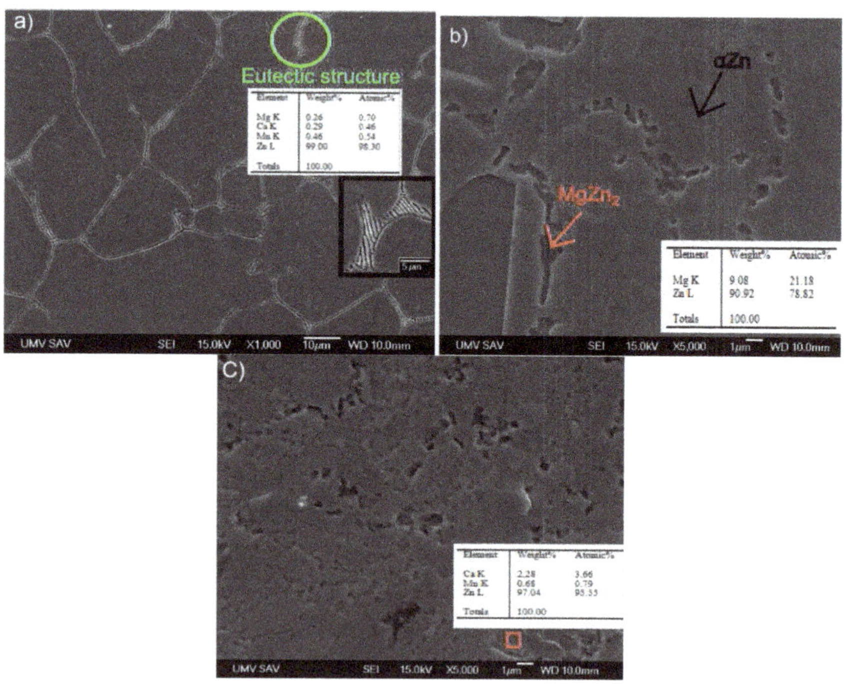

Figure 2. SEM micrographs (**a**) Zn-0.4Mg-0.4Ca-0.4Mn as-cast alloy with detail of eutectic structure, (**b**) Zn-0.4Mg-0.4Ca-0.4 Mn extruded alloy, (**c**) Zn-0.4Mg-0.4Ca-0.8Mn extruded alloy, red box represents region with precipitates.

The synchrotron X-ray diffraction patterns provided confirmation of the presence of the mentioned phases in the alloy. The Zn matrix phase (ICDD: 00-004-0831 [23]) exhibited a hexagonal structure with the space group P63/mmc. Similarly, the $MgZn_2$ phase (ICDD: 00-077-1177 [24]) also displayed a hexagonal structure with the same space group. The $CaZn_{13}$ phase (ICDD: 00-028-0258 [25]) demonstrated a cubic structure with the space group Fm3c, as depicted in Figure 3. These structural characterizations contribute to our understanding of the alloy's crystallographic properties and aid in further analysis of its behavior and performance.

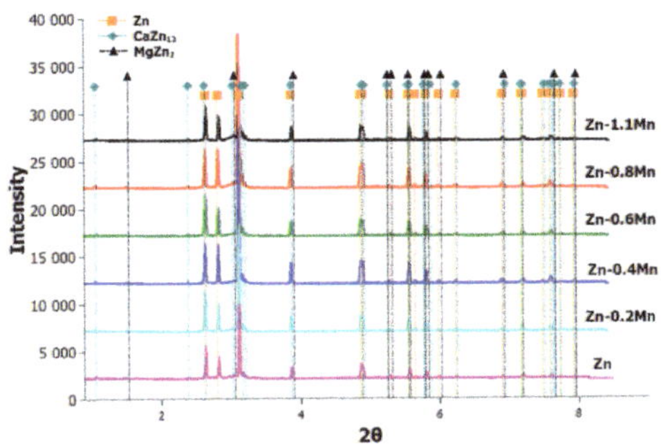

Figure 3. Synchrotron patterns of Zn-Mg-Ca-Mn extruded alloys, **Zn** represents Zn-0.4Mg-0.4Ca alloy.

Rietveld refinement of the synchrotron XRD data provided additional valuable information about the unit cell volumes of each phase (Figure 4a) and quantitative fraction of these phases, see Figure 4b).

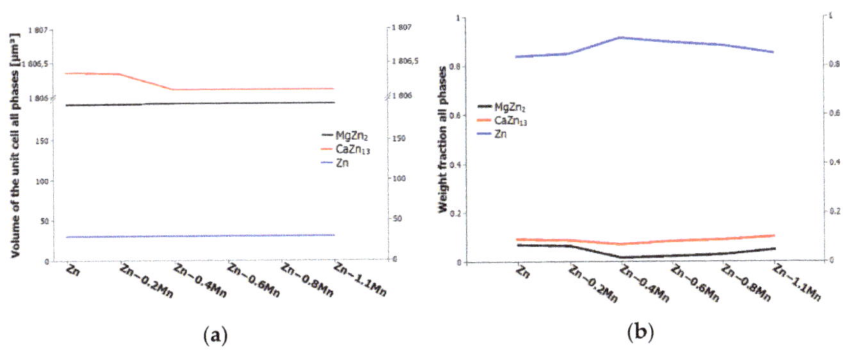

Figure 4. Synchrotron analysis: (**a**) volume of unit cell, (**b**) weight fraction all phases. **Zn** represents Zn-0.4Mg-0.4Ca alloy.

Figure 5b shows TEM bright field image of the Zn-0.4Mg-0.4Ca sample. The TEM analysis confirms presence of the above-mentioned intermetallic phases. The larger grains were identified as the pure Zn phase with a space group of P63/mmc, while the smaller grains were identified as the $CaZn_{13}$ phase with a space group of Fm-3c.

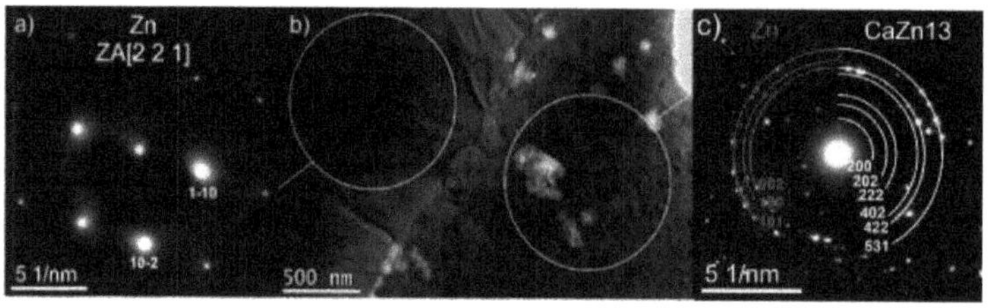

Figure 5. (**a**) SAED of the areas selected in the image (**b**), (**b**) TEM image of two different phases Zn and $CaZn_{13}$ with magnification 10 kx, and (**c**) SAED of the areas selected in the image (**b**).

3.2. Mechanical Properties of Zn-Mg-Ca and Zn-Mg-Ca-Mn Alloys

Tensile, compressive, and three-point bending tests, as well as measurements of elastic moduli and microhardness, were performed in order to evaluate the influence of adding manganese to the Zn-Mg-Ca alloys. Based on the performed tests, it was confirmed that there is a direct correlation between the content of manganese and the maximum stresses in both tension and compression. As the Mn content increased, the maximum stresses observed during both tensile and compressive testing also increased. This indicates that the addition of Mn increases mechanical strength of the Zn-Mg-Ca alloys, resulting in higher stress tolerance in both tension and compression (Table 2). Among the tested alloys, the Zn-0.4Mg-0.4Ca-1.1Mn alloy exhibited the highest strength in tension having a UTS (ultimate tensile strength) of 379 MPa (Table 2) and a YS (yield strength) of 299 MPa, which are considerably higher values compared the pure Zn, UTS and YS of 30 MPa and 23 MPa, respectively [15]. However, the Zn-0.4Mg-0.4Ca-1.1Mn alloy exhibited the lowest elongation to fracture among the all the tested samples, only 1.6%. Despite its high tensile and yield strength, this alloy displayed reduced ductility, indicating a limited ability to undergo plastic deformation before fracture. The elastic moduli of the alloys were compared to the elastic modulus range of human bone, having elastic modulus between 5 and 23 GPa [26]. The results showed that the alloys had significantly higher values compared to human bone, Table 2. This suggests that the tested alloys possess a greater stiffness or rigidity than human bone, which may have implications for their use in load-bearing applications where matching or mimicking the mechanical properties of bone is desired. In compressive testing, the Zn-0.4Mg-0.4Ca-1.1Mn alloy demonstrated the highest UCS (ultimate compressive strength), reaching a maximum value of 645 MPa (Table 2). This indicates a significant improvement in the ultimate strength compared to pure zinc, which exhibited a value of 103 MPa [27]. Regarding mechanical stiffness, the alloys meet the requirements to be considered as potential candidates for load-bearing biodegradable implants and stents [28]. The stress–strain curves of tensile and compression tests are shown in Figure 6a,b.

Table 2. Summarized mechanical properties of experimental materials.

As-Extruded Alloys	Label	Tensile Test				Compression Test		Bending Test
		YS [MPa]	σ_{UTS} [MPa]	ε [%]	E [GPa]	σ_{UCS} [MPa]	ε [%]	σ_{UBS} [MPa]
Pure Zn	Pure Zn	22.85 [15]	28	3.3 ± 0.1	96.5 [29]	102.92 ± 6.73 [27]	-	-
Zn-0.4Mg-0.4Ca	Zn-0Mn	253 ± 10	269 ± 6	4.2 ± 0.8	147 ± 6	486 ± 0	32 ± 1	137
Zn-0.4Mg-0.4Ca-0.2Mn	Zn-0.2Mn	277 ± 19	289 ± 14	2.7 ± 0.6	129 ± 2	558 ± 0	48 ± 1	140
Zn-0.4Mg-0.4Ca-0.4Mn	Zn-0.4Mn	289 ± 13	322 ± 7	3.6 ± 0.1	148 ± 2	566 ± 1	28 ± 1	140
Zn-0.4Mg-0.4Ca-0.6Mn	Zn-0.6Mn	287 ± 15	332 ± 7	3.5 ± 0.3	145 ± 3	626 ± 6	33 ± 1	153
Zn-0.4Mg-0.4Ca-0.8Mn	Zn-0.8Mn	275 ± 21	326 ± 8	2.5 ± 0.3	124 ± 3	616 ± 3	33 ± 1	159
Zn-0.4Mg-0.4Ca-1.1Mn	Zn-1.1Mn	299 ± 8	379 ± 2	1.6 ± 0.2	138 ± 6	645 ± 0	31	162

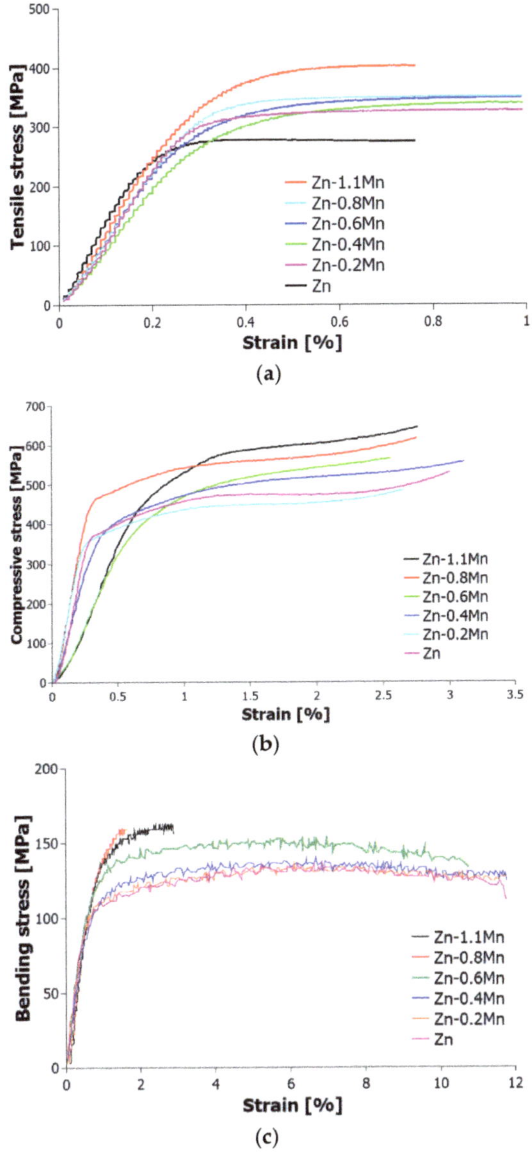

Figure 6. Ultimate strength and stress–strain curves of as-extruded alloys (**a**) tensile test, (**b**) compression test, and (**c**) three-point bending test.

In the three-point bending test, Zn-0.4Mg-0.4Ca-1.1Mn alloy exhibited the highest UBS (ultimate bending strength), with a maximum value of 162 MPa, which was 18.6% higher compared the Zn-0.4Mg-0.4Ca alloy. The stress–strain curves of the alloys showed that those with lower Mn concentrations had better ductility, whereas those with higher Mn concentrations became more brittle (Figure 6c).

Figure 7 and Table 3 present the results obtained from microhardness testing. The data clearly demonstrate that the microhardness (HV0.1) of the extruded materials increases with an increase in weight percentage of manganese. In general, the microhardness measurements indicate that the Zn-0.4Mg-0.4Ca-1.1Mn alloy exhibits a microhardness (HV0.1) that is 1.5 times higher compared to the Zn-0.4Mg-0.4Ca alloy (as shown in Figure 7). The observed increase in microhardness can be attributed to two factors: the formation of a harder solid solution matrix and the grain refinement caused by the presence of manganese.

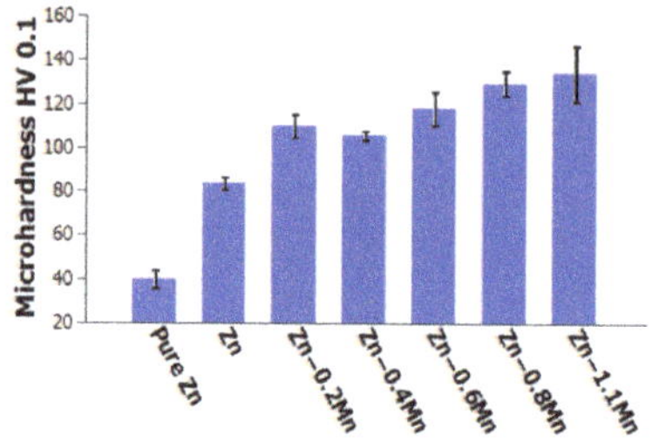

Figure 7. Microhardness HV0.1 of experimental materials.

Table 3. Microhardness HV0.1 of experimental materials.

As-Extruded Alloys	Microhardness HV0.1
Pure Zn	40 ± 2
Zn-0.4Mg-0.4Ca	83 ± 3
Zn-0.4Mg-0.4Ca-0.2Mn	110 ± 5
Zn-0.4Mg-0.4Ca-0.4Mn	105 ± 2
Zn-0.4Mg-0.4Ca-0.6Mn	118 ± 8
Zn-0.4Mg-0.4Ca-0.8Mn	129 ± 6
Zn-0.4Mg-0.4Ca-1.1Mn	134 ± 13

3.3. Fractography Analysis

Fracture surface analysis was performed on the specimens used for the tensile tests, comparing alloys without Mn and with the maximum Mn content.

Figure 8a displays the fracture surface of the Zn-0.4Mg-0.4Ca alloy, while Figure 8b provides a closer view of a specific area on the fracture surface. The images clearly document that the fracture is primarily transcrystalline, characterized by cleavage planes, with limited evidence of plastic deformation. A macro image of the fracture surface documents a wrinkled surface which is a manifestation of some degree of macroplasticity of the material. The detailed image of the fracture surface reveals the presence of a network of parallel cracks that propagated within the grains. These cracks have originated at the grain boundaries. This indicates that the fracture process involved intergranular cracking, where cracks initiated and propagated along the interfaces between adjacent grains.

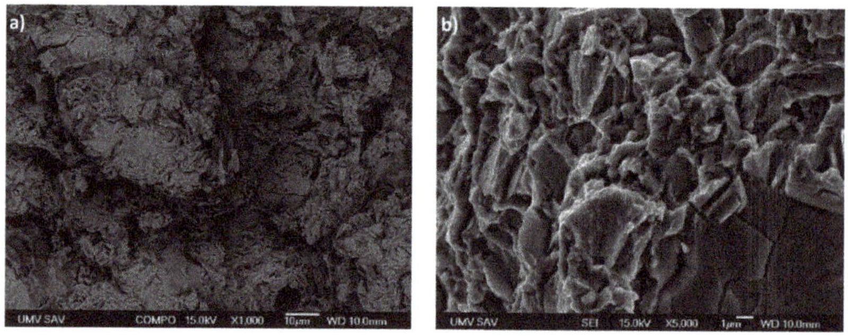

Figure 8. (**a**) Characteristic fracture surface of the alloy Zn-0.4Mg-0.4Ca after tensile test, (**b**) detail of the fracture surface.

Figure 9a shows the fracture surface of the Zn-0.4Mg-0.4Ca-1.1Mn alloy, while Figure 9b provides a detailed view on it. The fine particle morphology observed in the eutectic mixture of this alloy indicates its ability to withstand higher loads before fracturing. This is attributed to the presence of Mn which enhances the strength and toughness of the material. The refined morphology of the eutectic mixture, particularly around the cleavage planes, suggests that the material underwent plastic deformation before fracturing. The presence of Mn in the alloy contributes to its enhanced mechanical properties, allowing it to withstand higher tensile, Table 2.

Figure 9. (**a**) Characteristic fracture surface of the alloy Zn-0.4Mg-0.4Ca-1.1Mn after tensile test, (**b**) detail of fracture surface.

The correlation between the mixed fracture morphology and the mechanical properties of the examined alloys is indeed a crucial finding. The observation of thick and transcrystalline fractures in the Zn-0Mn alloy contrasts with the refined fracture surfaces and the presence of plastic deformation in the Zn-1.1Mn alloy having enhanced tensile strength. This correlation underscores the significance of alloy design in improving the mechanical properties of zinc-based materials. By carefully selecting and incorporating alloying elements, such as Mn, it is possible to enhance the strength and toughness of zinc-based alloys, resulting in improved mechanical performance. These findings provide valuable insights for alloy development and can guide the design of materials with desired mechanical properties.

4. Conclusions

The results of this experimental study can be summarized as follows:

1. The melt-stirring combined with hot-extrusion proved to be a suitable technique for preparation of the Zn-Mg-Ca-Mn alloy.
2. The addition of manganese was found to have a significant influence on both the microstructure and mechanical properties of the newly investigated biodegradable Zn-Mg-Ca-Mn material system.
3. The microstructure analysis revealed a noticeable difference between the as-cast Zn materials showing polyhedric grains of diameter 40 μm and after extrusion with grains of size 10 μm. The weight fraction of manganese in alloy has a significant impact on both the mechanical properties and microstructure of the alloys.
4. The microstructure and phase analysis confirms presence of the Zn, $CaZn_{13}$, $MgZn_2$, and Mn-containing phases in the eutectic structures of the alloys, with fraction varying depending on the composition.
5. The addition of Mn resulted in an increase in the maximum ultimate strength (US) in both tension and compression, as well as an increase in the microhardness (HV0.1) of the materials. Among all those tested, the Zn-0.4Mg-0.4Ca-1.1Mn alloy exhibited the highest values for ultimate tensile strength (UTS) at 379 MPa, ultimate compressive strength (UCS) at 645 MPa, ultimate bending strength (UBS) at 162 MPa, and

microhardness at 133 HV. These findings demonstrate the beneficial effect of Mn on enhancing the mechanical strength of the alloys. However, the incorporation of Mn also resulted in a decrease in ductility, as evidenced by the lowest elongation at fracture observed in the Zn-0.4Mg-0.4Ca-1.1Mn alloy.

6. The results also showed that the mechanical properties of the alloys can be tailored by varying alloy composition and heat treatment, making them suitable for different applications. In conclusion, this study offers significant insights into the microstructural modifications and mechanical performance of zinc-based alloys. These findings serve as a valuable resource for guiding the development of novel alloys with enhanced properties.

Author Contributions: Conceptualization, W.M., Z.M., B.B. and K.S.; methodology W.M., Z.M. and B.B.; Writing—original draft preparation W.M.; investigation, W.M., Z.M., B.B., M.D., R.D., D.C., M.L. and Š.M.; validation, K.S.; writing—review and editing, W.M., Z.M., B.B., Š.M. and K.S.; supervision, K.S. All authors have read and agreed to the published version of the manuscript.

Funding: This work was supported by the Slovak Research and Development Agency under contract No. APVV-20-0205, APVV-20-0068, APVV-20-0138 and APVV-21-0274. Authors are grateful also to the Scientific Grant Agency of the Ministry of Education, Science, Research and Sport of the Slovak Republic and the Slovak Academy of Sciences (VEGA project No. 2/0039/22 and VEGA2/0039/23). The Diamond Light Source is acknowledged for access to its beamline I12-JEEP under proposal mg34117-1.

Institutional Review Board Statement: Not applicable.

Informed Consent Statement: Not applicable.

Data Availability Statement: Not applicable.

Conflicts of Interest: The authors declare no conflict of interest.

References

1. Jamesh, M.; Kumar, S.; Narayanan, T.S.N. Corrosion behavior of commercially pure Mg and ZM21 Mg alloy in Ringer's solution–Long term evaluation by EIS. *Corros. Sci.* **2011**, *53*, 645–654. [CrossRef]
2. Chiu, K.Y.; Wong, M.H.; Cheng, F.T.; Man, H.C. Characterization and corrosion studies of fluoride conversion coating on degradable Mg implants. *Surf. Coat. Technol.* **2007**, *202*, 590–598. [CrossRef]
3. Zheng, Y.F.; Gu, X.N.; Witte, F. Biodegradable metals. *Mater. Sci. Eng. R Rep.* **2014**, *77*, 1–34. [CrossRef]
4. Hermawan, H. Updates on the research and development of absorbable metals for biomedical applications. *Prog. Biomater.* **2018**, *7*, 93–110. [CrossRef]
5. Agarwal, S.; Curtin, J.; Duffy, B.; Jaiswal, S. Biodegradable magnesium alloys for orthopaedic applications: A review on corrosion, biocompatibility and surface modifications. *Mater. Sci. Eng. C* **2016**, *68*, 948–963. [CrossRef] [PubMed]
6. Yang, J.X.; Koons, G.L.; Cheng, G.; Zhao, L.H.; Mikos, A.G.; Cui, F.Z. A review on the exploitation of biodegradable magnesium-based composites for medical applications. *Biomed. Mater.* **2018**, *13*, 022001. [CrossRef]
7. Zhang, S.X.; Zhang, X.N.; Zhao, C.L.; Li, J.N.; Song, Y.; Xie, C.Y.; Tao, H.R.; Zhang, Y.; He, Y.H.; Jiang, Y.; et al. Research on an Mg–Zn alloy as a degradable biomaterial. *Acta Biomater.* **2010**, *6*, 626–640. [CrossRef] [PubMed]
8. Duan, J.; Li, L.; Liu, C.; Suo, Y.; Wang, X.; Yang, Y. Novel Zn-2Cu-0.2Mn-xLi (x = 0, 0.1 and 0.38) alloys developed for potential biodegradable implant applications. *J. Alloys Compd.* **2022**, *916*, 165478. [CrossRef]
9. Naik, M.V.; Narasaiah, N.; Chakravarthy, P.; Kumar, R.A. Hot-extrusion behavior of biodegradable Zn-Mg alloys. *Mater. Today Proc.* **2022**, *56*, 1432–1439. [CrossRef]
10. Vojtěch, D.; Kubásek, J.; Šerák, J.; Novák, P. Mechanical and corrosion properties of newly developed biodegradable Zn-based alloys for bone fixation. *Acta Biomater.* **2011**, *7*, 3515–3522. [CrossRef] [PubMed]
11. Bakhsheshi-Rad, H.R.; Idris, M.H.; Abdul-Kadir, M.R.; Ourdjini, A.; Medraj, M.; Daroonparvar, M.; Hamzah, E. Mechanical and bio-corrosion properties of quaternary Mg–Ca–Mn–Zn alloys compared with binary Mg–Ca alloys. *Mater. Des.* **2014**, *53*, 283–292. [CrossRef]
12. Hagelstein, S.; Zankovic, S.; Kovacs, A.; Barkhoff, R.; Seidenstuecker, M. Mechanical Analysis and Corrosion Analysis of Zinc Alloys for Bioabsorbable Implants for Osteosynthesis. *Materials* **2022**, *15*, 421. [CrossRef]
13. Venezuela, J.; Dargusch, M. The influence of alloying and fabrication techniques on the mechanical properties, biodegradability and biocompatibility of zinc: A comprehensive review. *Acta Biomater.* **2019**, *87*, 1–40. [CrossRef]

14. Yang, L.; Guo, P.; Niu, Z.; Li, F.; Song, Z.; Xu, C.; Liu, H.; Sun, W.; Ren, T. Influence of Mg on the mechanical properties and degradation performance of as-extruded Zn Mg Ca alloys: In vitro and in vivo behavior. *J. Mech. Behav. Biomed. Mater.* **2019**, *95*, 220–231. [CrossRef]
15. Liu, X.; Sun, J.; Zhou, F.; Yang, Y.; Chang, R.; Qiu, K.; Pu, Z.; Li, L.; Zheng, Y. Micro-alloying with Mn in Zn–Mg alloy for future biodegradable metals application. *Mater. Des.* **2016**, *94*, 95–104. [CrossRef]
16. Panaghie, C.; Cimpoesu, N.; Benchea, M.; Roman, A.-M.; Manole, V.; Alexandru, A.; Cimpoesu, R.; Cazacu, M.M.; Wnuk, I.; Zegan, G. 'In-Vitro' Tests on New Biodegradable Metallic Material Based on ZnMgY-Archives of Metallurgy and Materials-Pas Journals. *Arch. Metall. Mater.* **2022**, *67*, 587–594.
17. Shi, Z.-Z.; Yu, J.; Liu, X.-F.; Zhang, H.-J.; Zhang, D.-W.; Yin, Y.-X.; Wang, L.-N. Effects of Ag, Cu or Ca addition on microstructure and comprehensive properties of biodegradable Zn-0.8Mn alloy. *Mater. Sci. Eng. C* **2019**, *99*, 969–978. [CrossRef]
18. Drakopoulos, M.; Connolley, T.; Reinhard, C.; Atwood, R.; Magdysyuk, O.; Vo, N.; Hart, M.; Connor, L.; Humphreys, B.; Howell, G.; et al. I12: The Joint Engineering, Environment and Processing (JEEP) beamline at Diamond Light Source. *J. Synchrotron Rad.* **2015**, *22*, 828–838. [CrossRef]
19. Filik, J.; Ashton, A.W.; Chang, P.C.Y.; Chater, P.A.; Day, S.J.; Drakopoulos, M.; Gerring, M.W.; Hart, M.L.; Magdysyuk, O.V.; Michalik, S.; et al. Processing two-dimensional X-ray diffraction and small-angle scattering data in DAWN 2. *J. Appl. Cryst.* **2017**, *50*, 959–966. [CrossRef]
20. *ASTM E8/E8M*; Standard Test Methods for Tension Testing of Metallic Materials. ASTM International: West Conshohocken, PA, USA, 2016.
21. *ASTM E9-89*; Standard test methods of compression testing of metallic materials at room temperature. ASTM International: West Conshohocken, PA, USA, 2019.
22. *ISO 6507-2:2017*; Metallic materials—Vickers hardness test. ISO: Geneva, Switzerland, 2017.
23. Joint Committee on Powder Diffraction Standards, Powder Diffraction File (JCPDS-ICDD), #00-004-0831. Available online: https://id.loc.gov/authorities/names/n78034813.html (accessed on 14 June 2023).
24. Joint Committee on Powder Diffraction Standards, Powder Diffraction File (JCPDS-ICDD), #00-077-1177. Available online: https://id.loc.gov/authorities/names/n78034813.html (accessed on 14 June 2023).
25. Joint Committee on Powder Diffraction Standards, Powder Diffraction File (JCPDS-ICDD), #00-028-0258. Available online: https://id.loc.gov/authorities/names/n78034813.html (accessed on 14 June 2023).
26. Kabir, H.; Munir, K.; Wen, C.; Li, Y. Recent research and progress of biodegradable zinc alloys and composites for biomedical applications: Biomechanical and biocorrosion perspectives. *Bioact. Mater.* **2020**, *6*, 836–879. [CrossRef]
27. Li, H.F.; Xie, X.H.; Zheng, Y.F.; Cong, Y.; Zhou, F.Y.; Qiu, K.J.; Wang, X.; Chen, S.H.; Huang, L.; Tian, L.; et al. Development of biodegradable Zn-1X binary alloys with nutrient alloying elements Mg, Ca and Sr. *Sci. Rep.* **2015**, *5*, srep10719. [CrossRef] [PubMed]
28. Bowen, P.K.; Drelich, J.; Goldman, J. Zinc exhibits ideal physiological corrosion behavior for bioabsorbable stents. *Adv. Mater.* **2013**, *25*, 2577–2582. [CrossRef] [PubMed]
29. Available online: https://www.matweb.com/search/DataSheet.aspx?MatGUID=8909140a76074049809ad74d536ed606&ckck=1 (accessed on 3 April 2023).

Disclaimer/Publisher's Note: The statements, opinions and data contained in all publications are solely those of the individual author(s) and contributor(s) and not of MDPI and/or the editor(s). MDPI and/or the editor(s) disclaim responsibility for any injury to people or property resulting from any ideas, methods, instructions or products referred to in the content.

materials

MDPI

Article

The Synergistic Lubrication Effects of h-BN and g-C_3N_4 Nanoparticles as Oil-Based Additives for Steel/Steel Contact

Wen Zhong [1,2], Jiazhi Dong [1,2], Siqiang Chen [1,2] and Zhe Tong [3,*]

1 The Key Laboratory of Fluid and Power Machinery, Ministry of Education, Xihua University, Chengdu 610039, China; zw1019@126.com (W.Z.); djz643071849@163.com (J.D.); wdtkbsn@126.com (S.C.)
2 Luzhou Laojiao Group Co., Ltd., Luzhou 646000, China
3 School of Mechanical Engineering, North University of China, Taiyuan 030051, China
* Correspondence: zhetong@nuc.edu.cn

Abstract: The synergistic effect of different types of solid particles in liquid lubricants is of great interest. In this work, g-C_3N_4 nanosheets were initially prepared using a calcination method and then as-prepared, and h-BN were used as lubricating additives to the white oil. A comparison between the mixed additives and the single g-C_3N_4 or h-BN additives revealed that the base oil with the addition of g-C_3N_4 and h-BN showed the best lubricating properties. The results show a 12.3% reduction in friction coefficient, resulting in a 68.6% reduction in wear rate compared to the white oil when filled with 0.5 wt% g-C_3N_4 and h-BN (1:1 by weight). Moreover, the addition of g-C_3N_4 and h-BN improves the high-temperature lubrication properties of the white oil. However, the friction coefficient and wear rate increase with increasing oil temperature. The large contact area between g-C_3N_4 and its sliding counterpart and the strong adhesive force between h-BN and its sliding counterpart improve the film formation efficiency, leading to enhanced tribological properties under oil lubrication conditions.

Keywords: oil lubrication conditions; high-temperature tribology; g-C_3N_4 and h-BN; synergistic effect

Citation: Zhong, W.; Dong, J.; Chen, S.; Tong, Z. The Synergistic Lubrication Effects of h-BN and g-C_3N_4 Nanoparticles as Oil-Based Additives for Steel/Steel Contact. *Materials* **2023**, *16*, 4979. https://doi.org/10.3390/ma16144979

Academic Editors: Pavol Liptai, Jaroslav Briančin and Maroš Halama

Received: 5 June 2023
Revised: 3 July 2023
Accepted: 5 July 2023
Published: 13 July 2023

Copyright: © 2023 by the authors. Licensee MDPI, Basel, Switzerland. This article is an open access article distributed under the terms and conditions of the Creative Commons Attribution (CC BY) license (https://creativecommons.org/licenses/by/4.0/).

1. Introduction

Oil lubrication has been widely utilized in various industrial sectors such as lubrication, machining, and heat dissipation, etc., because of its superior film-forming property and liquidity [1–4]. However, the low carrying capacity of the oil film prevents its low-speed application in heavy duty conditions. Adding solid lubricating particles to the oil lubricant can effectively improve the bearing capacity and lubrication conditions due to the formation of solid tribo-film [5–7].

Two-dimensional (2D) graphite-like materials have excellent physical and chemical properties and are usually used as solid additives to improve the tribological properties of oil-based or bulk materials [8–10]. The interfacial friction of h-BN can be adjusted and controlled by applying a voltage due to its semiconducting nature and low-friction properties, thus possessing potential as a lubricant additive [11,12]. The formation of tribo-film is considered the main mechanism for enhancing the tribological properties when using a 2D additive [13]. The additive will be absorbed on the surface of the counterpart and grow a protective solid tribo-film, preventing the direct contact of the friction interface during the sliding progress, which is significant in reducing wear and friction, especially under the boundary lubrication conditions [14–16]. Graphite carbon nitride (g-C_3N_4) is a typical layered structure that has been investigated as a prospective material in varied areas, such as catalysis, CO_2 reduction, and lubrication [17–19]. The film formation efficiency and quality at the sliding interface are the main factors affecting the tribological properties of the lubricant. Additionally, the size, shape, and bonding force between the additives and interface have important effects on the formation of the lubricating film [20].

In recent years, numerous studies have shown that the addition of individual additives does not satisfy engineering requirements. The use of hybrids or multiple particles has attracted attention in tribology design [21,22]. He et al. reported that the use of core–shell Cu@rGO hybrids can significantly improve the mechanical and self-lubricating properties. This was attributed to the prominent load-bearing capacity of Cu NPs and excellent lubrication performance of rGO nanosheets [23]. Li et al. found that the hybridization of CNTs and ZnS improves the dispersion of hybrids in epoxy-based composites. Additionally, the "micro-roller" of CNTs and stress distribution improvement of ZnS show a synergetic effect regarding reductions in friction and wear [24]. Nevertheless, the hybrids typically require chemical or physical reactions to achieve the binding of different particles, which is time-consuming. The direct mixing of different particles in the matrix can realize friction-reducing and wear-resistant properties [25]. Ye et al. reported that MoS_2 and Zr coating exhibits high hardness and good adhesion strength, improving the tribological behavior and extending the endurance life of the coating [26]. Su et al. found that using graphene oxide (GO) and onion-like carbon (OLC) as lubricating additives can significantly reduce the friction coefficients and wear rates of steel discs. This effect was attributed to the rolling action of OLC and excellent lubricating ability of the GO [27].

For the present study, g-C_3N_4 nanosheets were synthesized via a facile one-step calcination method and used together with h-BN to cooperatively modify the tribological properties of white oil under ball–disk sliding conditions at different temperatures. In this paper, the microstructure, components of g-C_3N_4- and h-BN, and the underlying lubrication mechanism behind the g-C_3N_4- and h-BN-induced enhancements with respect to its tribological properties are systematically described and discussed. Hopefully, this work can provide some guidance for extending the synergetic application of multiple additives in the lubrication field.

2. Materials and Experiment

2.1. Materials

h-BN with an average size of 2 μm was obtained from Aladdin Reagent (Shanghai) Co., Ltd., Shanghai, China. Urea (CSN_2H_4) was purchased from Sinopharm Group, China. White oil was purchased from Aladdin Reagent (Shanghai) Co., Ltd., Shanghai, China. All chemical reagents were used as received without further purification.

2.2. Preparation of g-C_3N_4

g-C_3N_4 nanosheets were synthesized by calcinating urea. Following typical synthesis, 10 g CSN_2H_4 was placed in quartz crucible and covered. The reaction was carried out at 550 °C for 5 h in a Muffle furnace with a heating rate of 20 °C/min. Finally, the product was obtained after natural cooling.

2.3. Characterization

The crystal phase and microtopography of h-BN and g-C_3N_4 nanoparticles were characterized to confirm the synthesis effect. The morphologies of h-BN and g-C_3N_4 nanoparticles were characterized by using scanning electron microscopy (SEM, Gemini SEM 500 at 10 kV accelerating voltage, Jena, Germany) and transmission electron microscopy (TEM, Hitachi H800 at 200 kV accelerating voltage, Tokyo, Japan). The crystal phase structures of h-BN and g-C_3N_4 nanoparticles were obtained by X-ray diffraction measurements (D8 Advance A25 at 40 kV and 40 mA with Cu Ka). The functional group of h-BN and g-C_3N_4 were characterized by Fourier-transform infrared spectroscopy (FTIR Nicolet iS50, Waltham, MA, USA).

2.4. Tests and Analysis

A certain quality of particles was added into the white oil and ultrasonically treated for 30 min to obtain a uniform oil-based lubricant, the concentration of soliditives for sliding test was 0.5 wt%. The tribological properties of the oil-based lubricant were evaluated by using a UMT-2 ball-on-disk tribometer (CETR-2, Campbell, CA, USA) with a reciprocating sliding length of 6 mm. The schematic diagram of the contact configuration and the sliding mode is shown in Figure 1. A heating module was placed at the bottom of the plate to achieve different ambient temperatures. The GCr15 bearing steel ball with a diameter of 9.6 mm with surface roughness of Ra = 0.22 μm and bearing steel disk with a diameter of 30 mm, roughness of Ra = 0.1 μm were used as sliding counterparts. The friction tests were carried out at a sliding frequency of 2 Hz with an applied load of 2 N under different temperatures (Room temperature, 50 ± 1 °C and 80 ± 1 °C, 100 ± 1 °C); the sliding duration was set to 1800 s. The ball and disk were cleaned with absolute alcohol prior to testing. Each test was repeated three times, and the average value was used as the final result.

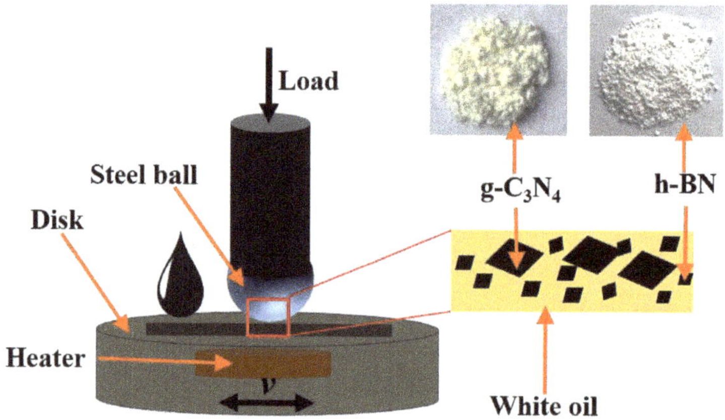

Figure 1. Schematic diagram of contact configuration.

After testing, the wear volume and wear scar were measured and evaluated using a laser scanning confocal microscope (Olympus, Tokyo, Japan, OLS4000) and SEM.

3. Results and Discussion

3.1. Microstructure of h-BN and g-C_3N_4 Particles

Figure 2 shows the microstructural properties of the h-BN and g-C_3N_4 particles. It is evident the size of h-BN range from 1 to 3 μm, and the g-C_3N_4 exhibit a large lamellar structure with obvious aggregation. Meanwhile, the typical laminated structure of h-BN and g-C_3N_4 nanosheets can be seen from the TEM images (Figure 2c,d).

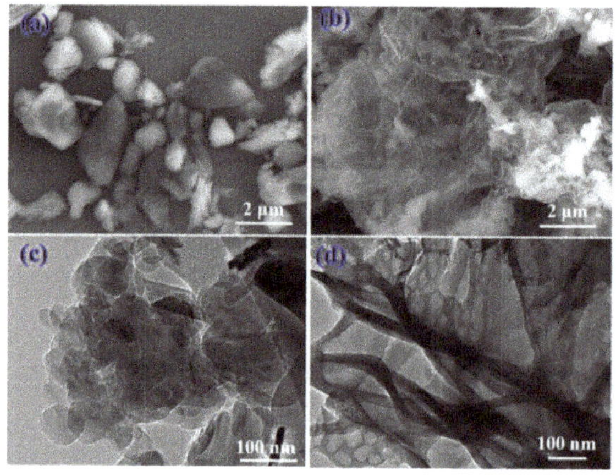

Figure 2. SEM micrographs of (**a**) h-BN nanoparticles and (**b**) g-C_3N_4 nanosheets; TEM images of (**c**) h-BN nanoparticles and (**d**) g-C_3N_4 nanosheets.

Figure 3 shows the FTIR spectrum of g-C_3N_4 and h-BN particles for g-C_3N_4; the peaks at 3400–3100 cm^{-1} correspond to the stretching vibrations of N-H. The absorption band of 1700–1100 cm^{-1} correspond to the main characteristic band of g-C_3N_4, while the 1650–1630 cm^{-1} band is attributed to N-H shear vibrations and the 1570–1515 cm^{-1} and 1310–1200 cm^{-1} bands are attributed to the N-H bending vibration and C-N stretching vibrations. In addition, the peaks at 880–820 cm^{-1} are attributed to C-H and N-H bending vibrations, indicating the generation of g-C_3N_4 [28,29]. For the h-BN, two significant infrared absorption peaks can be seen at 1364 cm^{-1} and 768 cm^{-1}, which are attributed to the B-N reticular and B-N-B bending modes, respectively.

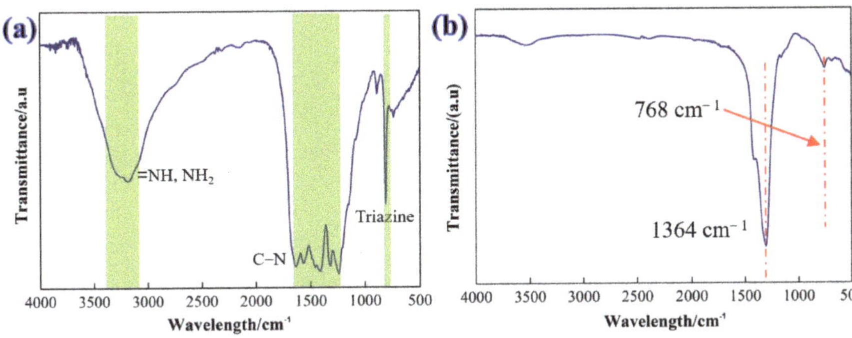

Figure 3. FTIR spectrum of (**a**) as-synthesized g-C_3N_4 nanosheets and (**b**) h-BN nanoparticles.

XRD was used to characterize the chemical composition and crystal structure of g-C_3N_4 and h-BN particles. Figure 4 shows the XRD pattern of the g-C_3N_4 nanosheets and the h-BN particles. For the g-C_3N_4 nanosheets, the two peaks located at 13.2° and 27.6° are attributed to the (100) and (002) planes, which correspond to the in-plane structural packing and inter-layer structural packing, respectively [30]. Additionally, regarding h-BN particles, the typical diffraction peak located at 26.5° corresponds to the (002) crystal plane of h-BN, which is consistent with the (002) reflection of the hexagonal graphitization [31]. Furthermore, the characteristic peaks located at 41.7° and 76.2° are assigned to the (100) and (110) crystallographic planes of h-BN [32].

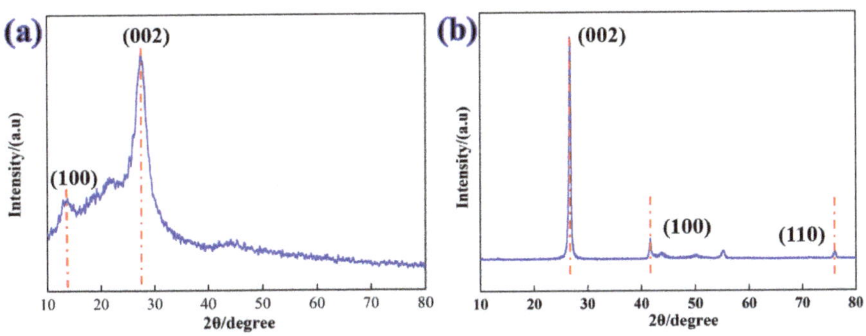

Figure 4. XRD pattern of (**a**) as-synthesized g-C_3N_4 nanosheets and (**b**) h-BN nanoparticles.

3.2. Tribological Properties

Figure 5 shows the real-time friction coefficients of oil-based lubricants containing different kinds of solid additives at room temperature. The friction coefficient of the pure oil starts at a very low level and then decreases rapidly with a relatively long running time before reaching final stability. In contrast, the addition of solid additives leads to different levels of reduction in both the friction and running-in time of oil-based lubricant. In particular, the oil-based lubricant shows the lowest friction coefficient with the addition of both h-BN and g-C_3N_4, simultaneously. Oil film can form quickly for pure white oil, resulting in a low friction coefficient at the early stages. However, the damage of tribo-film had occurred as the sliding went on due to the low bearing capacity of oil film. With the addition of solid additives, the relative sliding of counterparts was impeded due to an obstacle effect before the additives sufficiently dispersed [33], thus causing an increase in frictional force. However, as the sliding progressed, the solid lubricating film was formed on the surface of the counterparts, therefore reducing the friction coefficient. In addition, the simultaneous presence of h-BN and g-C_3N_4 promotes the formation of the solid lubricating film.

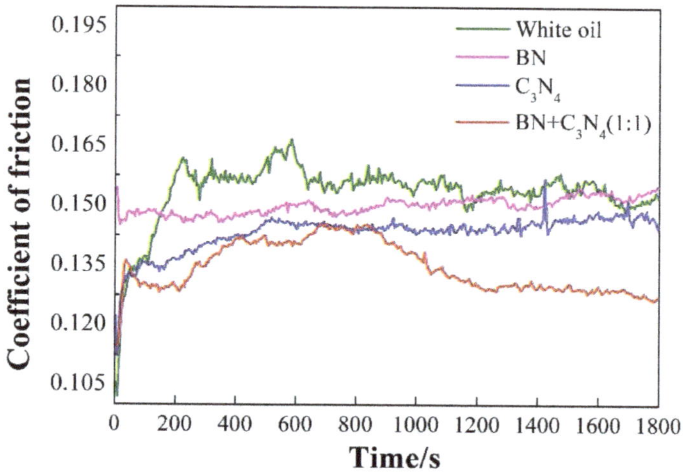

Figure 5. Effect of nano additives on the real-time friction curves of the oil-based lubricant.

Figure 6 shows the average friction coefficient and the corresponding wear rate for different types of oil-based lubricants. It can be established that both the average friction coefficient and wear rate of the three lubricants decrease with the addition of the additive in the white oil, evidencing the excellent lubricating effect of the solid additive. In contrast to h-BN, the counterparts have a lower friction coefficient and wear rate. In addition, the synergistic lubrication effect of h-BN and g-C_3N_4 are found to be where the average friction coefficient and wear rate are maximally reduced by 12.3% and 68.6%, respectively, in the presence of both h-BN and g-C_3N_4.

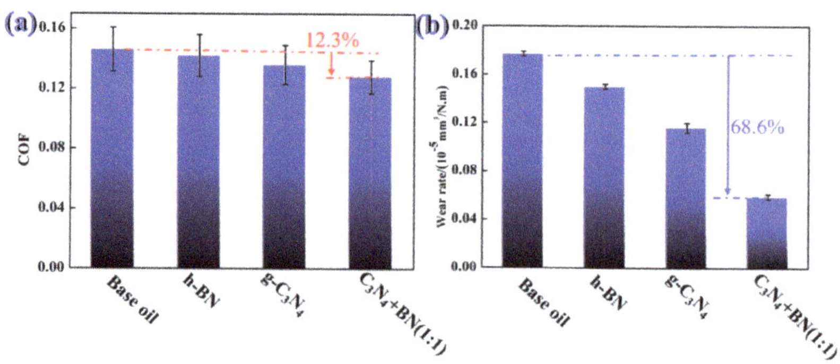

Figure 6. (**a**) Average friction coefficient and (**b**) wear rate of white oil containing different types of additives.

Figure 7 shows the SEM images of wear tracks of the disk lubricated with four types of oil-based lubricants. The wear surface was severely damaged with the addition of base oil, as were many deep groves and pits on the disk surface (Figure 7a). The oil film is prone to breaking at a high contact pressure, and direct contact between the counterpart ball and disk occurred, resulting in furrow wear. The width and depth of the wear track and pits decrease with the addition of h-BN to the base oil (Figure 7b). In addition, the g-C_3N_4 imbues the base oil with a wear-resistant effect. However, many shallow grooves on the disk surface can still be observed (Figure 7c). The wear surface with g-C_3N_4 and h-BN is significantly improved and shows reduced wear width and depth. Additionally, the surface is smoother than those under other lubricants (Figure 7d). In addition, it can be seen from Figure 7d that the main diffraction peak of the XRD pattern indicates the presence of g-C_3N_4 and h-BN in the tribo-film, which shows a synergistic effect on friction reduction and improves the wear resistance of g-C_3N_4 and h-BN.

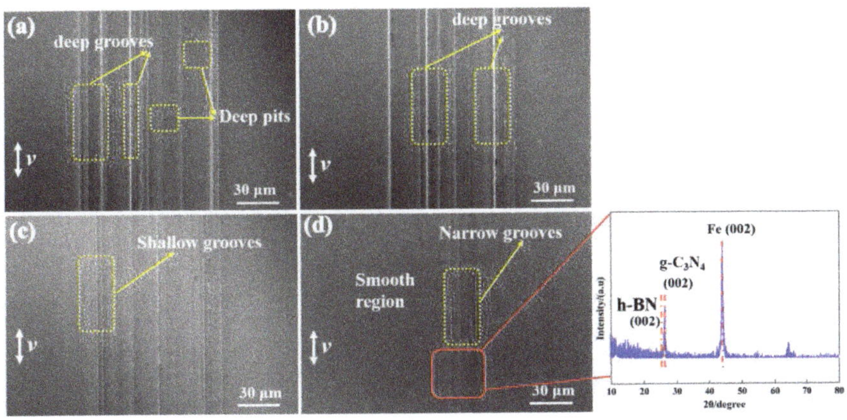

Figure 7. SEM images of the worn surfaces of the disk lubricated by (**a**) pure white oil, (**b**) white oil with h-BN, (**c**) white oil with g-C_3N_4, and (**d**) white oil with h-BN and g-C_3N_4, as well as the corresponding XRD pattern.

Figure 8 shows the 3D morphology of oil lubricants by adding different types of solid additives. The surface of the disk has characteristics that indicate rough wear and shows the deep grooves (Figure 8a). When the h-BN particles were added to the base oil, the width of the wear track decreases significantly. The particles will adhere to the friction surface to prevent direct contact with the sliding counterparts (Figure 8b). As the sliding of the interface continues, the solid lubricating film is formed, thus improving furrow. Compared to that lubricated by h-BN, the addition of g-C_3N_4 significantly decreases the width and depth of the wear track, showing an improved anti-wear effect compared to h-BN nanosheets (Figure 8c). The worn surface lubricated by h-BN and g-C_3N_4 exhibits a smooth character with only slight grooves being observable; these results are in good agreement with those shown in Figure 7.

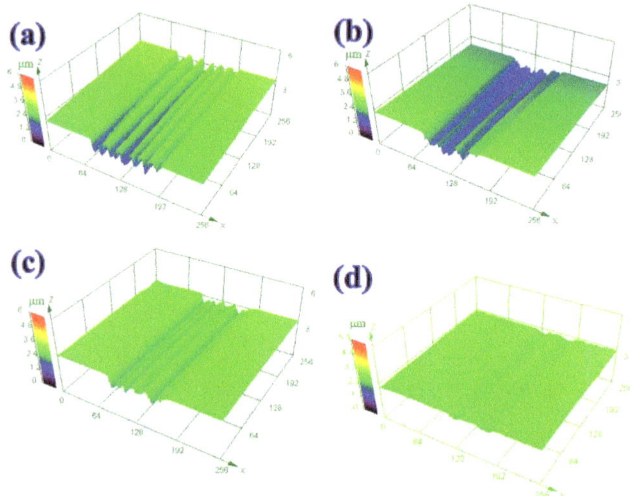

Figure 8. Three-dimensional morphologies of worn tracks lubricated by: (**a**) Pure white oil, (**b**) white oil with h-BN, (**c**) white oil with g-C_3N_4, and (**d**) white oil with h-BN and g-C_3N_4.

3.3. Effect of Temperature on the Tribological Properties of Oil Lubricants

Frictional heat accompanies the sliding process, and the temperature has a significant effect on the tribological properties of the liquid lubrication system [34,35]. The above results show that lubricants supplemented with h-BN and g-C_3N_4 exhibit the best friction reduction and wear resistance. Therefore, the effect of temperature on the tribological properties of lubricants containing h-BN and g-C_3N_4 was further investigated, and the results are shown in Figure 9. The friction coefficients of the two lubricating systems experience a slight increase with elevated temperature, and the lubricant containing h-BN and g-C_3N_4 additives still shows a lower friction coefficient than pure white oil at identical temperatures (Figure 9a,b), suggesting that a boundary oil film might be formed on the sliding surface. The bearing capacity of the oil film decreases with an increase in temperature, causing an increase in the friction coefficient [36,37]. Additionally, the wear rate shows the same trend with the friction coefficient (Figure 9c). This seems to indicate that raising the temperature may play a passive role in improving the tribological performances of white oil with additives for ball-disk contact.

SEM images of the worn surface are shown in Figure 10. Clearly, many deep grooves can be found in Figure 10a, and the dominant friction and wear mechanism is typical furrow wear. From Figure 10c,d, it can be seen that the width and depth of the wear tracks increase with increasing temperature. Moreover, plastic deformation can be observed on the surface at 80 °C and 100 °C. In addition to changes in the lubrication conditions, the hardness of the counterparts is an important influencing factor on wear behavior. The hardness of the counterparts decreases with increasing temperature, therefore resulting in severe wear, which may be caused by adhesive wear.

Figure 11 shows the 3D morphologies of the oil lubricants supplemented with h-BN and g-C_3N_4 additives at different temperatures. The worn surfaces show increasingly apparent wear tracks with increasing temperature. The wear track at 100 °C shows a wide and shallow profile compared with that at 80 °C. This can be attributed to variations in the contact area of the sliding interface. The hardness decreases as the temperature increases from 80 °C to 100 °C, thus leading to an increase in contact area and resulting in low contact pressure.

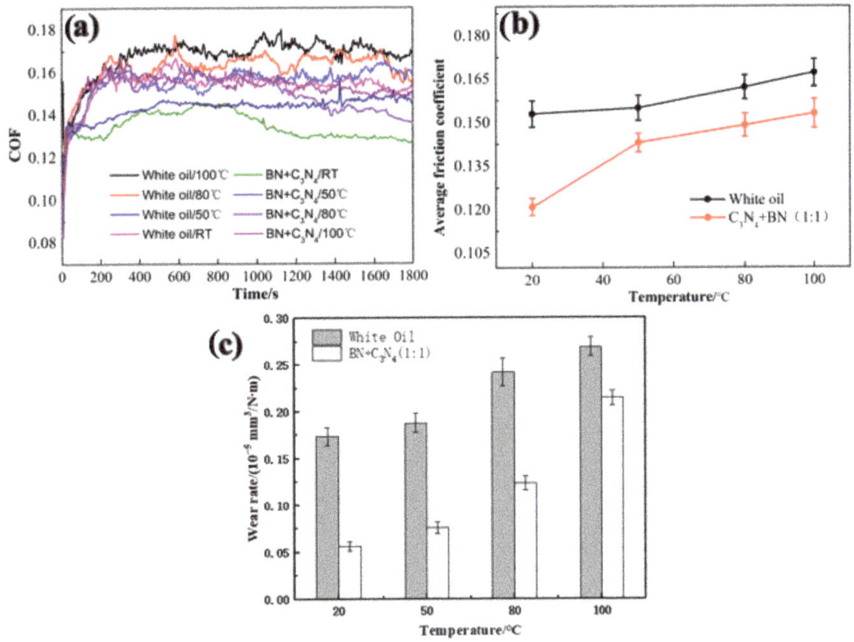

Figure 9. (**a**) Real-time friction curve, (**b**) average coefficient of friction, and (**c**) wear rate of pure white oil and white oil with adding of h-BN and g-C_3N_4 at different temperatures.

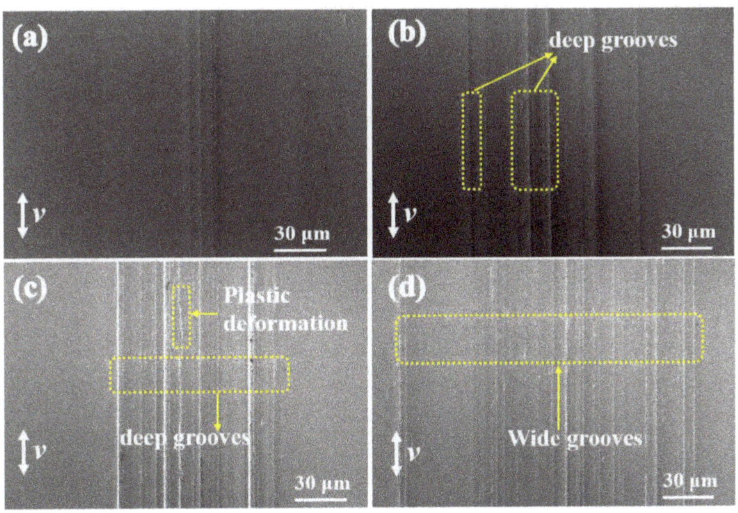

Figure 10. SEM images of the worn surfaces of the disk lubricated by white oil with h-BN and g-C_3N_4 additives at (**a**) room temperature, (**b**) 50 °C, (**c**) 80 °C, and (**d**) 100 °C.

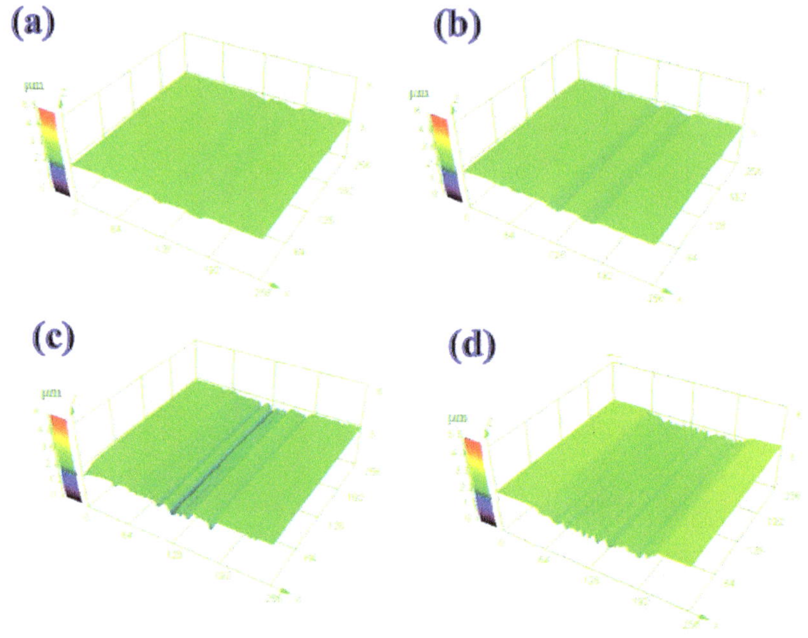

Figure 11. Three-dimensional morphologies of worn tracks lubricated by white oil with h-BN and g-C_3N_4 additives at (**a**) room temperature, (**b**) 50 °C, (**c**) 80 °C, and (**d**) 100 °C.

3.4. Tribological Mechanism

Based on the above-described results and analysis, the reinforcing mechanisms of g-C_3N_4 and h-BN additives on the tribological properties of oil-based lubricants is shown in Figure 12. When h-BN particles are used as additives in an oil-based lubricant, the particles tend to move with the flow of oil, resulting in less efficient solid film formation, and the lubricant with h-BN exhibits a relatively high friction coefficient and wear loss. The film formation efficiency may be enhanced when g-C_3N_4 nanosheets are added to the lubricant due to their large area. At the same time, the poor adhesion of g-C_3N_4 to metal may restrict further improvements in film formation efficiency. When g-C_3N_4 and h-BN are added to an oil-based lubricant at the same time, the film formation efficiency and properties can be effectively enhanced compared to single-type addition.

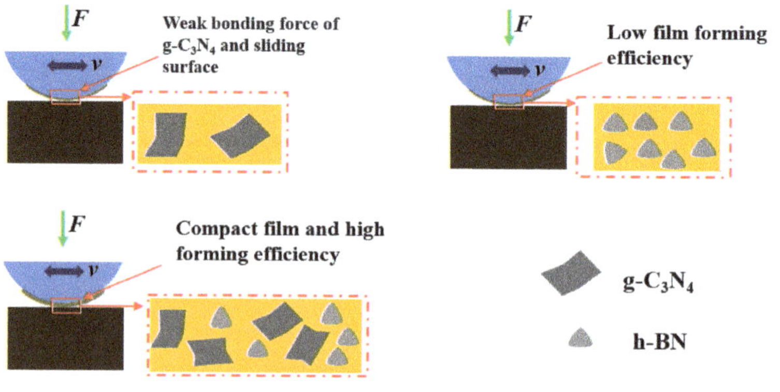

Figure 12. Reinforcing mechanism of g-C_3N_4 and h-BN additives on the tribological properties of oil-based lubricants.

4. Conclusions

In summary, g-C_3N_4 nanosheets were successfully prepared and subsequently introduced into white oil with h-BN particles to enhance the efficiency of solid lubrication film formation at the sliding interface. The tribological properties of oil-based lubricants after the addition of various additives were investigated at various temperatures, and the corresponding tribological mechanisms were discussed. The main conclusions we have drawn from this study are as follows:

(a) g-C_3N_4 nanoparticles were successfully prepared, and the nanosheets possess a high specific surface area; the diameter of the g-C_3N_4 nanoparticles was microscale.
(b) After the addition of the solid additives, both the friction coefficient and wear loss of the disk decrease with the addition of various solid additives compared to the base oil. Additionally, the lubricant supplemented with g-C_3N_4 exhibits better tribological properties than that supplemented with h-BN. Furthermore, the oil-based lubricant supplemented with both g-C_3N_4 and h-BN (1:1 by weight) showed the lowest friction coefficient of 0.13, and the corresponding disk achieved the lowest wear rate of 0.06×10^{-5} $mm^3/(N{\cdot}m)$. This can be attributed to the increased formation efficiency of the solid tribo-film due to the synergistic effect of g-C_3N_4 and h-BN.
(c) However, increasing the temperature resulted in an increase in both wear and the friction coefficient for both the base oil the oil lubricant supplemented with g-C_3N_4 and h-BN. The loss of mechanical properties suffered by the sliding counterpart and reduction in oil film bearing capacity with increasing temperature are believed to be the major contributors for this.

Author Contributions: Conceptualization, W.Z.; Formal analysis, W.Z.; Investigation, Z.T.; Methodology, J.D.; Resources, J.D.; Data Curation, S.C.; Writing—original draft, S.C.; Writing—review & editing, Z.T. All authors have read and agreed to the published version of the manuscript.

Funding: The authors are grateful for the support they received from the Sichuan Provincial Natural Science Foundation Project (Grant No. 2023NSFSC0871), Key Research and Development Project of Science and Technology Plan in Luzhou City (Grant No. 2020-GYF-1), Chunhui Project from Education Ministry of China (Grant No. RZ1900010894).

Institutional Review Board Statement: Not applicable.

Informed Consent Statement: Not applicable.

Data Availability Statement: The data presented in this study are available on request from the corresponding author.

Conflicts of Interest: The authors declare no conflict of interest.

References

1. Singh, N.; Sinha, S.K. Tribological and mechanical analysis of hybrid epoxy based polymer composites with different in situ liquid lubricants (silicone oil, PAO and SN150 base oil). *WEAR* **2022**, *504–505*, 204404. [CrossRef]
2. Emmanual, L.; Loganathan, M.; Karthikeyan, T. Machining studies on Monel K–500 using TiAlN coated tungsten carbide inserts under Ag nanoparticles incorporated modified pongamia pinnata oil lubrication. *Mater. Res. Express* **2022**, *9*, 076512. [CrossRef]
3. Adachi, N.; Matsuo, Y.; Todaka, Y.; Fujimoto, M.; Hino, M.; Mitsuhara, M.; Oba, Y.; Shiihara, Y.; Umeno, Y.; Nishida, M. Effect of grain boundary on the friction coefficient of pure Fe under the oil lubrication. *Tribol. Int.* **2021**, *155*, 106781. [CrossRef]
4. Thirumalai, R.; Senthilkumaar, J.S.; Selvarani, P.; Ramesh, S. Machining characteristics of Inconel 718 under several cutting conditions based on Taguchi method. *Proc. Inst. Mech. Eng. Part C J. Mech. Eng. Sci.* **2013**, *227*, 1889–1897. [CrossRef]
5. Conradi, M.; Podgornik, B.; Kocijan, A.; Remškar, M.; Klobčar, D. Water versus Oil Lubrication of Laser-Textured Ti6Al4V Alloy upon Addition of MoS_2 Nanotubes for Green Tribology. *Materials* **2022**, *15*, 2974. [CrossRef]
6. Raina, A.; Irfan Ul Haq, M.; Anand, A.; Mohan, S.; Kumar, R.; Jayalakshmi, S.; Arvind Singh, R. Nanodiamond Particles as Secondary Additive for Polyalphaolefin Oil Lubrication of Steel-Aluminium Contact. *Nanomaterials* **2021**, *11*, 1438. [CrossRef]
7. Gulzar, M.; Masjuki, H.H.; Kalam, M.A.; Varman, M.; Zulkifli, N.W.M.; Mufti, R.A.; Zahid, R. Tribological performance of nanoparticles as lubricating oil additives. *J. Nanopart. Res.* **2016**, *18*, 223–231. [CrossRef]

8. Trucco, D.; Vannozzi, L.; Teblum, E.; Telkhozhayeva, M.; Nessim, G.D.; Affatato, S. Graphene Oxide-Doped Gellan Gum–PEGDA Bilayered Hydrogel Mimicking the Mechanical and Lubrication Properties of Articular Cartilage. *Adv. Healthc. Mater.* **2021**, *10*, 873–881. [CrossRef]
9. Kogovšek, J.; Kalin, M. Lubrication performance of graphene-containing oil on steel and DLC-coated surfaces. *Tribol. Int.* **2019**, *138*, 59–67. [CrossRef]
10. Zhong, W.; Chen, S.; Ma, L.; Tong, Z. Tribological Properties of Carbon Fabric/Epoxy Composites Filled with FGr@MoS_2 Hybrids under Dry Sliding Conditions. *Materials* **2022**, *15*, 7951. [CrossRef]
11. Yu, K.; Zou, K.; Lang, H.; Peng, Y. Nanofriction characteristics of h-BN with electric field induced electrostatic interaction. *Friction* **2020**, *9*, 1–12. [CrossRef]
12. Nasser, K.I.; Liñeira del Río, J.M.; Mariño, F.; López, E.R.; Fernández, J. Double hybrid lubricant additives consisting of a phosphonium ionic liquid and graphene nanoplatelets/hexagonal boron nitride nanoparticles. *Tribol. Int.* **2021**, *163*, 189–199. [CrossRef]
13. Erdemir, A.; Ramirez, G.; Eryilmaz, O.L.; Narayanan, B.; Liao, Y.; Kamath, G.; Sankaranarayanan, S.K.R.S. Carbon-based tribofilms from lubricating oils. *Nature* **2016**, *536*, 67–71. [CrossRef]
14. Meng, F.; Han, H.; Gao, X.; Yang, C.; Zheng, Z. Experiment study on tribological performances of GNPs/MoS_2 coating. *Tribol. Int.* **2018**, *118*, 400–407. [CrossRef]
15. Bobby, S.; Samad, M.A. Enhancement of tribological performance of epoxy bulk composites and composite coatings using micro/nano fillers: A review. *Polym. Adv. Technol.* **2017**, *28*, 633–644. [CrossRef]
16. Furlan, K.P.; de Mello, J.D.B.; Klein, A.N. Self-lubricating composites containing MoS_2: A review. *Tribol. Int.* **2018**, *120*, 280–298. [CrossRef]
17. Jayashree, P.; Matějka, V.; Foniok, K.; Straffelini, G. Comparative Studies on the Dry Sliding Behavior of a Low-Metallic Friction Material with the Addition of Graphite and Exfoliated g-C_3N_4. *Lubricants* **2022**, *10*, 27. [CrossRef]
18. Kumar, A.; Thakre, D.G.; Arya, K.P.; Jain, A.K. 2D Structured Nano-Sheets of Octadecylamine Grafted Graphitic-Carbon Nitride (g-C_3N_4) as Lubricant Additives. *Macromol. Symp.* **2017**, *376*, 9–17. [CrossRef]
19. Richa, R.; Rose, A.C.; Thomas, A.; Korah, B.K. Recent Advances in Graphitic Carbon Nitrides (g-C_3N_4) as Photoluminescence Sensing Probe: A Review. *ChemistrySelect* **2022**, *7*, e202200876.
20. Ye, J.; Khare, H.S.; Burris, D.L. Quantitative characterization of solid lubricant transfer film quality. *Wear* **2014**, *316*, 133–143. [CrossRef]
21. Sharma, K.A.; Tiwari, K.A.; Dixit, R.A.; Singh, R.K. Novel uses of alumina/graphene hybrid nanoparticle additives for improved tribological properties of lubricant in turning operation. *Tribol. Int.* **2018**, *119*, 99–111. [CrossRef]
22. Tucker, Z. The Performance of Translucent Silicon-Oxide Nanoparticle Lubricant Additives. *Tribol. Lubr. Technol.* **2017**, *73*, 32–34.
23. He, G.; Li, Y.; Wu, L.; Wang, Y.; Liu, M.; Yuan, J.; Men, X. Synergy of core-shell Cu@rGO hybrids for significantly improved thermal and tribological properties of polyimide composites. *Tribol. Int.* **2021**, *161*, 91–105. [CrossRef]
24. Li, X.; Chen, B.; Jia, Y.; Li, X.; Yang, J.; Li, C.; Yan, F. Enhanced tribological properties of epoxy-based lubricating coatings using carbon nanotubes-ZnS hybrid. *Surf. Coat. Technol.* **2018**, *344*, 154–162. [CrossRef]
25. Ren, Y.; Zhang, L.; Xie, G.; Li, Z.; Chen, H.; Gong, H.; Xu, W.; Guo, D.; Luo, J. A review on tribology of polymer composite coatings. *Friction* **2021**, *9*, 429–470. [CrossRef]
26. Ye, M.; Zhang, G.; Ba, Y.; Wang, T.; Wang, X.; Liu, Z. Microstructure and tribological properties of MoS_2+Zr composite coatings in high humidity environment. *Appl. Surf. Sci.* **2016**, *367*, 140–146. [CrossRef]
27. Su, F.; Chen, G.; Huang, P. Lubricating performances of graphene oxide and onion-like carbon as water-based lubricant additives for smooth and sand-blasted steel discs. *Friction* **2020**, *8*, 47–57. [CrossRef]
28. Dos Santos, D.F.; Santiago, A.A.G.; Teodoro, M.D.; Motta, F.V.; Bomio, M.R.D. Investigation of the photocatalytic and optical properties of the $SrMoO_4$/g-C_3N_4 heterostructure obtained via sonochemical synthesis with temperature control. *J. Environ. Manag.* **2023**, *325*, 116396. [CrossRef]
29. Van, K.N.; Huu, H.T.; Nguyen Thi, V.N.; Le Thi, T.L.; Truong, D.H.; Truong, T.T.; Dao, N.N.; Vo, V.; Tran, D.L.; Vasseghian, Y. Facile construction of S-scheme SnO_2/g-C_3N_4 photocatalyst for improved photoactivity. *Chemosphere* **2022**, *289*, 133120. [CrossRef]
30. Hong, Y.; Shi, J.; Shi, W.; Fang, Z.; Chen, R.; Huang, Y. A facile and scalable route for synthesizing ultrathin carbon nitride nanosheets with efficient solar hydrogen evolution. *Carbon* **2018**, *136*, 160–167. [CrossRef]
31. Kostoglou, N.; Polychronopoulou, K.; Rebholz, C. Thermal and chemical stability of hexagonal boron nitride (h-BN) nanoplatelets. *Vacuum* **2015**, *112*, 42–45. [CrossRef]
32. Chen, T.; Zhang, Q.; Xie, Z.; Tan, C.; Chen, P.; Zeng, Y.; Wang, F.; Liu, H.; Liu, Y.; Liu, G.; et al. Carbon nitride modified hexagonal boron nitride interface as highly efficient blue LED light-driven photocatalyst. *Appl. Catal. B Environ.* **2018**, *238*, 410–421. [CrossRef]
33. Mao, J.; Zhao, J.; Wang, W.; He, Y.; Luo, J. Influence of the micromorphology of reduced graphene oxide sheets on lubrication properties as a lubrication additive. *Tribol. Int.* **2018**, *119*, 614–621. [CrossRef]
34. Decrozant-Triquenaux, J.; Pelcastre, L.; Courbon, C.; Prakash, B.; Hardell, J. High temperature tribological behaviour of PVD coated tool steel and aluminium under dry and lubricated conditions. *Friction* **2021**, *9*, 802–821. [CrossRef]
35. Gong, K.; Wu, X.; Zhao, G.; Wang, X. Nanosized MoS_2 deposited on graphene as lubricant additive in polyalkylene glycol for steel/steel contact at elevated temperature. *Tribol. Int.* **2017**, *110*, 1–7. [CrossRef]

36. Wang, Z.; Ren, R.; Song, H.; Jia, X. Improved tribological properties of the synthesized copper/carbon nanotube nanocomposites for rapeseed oil-based additives. *Appl. Surf. Sci.* **2018**, *428*, 630–639. [CrossRef]
37. Zhang, Y.; Wang, W.; Liang, H.; Zhao, Z. Layered oil slip model for investigation of film thickness behaviours at high speed conditions. *Tribol. Int.* **2019**, *131*, 137–147. [CrossRef]

Disclaimer/Publisher's Note: The statements, opinions and data contained in all publications are solely those of the individual author(s) and contributor(s) and not of MDPI and/or the editor(s). MDPI and/or the editor(s) disclaim responsibility for any injury to people or property resulting from any ideas, methods, instructions or products referred to in the content.

materials

MDPI

Article

Removal of Impurities from EAFD Ammonium Carbonate Leachate and Upgrading the Purity of Prepared ZnO

Zita Takacova *, Jana Piroskova, Andrea Miskufova, Tomas Vindt, Maria Hezelova and Dusan Orac

Institute of Recycling Technologies, Faculty of Materials, Metallurgy and Recycling, Technical University of Kosice, Letna 1/9, 04200 Kosice, Slovakia; jana.piroskova@tuke.sk (J.P.); andrea.miskufova@tuke.sk (A.M.); tomas.vindt@tuke.sk (T.V.); maria.hezelova@tuke.sk (M.H.); dusan.orac@tuke.sk (D.O.)

* Correspondence: zita.takacova@tuke.sk; Tel.: +421-55-6022493

Abstract: The paper describes cementation as a suitable method applied in the refining of EAFD leachates in order to obtain required purity of ZnO for specific industrial application. For study of cementation conditions, the leachate from alkaline leaching with $(NH_4)_2CO_3$ was used. The leachates contained a high amount of zinc (8000–12,000 μg/mL) and a low content of impurities such as iron, lead, copper, chromium and manganese in the range of 1–21 μg/mL. Cementation conditions were predicted by thermodynamic study, theoretically confirming the viability of the proposed experiments at the considered pH = 8–9. Cementation experiments were carried out using powdered zinc and aluminium (5 g/L) as cementation agents in the first phase separately. To increase the cementation efficiency, their combination was used (2.5:2.5 g/L or 5:5 g/L) at temperatures of 20, 40, 60 and 80 °C for 30 min. The optimum cementation conditions were established as: Zn:Al = 5:5 g/L, 60 °C. Under the given conditions, 100% of Fe, Cu, Pb and Mn were removed from the leachate. The ZnO with the maximum purity of 96.67% was obtained by crystallization of cemented leachate at 105 °C, followed by calcination at 900 °C for 4 h. ZnO with such a purity is suitable for use in the electrical or rubber industries.

Keywords: electric arc furnace dust (EAFD); hydrometallurgy; cementation; zinc; iron; lead

Citation: Takacova, Z.; Piroskova, J.; Miskufova, A.; Vindt, T.; Hezelova, M.; Orac, D. Removal of Impurities from EAFD Ammonium Carbonate Leachate and Upgrading the Purity of Prepared ZnO. *Materials* **2023**, *16*, 5004. https://doi.org/10.3390/ma16145004

Academic Editor: Chih-Chun Hsieh

Received: 23 June 2023
Revised: 10 July 2023
Accepted: 13 July 2023
Published: 14 July 2023

Copyright: © 2023 by the authors. Licensee MDPI, Basel, Switzerland. This article is an open access article distributed under the terms and conditions of the Creative Commons Attribution (CC BY) license (https://creativecommons.org/licenses/by/4.0/).

1. Introduction

Currently, the recycling of electric arc furnace dust (EAFD) is a widely discussed topic, because of its high production (world production of EAFD in 2020 = 37 million tonnes [1]), its hazardous nature and its economic value due to the zinc content. Nowadays, EAFD with a low Zn content (around 3%) is mixed into the pellet/briquette for iron production. For EAFD with a higher Zn content (25–46% [2]), pyrometallurgical processing is preferred, mainly the Waelz Kiln Process (WKP). Almost 90% of produced EAFD is processed by WKP. Crude ZnO as a product of the WKP is subsequently refined by sulphate electrowinning or by the Imperial Smelting Process [3,4]. Hydrometallurgical processing of EAFD is rarely used in industry. However, on a research scale, there are a number of studies describing acidic leaching of EAFD using H_2SO_4 [5–13], HCl [2,14–16], HNO_3 [2,17–19] and other leaching agents [20–22]. Leaching with NH_4Cl is also an option [23].

The problem in acidic leaching of zinc from EAFD is mainly the co-leaching of iron (from ferrite and other Fe oxides), calcium and miscellaneous contaminants such as Al, SiO_2 and others [5,11,24,25]. Therefore, the obtained leachates should be refined. Precipitation, cementation, ion exchange and other available methods can be used for impurities removal from acidic leachates [25]. Subsequently, zinc can be extracted from the leachate by precipitation in the form of hydroxide or carbonate, or by combination of solvent extraction and electrowinning [2,5,6,8,11,18,21]. Prior to leaching, vortex layer apparatuses [26] can be successfully used to separate the iron phases from the EAFD.

In addition to acidic leaching, alkaline media such as $(NH_4)_2CO_3$ and NaOH can be used in EAFD processing. The advantage of alkaline leaching agents is the mutual separation of iron and zinc. In alkaline leaching agents, the zinc passes into the leachate and the majority of the iron is concentrated in the insoluble residue. This is mainly because zinc is extracted from ZnO, while zinc ferrite ($ZnFe_2O_4$) is not leached [4,10,18,27–29]. Unfortunately, in addition to zinc, a number of accompanying metals are also extracted into the leachate. For example, the most common and frequently identified impurities in EAFD alkaline leachates are Fe, Pb, Cu, Cr, Mn, Cd, Sn, Al, Ni and Si. Potentially Na, K, Ca, etc. may also be present [30–34]. As with acidic leachates, alkaline zinc leachates need to be refined. The reason is to ensure a suitable purity of the leachate for Zn electrowinning and the high purity of Zn compounds for further applications, for instance ZnO. The production of pure ZnO from EAFD has a high potential for industry. The largest consumers of ZnO are the rubber industry (nearly 45%) and ceramics (about 15%) [35]. Possible methods for alkaline leachate refining prior to subsequent precipitation or zinc electrowinning include cementation, sorption, solvent extraction or membrane processes. Cementation is the most common and available option, using zinc as a cementation agent, which also enriches the leachate.

Several studies [5–8,11,16,21] investigate the cementation of the accompanying metals, but the input is the acidic leachates of EAFD. In contrast, cementation as a refining method for alkaline leachates has not been sufficiently investigated so far, or only marginally so, and detailed cementation conditions are not known. The refining of alkaline leachates by cementation with zinc and the recovery of zinc in the form of ZnO were partially described in [36]. EAFD was leached in $(NH_4)_2CO_3$ after washing with water. Removal of the present contaminants (Cu, Cd and Pb) was carried out by cementation with zinc dust (5 g/L) for 10 min at 20 °C. After crystallization and calcination, high purity ZnO (99.5%) was obtained. The cementation efficiencies for each contaminant were: Pb—97.1%; Cd—97.3% and Cu—97.1%. In the study of [31], the cementation of Pb and Mn from alkaline leachate of EAFD with zinc was carried, but the other impurities were not discussed. The optimum conditions for reducing the content of impurities to an acceptable concentration are given by the authors as: addition of 2 g/L Zn dust, cementation time = 1 h, temperature = 35 °C and pH of the ammonia solution = 9.03.

The aim of this work was to study the conditions and optimization of Fe, Pb, Cu, Cr, Mn cementation and the way of removal of other present metals (mainly Ca) from ammonium carbonate leachates formed during EAFD leaching and final recovery of ZnO. Based on preliminary experiments, powdered zinc and aluminium were selected as suitable cementation agents, first separately and subsequently in combination. The advantage of using these cementators is their availability and relatively low costs. In addition, zinc is the main metal of interest in the leachate and, by using zinc as a cementing agent, its concentration in the leachate will increase.

The goal was to observe the behaviour of the impurities during cementation at different temperatures and to verify the possibilities and limits for achieving a suitable purity of Zn solution before the subsequent crystallization of the Zn precursor. The ultimate goal was to obtain pure ZnO with suitable properties for its use, for example, in the ceramic, electrical or rubber industries.

2. Thermodynamic Study of Cementation of Impurities by Zinc and Aluminium

For the study of thermodynamics, four main contaminants—Fe, Pb, Cu and Cr were chosen. The probable chemical reactions of cementation of these contaminants from zinc ammonium carbonate leachate using two cementation agents—Zn and Al at two considered temperatures (20 and 80 °C) were calculated using HSC Chemistry v6.1 software [37], Tables 1 and 2. Temperatures of 20 and 80 °C were chosen as the two boundary temperatures to be considered for the cementation experiments, to show the tendency of the system behaviour with increasing temperature.

Table 1. Predicted chemical reactions of cementation of accompanying metals from ammonium carbonate leachate using zinc at 20 and 80 °C.

Chemical Reaction	ΔG^0 [kJ/mol]		N°
	20 °C	80 °C	
$Cu^{+2} + 1.5Zn + 3\ CO_3{}^{-2} + 3H_2O = 1.5Zn(OH)_2 + Cu + 3HCO_3{}^- + e^-$	−362.18	−395.26	(1)
$Cu^{+2} + Zn + 4CO_3{}^{-2} + 4NH_4{}^+ = [Zn(NH_3)_4]^{+2} + Cu + 4HCO_3{}^-$	−282.44	−318.09	(2)
$Pb^{+2} + 1.5Zn + 3CO_3{}^{-2} + 3H_2O = 1.5Zn(OH)_2 + Pb + 3HCO_3{}^- + e^-$	−273.33	−301.53	(3)
$Fe^{+2} + 1.5Zn + 3CO_3{}^{-2} + 3H_2O = 1.5Zn(OH)_2 + Fe + 3HCO_3{}^- + e^-$	−218.10	−253.33	(4)
$Pb^{+2} + Zn + 4CO_3{}^{-2} + 4NH_4{}^+ = [Zn(NH_3)_4]^{+2} + Pb + 4HCO_3{}^-$	−193.58	−224.34	(5)
$2Pb^{+2} + Zn + 4CO_3{}^{-2} + H_2O = ZnO + 2PbCO_3 + 2HCO_3{}^- + 2e^-$	−171.20	−195.41	(6)
$Cr^{+3} + 1.5Zn + 6CO_3{}^{-2} + 6NH_4{}^+ = 1.5[Zn(NH_3)_4]^{+2} + Cr + 6HCO_3{}^-$	−118.94	−181.68	(7)
$Fe^{+2} + Zn + 4CO_3{}^{-2} + 4NH_4{}^+ = [Zn(NH_3)_4]^{+2} + Fe + 4HCO_3{}^-$	−125.86	−161.75	(8)
$2Fe^{+2} + Zn + 4CO_3{}^{-2} + H_2O = ZnO + 2FeCO_3 + 2HCO_3{}^- + 2e^-$	−157.27	−188.45	(9)
$2Cu^{+2} + Zn + 4CO_3{}^{-2} + H_2O = ZnO + 2CuCO_3 + 2HCO_3{}^- + 2e^-$	−153.81	−182.21	(10)
$Cr^{+3} + 1.5Zn + 3CO_3{}^{-2} + 3H_2O = 1.5(ZnOH)_2 + Cr + 3HCO_3{}^-$	−98.13	−140.49	(11)

Table 2. Predicted chemical reactions of cementation of accompanying metals from ammonium carbonate leachate using aluminium at 20 and 80 °C.

Chemical Reaction	ΔG^0 [kJ/mol]		N°
	20 °C	80 °C	
$Cu^{+2} + Al + 3CO_3{}^{-2} + 3H_2O = Al(OH)_3 + Cu + 3HCO_3{}^- + e^-$	−670.50	−703.65	(12)
$Pb^{+2} + Al + 3CO_3{}^{-2} + 3H_2O = Al(OH)_3 + Pb + 3HCO_3{}^- + e^-$	−578.00	−604.32	(13)
$Fe^{+2} + Al + 3CO_3{}^{-2} + 3H_2O = Al(OH)_3 + Fe + 3HCO_3{}^- + e^-$	−526.41	−561.71	(14)
$Cr^{+3} + Al + 3CO_3{}^{-2} + 3H_2O = Al(OH)_3 + Cr + 3HCO_3{}^-$	−394.66	−435.15	(15)
$6Fe^{+2} + 2Al + 12CO_3{}^{-2} + 3H_2O = Al_2O_3 + 6FeCO_3 + 6HCO_3{}^- + 6e^-$	−261.91	−292.99	(16)
$6Pb^{+2} + 2Al + 12CO_3{}^{-2} + 3H_2O = Al_2O_3 + 6PbCO_3 + 6HCO_3{}^- + 6e^-$	−275.85	−299.94	(17)
$6Cu^{+2} + 2Al + 12CO_3{}^{-2} + 3H_2O = Al_2O_3 + 6CuCO_3 + 6HCO_3{}^- + 6e^-$	−258.27	−285.73	(18)

On the basis of the calculated ΔG^0 values, the cementation order of the individual metals was determined using zinc and aluminium cementators as follows: Cu → Pb → Fe → Cr. This order is consistent with the values of standard electrode redox potential pairs, Figure 1.

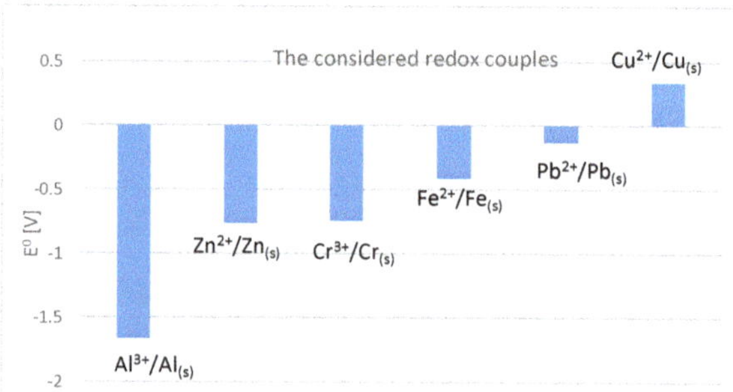

Figure 1. E° values of redox pairs of the present metals [38].

In order to predict the behaviour of the impurities in the leachate during cementation, E–pH diagrams for the Me–Zn–Al–C–N systems (where Me = Fe, Pb, Cu, Cr) were constructed with the HSC Chemistry 6.1 [37] at both considered temperatures (20 and 80 °C), Figure 2a–h.

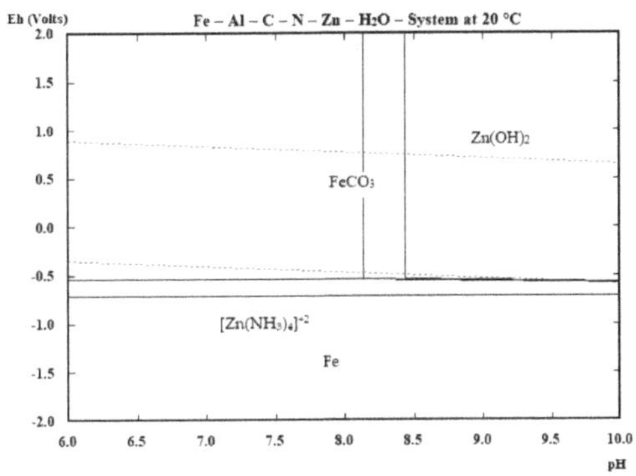

(**a**)

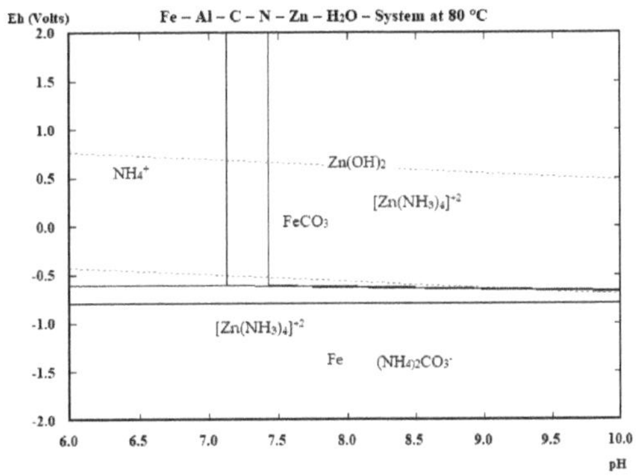

(**b**)

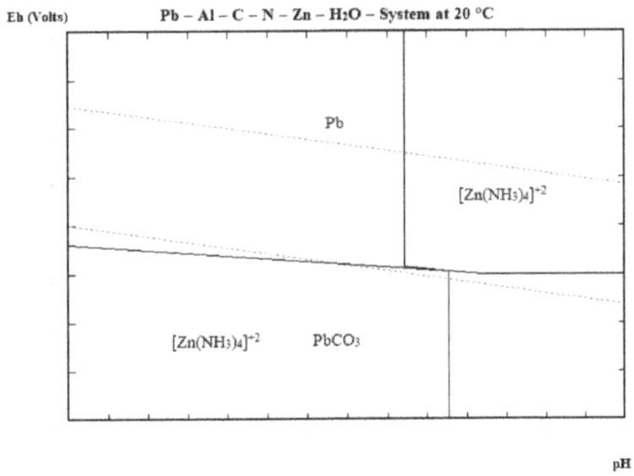

(**c**)

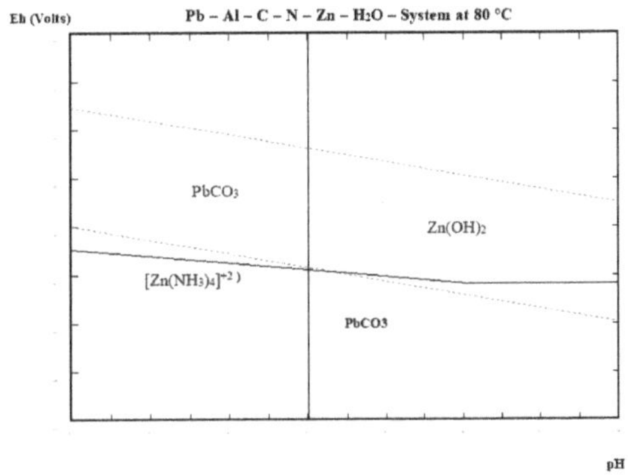

(**d**)

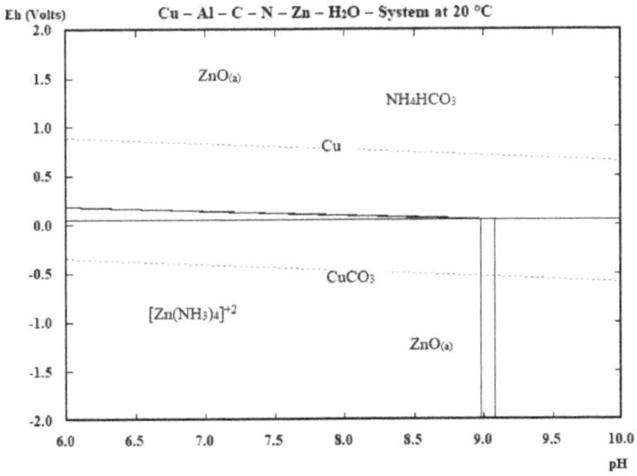

(**e**)

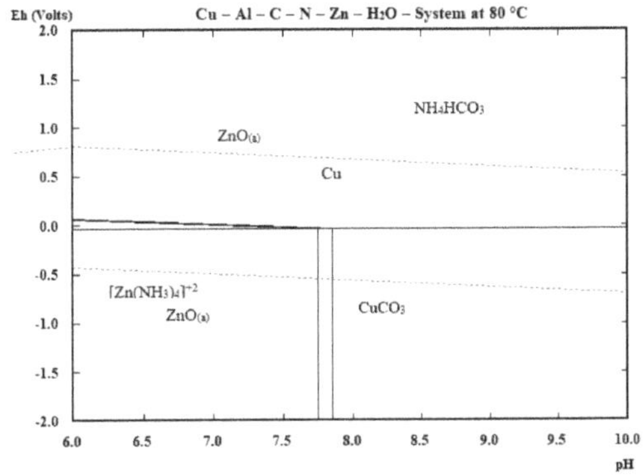

(**f**)

Figure 2. *Cont.*

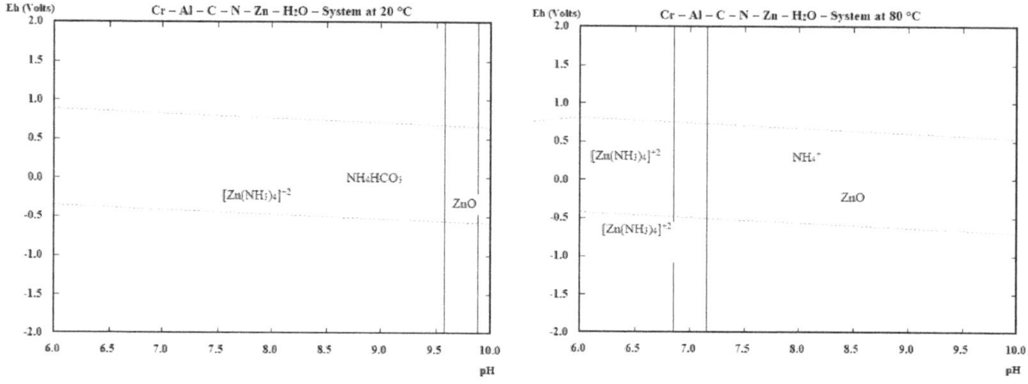

(g) (h)

Figure 2. E–pH diagrams for Me–Zn–Al–C–N–H_2O system at 20 and 80 °C at atmospheric pressure, where Me = Fe (**a**,**b**), Pb (**c**,**d**), Cu (**e**,**f**), Cr (**g**,**h**) with molality of 1×10^{-4} mol/kg for Fe, Pb, Cu and 8×10^{-5} mol/kg for Cr.

For the construction of the diagrams, a pH range of 6–10 was selected and the presence of the solid phases of the examined metals was evaluated at pH = 8–9, which represents the real pH of the input leachate. At the same time, the maximum concentrations of the cemented metals in the leachates were also taken into account (expressed in mol/kg).

Stability region of metallic iron in the presence of $(NH_4)_2CO_3$ and other dissolved impurities is located in the areas outside the stability of the water over the whole considered pH range, therefore it can be expected that, in the presence of zinc and aluminium at the considered pH = 8–9, iron can be removed from solution as $FeCO_3$ rather than as metallic at both considered temperatures. On the other hand, Fe is removed by Zn and Al from the investigated metals with a lower probability from a thermodynamic point of view. Higher temperatures only slightly extend the region of existence of elemental Fe in the alkaline pH beyond the water stability limit. Lead exists at pH = 8–9 as Pb and $PbCO_3$ at 20 °C. At 80 °C there is no area of metallic Pb and it could probably be removed from solution as carbonate. Copper can be removed out of solution as Cu or $CuCO_3$ at pH = 8–9 at both studied temperatures. In the case of chromium, there is no phase in E–pH that would predict the cementation of chromium, e.g., in metallic form, although the cementation reactions are thermodynamically viable.

The thermodynamic study confirmed the probability of the proposed cementation reactions of accompanying metals such as Fe, Cu, Pb and Cr using powdered zinc and aluminium, where ΔG^0 of most of the expected reactions takes a negative value at both considered temperatures, and hence should proceed in the direction of product formation. The E–pH diagrams confirmed the presence of the solid phases of the studied metals at the considered pH = 8–9, with the exception of chromium. At 80 °C, the cementation process may differ from 20 °C, also due to the decomposition of $(NH_4)_2CO_3$ and the presence of other nitrogen, carbon and water based species and the change in the activity of the individual species, since at higher temperatures (ca. 50 °C) the decomposition of $(NH_4)_2CO_3$ and the release of gaseous NH_3 and CO_2 occur according to the Reaction (19)

$$(NH_4)_2CO_3 = 2NH_{3(g)} + CO_{2(g)} + H_2O, \Delta G^0{}_{353} = -6.2 \text{ kJ} \quad (19)$$

The question on a practical scale is also the passivation of powdered Zn and Al and its influence on the reactions proceeding, form of cementation products and their removal rate under specific conditions and finally the practical efficiency of accompanying metals' cementation. The passivation of the Zn and Al is illustrated via E–pH diagrams at 20 and 80 °C in Figures 3 and 4.

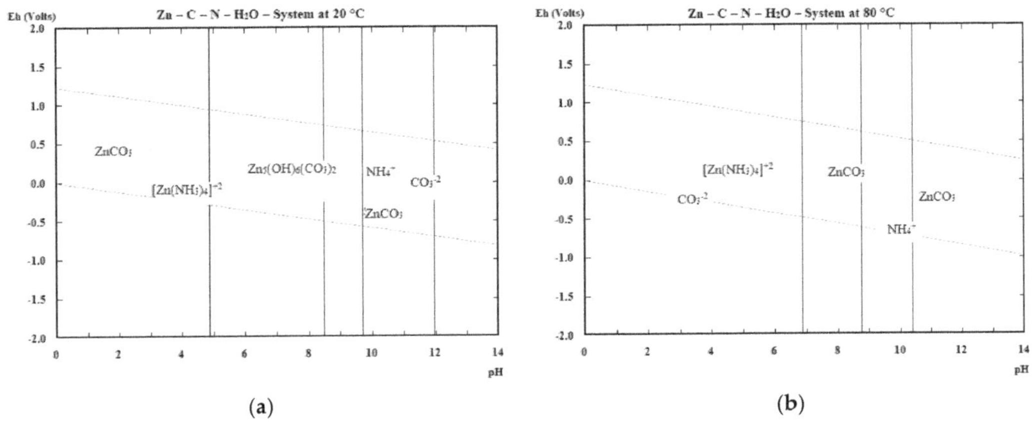

Figure 3. Passivation of zinc in the Zn–$(NH_4)_2CO_3$ system (**a**) at 20 °C (**b**) at 80 °C.

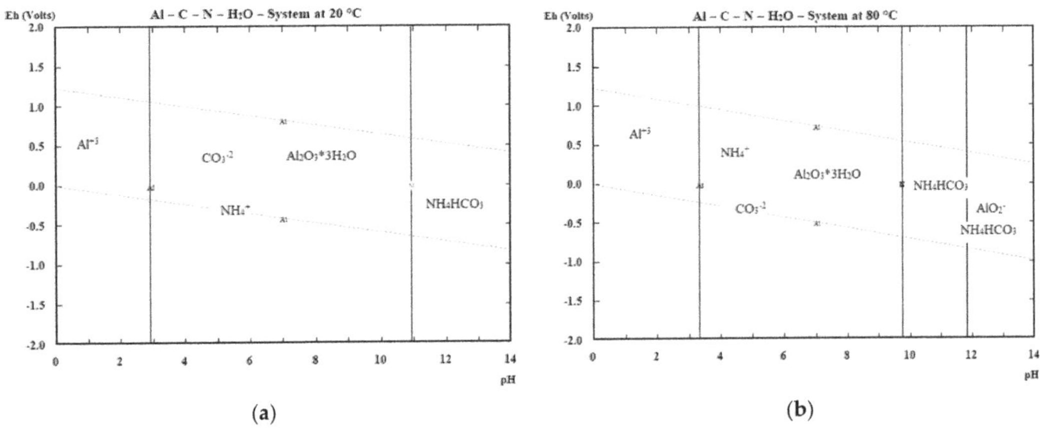

Figure 4. Passivation of aluminium in the Al–$(NH_4)_2CO_3$ system (**a**) at 20 °C, (**b**) at 80 °C.

The passivation of zinc in the cementation process at pH = 8–9 seems to take place by $ZnCO_3$ formation mainly at 20 °C. At higher temperatures, $Zn_5(OH)_6(CO_3)_2$ exists in addition to $ZnCO_3$. In case of aluminium the passivation occurs mostly through the formation of hydrated alumina at 20 °C in the considered region of pH = 8–9, up to pH~11. At 80 °C, the hydrated alumina presence range is reduced to below pH 10. The presence of alumina and hydrated alumina is also predicted by the cementation reactions given in Table 2. In case of using both cementing agents at the same time, the formation of a complex compound based on $ZnO.Al_2O_3$ is more probable (Reaction (20)) than individual Zn and Al oxides/hydroxides/carbonates from the thermodynamic point of view.

$$2Al + Zn + 4H_2O + 8CO_3^{-2} = ZnO{*}Al_2O_3 + 8HCO_3^{-} + 8e^{-}, \Delta G^0{}_{353} = -1555 \text{ kJ} \quad (20)$$

3. Materials and Methods

For the experimental study of cementation, the leachates produced by leaching the EAFD in ammonium carbonate were used. Previously, a neutral leaching of the EAFD was carried out to remove the chlorides present.

For cementation experiments powdered zinc and aluminium were used, both in analytical grade p.a. (Centralchem, Bratislava, Slovakia).

The chemical analysis of EAFD, leachates and obtained product were carried out by the AAS method using Varian AA240+ or Thermo Scientific (iCE 3000 series, London, UK) with the stock solutions at a concentration of 1 g/L ± 0.002 or 0.005 for metals such as Fe, Cu, Pb, Cr, Al, Si and others, and at a concentration of 10 g/L ± 0.020 for zinc. The results from the AAS analysis are acceptable for this method with a relative standard deviation value below 6% for three replicate measurements. By AAS, the content of Zn, Fe, Pb, Cd, Cu, Cr, Mn, Ca, Si and as well Al in the case of cementation with Al was determined. Chloride content was determined by titration and sulphate by a UV/VIS spectrophotometer HI83099 Hanna Instruments (Smithfield, RI, USA).

The chemical composition of the input EAFD and the dust after neutral leaching is given in Table 3. Mineralogical study of EAFD showed that phases such as franklinite $ZnFe_2O_4$, zincite ZnO, magnetite Fe_3O_4, limestone $CaCO_3$ and silica SiO_2 are predominantly present in the sample.

Table 3. Chemical composition of input dusts.

(wt.%)	Zn	Fe	Pb	Cd	Cu	Cr	Mn	Ni	Ca	Si	Mg	Cl^-
Input EAFD	28.01	26.34	0.73	0.02	0.12	0.31	1.90	0.013	5.25	1.44	1.94	1.77
Washed	28.06	27.17	0.70	0.02	0.11	0.16	1.79	0.013	3.66	1.44	1.83	1.48

Analysis of the tap water intended for neutral leaching from an external source is given in Table 4.

Table 4. Tap water analysis [39].

Analysed Parameter	Total Hardness (mmol/L)	NO_3^- (mg/L)	Fe (mg/L)	Mn (mg/L)	Free Chlorides (mg/L)	pH
Tap water	1.30	3.50	0.080	<0.030	<0.10	8.0
Limit	1.1–5.0	50.0	0.20	0.05	0.30	6.50–9.50

The chemical composition of the leachates from the ammonium carbonate leaching of the EAFD as an input for the cementation experiments is given in Table 5. Leachate 1 was obtained by a semi-operating experiment, (S/L = 10, 100 g/L $(NH_4)_2CO_3$), 50 °C, 30 min). Leachates 2 and 3 were obtained by repeated leaching the washed EAFD in the initial leachate with the addition of 25 g/L $(NH_4)_2CO_3$ (analytical grade, p.a., microCHEM, Pezinok, Slovakia) for 30 min, S/L = 10 and 50 °C.

Table 5. Chemical composition of leachates used for cementation.

Leachate N°	Zn	Fe	Pb	Cd	Cu	Cr	Mn	Ca	Cl^-	SO_4^{2-}	pH
	(μg/mL)								(g/L)		
1	4552	7.642	0.464	0	0	0.052	0.494	44.952	5.32	-	8.28
2	8250	4.460	21.39	0	7.590	1.320	1.430	57.200	3.90	0.57	9.02
3	11,750	2.670	1.510	0	9.310	4.100	0.945	88.800	2.98	0.62	8.89

Cementation of enriched leachates (leachates 2 and 3) was carried out in a glass laboratory reactor, which was placed in a thermostatically controlled water bath. The input volume of the cemented leachates was 300 mL. The pH of the leachates was measured using a WTW inoLab pH meter (pH/ION7320, Burladingen, Germany) before and after the cementation experiments. Cementation with zinc (5 g/L) or aluminium (5 g/L) separately was carried out under the following conditions: 20 and 60 °C, 30 min, 300 rpm.

To study the effect of the combination of cementation agents (Zn:Al = 2.5:2.5 g/L or 5:5 g/L) on the process efficiency, the experiments were carried out at 20, 40, 60 and 80 °C, 30 min at 300 rpm. Preliminary experiments were also carried out at 20 °C with the combination Zn:Al = 7.5:7.5 g/L, but without a significant increase in cementation efficiency.

The cementation efficiency was calculated according (21)

$$\mu = \frac{c_{(0)} - c_{(1)}}{c_{(0)}} \cdot 100\ (\%) \quad (21)$$

where $c_{(0)}$ is the initial metal concentration in the leachate and $c_{(1)}$ is the metal concentration in the leachate after cementation. The resulting efficiencies represent the average value obtained from three measurements (three samples taken and analysed after cementation).

The solid residues after washing and drying (100 °C, 24 h) were analysed by XRD using a Philips X'Pert PRO MRD (Co-Kα), range of measuring (10–120° 2theta), scan step (0.0170°) diffractometer (Philips, Amsterdam, The Netherlands). The phases were identified using X'Per HighScore plus software, v3.0a (3.0.1). Selected samples were subjected to XRF (Shimadzu EDX 7000 (Tokyo, Japan) and scanning electron microscopy (SEM) analysis along with energy-dispersive spectrometry, MIRA3 FE-SEM (resolution: 1.2 nm at 30 kV; 2.3 nm at 3 kV, TESCAN, Warrendale, PA, USA). Selected cementation residues were subjected to X-ray photoelectron spectroscopy (XPS) using a Kratos AXIS ULTRA DLD spectrometer (Kratos Analytical, Manchester, UK) with: X ray monochromatic Al Kα (1486.6 eV), 150 W, charge neutraliser system, analysis area: 300 × 700 μm^2, energy calibration C 1s = 284.7 Ev.

The leachates after cementation were further subjected to crystallization and calcination in order to obtain pure zinc oxide. Crystallization was carried out at 105 °C. The crystallization products were subjected to calcination at 900 °C for 4 h. The calcinates were analysed for the content of the zinc and accompanying metals by AAS. In addition, SEM—EDX analysis and XRD were also carried out on selected ZnO samples.

4. Results and Discussion

4.1. Metal Cementation from Ammonium Carbonate Leachate

The results of cementation of the accompanying metals using zinc are shown in Table 6 and using aluminium in Table 7.

Table 6. Results of cementation with pure Zn 5 g/L, 20 and 60 °C, 30 min.

N°	(°C)	Leachate 3	Zn	Fe	Pb	Cu	Cr	Mn	Ca	pH
			(μg/mL)							
1	20	Before cementation	11,750	2.67	1.51	9.31	4.10	0.95	88.8	8.82
		After cementation	12,100	2.56	0.78	0.30	0	0	28.4	8.81
		η(%)	-	**2.99**	**48.34**	**96.78**	**100**	**100**	**68.01**	
2	60	Before cementation	11,750	2.67	1.51	9.31	4.10	0.95	88.8	8.82
		After cementation	12,460	0	0.46	0.41	0	0	26.9	8.9
		η(%)	-	**100**	**69.54**	**95.49**	**100**	**100**	**69.71**	

The results of cementation with the combination of Zn:Al = 2.5:2.5 g/L and 5:5 g/L at 20, 40, 60 and 80 °C are given in Table 8.

The achieved results of cementation with Zn (Table 6) show that, even at 20 °C, almost complete removal of Cu, Cr and Mn occurred, but only 3% of Fe was removed. Increasing the temperature had a positive effect on the cementation of iron, because, at 60 °C, 100% Fe cementation efficiency was achieved. In the case of Pb, only partial removal occurred at both used temperatures (20 °C—48.3%, 60 °C—70%). When aluminium was used as cementation agent (Table 7), about 25% of Fe was cemented at 20 °C. A better but not completely satisfactory result was obtained at 60 °C—61%. Cr and Mn cemented with

100% efficiency already at 20 °C. Cu reached a maximum cementation of 90% at 60 °C, Pb was cemented in the range of 64–68%. At the same time, almost 70% of the calcium was removed in the process when both cementators had been used at both temperatures.

Table 7. Results of cementation with pure Al 5 g/L, 20 and 60 °C, 30 min.

N°	(°C)	Leachate 3	Zn	Fe	Pb	Cu	Cr	Mn	Ca	pH
						(μg/mL)				
3	20	Before cementation	11,750	2.67	1.51	9.31	4.10	0.95	88.8	8.82
		After cementation	11,500	2.02	0.55	1.63	0	0	28.6	8.82
		η [%]	-	**24.35**	**63.56**	**82.49**	**100**	**100**	**67.79**	
4	60	Before cementation	11,750	2.67	1.51	9.31	4.10	0.95	88.8	8.82
		After cementation	11,140	1.04	0.48	0.88	0	0	28.0	8.86
		η [%]	-	**61.05**	**68.21**	**90.55**	**100**	**100**	**68.47**	

Table 8. Results of cementation with Zn:Al = 2.5:2.5 g/L and 5:5 g/L at 20, 40, 60 and 80 °C.

N°	t (°C)	Leachate 2 and 3		Zn	Fe	Pb	Cu	Cr	Mn	Ca	pH
		Cementation Agent	Cementation				(μg/mL)				
5	20 °C	2.5 g/L Zn 2.5 g/L Al	Before	8250	5.54	21.39	7.59	1.32	1.43	57.2	9.02
			After	8180	3.39	0	0	0.62	0	51	9.01
			η [%]	-	**38.8**	**100**	**100**	**53**	**100**	**10.8**	-
6	20 °C	5 g/L Zn 5 g/L Al	Before	8250	5.54	21.39	7.59	1.32	1.43	57.2	9.02
			After	9000	2.76	0	0	0.64	0	54	9.02
			η [%]	-	**50.2**	**100**	**100**	**52**	**100**	**5.6**	-
7	40 °C	2.5 g/L Zn 2.5 g/L Al	Before	11,750	2.67	1.51	9.31	4.1	0.945	88.8	8.89
			After	11,180	0.71	0	0	3.25	0	77.4	8.89
			η [%]	-	**71**	**100**	**100**	**26**	**100**	**13**	-
8	40 °C	5 g/L Zn 5 g/L Al	Before	11,750	2.67	1.51	9.31	4.1	0.945	88.8	8.89
			After	10,560	0	0	0	2.07	0	86.6	8.87
			η [%]	-	**100**	**100**	**100**	**49.5**	**100**	**2.5**	-
9	60 °C	2.5 g/L Zn 2.5 g/L Al	Before	8250	5.54	21.39	7.59	1.32	1.43	57.2	9.02
			After	9500	1.13	0	0	0.68	0	48	9.00
			η [%]	-	**79.6**	**100**	**100**	**49**	**100**	**16.1**	-
10	60 °C	5 g/L Zn 5 g/L Al	Before	8250	5.54	21.39	7.59	1.32	1.43	57.2	9.02
			After	9640	0	0	0	0.57	0	49	9.01
			η [%]	-	**100**	**100**	**100**	**57**	**100**	**14.34**	-
11	80 °C	2.5 g/L Zn 2.5 g/L Al	Before	11,750	2.67	1.51	9.31	4.1	0.945	88.8	8.89
			After	10,600	0	0	0	3.19	0	68.6	8.80
			η [%]	-	**100**	**100**	**100**	**22**	**100**	**23**	-
12	80 °C	5 g/L Zn 5 g/L Al	Before	11,750	2.67	1.51	9.31	4.1	0.945	88.8	8.89
			After	11,700	0	0	0	3.58	0	65.0	8.88
			η [%]	-	**100**	**100**	**100**	**13**	**100**	**27**	-

The results correlate with the thermodynamic study. In the case of Cr and Mn, it is shown that, in a given leaching medium, it does not depend on the type of cementation agent and complete cementation occurs under the given conditions.

The insufficient cementation efficiencies of Fe, Pb and partially of Cu resulted in experiments that combined both cementation agents (Table 8). Two leachates were applied for cementation, namely leachate N° 2 at 20 and 60 °C and leachate N° 3 at 40 and 80 °C (the chemical composition of these leachates is in Table 5).

In addition to the elements listed in Table 8, the Cd, Si and Al contents were also determined but were under the detection limit of the AAS method. In all leachates where aluminium was used as a cementation agent there was no transfer of aluminium to the leachate. probably indicating its significant passivation in the form of insoluble Al_2O_3 or its precipitation from the leachate as $Al(OH)_3$. Aluminium may act as an electron donor with subsequent precipitation from the leachate [40,41]. However, its absence in the leachates can be considered as a large advantage from the point of view of their further processing.

The observed cementation results confirm the significant effect of temperature in the cementation of iron. While at 20 °C iron cementation efficiency of 38.8% was achieved using Zn:Al = 2.5:2.5 g/L, at 60 °C, the efficiency increased to almost 80% on otherwise identical terms.

At 80 °C, complete removal of iron from the leachate was achieved even at lower weights of cementators. The cementation of Pb and Cu is not significantly affected by temperature; at 20 °C maximum removal efficiency (100%) of both metals was achieved. For chromium cementation, the efficiency even decreases with temperature in some experiments. In addition to the main impurities, the calcium content of the leachates after cementation also decreased. The highest efficiency of calcium removal was obtained at 80 °C at Zn:Al = 5:5 g/L.

A temperature of 60 °C can be considered sufficient to remove the majority of the impurities, where 100% removal efficiency of Fe, Pb, Cu and 57% removal efficiency of Cr can be achieved using a combination of Zn:Al = 5:5 g/L. A lower proportion of cementators leads to a reduction in the cementation efficiency of Fe under otherwise identical conditions, but Pb and Cu are removed to 100%.

Interestingly, while the efficiency of chromium cementation using zinc and aluminium as cementation agents separately was 100%, using their combination it dropped to a maximum of 57%. The reasons for the preferential cementation and removal of Fe, Cu and Pb over Cr are probably due to the change in the redox potential value of Al−Zn pair in ammonium carbonate solution due to competitive interaction of those cementation agents and at the same time the change of the overall redox potential of Al−Zn−Me system (Me = accompanying metal ion in ammonium carbonate solution) after reaching the steady state (quasi-equilibrium under given conditions) in comparison to the Al−Me system and Zn−Me system and extending the cementation time >30 min could lead to complete removal of chromium.

4.2. *Characterization of Cementation Residue*

The aim of XRD qualitative phase analysis of cementators under different conditions was to describe or indicate the probable presence of cementation products or the possible formation of by-products during cementation. However, due to the low concentrations of metals in solution, the results of the XRD qualitative phase analysis should be taken as a guideline.

In the case of the use of pure Al (5 g/L) as the cementator at 60 °C, the possible presence of Fe and Cr in elemental form was identified, while the presence of Pb and Ca was not confirmed by the given measurement. For Ca, due to its electronegativity, its removal in elemental form was not expected. However, due to the decrease in its concentration during cementation, it was assumed that Ca could be removed as solid $CaCO_3$, but this was not confirmed under the given conditions.

Copper was identified as CuCl and CuO and manganese as Mn_3Si. Moreover, manganese has the thermodynamic ability to bind with the present silicon (either in solution or as an impurity in the Al cementator). Zinc from the carbonate solution reacted with the Al cementator to form the most likely compound $ZnAl_2O_4$, which is consistent with the thermodynamic prediction in the Zn-Al-H_2O (or $(NH_4)_2CO_3$) system.

The aluminium remained mostly in the elemental form after cementation at 60 °C (as well as at 20 °C) and no other corrosion products except $ZnAl_2O_4$ based on aluminium like hydroxides or oxides could be identified. The cemented metals were mostly subject to oxidation and reactions with the cementation matrix as well as other present impurities. This depended on the actual process conditions and the affinity of the metal ions in solution for the given concentration and temperature.

In the case of zinc cementation at 60 °C, the presence of a ZnO corrosion product is evident, which also agrees with the thermodynamic calculation. At the same time, the presence of hydrozincite $Zn_5(OH)_6(CO_3)_2$ and $Zn_3Cu_2(OH)_6(CO_3)_2$, respectively, was indicated. Copper, in addition to Cu-hydrozincite, can be bound as CuO or $CaCuO_2$. Iron could be present as $Ca_2Fe_2O_5$ rather than the elemental Fe and manganese as Mn_2O_3. Calcium was identified as CaO and $Ca(OH)_2$ in addition to the above mentioned compounds. Lead could also be present as $CaPb_3$.

In the case of a mixture cementators, the situation is slightly different. The XRD pattern of the cementation residue from experiment N° 10 where the mixture Zn and Al was used (Zn:Al = 5:5 g/L, 60 °C), is shown in Figure 5.

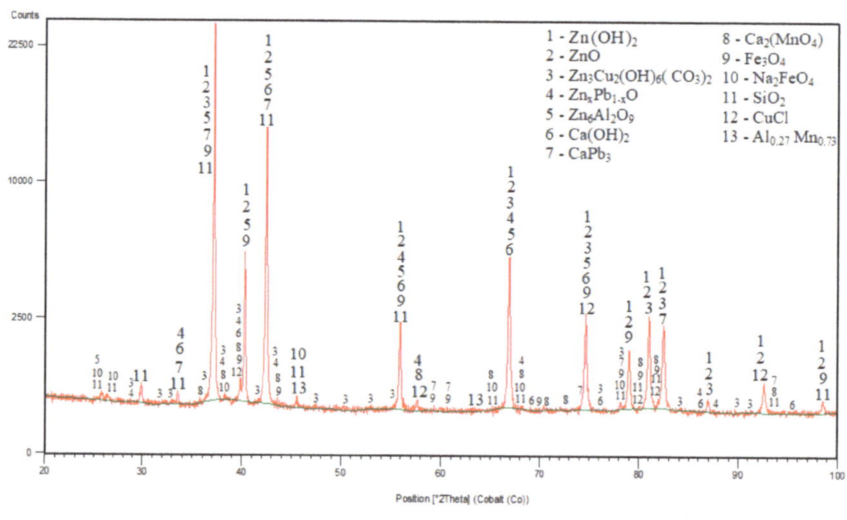

Figure 5. Cementation residue from experiment N° 10 (Zn:Al = 5:5 g/L, 60 °C).

After 30 min of cementation, practically no elemental aluminium or zinc was identified. XRD analysis indicates an almost complete surface reaction of aluminium with zinc and their oxidation in the studied leaching system during cementation experiment to form the compound $6ZnO.Al_2O_3$. Zinc was identified mainly as ZnO and $Zn(OH)_2$. Iron was identified as Fe_3O_4 and Na_2FeO_4. Copper was possibly present as CuCl and Cu-hydrozincite and lead as $Zn_xPb_{1-x}O$ and $CaPb_3$. Moreover, calcium was identified as $Ca(OH)_2$ similarly to the pure Zn cementators at 60 °C and together with manganese as Ca_2MnO_4. In addition to the investigated metals, minor concomitant elements such as chlorine, sodium and potassium (e.g., identified in CuCl, Na_2FeO_4) may also be present in the cementation residue.

From Figure 5 it is visible coincidence or overlapping of some diffractions (peaks) which belong to or can be associated with diffractions of more than one individual phase in the diffraction pattern of the whole mixture. Based on the standard qualitative diffraction phase analysis procedure by using the HighScore software, v3.0a (3.0.1), taking into account also that the phase diffractions coincide together with the thermodynamic assumption and probability calculations, the most probable phases have been identified and selected for the given cementation system.

The complete surface oxidation of Al during cementation with the mixture (Zn:Al = 5:5 g/L, 60 °C) is also confirmed by XPS analysis—spectrum 2 (Figure 6). Spectrum 1 represents the sample after cementation using pure Al (5 g/L, 60 °C) for comparison. The XPS spectra indicated that both samples contained zinc, confirming the participation of zinc ions in the reaction during using pure Al as cementator in carbonate solution and its deposition and further reactions on the Al surface.

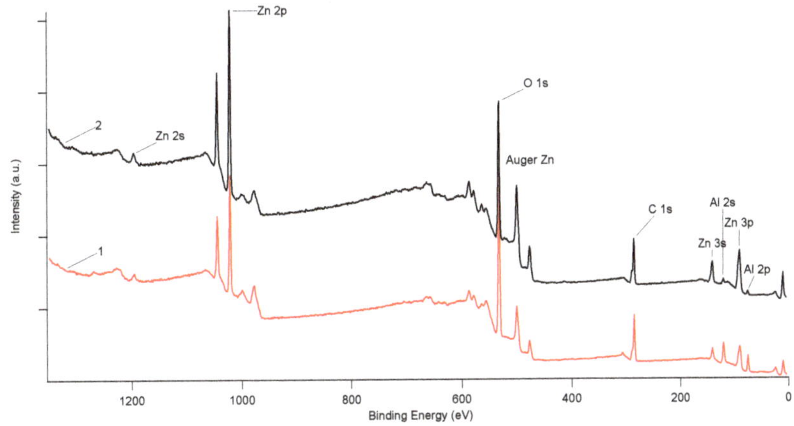

Figure 6. XPS spectra of cementation residues (from experiment N° 4—spectrum 1, from experiment N° 10—spectrum 2).

The obtained results showed that the combination of the two cementation agents (Zn + Al) is a more complex process and caused more significant oxidation of the cementators as well as cemented impurities on the surface of the cementators. This was in contrast to the use of the cementation agents separately, where the oxidation rate of the cementators on their surface after 30 min was considerably lower and moreover created different Zn phases (ZnO and hydrozincite). On the contrary, in the cementation residue, when both of Zn and Al have been used, zinc is probably preferentially bound with Al as $Zn_6Al_2O_9$. Subsequently, ZnO and $Zn(OH)_2$ are generated, but the formation and ratio of individual Zn phases during cementation will be influenced at least by temperature, overall redox potentials of the solution or potentials of individual metal pairs, content and concentration of dissolved impurities in the carbonate solution.

The cementation process seems to be quite easy to perform but, on the other hand, it is rather complicated from a mechanism point of view due to the complexity of the leaching system. Moreover, the effectivity and kinetics of the cementation will play roles and also other phenomena like reactions on the three phase boundary (liquid, solid, gas), where there are also dissolved gases (e.g., $H_{2(g)}$) due to the reaction of Al and Zn with water and $NH_{3(g)}$ from decomposition of ammonium carbonate. XRD analysis also indicated that calcium is removed from the solution mostly as $Ca(OH)_2$ rather than $CaCO_3$, which is more probable from a thermodynamic point of view. On the other hand, it was also proved that calcium could be separated from the solution not only as precipitated solid $Ca(OH)_2$ but also by reaction with other cemented impurities such as Fe, Mn and Pb which form specific compounds.

4.3. Characterization of Obtained Calcinates

The chemical composition of the calcinates obtained by crystallization and subsequent calcination of the refined leachates from experiments N° 1–4 are given in Table 9.

Table 9. Chemical composition of obtained calcinates from experiments N° 1–4.

N°	Zn in Leachate (µg/mL)	Input for Calcination (g)	Calcinates (g)	Content (wt.%)				Purity of ZnO (wt.%)
				Zn	Fe	Ca	Si	
1	12,100	2.0	1.27	73.13	0.08	0.25	0	91.02
2	12,460	2.0	1.26	66.51	0.06	0.16	0	82.78
3	11,500	2.0	1.23	62.74	0.08	0.17	0	78.1
4	11,140	2.0	1.24	49.10	0.06	0.14	0	61.1

The XRD patterns of the selected calcinates are shown in Figures 7 and 8.

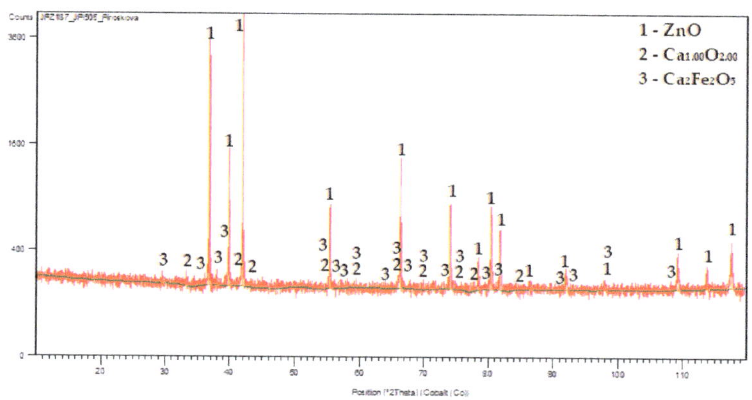

Figure 7. Calcinate (73.13 wt.% of Zn) from experiment N° 1 (5 g/L Zn at 20 °C).

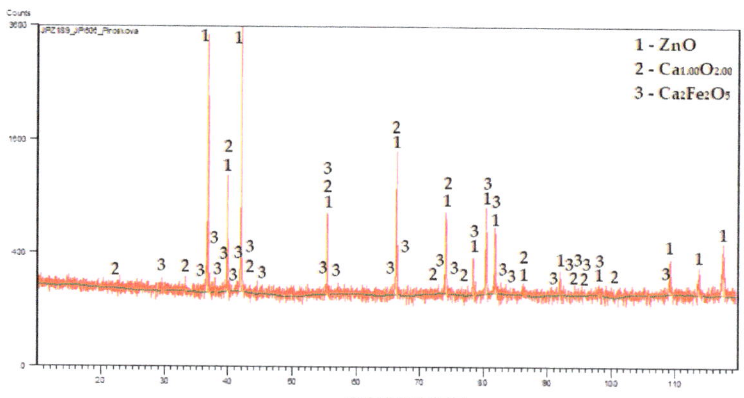

Figure 8. Calcinate (62.74 wt.% of Zn) from experiment N° 3 (5 g/L Al at 20 °C).

The chemical composition of the calcinates obtained by crystallization and subsequent calcination of the refined leachates from cementation experiments N° 5–12 are given in Table 10.

Table 10. Chemical composition of obtained calcinates from experiments N° 5–12.

N°	Zn in Leachate (μg/mL)	Input for Calcination (g)	Calcinates (g)	Content (wt.%)				Purity of ZnO (wt.%)
				Zn	Fe	Ca	Si	
5	8180	4.30	2.60	75.67	0.05	0.31	0.46	94.31
6	9000	3.00	1.86	76.33	0.04	0.26	0.37	95.00
7	11,180	4.68	2.94	75.36	0.06	0.41	0.55	93.80
8	10,560	5.60	3.79	77.61	0.04	0.37	0.46	96.60
9	9640	2.80	1.28	77.21	0.00	0.27	0.45	96.10
10	9500	3.00	1.80	77.67	0.04	0.27	0.44	96.67
11	10,600	5.82	3.74	76.35	0.03	0.40	0.61	95.19
12	11,700	3.64	2.44	74.47	0.03	0.50	0.62	93.03

ZnO with the highest zinc content (from Experiment N° 10) is shown in Figure 9. The results of its XRD analysis, SEM at magnifications of 3000×, 10,000× and 25,000× and EDX analysis are shown in Figures 10–12.

Figure 9. ZnO with the highest Zn content (from Experiment N° 10).

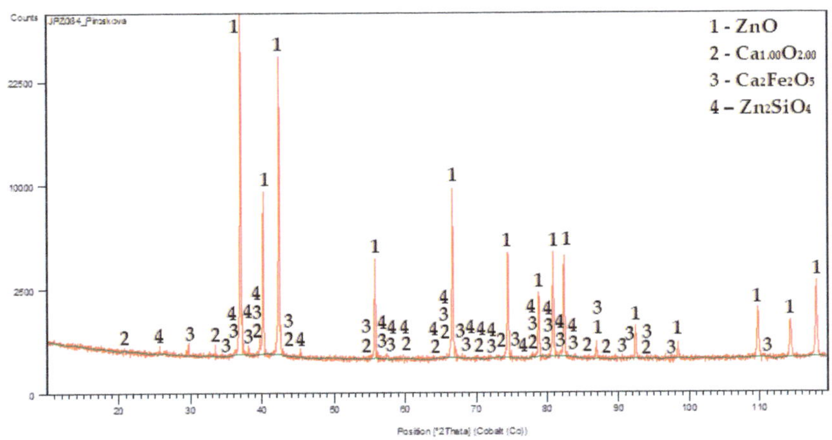

Figure 10. XRD pattern of ZnO with the highest Zn content.

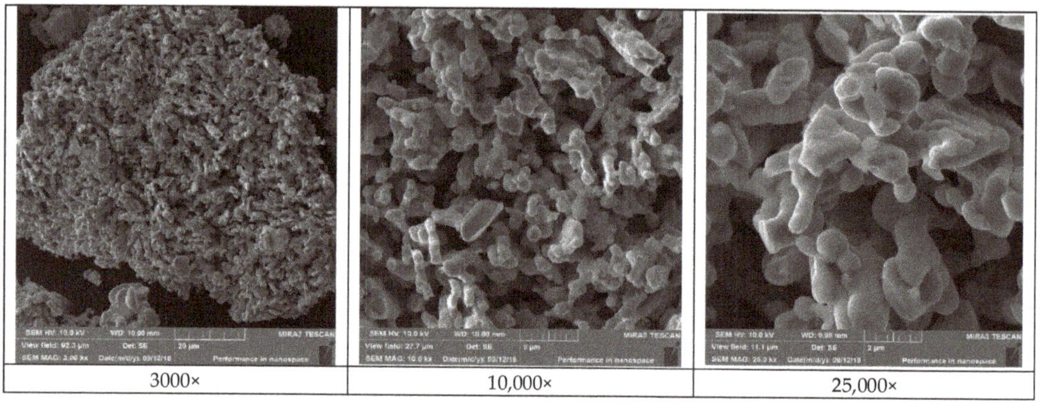

Figure 11. Morphology of ZnO with the highest Zn content.

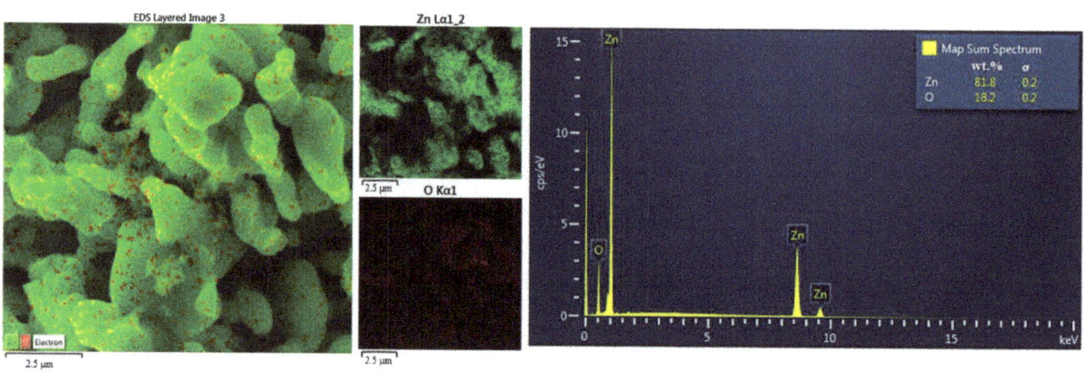

Figure 12. EDX analysis of ZnO with the highest Zn content.

The Zn, Fe, Pb, Cu, Cr, Mn and Si contents were determined in selected calcinates prepared by crystallization of the refined leachate and subsequent calcination of the obtained product. It was confirmed that, at a cementation efficiency of 100% of accompanying metals such as Cu, Pb and Mn, calcinates did not contain these metals or their content was below the limit of detection.

From the obtained results (Table 9), it can be concluded that the zinc content in the ZnO ranged from 74.47% to 77.67%. The highest zinc content (77.67%), which corresponds to ZnO content of 96.67%, was obtained from Experiment N° 10 (Zn:Al = 5.0:5.0 g/L, 60 °C). From the XRD pattern shown in Figure 10, the majority presence of the ZnO phase can be observed. The minor contaminant is the dicalciumferrite $Ca_2Fe_2O_5$, which corresponds to the chemical analysis presented in Table 5. The presence of willemite—Zn_2SiO_4 was also confirmed, while silica can be considered as a stable contaminant with no significant influence on the quality and applicability of the final product. The ZnO particles (Figure 11) are of the mostly elongated or irregular shape, less of a globular shape with a dimension range mostly from to 5 μm in length and 1 micrometre in diameter for the individual particles which are formed to the bigger clusters of around 100 μm in diameter. The form of the particles indicates the sintering of individual particles due to the high calcination temperature. It can be seen that the EDX mapping analysis confirmed the major content of zinc and oxygen. From the economic point of view, and as further investigations of the research team showed [42], for selected ZnO applications (e.g., tyre production), a lower calcination temperature (around 400–500 °C) can be also considered while maintaining sufficient purity and the required properties, since at 700–900 °C the particles are sintered and the specific surface area is reduced.

On the basis of the results from neutral and alkaline leaching published previously by us as the authors [43–47], and according to the results of cementation, a comprehensive processing for EAFD was proposed, Figure 13.

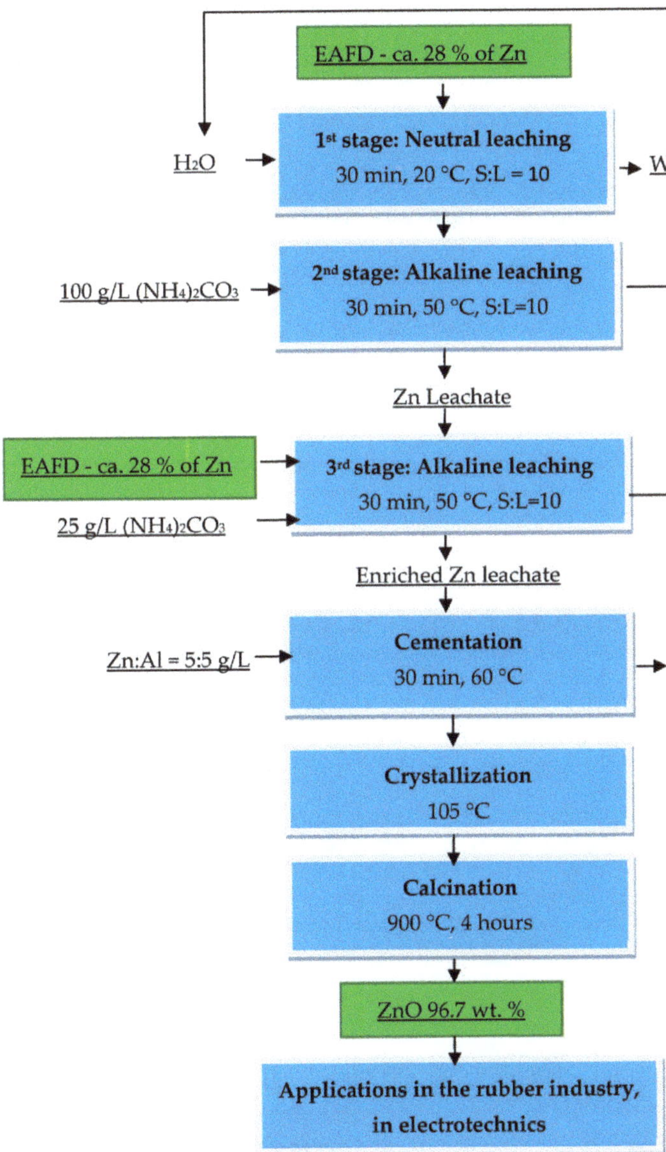

Figure 13. Proposal of comprehensive hydrometallurgical treatment of EAFD with the aim to recover pure ZnO product.

The aim is to obtain a high-purity ZnO with minimal impurities of Fe, Ca and Si, which can be used, for example, in the rubber industry. In the literature [48], the minimum purity of ZnO for application in the rubber industry is reported to be 93%, with lead and cadmium contents of 3 and 2 ppm, respectively. The final product of ZnO meets the required purity. The obtained ZnO has particles of irregular elongated shape, while according to [49] the rubber industry prefers a spherical shape of the ZnO particles. The practical application of the obtained ZnO in the rubber industry could be the subject of further investigation as well as the modification of the ZnO particles' shape by changing the parameters of crystallization and the calcination process of the intermediate Zn product.

According to [50,51], ZnO with such purity can also be used in the production of varistors. The impurities in the form of Si, Ca and Fe do not reduce the quality of the component in terms of electrical properties, moreover, they could even act as a required dopant.

5. Conclusions

The efficient treatment of EAFDs is a highly topical issue due to their quantity, hazardous nature and high zinc content as the main metal of interest (about 30 wt.%). In this work, a possible procedure for the refining of alkaline leachate was proposed and verified, which represents one of the steps in the comprehensive processing of the EAFD presented in the scheme, Figure 13. The alkaline leaching process ensures the selective transfer of zinc from the ZnO phase to the leachate without leaching the major amount of iron, the content of which in the leachate is at a similar level to the zinc content (approximately 30 wt.%). The insoluble ZnO-depleted residue is subsequently treated by acidic leaching to obtain iron and residual zinc from the $ZnFe_2O_4$, which represents the second treatment stream.

The zinc-containing leachates obtained by alkaline leaching have to be refined from the accompanying metals, which are iron, lead, copper, chromium, manganese, etc. Cementation is a simple, cheap and reliable method of refining these leachates and, in addition, in the case of zinc as a cementation agent, the leachate is enriched in zinc.

The combination of Zn and Al as cementation agents proved to be the most effective for the experiments carried out, where, under optimum conditions (Zn:Al = 5:5 g/L, 60 °C, 30 min), a substantial part of the contaminants was removed, in some cases with 100% efficiency (Fe, Cu, Pb and Mn). At the same time, aluminium was not transferred to the leachate. In contrast to cementation with zinc and aluminium separately, in the case of zinc−aluminium mixtures, both of them were significantly oxidized. The impurities present on their surface were also oxidized. The type and amount of phases formed during cementation depends on the temperature as well as on the total redox potential of the solution or individual metal pairs, the content and concentration of dissolved impurities in the carbonate solution, etc.

Zinc oxide with a purity of almost 97% can be obtained by subsequent treatment of the leachates via crystallization at 105 °C and calcination at 900 °C for 4 h. The remaining fraction consists of phases such as dicalcium ferrite, willemite and calcium oxides. Elements such as silicon and calcium are contaminants which are removed by the cementation only partially, but their minimal presence does not affect the quality of the product. ZnO with the achieved purity can be used in the rubber industry or in the production of semiconductor components called varistors.

Author Contributions: Conceptualization, Z.T., A.M. and J.P.; methodology, A.M., J.P., D.O. and Z.T.; investigation, Z.T., J.P., T.V., M.H. and D.O.; resources, J.P., T.V.; writing—original draft preparation, A.M., J.P. and Z.T.; writing—review and editing, Z.T. and J.P.; project administration, J.P. and Z.T. All authors have read and agreed to the published version of the manuscript.

Funding: This work was funded by the Ministry of Education of the Slovak Republic under grant VEGA 2/0080/23 and VEGA 1/0678/23.

Institutional Review Board Statement: Not applicable.

Informed Consent Statement: Not applicable.

Data Availability Statement: Not applicable.

Acknowledgments: Authors would like also to express their gratitude to Institut Français de Slovaquie, Jean-Christophe P. Gabriel and Jocelyne Leroy from of CEA Saclay (IRAMIS/NIMBE/LICSEN), France, for their research support in France and the possibility to perform the XPS measurements.

Conflicts of Interest: The authors declare no conflict of interest.

References

1. 2021 World Steel in Figures. World Steel Association. Available online: https://worldsteel.org/wp-content/uploads/2021-World-Steel-in-Figures.pdf (accessed on 25 November 2022).
2. Al-Harahsheh, M.; Al-Nuairat, J.; Al-Otoom, A.; Al-Hammouri, I.; Al-Jabali, H.; Al-Zoubi, M.; Alasal, S.A. Treatments of electric arc furnace dust and halogenated plastic wastes. *J. Environ. Chem. Eng.* **2019**, *7*, e102856. [CrossRef]
3. Antunano, N.; Cambra, J.F.; Arias, P.L. Hydrometallurgical processes for Waelz oxide valorisation. *Process Saf. Environ. Prot.* **2019**, *129*, 308–320. [CrossRef]
4. Palimaka, P.; Pietrzyk, S.; Stepien, M.; Ciecko, K.; Nejman, I. Zinc Recovery from Steelmaking Dust by Hydrometallurgical Methods. *Metals* **2018**, *8*, 547. [CrossRef]
5. Kaya, M.; Hussaini, S.; Kursunoglu, S. Critical review on secondary zinc resources and their recycling Technologies. *Hydrometallurgy* **2020**, *195*, e105362. [CrossRef]
6. Oustadakis, P.; Tsakiridis, P.E.; Katsiapi, A.; Agatzini-Leonardou, S. Hydrometallurgical process for zinc recovery from electric arc furnace dust (EAFD) Part I: Characterization and leaching by diluted sulphuric acid. *J. Hazard. Mater.* **2010**, *179*, 1–7. [CrossRef] [PubMed]
7. Hazaveha, P.K.; Karimia, S.; Rashchia, F.; Sheibania, S. Purification of the leaching solution of recycling zinc from the hazardous electric arc furnace dust through an as-bearing jarosite. *Ecotox. Environ. Saf.* **2020**, *202*, 110893. [CrossRef]
8. Xanthopoulos, P.; Agatzini-Leonardou, S.; Oustadakis, P.; Tsakiridis, P.E. Zinc recovery from purified electric arc furnace dust leach liquors by chemical precipitation. *J. Environ. Chem. Eng.* **2017**, *5*, 3550–3559. [CrossRef]
9. Havlik, T.; Turzakova, M.; Stopic, S.; Friedrich, B. Atmospheric leaching of EAF dust with diluted sulphuric acid. *Hydrometallurgy* **2005**, *77*, 41–50. [CrossRef]
10. Havlik, T.; e Souza, B.V.; Bernardes, A.M.; Schneider, I.A.H.; Miskufova, A. Hydrometallurgical processing of carbon steel EAF dust. *J. Hazard. Mater.* **2006**, *B135*, 311–318. [CrossRef]
11. Tsakiridis, P.E.; Oustadakis, P.; Katsiapi, A.; Agatzini-Leonardou, S. Hydrometallurgical process for zinc recovery from electric arc furnace dust (EAFD). Part II: Downstream processing and zinc recovery by electrowinning. *J. Hazard. Mat.* **2010**, *179*, 8–14. [CrossRef]
12. Yan, H.; Chai, L.Y.; Peng, B.; Li, M.; Peng, N.; Hou, D.K. A novel method to recover zinc and iron from zinc leaching residue. *Miner. Eng.* **2014**, *55*, 103–110. [CrossRef]
13. Laubertova, M.; Havlik, T.; Parilak, L.; Derin, B.; Trpcevska, J. The Effects of Microwave-assisted leaching on the treatment of electric arc furnace dusts (EAFD). *Arch. Metall. Mater.* **2020**, *65*, 321–328. [CrossRef]
14. Yao, Y.T.; Hwang, S.L. Improved hydrometallurgical extraction of zinc and iron from electric arc furnace (EAF) dust waste using hydrochloric acid. *AIP Conf. Proc.* **2019**, *2157*, 020017. [CrossRef]
15. Langova, S.; Lesko, J.; Matysek, D. Selective leaching of zinc from zinc ferrite with hydrochloric acid. *Hydrometallurgy* **2009**, *95*, 179–182. [CrossRef]
16. Hui-gang, W.; Jian-ming, G.; Wenwu, L.; Mei, Z.; Min, G. Recovery of metal-doped zinc ferrite from zinc-containing electric arc furnace dust: Process development and examination of elemental migration. *Hydrometallurgy* **2016**, *166*, 1–8. [CrossRef]
17. Zoraga, M.; Ilhan, S.; Kalpakli, A.O. Leaching kinetics of electric arc furnace dust in nitric acid solutions. *Int. J. Chem. Kinet.* **2020**, *52*, 933–942. [CrossRef]
18. Binnemans, K.; Jones, P.T.; Fernandez, A.M.; Torres, V.M. Hydrometallurgical Processes for the Recovery of Metals from Steel Industry By-Products. *J. Sustain. Metall.* **2020**, *6*, 505–540. [CrossRef]
19. Yucel, T.; Zoraga, M.; Ilhan, S.; Kalpakli, A.O. Hydrometallurgical Extraction of Zinc from EAF Dust in Nitric Acid Solution. In *The Congress Proceedings Book, 19th International Metallurgy & Materials Congress-IMMC 2018, TÜYAP Fair, Convention & Congress Center, Istanbul, Turkey 25–27 October 2018*; Metem: Parsippany, NJ, USA, 2018.
20. Halli, P.; Hamuyuni, J.; Leikola, M.; Lundström, M. Developing a sustainable solution for recycling electric arc furnace dust via organic acid leaching. *Miner. Eng.* **2018**, *124*, 1–9. [CrossRef]
21. Halli, P.; Agarwal, V.; Partinen, J.; Lundstrom, M. Recovery of Pb and Zn from a citrate leach liquor of a roasted EAF dust using precipitation and solvent extraction. *Sep. Purif. Technol.* **2020**, *236*, 116264. [CrossRef]
22. Leclerc, N.; Meux, E.; Lecuire, J.M. Hydrometallurgical extraction of zinc from zinc ferrites. *Hydrometallurgy* **2003**, *70*, 175–183. [CrossRef]
23. Miki, T.; Chairaksa-Fujimoto, R.; Maruyama, K.; Nagasaka, T. Hydrometallurgical extraction of zinc from CaO treated EAF dust inammonium chloride solution. *J. Hazard. Mat.* **2016**, *302*, 90–96. [CrossRef] [PubMed]
24. Nakamura, T.; Shibata, E.; Takasu, T.; Itou, H. Basic Consideration on EAF Dust Treatment Using Hydrometallurgical Processes. *Resourc. Process.* **2008**, *55*, 144–148. [CrossRef]

25. Kukurugya, F.; Havlík, T.; Parilák, L. Iron removal from solutions after leaching steelmaking dust. In Proceedings of the WASTE—Secondary Raw Materials 5, Liptovsky Jan, Slovakia, 4–7 June 2013.
26. Bizhanov, A.M. Use of vortex layers apparatuses for preparation of oily scale and zinc-containing dust and sludge for briquetting. *Metallurgist* **2023**, *5*, 98–107. (In Russian)
27. Krishnan, S.; Zulkapli, N.S.; Kamyab, H.; Taib, S.M.; Din, M.F.B.M.; Majid, Z.A.; Chaiprapat, S.; Kenzo, I.; Ichikawa, Y.; Nasrullah, M.; et al. Current technologies for recovery of metals from industrial wastes: An overview. *Environ. Technol. Innovat.* **2021**, *22*, 101525. [CrossRef]
28. Havlik, T.; Maruskinova, G.; Miskufova, A. Determination of ZnO amount in electric arc furnace dust and temperature dependence of leaching in ammonium carbonate by using of X-ray diffraction. *Arch. Metall. Mater.* **2018**, *63*, 653–658. [CrossRef]
29. Maccagni, M.G. INDUTEC/EZINEX Integrate Process on Secondary Zinc-Bearing Materials. *J. Sustain. Metall.* **2016**, *2*, 133–140. [CrossRef]
30. Yu, B.S.; Wang, Y.R.; Chang, T.C. Hydrothermal treatment of electric arc furnace dust. *J. Hazard. Mater.* **2011**, *190*, 397–402. [CrossRef]
31. Li, J.; Chen, Q.; Hu, H.-P.; Wang, W.-Z. Deep purification of zinc ammoniacal leaching solution by cementation with zinc dust. *Chin. J. Process Eng.* **2010**, *10*, 799–885.
32. Lenz, D.M.; Martins, F.B. Lead and zinc selective precipitation from leach electric arc furnace dust solutions. *Rev. Matéria* **2007**, *12*, 503–509. [CrossRef]
33. Lia, Q.; Zhao, Y.; Jiang, J.; Zhang, C. Optimized hydrometallurgical route to produce ultrafine zinc powder from industrial wastes in alkaline medium. *Procedia Environ. Sci.* **2012**, *16*, 674–682. [CrossRef]
34. Halli, P.; Hamuyuni, J.; Revitzer, H.; Lundstrom, M. Selection of leaching media for metal dissolution from electric arc furnace dust. *J. Clean. Prod.* **2017**, *164*, 265–276. [CrossRef]
35. Zinc Oxide Market Size, Share & Trends Analysis Report by Application (Rubber, Paints & Coatings, Chemicals, Ceramics), by Process (Wet Chemical, Direct, Indirect), by Region, And Segment Forecasts, 2022–2030. Available online: https://www.grandviewresearch.com/industry-analysis/zinc-oxide-market-report (accessed on 25 February 2023).
36. Ruiz, O.; Clemente, C.; Alonso, M.; Alguacil, F.J. Recycling of an electric arc furnace flue dust to obtain high grade ZnO. *J. Hazard. Mater.* **2007**, *41*, 33–36. [CrossRef] [PubMed]
37. Roine, A. *HSC Chemistry®*, version 6.1. Software Outotec Research Oy. HSC Chemistry: Tempere, Finland, 2006.
38. Comparing Strengths of Oxidants and Reductants. Available online: https://chem.libretexts.org/Bookshelves/Analytical_Chemistry/Supplemental_Modules_(Analytical_Chemistry)/Electrochemistry/Redox_Chemistry/Comparing_Strengths_of_Oxidants_and_Reductants (accessed on 6 July 2023).
39. Products and Services—Drinking and Waste Water Quality Quality Tap Water Shows Positive Effects on Human Health. Available online: https://www.vodarne.eu/produkty-a-sluzby/kvalita-pitnej-a-odpadovej-vody (accessed on 25 June 2022). (In Slovak)
40. Choi, S.; Jeon, S.; Park, I.; Ito, M.; Hiroyoshi, N. Enhanced Cementation of Co^{2+} and Ni^{2+} from Sulfate and Chloride Solutions Using Aluminum as an Electron Donor and Conductive Particles as an Electron Pathway. *Metals* **2021**, *11*, 248. [CrossRef]
41. Choi, S.; Jeon, S.; Park, I.; Tabelin, C.B.; Ito, M.; Hiroyoshi, N. Enhanced cementation of Cd^{2+}, Co^{2+}, Ni^{2+}, and Zn^{2+} on Al from sulfate solutions by activated carbon addition. *Hydrometallurgy* **2021**, *201*, 105580. [CrossRef]
42. Maruskinova, G.; Havlik, T.; Parilak, L.; Chomic, V.; Turna, S.; Balay, M. Hydrometallurgical processing of EAFD—Alkaline way. In Proceedings of the ŽP VVC 2018 Expert Conference, Tale, Horna Lehota, Slovakia, 4–6 September 2018; pp. 197–208, ISBN 978-80-973141-0-1. (In Slovak)
43. Vindt, T.; Havlik, T.; Miskufova, A.; Klein, D. Recycling of EAF dust at the pilot plant within the Laboratory of Processing Industrial Waste. In Proceedings of the TOP 2018—24th International Conference "Engineering for Environment Protection", High Tatras, Strbske Pleso, Slovakia, 19–21 September 2018; ISBN 978-80-227-4835-3.
44. Horvathova, H.; Miskufova, A.; Havlik, T. The study of removal contaminants from wastewater produced during eaf dust treatment. In Proceedings of the Material Recyling of Industrial Waste, Tale, Horna Lehota, Slovakia, 20–21 March 2018; pp. 23–34, ISBN 978-80-553-2951-2.
45. Vindt, T.; Havlik, T.; Parilak, L.; Chomic, V. Industrial Waste Processing Laboratory—Hydrometallurgical Treatment of EAF Dust. *Metall. J.* **2019**, *72*, 21–25.
46. Piroskova, J.; Hezelova, M.; Miskufova, A.; Takacova, Z. The refining of leachate after alkaline leaching EAF dust. In Proceedings of the Material Recycling of Industrial Waste, Tale, Horna Lehota, Slovakia, 20–21 March 2018; pp. 70–77, ISBN 978-80-553-2951-2. (In Slovak)
47. Horvathova, H.; Miskufova, A.; Kochmanova, A.; Halvik, T. Watewater Treatment produced in processing of EAF dust. In Proceedings of the Material Recycling of Industrial Waste, Tale, Horna Lehota, Slovakia, 22–23 September 2016; pp. 24–33, ISBN 978-80-553-2597-2. (In Slovak)
48. RheinChemie Additives. Specialty Rubber Products. Available online: https://rch.lanxess.com/product/zinkoxyd-aktiv-2?lang=en (accessed on 16 March 2023).
49. Anand Krishnamoorthy, K.; Varghese, S. Review—The Importance of Zinc Oxide (ZnO) in Rubber Technology. *Raw Mater. Appl.* **2018**, *10*, 33–39.

50. Liptai, P.; Dolnik, B.; Havlik, T.; Orac, D. *Semiconductor Component (Varistor) Made from ZnO Obtained by Recycling Industrial Waste;* Utility Model PUV 50037-2022; Technical University of Kosice: Kosice, Slovakia, 2022. (In Slovak)
51. Liptai, P.; Dolnik, B.; Briancin, J.; Havlik, T. Hydrometallurgical Recycling of Electric Arc Furnace Dust. *Waste Biomass Valor.* **2020**, *11*, 4419–4428. [CrossRef]

Disclaimer/Publisher's Note: The statements, opinions and data contained in all publications are solely those of the individual author(s) and contributor(s) and not of MDPI and/or the editor(s). MDPI and/or the editor(s) disclaim responsibility for any injury to people or property resulting from any ideas, methods, instructions or products referred to in the content.

materials

Article

Evolutionary Optimizing Process Parameters in the Induction Hardening of Rack Bar by Response Surface Methodology and Desirability Function Approach under Industrial Conditions

Grzegorz Dziatkiewicz [1,*], Krzysztof Kuska [2] and Rafał Popiel [3]

1 Department of Computational Mechanics and Engineering, Silesian University of Technology, ul. Konarskiego 18a, 44-100 Gliwice, Poland
2 HL Mando Corporation Poland Ltd., ul. Uczniowska 36, 58-306 Wałbrzych, Poland; krzysztof-kuska@o2.pl
3 PST Consulting Ltd., ul. Biblioteczna 29/14, 43-100 Tychy, Poland; rafal.popiel@pstconsulting.pl
* Correspondence: grzegorz.dziatkiewicz@polsl.pl

Abstract: Conditions of industrial production introduce additional complexities while attempting to solve optimization problems of material technology processes. The complexity of the physics of such processes and the uncertainties arising from the natural variability of material parameters and the occurrence of disturbances make modeling based on first principles and modern computational methods difficult and even impossible. In particular, this applies to designing material processes considering their quality criteria. This paper shows the optimization of the rack bar induction hardening operation using the response surface methodology approach and the desirability function. The industrial conditions impose additional constraints on time, cost and implementation of experimental plans, so constructing empirical models is more complicated than in laboratory conditions. The empirical models of nine system responses were identified and used to construct a desirability function using expert knowledge to describe the quality requirements of the hardening operation. An analysis of the hypersurface of the desirability function is presented, and the impossibility of using classical gradient algorithms during optimization is empirically established. An evolutionary strategy in the form of a floating-point encoded genetic algorithm was used, which exhibits a non-zero probability of obtaining a global extremum and is a gradient-free method. Confirmation experiments show the improvement of the process quality using introduced measures.

Keywords: induction hardening; thermal residual deformation; multiobjective optimization; response surface methodology; desirability function; evolutionary computations; genetic algorithm

Citation: Dziatkiewicz, G.; Kuska, K.; Popiel, R. Evolutionary Optimizing Process Parameters in the Induction Hardening of Rack Bar by Response Surface Methodology and Desirability Function Approach under Industrial Conditions. *Materials* **2023**, *16*, 5791. https://doi.org/10.3390/ma16175791

Academic Editors: Pavol Liptai, Jaroslav Briančin and Maroš Halama

Received: 27 July 2023
Revised: 19 August 2023
Accepted: 22 August 2023
Published: 24 August 2023

Copyright: © 2023 by the authors. Licensee MDPI, Basel, Switzerland. This article is an open access article distributed under the terms and conditions of the Creative Commons Attribution (CC BY) license (https://creativecommons.org/licenses/by/4.0/).

1. Introduction

Engineering requirements of gear parts that have specific properties lead to research on developing the induction hardening processes. Especially in the automotive industry, reaching a compromise between cost, quality and reliability by using semi-automatic heat treatment processes to ensure the mechanical and geometrical features is receiving more and more attraction. However, the induction hardening process is challenging due to thermal strains affecting the functionality and quality of crucial elements in the produced structures. Other limitation factors are the formation of non-hardened zones in parts and the change in the designed hardness of the treated elements. Moreover, the induction hardening process is quite complex; it could be represented as a coupled field problem with mechanical, electromagnetic and thermal fields, which are pretty hard to model in the industrial environment, especially in the serial mode of production.

Optimization of the induction hardening process is well-known in the literature. Two main approaches can be distinguished here: one using complex computational mechanics models and the other using empirical modeling. The following works represent the computational mechanics approach. Favennec et al. [1] show how to solve the problem of temperature distribution optimization using optimal control techniques, and similarly, Jakubovicowa et al. [2] optimized the process for the uniform surface temperature distribution criterion. Nemkov et al. [3] simulated the stress and distortion evolution during the induction hardening process of tubes using the finite element method; the hardness profile sensitivity in the induction hardening process with the finite element simulations was explored in [4]. The coupled electromagnetic, thermal and mechanical computational models of the heat treatment process using electromagnetic fields were shown in [5]. Fisk et al. [6] have presented the complex models of induction hardening in low alloy steels. Reference [7] shows the optimization of the edge effect of 4340 steel specimens heated by induction process with flux concentrators using finite element axis-symmetric simulation. The computational approach brings several advantages, such as universality and flexibility, as shown in [8,9], and also lower cost and faster results than an experimental and analytical approach. However, the computational approach requires tuned, complex constitutive models, which are hard to identify, especially in industrial conditions. That is why the empirical modeling approach was also successfully applied to solving various optimization problems [10], also arising during the induction hardening process.

The empirical modeling approach allows for establishing relations between process outputs and inputs using experimentation and measurements with statistical techniques for managing data [11]. Kohli and Singh [12] have used response surface methodology (RSM) to find the optimal values of process parameters for induction hardening of AISI 1040 steel. Various process parameters, such as feed rate, current, dwell time, and the gap between the workpiece and induction coil, are experimentally explored. In [13], the RSM and Taguchi method optimized the induction hardening process for maximum depth and minimum edge effect. The multiobjective optimization problem with the appropriate economic, environmental, and social metrics was analyzed in [14] to assure the sustainability of the induction hardening process using empirical models. The reduction of edge effect using the RSM and artificial neural network modeling of a spur gear treated by induction with flux concentrators was shown in [15]. Artificial intelligence modeling of induction contour hardening of 300M steel bar and C45 steel spur-gear was performed in [16]. Reference [17] shows that the central composite design, with a second-order response surface design, was employed to systematically estimate the empirical models of temperature and phase transformation geometry during the induction hardening. The effect of scanning speed and air gap on the uniformity of hardened depth and mechanical properties of large-size spur gears was investigated in [18]. Multi-response optimization using the desirability function approach of the induction hardening process using quality responses such as the effective case depth and hardness values were analyzed in [19] for different combinations of medium frequency power, feed rate, quench pressure and temperature. Asadzadeh et al. [20] have shown the hybrid model, integrating measurements and physics, of the induction hardening.

A gap in the literature can be identified based on the authors' best knowledge and the review presented. It concerns the solution to the problem of multi-criteria optimization of the induction hardening process in industrial conditions utilizing a hybrid approach using empirical modeling and computational intelligence tools. The novelty of this work lies in the application of multiple qualitative metrics regarding deformation, hardness profile and hardening depth in a complex structural component such as a steering gear rack bar to optimize the parameters of the induction hardening process, using empirical modeling, a desirability function and an evolutionary algorithm.

Some essential points characterize the present work, which are identified and presented in this paper:

- There is no evidence that evolutionary optimization using a genetic algorithm has been applied to induction hardening processes under complex industrial process requirements, modeled empirically;

- Lack of availability of studies showing the effect of the form of the desirability hypersurface on the effectiveness of optimization algorithms for the induction hardening process;
- Lack of availability of studies showing the effectiveness of global optimization algorithms, such as genetic algorithms, for obtaining a set of quasi-optimal solutions to the problem for the desirability function formulated for the induction hardening process;
- There is a lack of availability of studies showing the complex problem of conducting experiments to confirm the optimal parameter settings of the induction hardening process under industrial conditions and analyzing their results contained in small samples, which precludes the use of parametric tests.

In this work, RSM with the central composite design (CCD) of experiments was employed to establish the functional relationship between the three main process parameters that served as design variables of the induction hardening multi-criteria optimization problem. Several responses related to hardness profiles and geometrical measures of quenching quality of the heat treatment operation of the automotive steering gear were formulated. Among them, the most important in the following work is the problem of minimizing residual thermal deformations, which is enforced by an additional straightening operation, increasing the duration of the process and its costs. However, the desirability approach allowed for transforming the multiobjective optimization problem into a single-optimization problem using expert knowledge for setting the weights in the desirability function. The quadratic models for the process responses were quantitatively analyzed, and their significance and accuracy were confirmed statistically. Next, the reduction of the models is performed to consider significant terms of the regression models. This step introduces a specific consequence for the optimization problem formulation in a change of an optimized function form. That change affects the effectiveness of the applied optimization algorithms. The article shows that the global optimization technique as the evolutionary strategy in the form of GA allows to meet the difficulties resulting from the form of the optimized function. Finally, the confirmation experiments should be conducted to verify the optimal solution the GA method determines.

This article consists of six Sections. After the introduction in the present Section, Section 2 briefly describes the considered induction hardening process with its quality indicators and experimentation methods. Section 3 is devoted to a description of the empirical models that were obtained with RSM. Section 4 focuses on formulating the multi-criteria optimization problem and obtaining the solution using the computational intelligence technique, namely the GA algorithm. Section 5 analyzes confirmation experiments. The last Section, Section 6, briefly summarizes the conducted research.

2. Process, Quality Indicators, and Experimentation Methods

2.1. Induction Hardening Process

The process is conducted using a particular equipment. The automatic inductive hardening and tempering machine have a rotation table with three stations, as shown in Figure 1:

- For loading rack bars before the heat treatment process and unloading rack bars after the process is finished;
- Hardening station;
- Tempering station.

The machine is available for hardening and tempering rack bars with lengths of 500–900 mm and 22–32 mm diameters.

The hardening machine provides a possibility to control the hardening and tempering process condition by adjusting:

- The hardening power is in the range of 0–100%;
- The hardening feed rate ranges from 100–50,000 mm/min;
- The distance of the hardening coil to rack bar teeth is in the 1.5–5 mm range.

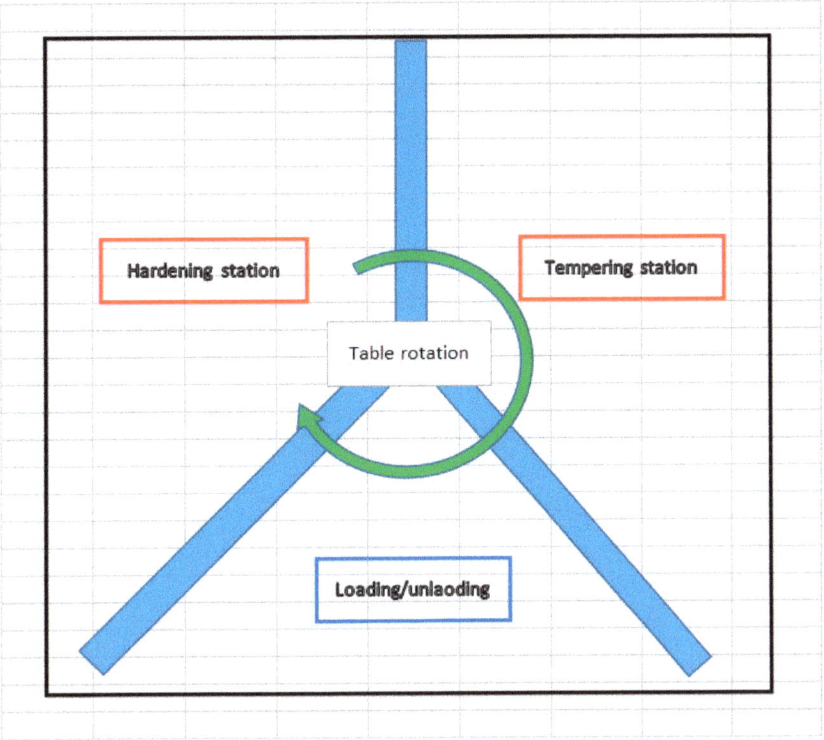

Figure 1. The automatic inductive hardening and tempering machine.

Moreover, the flow of quenching water could be controlled; however, due to the low accuracy of the machine flow meter and lack of control possibility in auto mode, the quenching flow rate was kept constant through the process with a value equal to 45 L/min. Moreover, the hardening station has sensors in order to provide a continuous measurement of a rack bar distance from the coil. It also provides a stable distance from the hardening coil during the hardening process. Furthermore, during the tempering process, which gave the required surface hardness, the power and the feed were kept constant through experiments with settings:

- The power equals 47%;
- The feed rate equals 1600 mm/min.

The measurements of the following quantities were conducted during the quality control of rack bars:

1. Hardness depth;
2. Surface hardness;
3. It is mandatory to use a straightening process after hardening due to deformation, which is an effect of the hardening process; the time of the straightening operation increases when rack bars deformation is over 1100 μm.

The main aim of the present study was to establish hardening process parameters in order to achieve hardening conditions that match quality requirements, i.e., minimize rack bar deformations with the proper hardening depth and hardness profiles and avoid increasing a total machine cycle time, mainly when it includes additional operations, e.g., straightening.

Additional ancillary factors, e.g., coolant state, coolant flow rate, environment temperature, humidity, hardening coil condition, deviation in allow composition of steel, etc., can influence the output variables; nevertheless, the performed experiments are restricted to the three hardening parameters, which are automatically controlled by the machine. Figure 2 shows the process setup.

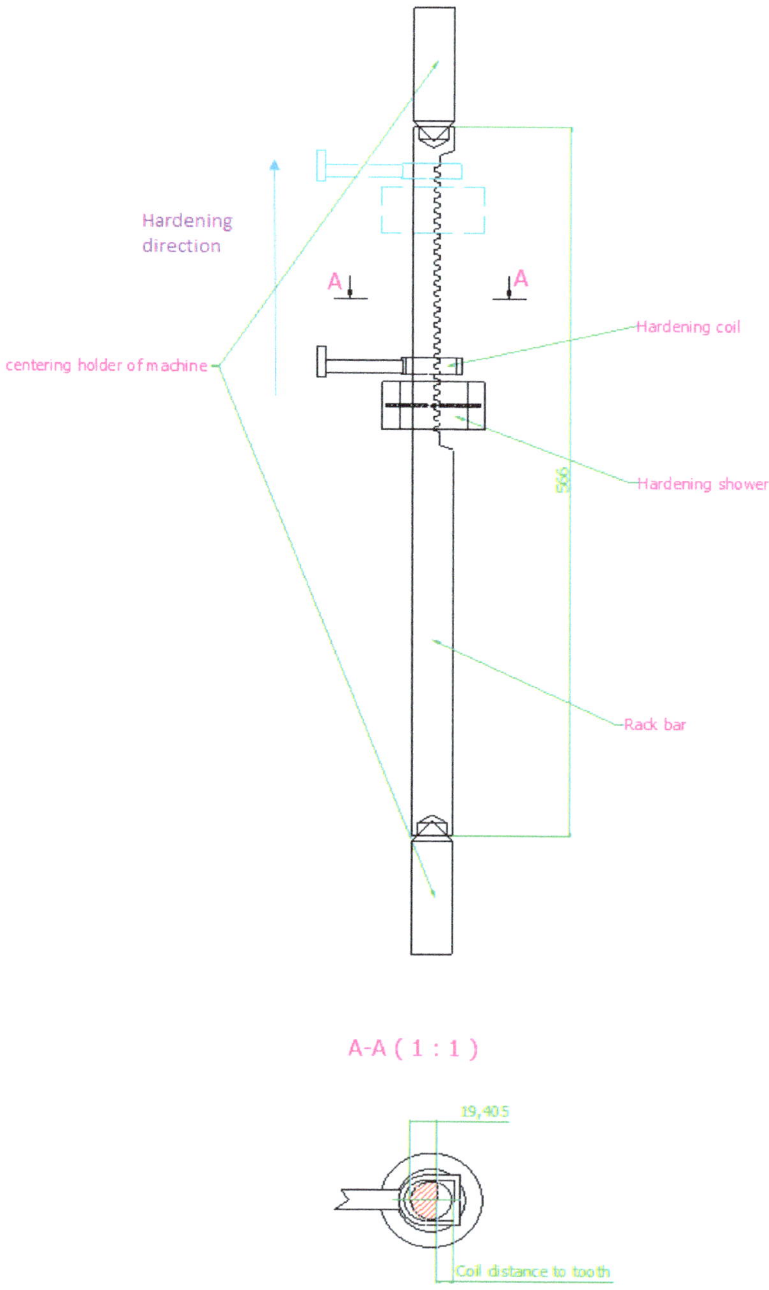

Figure 2. The process setup.

The hardening operation in automatic mode is as follows. After the hardening element is delivered to the loading station, it is mounted in holders, and the inductor is fixed in the initial position at the bottom, at a predetermined distance from the hardening piece, according to the plan of experiments. At a fixed power and feed rate, the quenching process begins, during which the coil moves upwards towards the upper holder with the coolant flow, as shown in Figure 2. After passing the set distance, the holder is released, the coil is discharged to the neutral position, and the hardened element is transferred to the tempering operation.

2.2. Process Quality Indicators and the Measurement Setup

Specific measurements are required to identify the quality of the hardening process qualitatively. Below, the list of performed measurements with the quality requirements is presented:

- The hardness depth on the teeth side with minimum requirements equal to 3.9 mm;
- The hardness depth on the back side (an opposite side to teeth) with requirements given by the range above 1 mm;
- The surface hardness on teeth with requirements given by the range 55–60 HRC;
- The surface hardness on the back side is within the 52–55 HRC range requirements;
- After finishing the hardening and tempering operations, all rack bars are straightened to obtain a maximal deflection of value not greater than 1000 μm before going to the next operation; in the presented study, the existence of thermal strains after induction hardening and tempering operations is a primary driving force for performing process optimization to reduce costs and time.

The general flow of the experimentation procedure is presented in Figure 3.

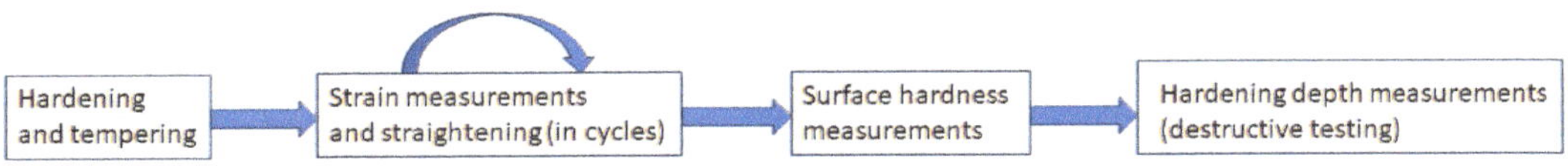

Figure 3. The general flow of the experimentation procedure.

The zones associated with the hardness measurements are shown in Figure 4. Figure 4 shows six hardness measurement zones: three zones for the tooth side and three for the back side of the rack bar. The numbers indicate which teeth the hardness measurements apply to.

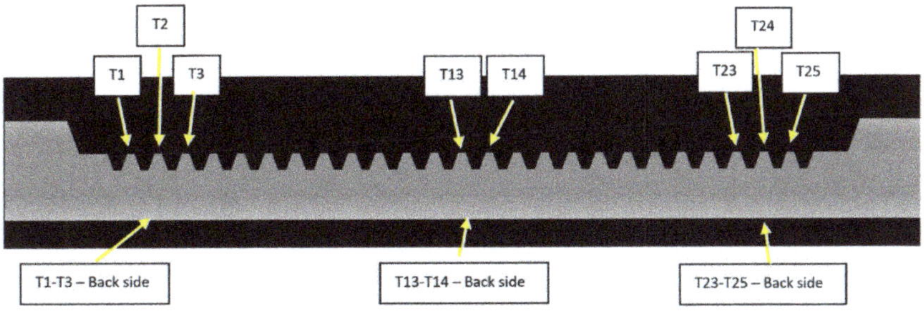

Figure 4. Zones for the hardness measurements of the rack bar after induction hardening operation.

Places of the hardness depth measurements are also presented on the cross-section of the rack bar in the middle of its length, as shown in Figure 5.

To measure the hardening depth, the micro Vickers method was used, and the FM-810 Micro Vickers Hardness Tester performed the inspection with an accuracy of ±1 HV. This measurement involves determining the microhardness profile along the rack bar's thickness for a tooth cross-section. The measurements assume that the hardening depth defines a point with a hardness of 400 HV. An example of the hardness profile is shown in Figure 6.

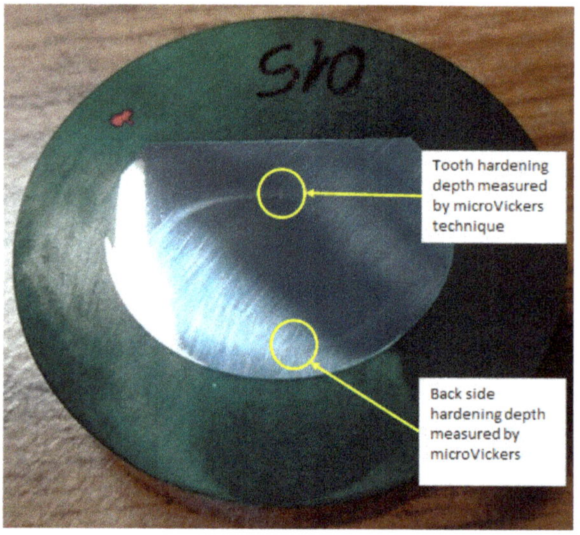

Figure 5. The setup of the hardness depth measurements.

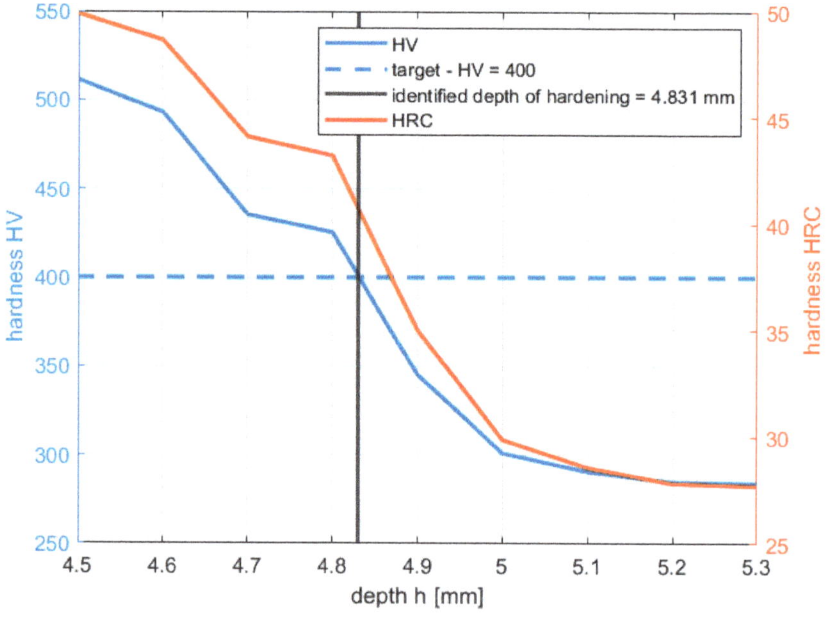

Figure 6. Example of hardness profile from the tooth side.

The Zwick/Roell ZHR 8150LK inspects surface hardness in the HRC scale with an accuracy of ±1 HRC. The mentioned methods are indirect; moreover, destroying the component (a rack bar) to perform measurements is necessary.

Runout inspection and straightening are performed by the automatic process of a rack bar straightening; the "Rack Bar Bend M/C" machine performs measurements by eight of the high accuracy digital displacement transducer gauge probe with an accuracy of ±1.2 μm. Figure 7 shows how deformation is measured and how the machine performs straightening during the automatic process.

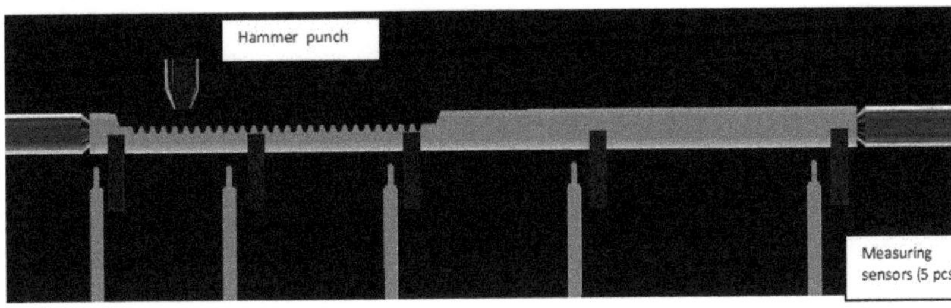

Figure 7. The setup of the deformation measurements.

2.3. Experimental Design

The central composite design (CCD) [11] as a plan of experiments is used. Three input factors (design variables) were set within the following ranges according to the knowledge of the process before optimization:

- The power x_1: 45–69%;
- The coil distance x_2: 1.5–5 mm;
- The feed rate x_3: 600–900 mm/min.

Due to the destructive inspection, it is necessary to scrap all components used in the experiments; therefore, due to cost reduction, there is no replication in the current plan of experiments. The results of the experiments are presented in Table 1.

Table 1. The CCD plan of experiments with process responses.

No.	x_1	x_2	x_3	y_1 [μm]	y_2 [mm]	y_3 [mm]	y_4 [HRC]	y_5 [HRC]	y_6 [HRC]	y_7 [HRC]	y_8 [HRC]	y_9 [HRC]
1	1	−1	−1	1721	10.358	5.098	47.9	50.0	53.3	47.4	52.7	54.5
2	−1	1	−1	720	4.008	1.416	56.1	57.6	58.0	55.0	55.1	55.6
3	1	1	1	1258	4.831	2.420	54.0	57.5	58.0	53.7	54.8	56.0
4	0	0	0	962	4.810	1.954	55.2	57.2	58.2	54.4	54.5	54.5
5	−1	−1	1	958	3.776	0.351	56.4	58.8	58.8	40.4	40.0	40.9
6	0	0	0	1160	4.767	1.957	56.0	58.6	58.9	54.0	54.9	55.9
7	−1	−1	−1	948	4.784	1.451	55.3	57.8	58.8	53.5	54.5	53.8
8	−1	1	1	1213	4.771	1.987	56.5	59.6	59.8	39.4	37.7	39.1
9	0	0	0	987	2.933	0.234	55.1	57.8	58.7	52.5	54.7	55.2
10	1	1	−1	2005	7.250	4.507	52.2	56.1	56.7	50.8	54.3	55.8
11	1	−1	1	981	5.778	2.512	53.4	56.5	57.1	52.2	55.0	54.7
12	0	0	0	1020	4.672	1.946	55.1	57.0	58.3	54.2	54.6	55.5
13	1.682	0	0	2131	7.704	4.579	46.2	51.2	56.1	46.3	51.6	51.5
14	0	0	1.682	745	4.059	1.174	56.4	58.3	58.9	54.6	54.3	54.5
15	0	0	0	1085	4.818	1.932	56.4	58.5	59.0	52.3	55.3	56.3
16	−1.682	0	0	1089	2.953	0.000	56.9	58.5	58.8	27.7	27.4	26.5
17	0	0	0	1013	4.766	1.969	55.6	57.4	58.0	54.0	55.2	56.1
18	0	1.682	0	821	4.331	1.938	55.7	58.2	58.6	55.1	56.0	56.0
19	0	−1.682	0	1435	5.669	1.991	55.0	57.3	57.8	53.0	55.1	56.0
20	0	0	−1.682	2057	6.132	3.267	52.1	55.4	56.0	52.9	55.0	56.4

Response y_1 denotes the maximal deflection of the rack bar after the induction hardening operation is finished. Responses y_2 and y_3 denote the hardness depths on the teeth and back-side, along the cross-section of the rack bar, in the middle of its length, respectively. Responses y_4–y_6 indicate the average hardness measured in the I, II and III zones on the teeth side and responses y_7–y_9 indicate the hardness measured in the I, II and III zones on the back side of the rack bar.

Grubb's test [11] was performed for data given in Table 1 to identify outliers. The significance level was 0.05, and the following outliers are identified in the collected data:

- In experiment no. 1—in responses 2, 5 and 6, the result indicates that the deep hardening was obtained without the required hardness of the teeth in the 2nd and 3rd zones;
- In experiment no. 13—response 4 indicates that the hardness of the teeth in the 1st zone is insufficient;
- In experiment no. 16—in responses 7, 8 and 9, the result indicates that the hardening of the back side of the rack bar was not achieved; moreover, in response 3—the hardening depth on the back side—is equal to 0, however, on the assumed significance level this response is not an outlier.

It is well known that outliers affect the response surface models, but in the presented study, no replications of experiments were assumed; hence all results were applied for modeling.

3. RSM for Empirical Modelling

Empirical models of process responses are assumed in the form of full quadratic models as follows:

$$
\begin{gathered}
y^{(i)} = b_0 + \mathbf{b}^T\mathbf{x} + \mathbf{x}^T\mathbf{B}\mathbf{x} + \varepsilon^{(i)}, \\
\mathbf{x} = \begin{bmatrix} x_1 & x_2 & x_3 \end{bmatrix}^T; \mathbf{b} = \begin{bmatrix} b_1 & b_2 & b_3 \end{bmatrix}^T; \\
\mathbf{B} = \begin{bmatrix} b_{11} & b_{12}/2 & b_{13}/2 \\ b_{12}/2 & b_{22} & b_{23}/2 \\ b_{13}/2 & b_{23}/2 & b_{33} \end{bmatrix}; \varepsilon^{(i)} \sim N\left(0, \sigma^{(i)}\right),
\end{gathered}
\tag{1}
$$

where $y^{(i)}$ is the i-th response, b_0 is the free term of the model, vector $\mathbf{x}$ denotes the design variables, vector $\mathbf{b}$ collects the linear terms coefficients, and matrix $\mathbf{B}$ includes the quadratic terms of the response surface model; it is assumed that the responses are uncorrelated and each model is under the normal noise with 0 mean and standard deviation $\sigma^{(i)}$. Models are linear with respect to the coefficients, which is why the least square method (LSM) [11] was applied to identify coefficients.

An analysis of variance (ANOVA) is carried out to estimate the validity of the identified mathematical models and the effect of every model term on the responses. The p-value is used to check the significance, which means the response is greatly determined by the model term whose p-value is sufficiently low (less than or equal to the significance level of 0.05). On the contrary, the higher the F-value, the stronger the significance of the model item. Symbol S indicates the significant term in the considered model.

The ANOVA of the mathematical model for the maximal deflection is shown in Table 2. The F-value (5.26) and p-value (0.008) imply that the model is significant, while the F-value (22.24) and p-value (0.002) imply that the lack of fit is significant. If the lack of fit is significant, the higher-order model terms should be taken into consideration to improve the accuracy of the model; however, in the present study, only the full quadratic models are considered. According to the F-value and p-value, there are four significant model terms, among which the power x_1 and the feed rate x_3 have the most significant influence, followed by the linear-by-linear interaction effect between the power x_1 and the feed rate x_3 and the quadratic effect of the power $x_1{}^2$. The coil distance x_2 and other model terms also containing it seem to have little influence on the maximal deflection of the rack bar.

Table 2. Analysis of variance for the mathematical model of maximal deflection response y_1.

Source	Sum of Squares	Sum of Squares Contribution [%]	*df*	Mean Square	*F*-Value	*p*-Value	Notes
Model	2,908,285.706	82.568	9	323,142.856	5.263	0.008	S
x_1	1,101,450.148	31.271	1	1,101,450.148	17.939	0.002	S
x_2	14,482.149	0.411	1	14,482.149	0.236	0.638	
x_3	745,417.899	21.163	1	745,417.899	12.140	0.006	S
x_1x_2	35,644.500	1.012	1	35,644.500	0.581	0.464	
x_1x_3	495,012.500	14.054	1	495,012.500	8.062	0.018	S
x_2x_3	28,322.000	0.804	1	28,322.000	0.461	0.512	
x_1^2	394,238.133	11.193	1	394,238.133	6.421	0.030	S
x_2^2	361.006	0.010	1	361.006	0.006	0.940	
x_3^2	120,678.769	3.426	1	120,678.769	1.965	0.191	
Residual	613,997.244	17.432	10	61,399.724			
Lack of fit	587,578.410	16.682	5	117,515.682	22.241	0.002	S
Pure error	26,418.833	0.750	5	5283.767			
Total	3,522,282.950	100.000	19	185,383.313			

The ANOVA of the mathematical model for the hardening depth of the teeth is shown in Table 3. The *F*-value and *p*-value of the lack of fit are 0.41 and 0.82, respectively, while the *F*-value and *p*-value of the model are 14.41 and less than 0.00013, respectively, which indicates that the model is significant enough and no higher-order model terms need to be considered. The power and the feed rate amount have the most significant influence on the hardening depth of the teeth, followed by the linear-by-linear interaction effect between the power and the feed rate, the coil distance x_2 and the quadratic effect of the power x_1^2. The linear-by-linear interaction effect between the power and the coil distance is also significant.

Table 3. Analysis of variance for the mathematical model of the hardening depth of teeth y_2.

Source	Sum of Squares	Sum of Squares Contribution [%]	*df*	Mean Square	*F*-Value	*p*-Value	Notes
Model	51.625	92.841	9	5.736	14.410	<0.001	S
x_1	26.068	46.881	1	26.068	65.488	<0.001	S
x_2	2.712	4.878	1	2.712	6.814	0.026	S
x_3	8.431	15.162	1	8.431	21.180	0.001	S
x_1x_2	2.283	4.106	1	2.283	5.736	0.038	S
x_1x_3	5.702	10.254	1	5.702	14.325	0.004	S
x_2x_3	1.933	3.476	1	1.933	4.855	0.052	
x_1^2	2.479	4.458	1	2.479	6.227	0.032	S
x_2^2	1.285	2.311	1	1.285	3.228	0.103	
x_3^2	1.592	2.863	1	1.592	3.999	0.073	
Residual	3.981	7.159	10	0.398			
Lack of fit	1.165	2.096	5	0.233	0.414	0.822	
Pure error	2.815	5.063	5	0.563			
Total	55.606	100.000	19	2.927			

The ANOVA of the mathematical model for the hardening depth on the back side is shown in Table 4. As can be seen from the table, the model is significant with an *F*-value (11.71) and a small *p*-value (<0.0004), while the lack of fit is still not significant with a smaller *F*-value (0.25) and larger *p*-value (0.92). The power x_1 has the most significant influence on the hardening depth on the back side, with the largest *F*-value (69.29), followed by the feed rate x_3 and the linear-by-linear interaction effect between the power and the feed rate. Other model terms seem to have little influence on the hardening depth on the back side of the rack bar.

Table 4. Analysis of variance for the mathematical model of the hardening depth on the back side y_3.

Source	Sum of Squares	Sum of Squares Contribution [%]	*df*	Mean Square	*F*-Value	*p*-Value	Notes
Model	32.331	91.339	9	3.592	11.718	<0.001	S
x_1	21.244	60.016	1	21.244	69.295	<0.001	S
x_2	0.050	0.142	1	0.050	0.164	0.694	
x_3	5.570	15.737	1	5.570	18.170	0.002	S
x_1x_2	0.652	1.842	1	0.652	2.127	0.175	
x_1x_3	2.147	6.064	1	2.147	7.002	0.024	S
x_2x_3	0.589	1.663	1	0.589	1.920	0.196	
x_1^2	1.127	3.185	1	1.127	3.677	0.084	
x_2^2	0.391	1.106	1	0.391	1.277	0.285	
x_3^2	0.939	2.654	1	0.939	3.064	0.111	
Residual	3.066	8.661	10	0.307			
Lack of fit	0.606	1.713	5	0.121	0.247	0.925	
Pure error	2.459	6.948	5	0.492			
Total	35.397	100.000	19	1.863			

The ANOVA of the mathematical model for the hardness of the teeth in the first zone is shown in Table 5. As can be seen from the table, the model is significant with an *F*-value of 20.72 and an extremely small *p*-value <0.0001, while the lack of fit is still not significant with an *F*-value of 4.42 and a *p*-value of 0.06. The power x_1 has the most significant influence on the hardness of the teeth in the first zone, with the largest *F*-value (112.54), followed by the quadratic effect of the power x_1^2 and the feed rate x_3. The linear-by-linear interaction effect between the power and the feed rate is also significant at the 0.05 significance level.

Table 5. Analysis of variance for the mathematical model of the hardness of the teeth in the first zone y_4.

Source	Sum of Squares	Sum of Squares Contribution (%)	*df*	Mean Square	*F*-Value	*p*-Value	Notes
Model	146.920	94.911	9	16.324	20.722	<0.001	S
x_1	88.654	57.271	1	88.654	112.539	<0.001	S
x_2	3.564	2.303	1	3.564	4.525	0.059	
x_3	18.820	12.158	1	18.820	23.890	0.001	S
x_1x_2	2.000	1.292	1	2.000	2.539	0.142	
x_1x_3	4.205	2.716	1	4.205	5.338	0.043	S
x_2x_3	2.420	1.563	1	2.420	3.072	0.110	
x_1^2	25.952	16.765	1	25.952	32.944	<0.001	S
x_2^2	0.000	0.000	1	0.000	0.000	0.995	
x_3^2	2.163	1.397	1	2.163	2.746	0.129	
Residual	7.878	5.089	10	0.788			
Lack of fit	6.424	4.150	5	1.285	4.420	0.064	
Pure error	1.453	0.939	5	0.291			
Total	154.798	100.000	19	8.147			

The ANOVA of the mathematical model for the hardness of the teeth in the second zone is shown in Table 6. The model is significant, with an *F*-value of 10.49 and a *p*-value < 0.0006, and the lack of fit is still insignificant. The most important factors are the power x_1, the feed rate x_3 and the quadratic effect of the power x_1^2. Similar significance levels show the coil distance x_2 and the linear-by-linear interaction effect between the power and the coil distance.

Table 6. Analysis of variance for the mathematical model of the hardness of the teeth in the second zone y_5.

Source	Sum of Squares	Sum of Squares Contribution (%)	*df*	Mean Square	*F*-Value	*p*-Value	Notes
Model	98.206	90.424	9	10.912	10.492	<0.001	S
x_1	49.412	45.497	1	49.412	47.512	<0.001	S
x_2	6.216	5.723	1	6.216	5.977	0.035	S
x_3	18.226	16.782	1	18.226	17.525	0.002	S
x_1x_2	5.281	4.863	1	5.281	5.078	0.048	S
x_1x_3	3.001	2.763	1	3.001	2.886	0.120	
x_2x_3	2.101	1.935	1	2.101	2.020	0.186	
x_1^2	13.157	12.114	1	13.157	12.651	0.005	S
x_2^2	0.070	0.065	1	0.070	0.067	0.800	
x_3^2	0.889	0.819	1	0.889	0.855	0.377	
Residual	10.400	9.576	10	1.040			
Lack of fit	8.125	7.481	5	1.625	3.571	0.094	
Pure error	2.275	2.095	5	0.455			
Total	108.606	100.000	19	5.716			

The ANOVA of the mathematical model for the hardness of the teeth in the third zone is shown in Table 7. The model is significant, with an *F*-value of 9.44 and a p-value < 0.0009, and the lack of fit is still insignificant. The most important factors are the power x_1, the feed rate x_3 and the coil distance x_2. Similar levels of significance, but with smaller values of the *F*-statistics than earlier considered factors, show the quadratic effects of the power and the feed rate.

Table 7. Analysis of variance for the mathematical model of the hardness of the teeth in the third zone y_6.

Source	Sum of Squares	Sum of Squares Contribution (%)	*df*	Mean Square	*F*-Value	*p*-Value	Notes
Model	36.465	89.467	9	4.052	9.438	<0.001	S
x_1	16.127	39.568	1	16.127	37.566	<0.001	S
x_2	2.502	6.138	1	2.502	5.828	0.036	S
x_3	10.156	24.918	1	10.156	23.658	0.001	S
x_1x_2	2.101	5.155	1	2.101	4.895	0.051	
x_1x_3	1.361	3.340	1	1.361	3.171	0.105	
x_2x_3	0.061	0.150	1	0.061	0.143	0.714	
x_1^2	2.257	5.536	1	2.257	5.256	0.045	S
x_2^2	0.246	0.603	1	0.246	0.572	0.467	
x_3^2	2.257	5.536	1	2.257	5.256	0.045	S
Residual	4.293	10.533	10	0.429			
Lack of fit	3.465	8.501	5	0.693	4.183	0.071	
Pure error	0.828	2.032	5	0.166			
Total	40.758	100.000	19	2.145			

The ANOVA of the mathematical model for the hardness of the back-side in the first zone is shown in Table 8. The model is significant with an *F*-value of 11.32 and p-value < 0.0004, and the lack of fit is significant with a p-value of 0.003. Only three factors are significant: the quadratic effect of the power x_1, the linear-by-linear interaction effect between the power and the feed rate and the pure quadratic effect of the power.

Table 8. Analysis of variance for the mathematical model of the hardness of the back-side in the first zone y_7.

Source	Sum of Squares	Sum of Squares Contribution (%)	*df*	Mean Square	*F*-Value	*p*-Value	Notes
Model	840.725	91.064	9	93.414	11.323	<0.001	S
x_1	162.321	17.582	1	162.321	19.676	0.001	S
x_2	5.841	0.633	1	5.841	0.708	0.420	
x_3	24.094	2.610	1	24.094	2.921	0.118	
x_1x_2	2.420	0.262	1	2.420	0.293	0.600	
x_1x_3	165.620	17.939	1	165.620	20.076	0.001	S
x_2x_3	2.420	0.262	1	2.420	0.293	0.600	
${x_1}^2$	455.825	49.373	1	455.825	55.254	<0.001	S
${x_2}^2$	2.348	0.254	1	2.348	0.285	0.605	
${x_3}^2$	1.276	0.138	1	1.276	0.155	0.702	
Residual	82.497	8.936	10	8.250			
Lack of fit	78.284	8.479	5	15.657	18.580	0.003	S
Pure error	4.213	0.456	5	0.843			
Total	923.222	100.000	19	48.591			

The ANOVA of the mathematical model for the hardness of the back-side in the second zone is shown in Table 9. The model is significant with an F-value of 13.27 and a p-value < 0.0002, and the lack of fit is very significant with a p-value < 0.00002. Only four factors are significant: the quadratic effect of the power ${x_1}^2$ followed by the power itself, the linear-by-linear interaction effect between the power and the feed rate, and the feed rate itself.

Table 9. Analysis of variance for the mathematical model of the hardness of the back-side in the second zone y_8.

Source	Sum of Squares	Sum of Squares Contribution (%)	*df*	Mean Square	*F*-Value	*p*-Value	Notes
Model	995.989	92.275	9	110.665	13.273	<0.001	S
x_1	360.856	33.432	1	360.856	43.280	<0.001	S
x_2	0.108	0.010	1	0.108	0.013	0.912	
x_3	67.118	6.218	1	67.118	8.050	0.018	S
x_1x_2	1.201	0.111	1	1.201	0.144	0.712	
x_1x_3	150.511	13.944	1	150.511	18.052	0.002	S
x_2x_3	2.761	0.256	1	2.761	0.331	0.578	
${x_1}^2$	395.712	36.662	1	395.712	47.461	<0.001	S
${x_2}^2$	2.716	0.252	1	2.716	0.326	0.581	
${x_3}^2$	0.194	0.018	1	0.194	0.023	0.882	
Residual	83.376	7.725	10	8.338			
Lack of fit	82.843	7.675	5	16.569	155.330	<0.001	S
Pure error	0.533	0.049	5	0.107			
Total	1079.366	100.000	19	56.809			

Similar conclusions could be formulated for the mathematical model of the hardness of the back-side in the third zone, as shown in Table 10. The model is significant with an F-value of 14.21 and a p-value < 0.0002, and the lack of fit is significant with a p-value < 0.0007. Moreover, only four factors are significant: the quadratic effect of the power ${x_1}^2$ followed by the power itself, the linear-by-linear interaction effect between the power and the feed rate, followed by the feed rate itself.

Table 10. Analysis of variance for the mathematical model of the hardness of the back-side in the third zone y_9.

Source	Sum of Squares	Sum of Squares Contribution (%)	*df*	Mean Square	*F*-Value	*p*-Value	Notes
Model	1053.639	92.749	9	117.071	14.213	<0.001	S
x_1	397.146	34.960	1	397.146	48.215	<0.001	S
x_2	0.495	0.044	1	0.495	0.060	0.811	
x_3	75.893	6.681	1	75.893	9.214	0.013	S
x_1x_2	0.845	0.074	1	0.845	0.103	0.755	
x_1x_3	111.005	9.771	1	111.005	13.477	0.004	S
x_2x_3	1.620	0.143	1	1.620	0.197	0.667	
${x_1}^2$	444.042	39.088	1	444.042	53.909	<0.001	S
${x_2}^2$	3.038	0.267	1	3.038	0.369	0.557	
${x_3}^2$	1.010	0.089	1	1.010	0.123	0.733	
Residual	82.369	7.251	10	8.237			
Lack of fit	80.161	7.056	5	16.032	36.299	0.001	S
Pure error	2.208	0.194	5	0.442			
Total	1136.008	100.000	19	59.790			

The lack of fit in the presented models indicates the need to use the higher-order models; however, in the present work, only quadratic models will be used. The higher-order models can be applied, e.g., in polynomials, neural networks and kriging.

The identified models are nonlinear, indicating room for improvement or even process optimization. Table 11 presents the essential statistical characteristics of the identified full quadratic models (1). The significance level was assumed as 0.05.

Table 11. Statistical characteristics of regression models for responses of the process.

No.	Model	Statistics				
		Multiple R	R^2	Adjusted R^2	*p*-Value	PRESS
1	The maximal deflection, y_1	0.909	0.826	0.669	0.008	4,540,000
2	Hardening depth of teeth, y_2	0.964	0.928	0.864	<0.001	12.891
3	The hardening depth on the back side, y_3	0.956	0.913	0.836	<0.001	8.402
4	The hardness of the teeth in the first zone, y_4	0.956	0.913	0.836	<0.001	52.643
5	The hardness of the teeth in the second zone, y_5	0.950	0.902	0.813	<0.001	72.291
6	The hardness of the teeth in the third zone, y_6	0.948	0.898	0.807	<0.001	31.142
7	The hardness of the back-side in the first zone, y_7	0.954	0.911	0.830	<0.001	600.672
8	The hardness of the back-side in the first zone, y_8	0.961	0.923	0.853	<0.001	630.241
9	The hardness of the back-side in the first zone, y_9	0.963	0.928	0.862	<0.001	614.364

The analysis shows that the identified models are strong and significant but with moderate predictive ability, as shown by the PRESS statistics. The backward elimination of the insignificant terms is performed to increase the predictive properties of the identified models. The reduced models will be used in the optimization process; models presented in Table 12 were shown with an accuracy of two significant digits. The values of the PRESS statistics are significantly smaller than for the full models, with a tiny drop in R^2 values.

Table 12. The reduced models of responses.

No.	Model	Equation	R^2	PRESS
1	The maximal deflection, y_1	$y_1 = 1107.86 + 283.98x_1 - 233.62x_3 + 157.55x_1^2 - 248.75x_1x_3$	0.771	1,520,000
2	Hardening depth of teeth, y_2	$y_2 = 4.45 + 1.38x_1 - 0.45x_2 - 0.79x_3 + 0.42x_1^2 + 0.30x_2^2 + 0.33x_3^2$ $-0.53x_1x_2 - 0.84x_1x_3 + 0.49x_2x_3$	0.928	12.891
3	The hardening depth on the back side, y_3	$y_3 = 1.66 + 1.25x_1 + 0.06x_2 - 0.64x_3 + 0.28x_1^2 + 0.17x_2^2 + 0.26x_3^2$ $-0.29x_1x_2 - 0.52x_1x_3 + 0.27x_2x_3$	0.913	8.402

Table 12. *Cont.*

No.	Model	Equation	R^2	PRESS
4	The hardness of the teeth in the first zone, y_4	$y_4 = 55.27 - 2.55x_1 + 1.18x_3 - 1.31x_1^2$	0.863	43.793
5	The hardness of the teeth in the second zone, y_5	$y_5 = 57.59 - 1.89x_1 + 0.68x_2 + 1.16x_3 - 0.94x_1^2$	0.801	41.591
6	The hardness of the teeth in the third zone, y_6	$y_6 = 58.41 - 1.09x_1 + 0.43x_2 + 0.85x_3 - 0.38x_1^2 - 0.38x_3^2 + 0.52x_1x_2$	0.862	17.054
7	The hardness of the back-side in the first zone, y_7	$y_7 = 54.05 + 3.45x_1 - 5.69x_1^2 + 4.55x_1x_3$	0.874	307.236
8	The hardness of the back-side in the second zone, y_8	$y_8 = 55.25 + 5.14x_1 - 2.22x_3 - 5.29x_1^2 + 4.34x_1x_3$	0.921	273.008
9	The hardness of the back-side in the third zone, y_9	$y_9 = 56.08 + 5.39x_1 - 2.36x_3 - 5.62x_1^2 + 3.725x_1x_3$	0.922	273.822

It is essential to emphasize that the reduced models change the background for optimization. Supposedly, we will consider the model for the maximal deflection y_1. In that case, the canonical analysis shows that for the full quadratic model, the stationary point $x_s = [-0.196 \;\; 0.764 \;\; 1.053]^T$ is a saddle point because the eigenvalues of matrix **B** {−31.45, 36.93, 269.55} are mixed in sign. The reduced model, according to Table 12, has a stationary point $x_s = [-1.032 \; 0 \; -0.377]^T$ which is also a saddle point, but from the ridge system because the eigenvalues are now {−63.10, 0, 2246.76}. Figure 8 shows contour plots of the response surface slices at $x_3 = 0.5$ for the full quadratic model (left) and the reduced model (right).

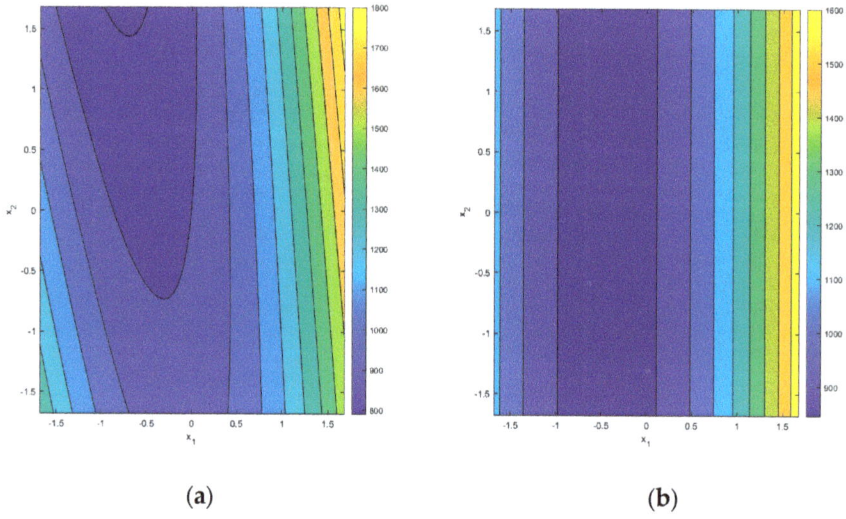

Figure 8. Contour plot for the slice $x_3 = 0.5$ of the maximal deflection response hypersurface y_1: (**a**) the full quadratic model; (**b**) the reduced model.

The change in the type of the response surface affects the optimization process, especially in the single-objective case. For the multiobjective case, the desirability approach allows taking into account various forms of response surfaces. However, the properties of the ridge systems will also be present in the objective function, as shown in the next Section.

4. Multiobjective Optimization Problem Formulation and Its Solution

The design variable vector **x** is as follows $\mathbf{x} = [x_1\ x_2\ x_3]^T$, where x_1 is the power, x_2 is the distance between the coil and the part, and x_3 is the feed rate. Then, the multiobjective optimization problem can be formulated using the desirability function D as given below:

$$\begin{aligned} \mathbf{x}^{opt} &= \underset{\mathbf{x}\in S}{\operatorname{argmax}} D(\mathbf{x}) \\ \text{s.t. } \mathbf{x}^T\mathbf{x} &\leq \alpha^2. \end{aligned} \tag{2}$$

The sphere S denotes the set of the acceptable solution, and this set is given by the constraint described by the radius α of the central composite plan of the experiments.

In the paper, the following form of the desirability function is used [11]:

$$\begin{aligned} D(\mathbf{x}) &= \left(\prod_{i=1}^{m} d_i(\mathbf{x})^{w_i}\right)^{1/\sum_{i=1}^{m} w_i} \quad \text{or} \\ D(\mathbf{x}) &= \left(\prod_{i=1}^{m} d_i(\mathbf{x})\right)^{1/m}, \end{aligned} \tag{3}$$

where d_i is the desirability function related to the optimization's i-th criterion, w_i denotes the weight of the i-th criterion, and m is the number of responses.

Two types of the desirability function are applied: the smaller—the better (STB) and the nominal—the better (NTB) [11] as follows:

$$\text{STB}: \quad d_i = \begin{cases} 1.0, & \hat{y}_i \leq L_i \\ \left(\frac{\hat{y}_i - U_i}{L_i - U_i}\right)^s, & L_i < \hat{y}_i < U_i \\ 0, & \hat{y}_i \geq U_i \end{cases} \tag{4}$$

and

$$\text{NTB}: \quad d_i = \begin{cases} 0, & (\hat{y}_i < L_i) \vee (\hat{y}_i > U_i) \\ \left(\frac{\hat{y}_i - L_i}{T_i - L_i}\right)^s, & L_i \leq \hat{y}_i \leq T_i \\ \left(\frac{\hat{y}_i - U_i}{T_i - U_i}\right)^s, & T_i < \hat{y}_i \leq U_i. \end{cases} \tag{5}$$

where for the STB function, L denotes the acceptable target for the response $\hat{y}_i$, and U is an acceptable upper limit of the response, s is the exponent which sets the sharpness of the desirability function. For the NTB function, T denotes the target for the response $\hat{y}_i$, and L and U are acceptable lower and upper limits of the response, respectively.

The responses $\hat{y}_i$ in the presented formulation are obtained using the identified RSM models. Table 13 shows the parameters of the identified models' desirability functions.

Table 13. Parameters of the desirability functions for the identified response models.

No.	Response $\hat{y}_i$	Type	L_i	T_i	U_i	s	w_i
1.	f_{max}	STB	1.0 mm	-	1.4 mm	2	1
2.	h_t	NTB	3.9 mm	4.1 mm	10.0 mm	2	0.8
3.	h_b	NTB	1.0 mm	1.4 mm	10.0 mm	2	0.8
4.	HRC_{t1}	NTB	55 HRC	57.5 HRC	60 HRC	2	0.6
5.	HRC_{t2}	NTB	55 HRC	57.5 HRC	60 HRC	2	0.6
6.	HRC_{t3}	NTB	55 HRC	57.5 HRC	60 HRC	2	0.6
7.	HRC_{b1}	NTB	52 HRC	53.5 HRC	55 HRC	2	0.6
8.	HRC_{b2}	NTB	52 HRC	53.5 HRC	55 HRC	2	0.6
9.	HRC_{b3}	NTB	52 HRC	53.5 HRC	55 HRC	2	0.6

Table 13 shows that the most critical response is the first one—related to the deflection of the rack bar.

Using the first of Equation (3) with Equations (4) and (5) and Table 12, the desirability function was calculated for the responses given in Table 1 obtained during the experimental phase of the study.

The values of the desirability function for the plan of experiments results are given in Figure 9.

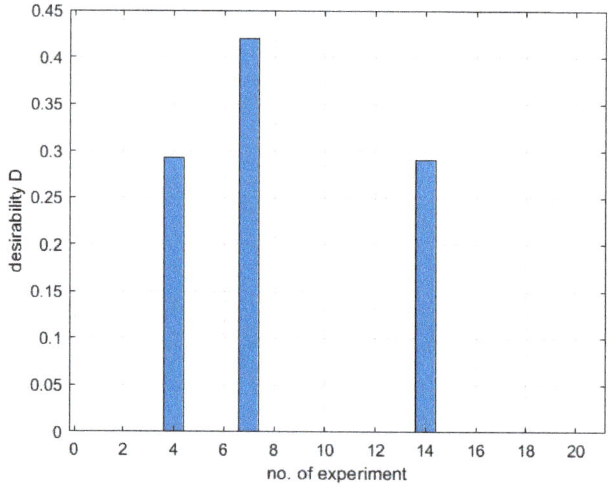

Figure 9. Desirability values for the experimental data obtained using CCD.

Figure 9 shows that only three out of twenty experiments gave non-zero desirability values with a maximum of 0.421 for the seventh and 0.292 and 0.291 for the fourth and fourteenth experiments, respectively. The results indicate that there is room for improvement in the process.

The classical optimization technique, namely the sequential quadratic programming (SQP) algorithm for constrained optimization problems, was applied in the first trial. One hundred tests were carried out with a random selection of the starting point of the optimization procedure. The results are presented in Figure 10.

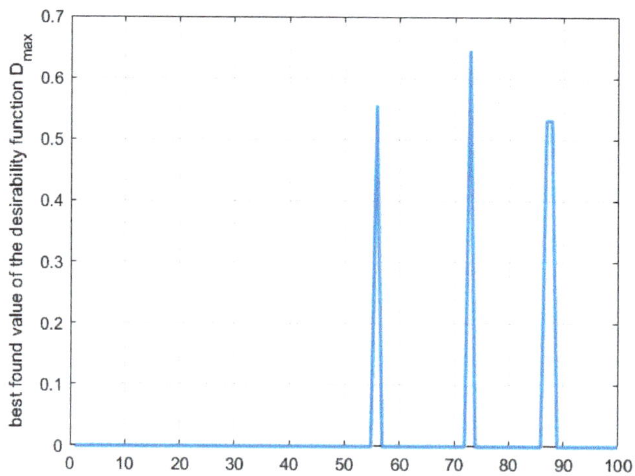

Figure 10. Best-found values of the desirability function for SQP optimization during N = 100 tests.

It is visible that only for four out of one-hundred trials, the best value of the desirability function is non-zero with a maximum of 0.646. This poor result can be explained by approximating the gradient of the objective function using difference quotients. However, one must also consider that the desirability functions (4) and (5) are not differentiable in the classical sense. In Table 14, the statistics for the SQP optimization process are presented. The results show that the process is not robust; the optimal design variables vector in uncoded values is $\mathbf{x}^{opt}$ = [56.4% 1.6 mm 796 mm/min]^T, which gives D^{opt} = 0.646. The second best result is $\mathbf{x}$ = [56.4% 3.1 mm 793 mm/min]^T which gives D^{opt} = 0.557. The latter result is significant from the point of view of the industrial conditions of the process; additional considerations will also be presented in this work.

Table 14. Statistics of the SQP optimization results for the desirability function values.

Statistics	Value
Sample mean value, $\overline{x}$	0.023
Sample standard deviation, σ	0.112
Sample median, M	0.000
Sample minimum, min(max(D))	0.000
Sample maximum, max(max(D))	0.646

Figure 11 shows slices of the desirability function hypersurface. The hypersurface of desirability has the form of narrow ribs with a wavy ridge, making it very difficult to solve the optimization problem using classical algorithms. In addition, for almost the entire set of admissible solutions, the values of the desirability function are equal to zero, which also justifies the inefficiency of the classic optimization algorithm SQP.

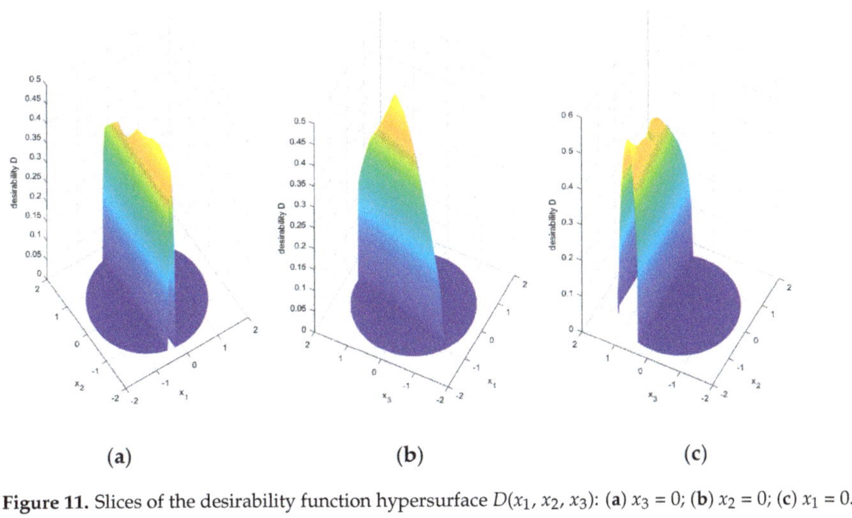

Figure 11. Slices of the desirability function hypersurface $D(x_1, x_2, x_3)$: (**a**) $x_3 = 0$; (**b**) $x_2 = 0$; (**c**) $x_1 = 0$.

Due to the presented features of the objective function, an evolutionary, floating-point coded genetic algorithm (GA) was used in the work.

A genetic algorithm (GA) [21] is an intelligent evolutionary procedure used to find the optimal solutions to problems based on natural genetics and natural selection principles. The foundation of this algorithm is the biologically inspired set of operations such as selection, combination or crossover, and mutation. In the present case, the individuals and the objective function are defined and coded at the real-coded strings. Then, the iterative process is carried up using the three operations, i.e., selection, crossover, and mutation. The selection operation is to choose individuals that start from a population generated randomly according to a probability distribution. The roulette wheel selection method is used to determine the selection of individuals. A crossover operation appears when two strings randomly picked from the populations exchange their substrings to create two new strings. The sum of individuals to apply the crossover operation is dominated by crossover probability, which is the ratio of all selected strings to the total population size. The mutation operation assures the diversity of the population. It is an occasional random alternative in one or more string positions where the small random number mutes the value. After successive iterative generations, the population evolves gradually toward an optimal solution. Finally, the evolutionary process is completed till the termination criterion is reached. This study proposed RSM with the desirability function to define the relationship between parameters and responses. A practical method for combining RSM and an evolutionary strategy in the form of a genetic algorithm (GA) is that RSM is utilized to build a functional relationship between parameters and responses, and then GA is used to optimize the given fitness function composed by the desirability approach.

The following parameters of the applied GA were listed based on the study to get optimal solutions with the lowest computational effort:

- Population size = 20;
- Maximum number of generations = 100;
- Crossover probability = 0.8;
- Mutation probability = 0.9;
- Stop criterion: max number of generations and average change in fitness function values (with tolerance 10^{-6}) and the average change of constraints values (with tolerance 10^{-3});
- The constraint: the first generation is chosen randomly according to the constraint.

The optimization process is conducted 100 times, and the best results after each trial are collected in Figure 12.

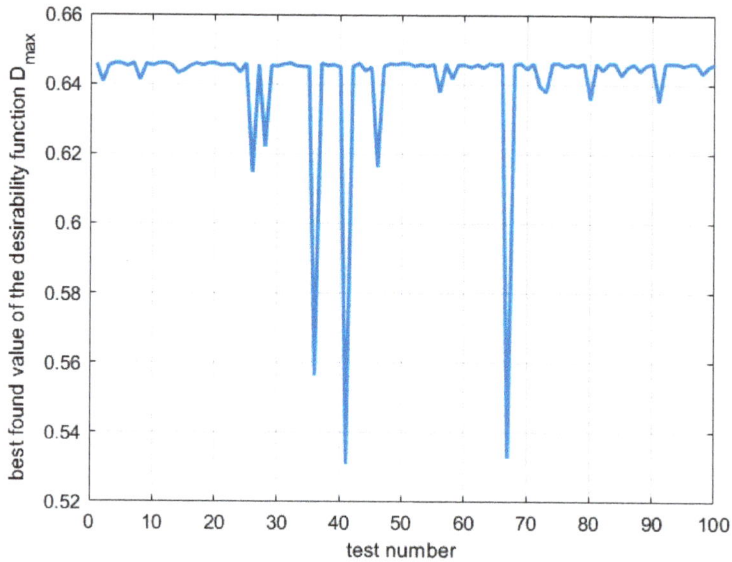

Figure 12. Best-found values of the desirability function for GA optimization during N = 100 tests.

It is visible that for all 100 trials, the best value of the desirability function is non-zero, with an overall maximum of 0.646. In a few cases, local minima were found, or the optimization process was interrupted due to the lack of a significant change in the value of the objective function. Most often, the optimization process requires five iterations of the algorithm, which is a relatively small value, in particular for methods of computational intelligence, such as the genetic algorithm. In Table 15, the statistics for the evolutionary optimization process are presented.

Table 15. Statistics of the evolutionary optimization results for the desirability function values.

Statistics	Value
Sample mean value, $\overline{x}$	0.641
Sample standard deviation, σ	0.019
Sample median, M	0.646
Sample minimum, min(max(D))	0.531
Sample maximum, max(max(D))	0.646

The results in Table 15 show that evolutionary optimization is a robust technique, especially for complex non-differentiable multidimensional fitness functions as the considered desirability function for the induction hardening process of the rack bar.

The global optimal design variables vector in uncoded values is $\mathbf{x}^{opt}$ = [56.4% 1.6 mm 796 mm/min]^T, which gives D^{opt} = 0.646; this result agrees with the one obtained for the SQP algorithm considering the rounding used. One of the results for local minima is also close to the second-best result found with the SQP algorithm, namely $\mathbf{x}$ = [56.4% 3.0 mm 795 mm/min]^T, which gives D^{opt} = 0.556. However, it should be noted that the GA found more local minima, as can be seen in Figure 12, but similar results are shown to compare the results of the two algorithms. Figure 13 shows a histogram of the optimal values of the design variables obtained during 100 tests of the GA. It can be seen that the most significant variation in optimal values occurs for the third component of the vector of design variables, that is, the feed rate. For the first and second components of the vector of design variables, the optimal values cluster with less variation than for the third component.

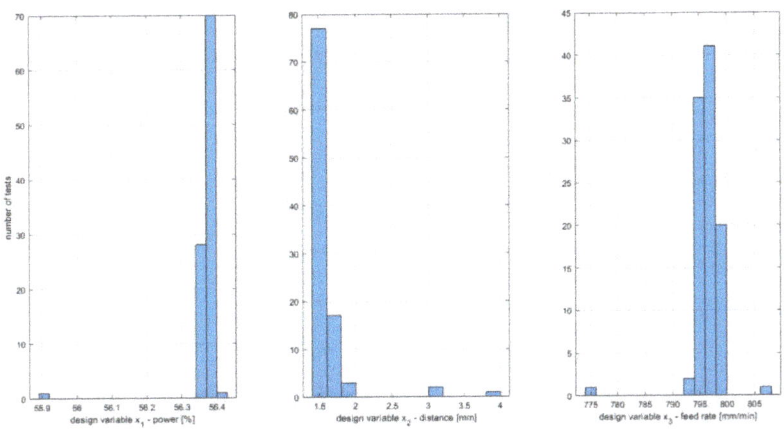

Figure 13. Histograms of optimal design variables for GA optimization results.

5. Confirmation Experiments and Discussion

The best and global optimal point obtained during the evolutionary optimization was found for the following values of the design variables: 56.4% of power, 1.6 mm of the distance of the hardening coil, and 804 mm/min of the feed rate. The second design variable was found close to its lower bound. Industrial conditions require a different approach to evaluating the presented optimal solution, which can be regarded with high probability as a global optimum. Preliminary tests have shown that the found optimal solution cannot be applied due to increased heating of the tool, which is a coil located close to the heated element, leading to its rapid wear, increasing the cost of the process. The original assumptions about the range of design variables included the possibility of fully controlling design variables without taking into account additional factors that arise in industrial conditions, especially in the conditions of mass production in an automatic cycle. Therefore, it was decided to adopt the following quasi-optimal values of design variables corresponding to the local minimum, $\mathbf{x} = [56.4\%\ 3.1\ \text{mm}\ 793\ \text{mm/min}]^T$, which gives $D^{opt} = 0.557$.

Finally, the confirmation experiment was conducted for the quasi-optimal setup, and the results are displayed in Table 16. Industrial cost and time constraints necessitate small sample sizes for experimentation to prove process optimization. Six experiments were performed, and Table 16 contains the measured responses and the corresponding values of the desirability function. In two out of six cases, the desirability function values are equal to zero. In experiments 2 and 4, responses y_9 (the hardness of the back side in the 3rd zone) are higher than the upper limit. The best-obtained result is ~8.9% higher than the calculated optimal desirability.

Table 16. Results of the confirmation experiment.

No.	y_1 [µm]	y_2 [mm]	y_3 [mm]	y_4 [HRC]	y_5 [HRC]	y_6 [HRC]	y_7 [HRC]	y_8 [HRC]	y_9 [HRC]	D
1	605	4.190	2.010	57.3	57.1	57.5	54.3	54.1	54.4	0.606
2	815	4.522	2.063	56.9	57.1	57.5	53.4	55.3	56.2	0.000
3	773	4.410	2.420	57.8	56.5	57.4	54.3	54.6	54.1	0.508
4	681	4.544	2.092	56.8	56.8	57.1	53.3	54.9	55.5	0.000
5	656	4.368	1.904	56.6	56.7	56.9	53.4	54.7	54.1	0.514
6	635	4.590	2.070	58.1	57.3	58.1	53.8	54.2	52.6	0.605

Due to the small sample size of the data from the confirmatory experiment, statistical testing of the results obtained requires special attention. This study used the non-parametric Wilcoxon signed rank test with a significance level of 0.05. Table 17 presents the results of several tests related to the obtained responses and desirability values. $\mathbf{X}^{conf}$ denotes the sample of the confirmatory experiment for $\mathbf{X}$ characteristic. In Table 17, the alternative hypotheses are presented. M denotes the median of the sample. For tests 5-8, it is assumed that the sample contains data from all appropriate measurement zones because the requirements are the same for each.

Table 17. Results of the Wilcoxon test for data from the confirmatory experiment.

No.	Hypothesis H_1	p-Value	Result
1	$M(\mathbf{D}^{conf}) \neq \max(D_{opt})$	0.031	the data in the sample come from a continuous distribution with a median different than the best optimal solution of 0.646
2	$M((\mathbf{y}_1)^{conf}) < L(y_1)$	0.016	the data in the sample come from a continuous distribution with a median less than the lower bound for the maximal deflection response of 1000 µm
3	$M((\mathbf{y}_2)^{conf}) > L(y_2)$	0.016	the data in the sample come from a continuous distribution with a median greater than the lower bound for the hardening depth of teeth of 3.9 mm
4	$M((\mathbf{y}_3)^{conf}) > L(y_3)$	0.016	the data in the sample come from a continuous distribution with a median greater than the lower bound for the hardening depth of the back side of 1.0 mm
5	$M([\mathbf{y}_4; \mathbf{y}_5; \mathbf{y}_6])^{conf}) > L(y_4)$	<0.001	the data in the sample come from a continuous distribution with a median greater than the lower bound for the hardness of the teeth in I, II and III zones of 55 HRC
6	$M([\mathbf{y}_4; \mathbf{y}_5; \mathbf{y}_6])^{conf}) < U(y_4)$	<0.001	the data in the sample come from a continuous distribution with a median less than the upper bound for the hardness of the teeth in the I, II and III zones of 60 HRC
7	$M([\mathbf{y}_7; \mathbf{y}_8; \mathbf{y}_9])^{conf}) > L(y_7)$	<0.001	the data in the sample come from a continuous distribution with a median greater than the lower bound for the hardness of the back side in the I, II and III zones of 52 HRC
8	$M([\mathbf{y}_7; \mathbf{y}_8; \mathbf{y}_9])^{conf}) < U(y_7)$	0.003	the data in the sample come from a continuous distribution with a median less than the upper bound for the hardness of the back side in the I, II and III zones of 55 HRC

The result for the first test shows that the obtained values of the desirability function are not statistically equal to the calculated optimal value, which is mainly influenced by the results of the second and fourth experiments, for which the values of the desirability function are zero. The rest of the test results indicate that the process response requirements are statistically met at a significance level of 0.05. Moreover, process quality improvement is observed, mainly if the deflection response is analyzed as the most critical response. Uncertainties of the process mean that it is impossible to talk about finding the optimal parameters of the induction hardening operation, both in the sense of single-criteria and multi-criteria optimization. However, it could be concluded that optimizing the desirability function based on the RSM models increases the quality of the process. It can be noted that if the confirmation sample had a larger size, the process capability analysis could have helped in assessing its quality improvement.

6. Conclusions

This work addressed the constrained multi-response optimization of processing conditions in induction hardening to decrease mechanical deformation during the hardening operation and improve the process's predefined geometrical and mechanical properties. The RSM models were used to model the highly nonlinear relationships between inputs (power, distance, feed rate) and technological responses. The GA was applied to determine the quasi-optimal values of performances measured and technological inputs.

The following important conclusions can be drawn. The induction hardening process can be significantly improved. The quadratic models of responses of the process were significant according to the quality measures of the least-square approximation. The developed RSM models act effectively in the optimization process. The models proposed can be used in industrial applications to predict technological responses with acceptable accuracy.

The optimization problem, in which maximal deflection, depths of the hardening and hardness of teeth and the back side of the rack bar are objectives, and the desirability function is defined using them, is practical and realistic in the induction hardening process optimization compared to a single objective or simultaneously optimizing nine responses in the Pareto sense. However, other formulations are possible, such as one in which some criteria act as constraints; this may provide directions for further research. Industrial conditions, on the other hand, favor the use of desirability functions. Statistically significant lack of fit of some response functions indicates directions for further research related to using other empirical modeling tools, such as neural networks or kriging. However, this forces larger training data sets, which are difficult to obtain in industrial conditions.

The optimization algorithms recommended the following quasi-optimal combination of process parameters: 56.4% power, 3.1 mm distance of the hardening coil, and 793 mm/min of the feed rate when the industrial conditions were considered. Industrial conditions in the considered case forced the rejection of the possible global optimum of the desirability function, and the quasi-optimal solution was recommended, related to the local optimum, indicated approximately by both the classical optimization algorithm, such as SQP and the genetic algorithm. Uncertainties in empirical modeling also indicate that industrial conditions with limited experimental data (also with a lack of replication) make it difficult to determine optimal solutions. However, the confirmation experiments show that the quasi-optimal process is reliable and high-quality. The use of the nonparametric Wilcoxon test allowed the evaluation of a confirmatory experiment that provided a small sample of data, which is due to the industrial conditions of the research conducted. The test results confirm a statistically significant improvement in process quality in terms of the measures introduced.

The proposed hybrid approach in this paper can be considered an effective technique in some academic research and industrial applications to identify the optimal parameters in the induction hardening process and decrease production costs and time.

Industrial conditions should be carefully applied when solving optimization problems to meet multiple requirements. However, it can be concluded that the influence of treatment conditions on other responses of the technological process, such as total energy consumption, tool wear, residual stresses, etc., has not been investigated. Therefore, more goals should be considered during the formulation of optimization problems.

Author Contributions: Conceptualization, G.D., K.K. and R.P.; Measurements and data acquisition, K.K.; Formal analysis, G.D., K.K. and R.P.; Programming and optimization, G.D.; Funding acquisition, G.D.; Resources, K.K.; Software, G.D.; Verification, Validation and Testing, G.D., K.K. and R.P.; Writing—original draft, G.D., K.K. and R.P; Writing—review and editing, G.D. All authors have read and agreed to the published version of the manuscript.

Funding: Publication supported by the rector's pro-quality grant. The Silesian University of Technology, grant number 10/040/RGJ22/0120.

Institutional Review Board Statement: Not applicable.

Informed Consent Statement: Not applicable.

Data Availability Statement: All data generated or analyzed in this research were included in this published article.

Conflicts of Interest: The authors declare that they have no known competing financial interest or personal relationships that could have appeared to influence the work reported in this paper.

References

1. Favennec, Y.; Labbé, V.; Bay, F. Induction Heating Processes Optimization a General Optimal Control Approach. *J. Comput. Phys.* **2003**, *187*, 68–94. [CrossRef]
2. Jakubovičová, L.; Andrej, G.; Peter, K.; Milan, S. Optimization of the Induction Heating Process in Order to Achieve Uniform Surface Temperature. *Procedia Eng.* **2016**, *136*, 125–131. [CrossRef]
3. Nemkov, V.; Goldstein, R.; Jackowski, J.; Ferguson, L.; Li, Z. Stress and Distortion Evolution during Induction Case Hardening of Tube. *J. Mater. Eng. Perform.* **2013**, *22*, 1826–1832. [CrossRef]
4. Barka, N.; Bocher, P.; Brousseau, J. Sensitivity Study of Hardness Profile of 4340 Specimen Heated by Induction Process Using Axisymmetric Modeling. *Int. J. Adv. Manuf. Technol.* **2013**, *69*, 2747–2756. [CrossRef]
5. Hachkevych, O.R.; Drobenko, B.D.; Vankevych, P.I.; Yakovlev, M.Y. Optimization of the High-Temperature Induction Treatment Modes for Nonlinear Electroconductive Bodies. *Strength Mater.* **2017**, *49*, 429–435. [CrossRef]
6. Fisk, M.; Lindgren, L.E.; Datchary, W.; Deshmukh, V. Modelling of Induction Hardening in Low Alloy Steels. *Finite Elem. Anal. Des.* **2018**, *144*, 61–75. [CrossRef]
7. Khalifa, M.; Barka, N.; Brousseau, J.; Bocher, P. Optimization of the Edge Effect of 4340 Steel Specimen Heated by Induction Process with Flux Concentrators Using Finite Element Axis-Symmetric Simulation and Experimental Validation. *Int. J. Adv. Manuf. Technol.* **2019**, *104*, 4549–4557. [CrossRef]
8. Lamura, M.D.P.; Ammarullah, M.I.; Hidayat, T.; Maula, M.I.; Jamari, J.; Bayuseno, A.P.; Pratama, M.D.; Ammarullah, M.I.; Hidayat, T.; Maula, M.I.; et al. Diameter Ratio and Friction Coefficient Effect on Equivalent Plastic Strain (PEEQ) during Contact between Two Brass Solids. *Cogent Eng.* **2023**, *10*, 2218691. [CrossRef]
9. Ammarullah, M.I.; Hartono, R.; Supriyono, T.; Santoso, G.; Sugiharto, S.; Permana, M.S. Polycrystalline Diamond as a Potential Material for the Hard-on-Hard Bearing of Total Hip Prosthesis: Von Mises. *Biomedicines* **2023**, *11*, 951. [CrossRef]
10. Farooq, M.U.; Anwar, S.; Bhatti, H.A.; Kumar, M.S.; Ali, M.A.; Ammarullah, M.I. Electric Discharge Machining of Ti6Al4V ELI in Biomedical Industry: Parametric Analysis of Surface Functionalization and Tribological Characterization. *Materials* **2023**, *16*, 4458. [CrossRef]
11. Myers, R.H.; Montgomery, D.C.; Anderson-Cook, C.M. *Response Surface Methodology: Process and Product Optimization Using Designed Experiments*, 4th ed.; John Wiley & Sons, Inc.: Hoboken, NJ, USA, 2016; ISBN 9786021018187.
12. Kohli, A.; Singh, H. Modeling and Microstructural Analysis of Induction Hardened Parts. *Mater. Manuf. Process.* **2012**, *27*, 278–283. [CrossRef]
13. Parvinzadeh, M.; Sattarpanah Karganroudi, S.; Omidi, N.; Barka, N.; Khalifa, M. A Novel Investigation into the Edge Effect Reduction of 4340 Steel Disc through Induction Hardening Process Using Magnetic Flux Concentrators. *Int. J. Adv. Manuf. Technol.* **2021**, *115*, 2959–2971. [CrossRef]
14. Eastwood, M.D.; Haapala, K.R. An Induction Hardening Process Model to Assist Sustainability Assessment of a Steel Bevel Gear. *Int. J. Adv. Manuf. Technol.* **2015**, *80*, 1113–1125. [CrossRef]
15. Khalifa, M.; Barka, N.; Brousseau, J.; Bocher, P. Reduction of Edge Effect Using Response Surface Methodology and Artificial Neural Network Modeling of a Spur Gear Treated by Induction with Flux Concentrators. *Int. J. Adv. Manuf. Technol.* **2019**, *104*, 103–117. [CrossRef]
16. Garois, S.; Daoud, M.; Traidi, K.; Chinesta, F. Artificial Intelligence Modeling of Induction Contour Hardening of 300M Steel Bar and C45 Steel Spur-Gear. *Int. J. Mater. Form.* **2023**, *16*, 26. [CrossRef]

17. Qin, X.; Gao, K.; Zhu, Z.; Chen, X.; Wang, Z. Prediction and Optimization of Phase Transformation Region After Spot Continual Induction Hardening Process Using Response Surface Method. *J. Mater. Eng. Perform.* **2017**, *26*, 4578–4594. [CrossRef]
18. Saputro, I.E.; Chen, C.P.; Jheng, Y.S.; Huang, H.C.; Fuh, Y.K. Mobile Induction Heat Treatment of Large-Sized Spur Gear—The Effect of Scanning Speed and Air Gap on the Uniformity of Hardened Depth and Mechanical Properties. *Steel Res. Int.* **2023**, *94*, 2200261. [CrossRef]
19. Misra, M.K.; Bhattacharya, B.; Singh, O.; Chatterjee, A. Multi Response Optimization of Induction Hardening Process—A New Approach. *IFAC Proc. Vol.* **2014**, *3*, 862–869. [CrossRef]
20. Asadzadeh, M.Z.; Raninger, P.; Prevedel, P.; Ecker, W.; Mücke, M. Hybrid Modeling of Induction Hardening Processes. *Appl. Eng. Sci.* **2021**, *5*, 100030. [CrossRef]
21. Badar, A.Q.H. *Evolutionary Optimization Algorithms*; Taylor and Francis Group: Boca Raton, FL, USA, 2022; ISBN 9780367750541.

Disclaimer/Publisher's Note: The statements, opinions and data contained in all publications are solely those of the individual author(s) and contributor(s) and not of MDPI and/or the editor(s). MDPI and/or the editor(s) disclaim responsibility for any injury to people or property resulting from any ideas, methods, instructions or products referred to in the content.

materials

MDPI

Article

Comparative Evaluation of Titanium Feedstock Powder Derived from Recycled Battlefield Scrap vs. Virgin Powder for Cold Spray Processing

Kiran G. Judd [1], Kyle Tsaknopoulos [1], Bryer C. Sousa [1], Marc Pepi [2] and Danielle L. Cote [1,*]

[1] Department of Mechanical and Materials Engineering, Worcester Polytechnic Institute, Worcester, MA 01609, USA; kgjudd@wpi.edu (K.G.J.); kltsaknopoulos@wpi.edu (K.T.); bcsousa@wpi.edu (B.C.S.)

[2] DEVCOM—Army Research Laboratory (ARL), Aberdeen Proving Ground, Aberdeen, MD 21005, USA; marc.s.pepi.civ@army.mil

* Correspondence: dlcote2@wpi.edu

Abstract: Gas-atomization is extensively used to produce metallic feedstock powders for additive manufacturing processes, including gas dynamic cold spray processing. This work explores the potential utility of on-demand recycled titanium scrap feedstock powder as a viable substitute for virgin powder sources. Three recycled titanium powders were atomized from different battlefield scrap sources using a mobile foundry developed by MolyWorks Materials Corporation. Recycled titanium alloy powders were compared against virgin Ti-6Al-4V powder to verify there were no significant variations between the recycled and virgin materials. Powder characterization methods included chemical analysis, particle size distribution analysis, scanning electron microscopy (SEM), Karl Fischer (KF) titration moisture content analysis, X-ray diffraction (XRD) phase analysis, microparticle compression testing (MCT), and nanoindentation. Results indicate that recycled titanium powder provides a viable alternative to virgin titanium alloy powders without compromising mechanical capabilities, microstructural features, or ASTM-specified composition and impurity standards. The results of this work will be used to aid future research efforts that will focus on optimizing cold spray parameters to maximize coating density, mechanical strength, and hardness of recycled titanium feedstock powders. "Cold spray" presents opportunities to enhance the sustainability of titanium component production through the utilization of recycled feedstock powder, mitigating issues of long lead times and high waste associated with the use of conventional virgin feedstock.

Keywords: gas atomization; titanium; recycled feedstock powder; cold spray; additive manufacturing

Citation: Judd, K.G.; Tsaknopoulos, K.; Sousa, B.C.; Pepi, M.; Cote, D.L. Comparative Evaluation of Titanium Feedstock Powder Derived from Recycled Battlefield Scrap vs. Virgin Powder for Cold Spray Processing. *Materials* **2024**, *17*, 1122. https://doi.org/10.3390/ma17051122

Academic Editor: Yong-Cheng Lin

Received: 4 February 2024
Revised: 23 February 2024
Accepted: 27 February 2024
Published: 29 February 2024

Copyright: © 2024 by the authors. Licensee MDPI, Basel, Switzerland. This article is an open access article distributed under the terms and conditions of the Creative Commons Attribution (CC BY) license (https://creativecommons.org/licenses/by/4.0/).

1. Introduction

Titanium and titanium (Ti) alloys grew to prominence in the late 1950s due to their high performance and unique structure-property relationships. Titanium alloys are advantageous in a versatile range of applications due to their high strength-to-weight ratio, high corrosion resistance, fracture toughness, high-temperature strength, biocompatibility, and processing capabilities [1–4]. These alloys are classified into five main categories, described as alpha (α), near α, beta (β), near β, and α-β alloys. Alpha and near-α titanium alloys contain low to moderate β-stabilizing elements, retaining the hexagonal close-packed (HCP) α phase as the dominant microconstituent, while metastable and stable β titanium alloys have sufficient β-stabilizer additions to form higher volumes of the body-centered cubic (BCC) β phase upon cooling from the β transus temperature [1,2,4,5]. This variation in alloying content and resultant microstructural constitution leads to differentiated mechanical properties and performances, with α and near-α alloys exhibiting superior fracture toughness and creep strength compared to β and near-β alloys which conversely display higher strength and formability [1,2,4,5]. Two of the most widely used titanium alloys are commercially pure (CP) titanium, an α alloy, and Ti-6Al-4V, an α- β alloy. For reasons previously described, Ti-6Al-4V is one of the most prominently used titanium alloys throughout aerospace, biomedical, automotive, naval, defense, and power generation industries [2,6,7].

Conventional production methods for titanium alloys (casting or forging), involving repeated thermo-mechanical processing and extensive machining, result in high costs, long lead times, and significant energy consumption [8]. In contrast, gas atomization is a commonly employed powder production method that involves the disintegration of a molten metal stream into fine droplets using a high-velocity inert gas jet; upon contact with the gas, the metal droplets rapidly solidify into spherical powder particles, resulting in refined microstructures and limited segregation [9]. Controlling gas atomization process parameters allows for tailored particle size distribution, morphology, and purity applicable to a range of additive manufacturing technologies, such as cold spray. Cold spray is a solid-state deposition process that accelerates feedstock powder particles via a heated carrier gas stream through a de Laval nozzle at supersonic velocities before depositing them onto a substrate. This process has been adopted for the remanufacture and repair of coatings and parts across a multitude of industries [10]. Cold spray offers opportunities to overcome the disadvantages of traditional titanium part manufacturing and enables improved sustainability and applicability for titanium alloy production.

The widespread use of titanium alloys in crucial industries, such as aerospace and defense, underlines the need to develop sustainable material sourcing and processing techniques that guarantee continued supply while minimizing environmental ramifications. As the global demand for materials like titanium continues to increase, the sustainability of their production methods becomes a crucial concern. The extraction and processing of virgin titanium are resource-intensive and have significant environmental impacts, including habitat destruction, energy consumption, and greenhouse gas emissions [11]. Recycling titanium scrap and other waste materials helps conserve natural resources by decreasing dependence on mining raw titanium ore. Additionally, recycling aids in securing a consistent titanium supply, which is critical for industries such as defense, aerospace, and biomedical where material availability can be rate-limiting for manufacturing operations.

Therefore, the present study investigates the viability of using recycled titanium feedstock powder atomized from battlefield scrap for cold spray applications. Three distinct recycled titanium alloy powders were characterized and compared against virgin Ti-6Al-4V powder to verify there were no significant variations between the recycled and virgin materials. Powder characterization methods included chemical analysis, particle size distribution analysis, scanning electron microscopy, Karl Fischer titration analysis, X-ray diffraction phase analysis, microparticle compression testing, and nanoindentation. The objective of this work is to demonstrate the feasibility of generating high-quality, on-demand, recycled titanium feedstock powder and establish a basis for optimizing cold spray deposition parameters using these materials.

2. Experimental Methods and Materials

2.1. Feedstock Powders

Plasma atomized grade 5 virgin Ti-6Al-4V powder was purchased from Advanced Powders and Coating (AP&C) (Boisbriand, QC, Canada). This powder was selected for its ideal characteristics for additive manufacturing applications due to the low levels of entrapped porosity, low satellite content, and notable flowability/packing density. Herein, this powder will be referred to as "virgin Ti-6Al-4V". Three varying alloys of titanium feedstock powder were produced using gas atomization by the MolyWorks Corporation (Cloverdale, CA, USA) from recycled battlefield titanium armor scrap and commercially pure titanium laser powder bed fusion (LPBF) build plate scrap. The gas atomization process utilized the Molyworks "Greyhound" mobile foundry system that is purposely built to be self-contained in an enclosed 20-foot-long shipping container. Scrap metal is melted in the Greyhound system by a 240-kW plasma torch by which melted material falls through a horizontal argon gas stream to form spheroidized additive manufacturing

(AM)-grade powder particles. Figure 1 shows the self-contained atomization unit that is housed inside a shipping container. The recycled titanium atomized feedstocks consisted of a CP Ti, Ti-6Al-4V, and an alloyed blend of the two (~Ti-4.4Al-2.6V). Henceforth, these powders will be referred to as "recycled CP Ti", "recycled Ti armor alloy", and "recycled Ti blend alloy". Titanium armor scrap was sourced from ballistically tested titanium armor plates from the ballistic test range at the DEVCOM—Army Research Laboratory (ARL) (Aberdeen Proving Ground, MD, USA). All ballistic damage and dirt residue were removed from the armor plates prior to atomization. The CP Ti LPBF AM build plates were similarly cleaned and sourced from the DEVOM-ARL Metals AM Laboratory (Aberdeen Proving Ground, MD, USA).

Figure 1. Interior of the Greyhound foundry encapsulated within a shipping container unit for ease of deployability.

2.2. *Powder Characterization*

The chemical composition of all titanium powders was measured via inert gas fusion (ASTM E 1409-13), combustion infrared detection (ASTM E 1941-10), and direct current plasma emission spectroscopy (ASTM E 2371-21) (Luvak Laboratories, Inc., Boylston, MA, USA). The oxygen (O), nitrogen (N), and hydrogen (H) contents will be compared between all powders in the Results and Discussion of this work.

Particle size distribution (PSD) and particle morphology were evaluated using a scanning electron microscope (SEM) and particle size/shape analyzer. The particle size distribution was measured in 3-dimensions and assuming spherical particles using a Microtrac TurboSync system (Microtrac Retsch GmbH, Haan/Duesseldorf, Germany). This system quantifies particle size and shape with a hybrid laser diffraction and dynamic image analysis method by an ensemble of particles in an air stream. Additionally, a Zeiss EVO MA-10 SEM (Carl Zeiss AG, Oberkochen, Germany) was used to qualitatively evaluate the size, morphology, and microstructure in 2-dimensions of all titanium powder samples. Loose powder particles were examined by mounting them to SEM stubs with carbon tape and were imaged at 10 kV at varied magnifications ranging from 100× to 2000× with secondary electron emission. The microstructures of all powders were investigated using mechanical and ion beam polishing techniques. A 6-step polishing procedure was used until a mirror finish was attained with a Buehler EcoMet 300 automatic grinder-polisher (Lake Bluff, IL, USA) and an ion-mill polish with argon gas in a JEOL IB-19530CP Cross Section Polisher (Tokyo, Japan) at 8 kV for 2 h to prepare powders for imaging, respectively. Both mechanical and ion beam polishing methods were utilized to ensure all potential microstructural features were identified and whether the observed features were inherent to the material or if artifacts were introduced during sample preparation. The cross-sectional microstructure of each powder was then imaged at 10 kV at varied magnifications with backscattered electron imaging.

The moisture content of all the titanium powders was measured using a coulometric Karl Fischer titration analysis method with a Mettler Toledo Coulometric Karl Fischer C30S Titrator with InMotion KF Oven Autosampler Pro (Columbus, OH, USA). This method places three separate sealed glass vials per titanium powder type, each containing 1–2 g of powder, onto the autosampler. These glass vials were then loaded into the system's oven where a needle would then pierce the plastic lid to flood the interior of the vial with desiccated air. Afterwards, the vial and encapsulated powder were heated to 220 °C, by which the temperature is sufficient to desorb the moisture from the titanium powder's surface for 5 min isothermally. The dried air, containing the desorbed moisture, was then carried out of the vial through tubing into the titrator cell. Once the moisture entered the titrator cell, air was bubbled and stirred into a Honeywell high-temperature rated methanol-based analyte "HYDRANAL-Coulomat AG Oven" (Charlotte, NC, USA). An electrode is used to complete the titration process between desorbed moisture introduced into the cell and iodine generated in the solution. The electrode is then used to measure the voltage change of the reaction, thereby converting the voltage measurement into a quantity of moisture introduced into the system (in ppm). This process was repeated for each titanium powder explored in this study, precursed by initial "blank" samples that were empty vials run at identical temperature and time conditions to account for environmental drift effects on the moisture contents. Any drift effects were then subtracted from the final sample moisture contents prior to averaging the results.

X-ray diffraction (XRD) was executed on all powder samples using a Malvern Panalytical Ltd. Empyrean X-ray Diffractometer (Almelo, Overijssel, The Netherlands), a Cu-Kα radiation source, and Ni filter. Additionally, the data analysis was completed using the HighScore Plus software (Version 4.9.0.27512).

Microparticulate compression testing (MCT) was performed to better understand the particle-scale mechanical property differences across differences in titanium powder compositions and material sourcing, i.e., virgin or recycled scrap feedstocks. Titanium particles were spread across the surface of a diamond compression plate, fixtured to a sliding stage enclosed within a Shimadzu MCT-510 Series system (Kyoto, Japan). Compression strengths were obtained by bringing individual powder particles into focus with two varied magnification objective lenses, in which the particle diameter was measured, and the stage height was focused to the top of the particle for compression. Afterwards, a 50 μm diamond flat-punch indenter probe applied a load at 20.74 mN/s until the target 1600 mN was obtained. As described in [12–14], the ultimate tensile strength (σ_u) (MPa) of microparticles was determined from load–displacement data at a displacement equal to 50% particulate compression ($\sigma_c(50\%)$). Equation (1) details this calculation:

$$\sigma_u = 1.4\left(\frac{aP}{\pi d^2}\right) \quad (1)$$

whereby, a is 2.48 [15], P is the test force (N) at a displacement of 50% of the particulate diameter, and d is the particle diameter (mm). An average of at least 15 measured powder particles compression strength values was taken for each titanium powder at 50% compression ($\sigma_c(50\%)$).

Nanoindentation testing was performed using an iMicro Pro from the KLA Corp. (Milpitas, CA, USA) on metallographically prepared epoxy-mounted powder particles. An InForce 1000 mN actuator was employed with a diamond Berkovich tip from Micro Star Technologies Inc. (Huntsville, TX, USA) and all indents were applied at the maximum specified load for 15 s. The nanoindenter system and data processing were operated using the InView Software program (Version 19.2.24). Twenty-five particles were tested for each titanium powder type; however, insufficient indents had to be removed during data analysis to ensure only valid indents were being measured. All hardness measurements were reported at depths averaging between 195 nm and 205 nm using a dynamic method, assuring that on average, all indents were compared at a relative depth of 200 nm. A Poisson's ratio of 0.34 was assumed for all titanium samples tested. Prior to testing the powders, the contact area function and the frame stiffness were determined by testing a fused silica reference standard to obtain the load-depth data. An operational check was also applied to the entire nanoindenter system to check that the actuator and stage mechanics were working properly. Final hardness measurements reported account for any thermal drift, pile-up, and creep-related phenomena that might have occurred during testing. Further information on nanoidentation of single powder particles can be explored here [16–18].

3. Results and Discussion

3.1. Chemical Analysis

Table 1 presents the chemical composition of the four titanium feedstock powders examined, as well as the ASTM F1580-18 standard specification for chemical compositions of CP Ti and Ti-6Al-4V powders [19]. Table 1 also compares the measured O, N, C, and H contents among all powders to ensure that both virgin and recycled powders conform to the maximum allowable impurity contents of the ASTM standards.

Table 1. Chemical composition of the four titanium feedstock powders compared with the ASTM F1580-18 standard for CP Ti and Ti-6Al-4V alloys.

Element (wt.%)	O	N	C	H	Al	Fe	V	Cr	Y	Ti
Recycled CP Ti	0.162	0.024	0.014	0.0027	0.2	0.093	0.069	0.02	N/A	Balance
Recycled Ti Armor Alloy	0.181	0.008	0.004	0.0028	6.04	0.17	4.13	0.014	N/A	Balance
Recycled Ti Blend Alloy	0.202	0.015	0.013	0.003	4.35	0.15	2.59	0.018	N/A	Balance
Virgin Ti-6Al-4V	0.18	0.01	0.01	0.002	6.35	0.19	3.98	N/A	<0.001	Balance
ASTM Standard CP Ti [19]	0.400 (Max)	0.050 (Max)	0.080 (Max)	0.050 (Max)	N/A	0.500 (Max)	N/A	N/A	N/A	Balance
ASTM Standard Ti-6Al-4V [19]	0.200 (Max)	0.050 (Max)	0.080 (Max)	0.015 (Max)	5.50–6.75	0.300 (Max)	3.50–4.50	N/A	0.005 (Max)	Balance

Recycled CP Ti, recycled Ti armor alloy, and virgin Ti-6Al-4V powders were all within the allowable composition and impurity specifications for their respective alloy compositions. Alternatively, the recycled Ti blend alloy did not adhere to the ASTM standard for Ti-6Al-4V powder composition, having a lower Al and V content. However, this can be explained by the fact that the blend alloy was a combination of Ti-6Al-4V and CP Ti material. The lack of additional alloying elements in CP Ti resulted in a decreased content of aluminum and vanadium when combined with the Ti-6Al-4V during atomization. Alloying elements aside, the Ti blend alloy was otherwise within the specification when comparing the measured content of O, N, C, and H to those maximum allowable contents for Ti-6Al-4V and CP Ti ASTM standards.

It was originally hypothesized that the recycled scrap powders would have increased O, N, C, and H contents when compared to their virgin powder counterpart due to the presence of dirt, grease, and environmental containments which could reduce their overall quality and mechanical properties. Nonetheless, the results of the chemical analysis showed that there was no significant difference in elemental composition between recycled and virgin (or ASTM standards) titanium powders. In some cases, the recycled powder, such as the Ti amor alloy, had a lower measured N and C content than the virgin Ti-6Al-4V powder. These results show titanium scrap material can successfully be atomized into usable recycled feedstock powders that are compositionally comparable to virgin powders when using the MolyWorks Greyhound foundry system.

3.2. Particle Size Distribution (PSD) Analysis

Particle size distribution and D10, D50, and D90 percentiles for all powders are shown in Figure 2. The D10, D50, and D90 percentile values refer to the particle diameter sizes (µm) below which 10%, 50%, and 90% of all particles are found, respectively. The channel percentage, located on the y-axis of Figure 2, refers to the probability density function of the data. This reports the percentage of powder particles that are present at a specific size and shows the breakdown of size percentiles. All powders demonstrated similar size distributions which is apparent by the overlapping distribution curves among recycled and virgin powders. Virgin Ti-6Al-4V powder was purchased within the 15–53 µm size range for its optimal size performance for the cold spray process. The PSD measurements confirmed that the virgin Ti-6Al-4V powder was within this desired range and matched the D10, D50, and D90 percentiles reported by the manufacturer. Recycled titanium powders were atomized to be within a similar desired size range as the virgin powder but were on average slightly smaller in size. Quantifiable differences in particle size can be compared by examining the D50 size values among powder types. Broadly, the recycled CP Ti powder was the smallest in size followed by the recycled Ti blend alloy, recycled Ti armor alloy, and the virgin Ti-6Al-4V, respectively. The minor differences between the average particle sizes are thought to have a negligible effect on the resultant deposition efficiency and consolidation quality when cold spraying the virgin and recycled powders.

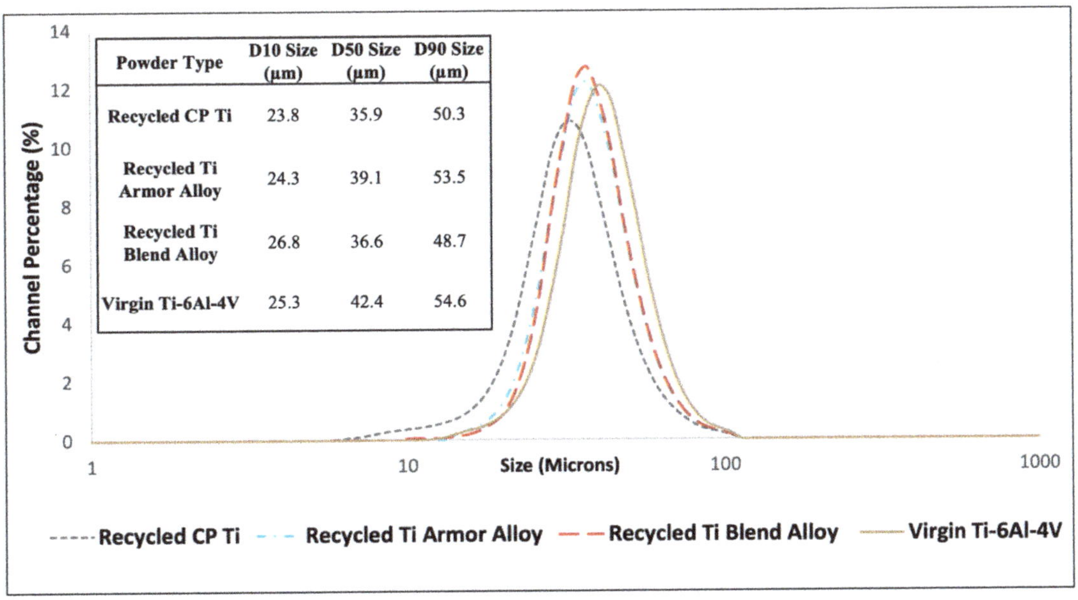

Figure 2. Cumulative particle size distribution and average percentiles of recycled and virgin titanium powders.

3.3. Karl Fischer Titration Moisture Content

The average moisture content of all titanium powders is displayed in Figure 3. The measured average moisture contents were 44.6 ppm, 25.4 ppm, 62.2 ppm, and 50.5 ppm for recycled CP Ti, recycled Ti amor alloy, recycled Ti blend alloy, and virgin Ti-6Al-4V powders, respectively. While all powders had low moisture contents, both the recycled CP Ti and recycled Ti armor alloy powders interestingly had lower moisture contents than the virgin Ti-6Al-4V. This was unexpected, because it was assumed that similarly to the predictions for O, N, C, and H content, recycled scrap material would have higher moisture content due to increased contaminants or exposure to atmospheric conditions.

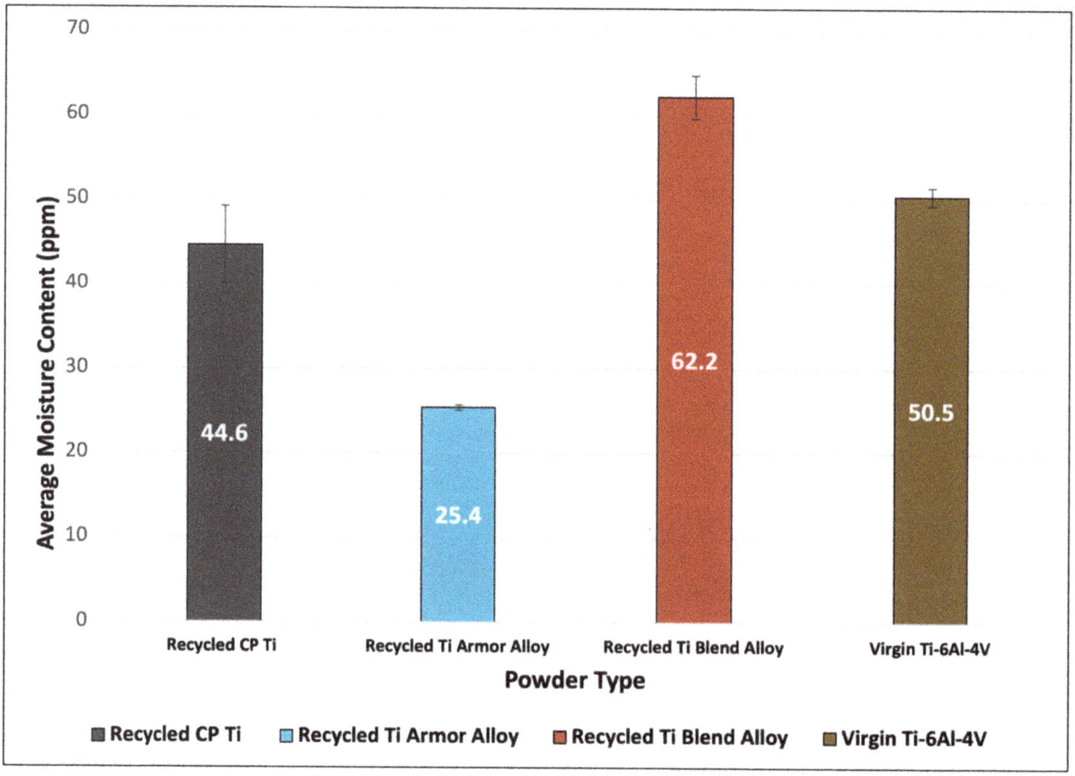

Figure 3. Average moisture content (ppm) of virgin and recycled titanium feedstock powders.

All powders were found to have suitable moisture contents for cold spray applications based on previous laboratory experiences where various material types containing higher moisture contents than those measured in this study were successfully deposited. Additionally, powder exposure studies by Grubbs et al. showed that environmentally exposed Al2024 powder (~320 ppm moisture content) similarly produced highly dense consolidations, <0.2% porosity, when compared to consolidations deposited using dried Al2024 powder (~175 ppm) [20,21]. The combination of these factors validates the usage of both virgin and recycled titanium powders for cold spray.

It should be noted that all powders (virgin and recycled alike) were stored in a laboratory setting after they were unsealed where there was little impact on the moisture content. Powder moisture content is not a static property and is subject to change if powders are not maintained at proper environmental conditions where the humidity and temperature is controlled. Powder storage plays a substantial role in making sure moisture content doesn't change due to variations in humidity and weather. These recycled titanium powders have shown to have suitably low moisture contents however, they must continue to be properly stored to ensure there are no substantial increases over time that may further deleteriously increase moisture and subsequently oxide contents.

3.4. Scanning Electron Microscopy (SEM)

The morphology and microstructure of the titanium powders were further explored through SEM analysis. Figure 4 displays SEM micrographs of loose feedstock powder

(a_1–a_4), mechanically polished cross-sections of powder particles (b_1–b_4), and ion beam polished cross-sections of powder particles (c_1–c_4). Superscripts 1, 2, 3, and 4 refer to the recycled CP Ti, recycled Ti armor alloy (Ti-6Al-4V), recycled Ti blend alloy, and virgin Ti-6Al-4V powders, respectively. Figure 4(a_1–a_4) loose powder SEM reveals that the plasma and gas atomized titanium powders were predominantly spherical with some smaller satellites on the surface of individual particles. Overall, the SEM reveals similar size, shape, and morphology across powder types, agreeing with PSD analysis. Polished cross-sections of powder particles displayed three unique microstructures with varying degrees of the typical α and β phases found in titanium alloys. The micrographs of recycled CP Ti powder, shown in Figure 4(b_1,c_1), highlight the single α-phase present in the microstructure. The α-phase in pure titanium remains stable until the β transus temperature is reached at 883 °C whereby the BCC β phase is formed [22]. The softer α-phase present in the recycled CP Ti powder will promote superior plastic deformation and mechanical interlocking during cold spray which in turn will improve porosity reduction and deposition efficiency. However, improved deposition and denseness of coatings will come at the expense of strength when compared to Ti-6Al-4V or other stronger titanium alloys. As expected, the microstructures of the recycled Ti armor alloy and the virgin Ti-6Al-4V were the same as they share the same alloy chemical composition. There were no visible morphological or microstructural differences between recycled and virgin Ti-6Al-4V powder samples. This was also apparent in the presence of primary α-phases and primary β-phases in Figure 4(b_2,c_2,b_4,c_4). The β-phases are identified by the lathy, needle-like appearance dispersed within the lighter colored α-phase matrix. This α-β phase combination allows for improved strength and toughness due to the presence β-phases. The increased strength of the recycled Ti armor alloy and virgin Ti-6Al-4V will make it more difficult to achieve dense coatings, requiring a higher gas pressure, temperature configurations, or thermal treatments to achieve the required plastic deformation during the cold spray process [23]. The microstructure of the recycled Ti blend alloy alternatively had a unique appearance because of its origins from both CP Ti and Ti-6Al-4V materials. Microstructural features of the recycled Ti blend alloy were more like that of the recycled CP Ti with fewer features matching that of the Ti-6Al-4V-based powders. It is hypothesized that the recycled Ti blend alloy will act as a "middle ground" between the recycled CP Ti and recycled Ti amor alloy/Ti-6Al-4V when depositing coatings using cold spray. A comparison of the ease of deposition between titanium powder types will be further discussed in Section 3.6 based on the mechanical properties provided by the phases identified.

3.5. X-ray Diffraction (XRD)

Figure 5 shows the X-ray diffraction spectra results for all four powder samples. These results indicate that there is no difference in secondary phases present within these powders, as all peaks indicate the structure of the titanium matrix. Slight shifting and broadening differences are likely due to differences in chemical composition between the powder samples.

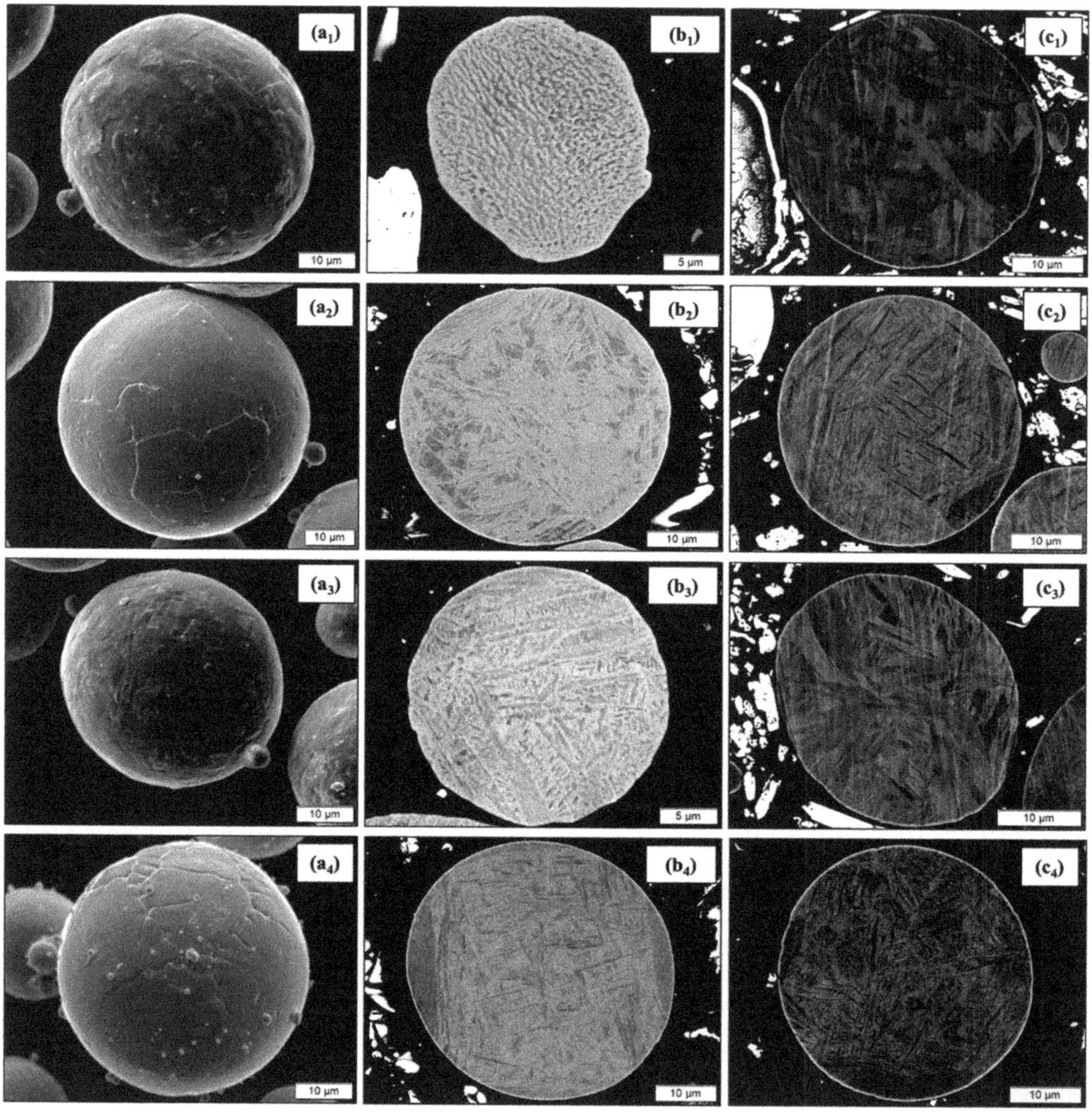

Figure 4. SEM micrographs of loose powder ($\mathbf{a_{1–4}}$) and polished cross-sections ($\mathbf{b_{1–4}}$,$\mathbf{c_{1–4}}$) of Ti CP, Ti armor alloy, Ti blend alloy, and virgin Ti-6Al-4V.

3.6. Micro-Particle Compression

Understanding the compression strength of feedstock powder is essential for predicting how the powder particles will behave during the cold spray process and can greatly improve particle impact simulations [13,14,24]. Figure 6 displays the results of microparticulate compression testing on each of the four titanium powders. The recycled Ti armor alloy, recycled Ti blend alloy, and virgin Ti-6Al-4V powders were all measured to have comparable compression strengths, especially when considering the standard deviations within average measurements. The recycled CP Ti powder had the lowest compression strength (~65% as strong as the other powders) due to a lack of alloying elements that would otherwise mechanically strengthen the powder. The recycled Ti armor alloy compression strength slightly exceeded that of the virgin Ti-6Al-4V powder, highlighting the similarity of mechanical properties between recycled and virgin titanium powders. The Ti blend alloy had the strongest average compression strength by a small margin but had more variation in measurements as reflected by the larger standard deviation. This is thought to be due to the mixed alloyed nature of the Ti blend alloy, leading some particles to be more compositionally and microstructurally analogous to that of the CP Ti or Ti-6Al-4V-based atomization material. The average ultimate tensile strength (UTS) calculated for the recycled CP Ti, recycled Ti armor alloy, recycled Ti blend alloy, and virgin Ti-6Al-4V was 1005 MPa, 1555 MPa, 1666 MPa, 1468 MPa, respectively. Analysis of the calculated powder UTS shows the same trends observed with the average compression strength measurements. Notably, all calculated UTS of titanium powder particles were shown to possess higher UTS values than those of identical wrought or cast titanium alloys [25,26]. This finding is in line with previous reporting that attributes increased powder particle strength to the rapidly solidified nature of atomized powders [13,14,24,27]. The rapid solidification during powder gas-atomization promotes the formation of fine grain sizes that are responsible for increased mechanical strength.

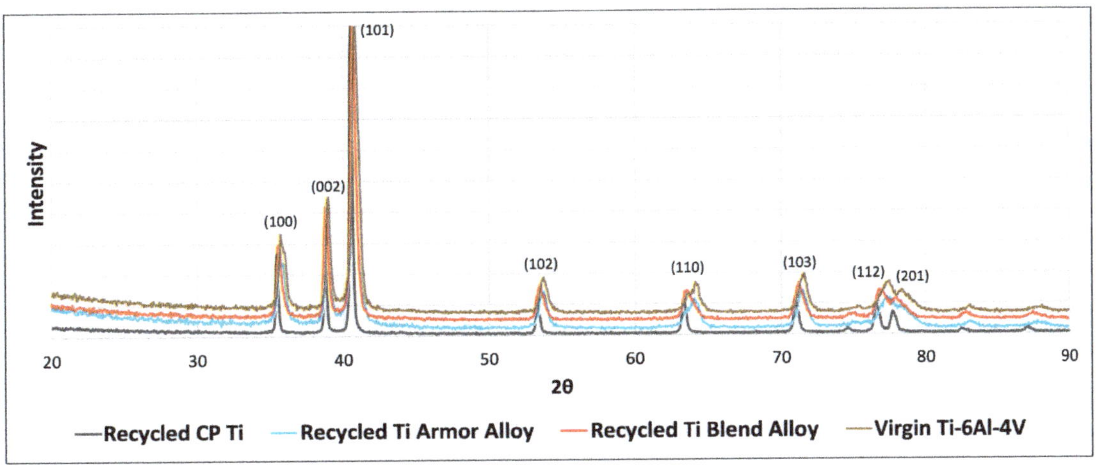

Figure 5. XRD spectra of the four powder samples.

3.7. Nanoindentation

Due to the importance of powder microstructure on mechanical properties and, in turn, impact deformation tendencies in cold spray, nanoindentation characterization was used to determine the hardness and modulus of all titanium powders. Figure 7a,b present the average hardness and modulus of elasticity, respectively. As with the average compression strength, the recycled Ti armor alloy, recycled Ti blend alloy, and virgin Ti-6Al-4V powders all had comparable properties with no significant differences in hardness between virgin or recycled powders. The recycled CP Ti powder exhibited the lowest hardness because it lacked hardening elements such as aluminum and vanadium found in Ti-6Al-4V. These elements provide Ti-6Al-4V and similar alloys with solute atoms that distort and interact with dislocations, resulting in improved hardness and bulk strength when compared with pure titanium. Despite containing lower concentrations of titanium and vanadium compared to Ti-6Al-4V, the recycled Ti blend alloy powder exhibited hardness values like those observed for the Ti-6Al-4V based powders. This suggests that the recycled Ti blend alloy has sufficient concentrations of alloying elements to provide comparable strengthening effects to that of Ti-6Al-4V. These results may suggest that factors beyond just titanium and vanadium content, such as distributions of additional alloying solutes, influence the hardness of the recycled Ti blend alloy.

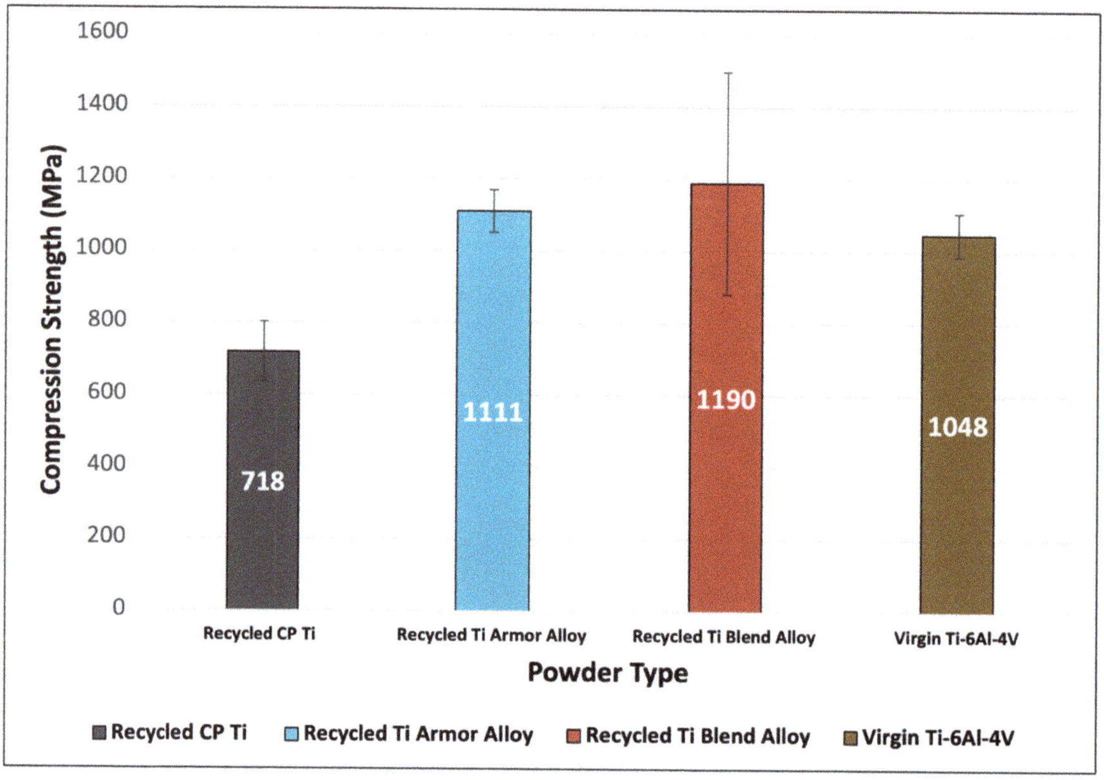

Figure 6. Average particle compression strength of virgin and recycled titanium feedstock powders.

A contrasting trend was observed between the hardness and modulus values across the recycled titanium powder variants. As shown in Figure 7a, the average hardness increased moving from the recycled CP Ti powder to the recycled Ti armor alloy powder to the recycled Ti blend alloy powder. However, Figure 7b indicates the inverse relationship for average modulus, with the recycled CP Ti powder having the highest average modulus followed by declining values for the recycled titanium armor alloy and recycled titanium alloy blend powders. The inverse relationship observed between hardness and modulus aligns with general materials science principles—namely, the difficulty in simultaneously achieving high strength/hardness and high elasticity in metal alloys. The strengthening mechanisms that increase hardness, such as solute atoms and precipitates obstructing dislocation motion, conversely tend to decrease modulus [28]. However, it is notable that the virgin Ti-6Al-4V powder exhibited the highest overall modulus of elasticity, which does not follow the trend of declining modulus with increasing hardness. This discrepancy suggests additional microstructural or compositional factors may be influencing the modulus versus hardness relationship in the virgin powder. Further investigation is required to understand the mechanisms resulting in the virgin Ti-6Al-4V's anomalously high modulus.

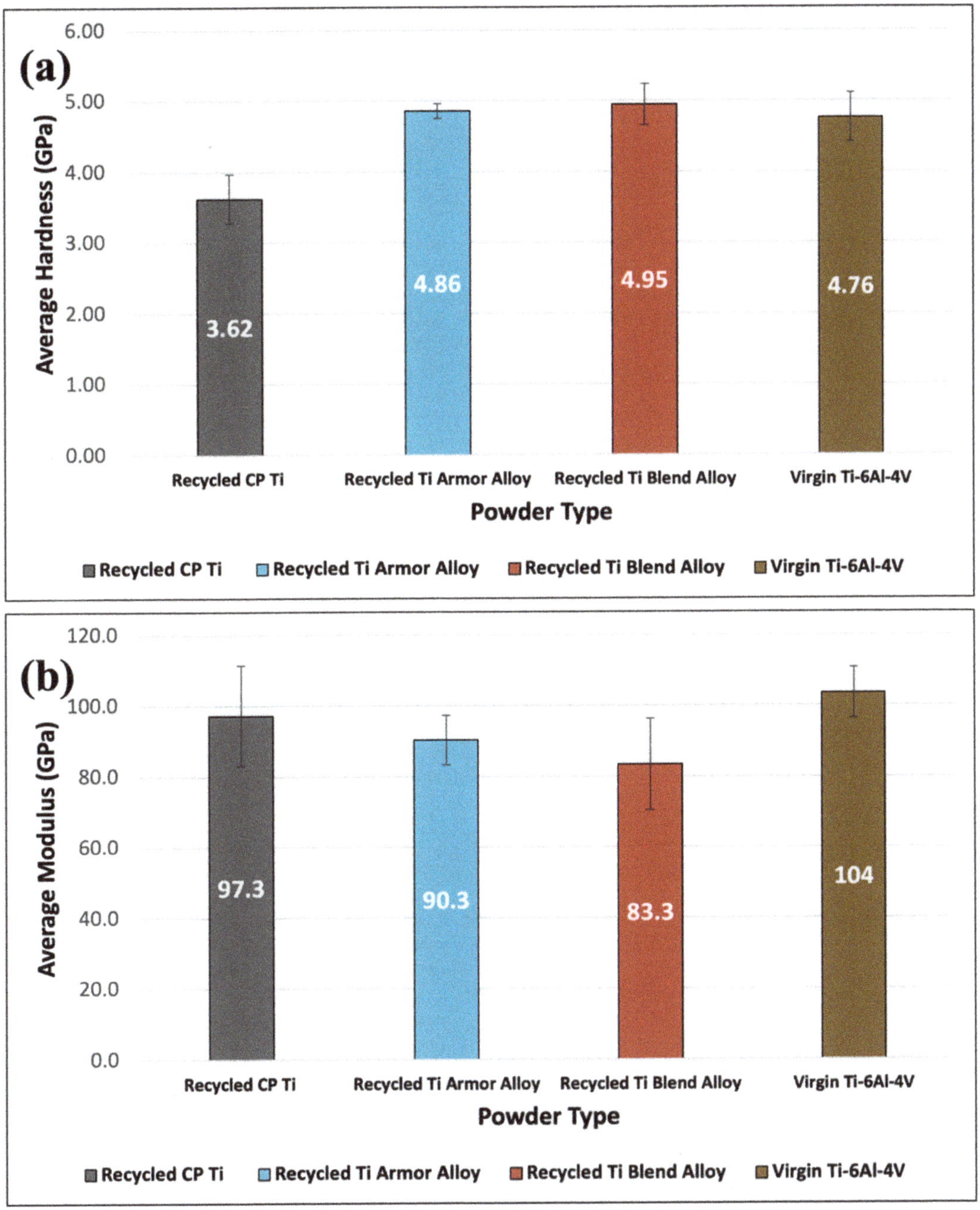

Figure 7. Average dynamic nanoindentation hardness (**a**) and modulus of elasticity (**b**) of virgin and recycled titanium feedstock powders.

4. Conclusions

This work characterized the microstructure and mechanical properties of four different titanium powders and explored the feasibility of using recycled titanium feedstock powders for cold spray applications. In this study, recycled CP Ti, recycled Ti armor alloy, and recycled Ti blend alloy powders were all analyzed and compared to virgin Ti-6Al-4V powder using chemical analysis, PSD, SEM, KF Titration, XRD, MCT, and nanoindentation. Analysis of the experimentation and results determined the following conclusions:

- Recycled CP Ti, recycled Ti armor alloy, and virgin Ti-6Al-4V powders all adhered to ASTM-specified composition and impurity standards for their respective alloy compositions. Compliance with the applicable compositional standards verifies the compositional quality of the recycled powder variants and their suitability for use in cold spray processing.
- PSD and SEM results reveal predominantly spherical particles with similar size, shape, and morphology across all powder types. Particle diameters were predominantly ranged between 20–50 μm, ideally sized for cold spray.
- Microstructural features vary between titanium alloys based on slight differences in primary phases identified from XRD.
- Moisture contents of all four titanium powders were found acceptable for cold spray.
- Mechanical testing demonstrated superior hardness and strength in alloyed titanium powders compared to CP titanium. There was no difference in mechanical strength between virgin and recycled powders.
- Recycled titanium powder produced through the atomization of scrap-sourced material provides a viable alternative to virgin titanium alloy powders without compromising mechanical capabilities. Implementing localized or mobile atomization systems to generate recycled titanium powders at a "point of need" enhances operational adaptability and sustainability while potentially decreasing material expenses. This approach provides an avenue to reduce waste, improve cost-effectiveness, and promote circular economy principles for advanced manufacturing.

Building on the findings from this initial study, future efforts will focus on optimizing cold spray parameters to maximize coating density, mechanical strength, and hardness. The four titanium powders analyzed, including the recycled variants, will be deposited using cold spray to assess the viability of applying on-demand recycled feedstock for part repair and replacement applications. Additional exploration into thermal treatments of the powders prior to spraying will also be performed. Specifically, annealing studies on the Ti-6Al-4V powder will be carried out to improve deposition efficiency and minimize porosity by softening the powder before spraying.

Author Contributions: Conceptualization, K.G.J., M.P. and D.L.C.; Methodology, K.G.J. and K.T.; Validation, K.G.J.; Formal analysis, K.T. and B.C.S.; Investigation, K.G.J.; Resources, D.L.C.; Data curation, K.G.J. and B.C.S.; Writing—original draft, K.G.J. and K.T.; Writing—review and editing, K.G.J., K.T., B.C.S. and D.L.C.; Visualization, K.G.J.; Supervision, K.T., M.P. and D.L.C.; Project administration, K.G.J.; Funding acquisition, M.P. and D.L.C. All authors have read and agreed to the published version of the manuscript.

Funding: This work was funded by OSD through LIFT—operated by the American Lightweight Materials Innovation Institute (ALMMII) under CS R&D Project #N00014-22-9-0014, with oversight from DEVCOM-Army Research Laboratory.

Institutional Review Board Statement: Not applicable.

Informed Consent Statement: Not applicable.

Data Availability Statement: Data will be made available upon request to author.

Conflicts of Interest: The authors declare no conflict of interest.

References

1. Pushp, P.; Dasharath, S.M.; Arati, C. Classification and applications of titanium and its alloys. *Mater Today Proc.* **2022**, *54*, 537–542. [CrossRef]
2. Leyens, C.; Peters, M. *Titanium and Titanium Alloys: Fundamentals and Applications*; Wiley-VCH: Weinheim, Germany, 2003.

3. Veiga, C.; Davim, J.P.; Loureiro, A.J.R. Properties and Applications of Titanium Alloys: A Brief Review. 2012. Available online: https://www.researchgate.net/publication/283863116 (accessed on 10 December 2023).
4. Gialanella, S.; Malandruccolo, A. Titanium and Titanium Alloys. In *Topics in Mining, Metallurgy and Materials Engineering*; Springer Science and Business Media Deutschland GmbH: Berlin/Heidelberg, Germany, 2020; pp. 129–189. [CrossRef]
5. Williams, J.C.; Boyer, R.R. Opportunities and issues in the application of titanium alloys for aerospace components. *Metals* **2020**, *10*, 705. [CrossRef]
6. Peters, M.; Kumpfert, J.; Ward, C.H.; Leyens, C. Titanium alloys for aerospace applications. *Adv. Eng. Mater.* **2003**, *5*, 419–427. [CrossRef]
7. Elshaer, R.N.; Ibrahim, K.M. Applications of Titanium Alloys in Aerospace Manufacturing: A Brief Review. *Bull. Tabbin Inst. Metall. Stud.* **2022**, *111*, 60–69. [CrossRef]
8. Tshephe, T.S.; Akinwamide, S.O.; Olevsky, E.; Olubambi, P.A. Additive manufacturing of titanium-based alloys—A review of methods, properties, challenges, and prospects. *Heliyon* **2022**, *8*, e09041. [CrossRef] [PubMed]
9. Beckers, D.; Ellendt, N.; Fritsching, U.; Uhlenwinkel, V. Impact of process flow conditions on particle morphology in metal powder production via gas atomization. *Adv. Powder Technol.* **2020**, *31*, 300–311. [CrossRef]
10. Champagne, V. *The Cold Spray Materials Deposition Process: Fundamentals and Applications*; Woodhead Publishing Limited: Thorston, UK, 2007.
11. Agboola, O.; Babatunde, D.E.; Fayomi, O.S.I.; Sadiku, E.R.; Popoola, P.; Moropeng, L.; Yahaya, A.; Mamudu, O.A. A review on the impact of mining operation: Monitoring, assessment and management. In *Results in Engineering*; Elsevier B.V.: Amsterdam, The Netherlands, 2020; Volume 8. [CrossRef]
12. Gleason, M.A.; Sousa, B.C.; Tsaknopoulos, K.; Grubbs, J.A.; Hay, J.; Nardi, A.; Brown, C.A.; Cote, D.L. Application of Mass Finishing for Surface Modification of Copper Cold Sprayed Material Consolidations. *Materials* **2022**, *15*, 2054. [CrossRef] [PubMed]
13. Tsaknopoulos, K.; Grubbs, J.; Sousa, B.C.; Siopis, M.; Nardi, A.; Cote, D.L. Evaluation of a laser powder bed fusion designer Al-Mg-Zr-Si alloy for cold spray additive manufacturing. *Mater. Des.* **2022**, *222*, 111105. [CrossRef]
14. Assadi, H.; Gärtner, F. Particle compression test: A key step towards tailoring of feedstock powder for cold spraying. *Coatings* **2020**, *10*, 458. [CrossRef]
15. *JIS R 1639-5*; Test Methods of Properties of Fine Ceramic Granules—Part 5: Compressive Strength of A Single Granule. Japanese Standards Association: Tokyo, Japan, 2007.
16. Sousa, B.C.; Hay, J.; Cote, D.L. Toward an Instrumented Strength Microprobe—Origins of the Oliver-Pharr Method and Continued Advancements in Nanoindentation: Part 1. In *Elasticity of Materials*; IntechOpen: London, UK, 2023. [CrossRef]
17. Sousa, B.C.; Aindow, M.; Lee, S.-W.; Lados, D.; Spangenberger, A.G.; Sample, C.M.; Cote, D.L. Characterization of cold-sprayed material consolidations. In *Advances in Cold Spray: A Coating Deposition and Additive Manufacturing Process*; Elsevier: Amsterdam, The Netherlands, 2023; pp. 205–298. [CrossRef]
18. Sousa, B.C.; Tsaknopoulos, K.; Cote, D.L. Microstructural and Micromechanical Characterization of Feedstock for Cold Spray Additive Manufacturing and Processing. *Powders* **2022**, *1*, 129–154. [CrossRef]
19. *ASTM F1580-18*; Standard Specification for Titanium and Titanium-6 Aluminum-4 Vanadium Alloy Powders for Coatings of Surgical Implants. ASTM International: West Conshohocken, PA, USA, 2019. [CrossRef]
20. Grubbs, J. *Exploring the Impact of Metallic Powder Quality on Cold Spray Processing and Consolidated Material Performance*; Worcester Polytechnic Institute: Worcester, UK, 2023.
21. Grubbs, J.; Sousa, B.C.; Cote, D. Exploration of the Effects of Metallic Powder Handling and Storage Conditions on Flowability and Moisture Content for Additive Manufacturing Applications. *Metals* **2022**, *12*, 603. [CrossRef]
22. Bieler, T.R.; Trevino, R.M.; Zeng, L. Alloys: Titanium. In *Encyclopedia of Condensed Matter Physics*; Elsevier: Amsterdam, The Netherlands, 2005.
23. Poirier, D.; Thomas, Y.; Guerreiro, B.; Martin, M.; Aghasibeig, M.; Irissou, E. Improvement of Tool Steel Powder Cold Sprayability Via Softening and Agglomeration Heat Treatments. *J. Therm. Spray Technol.* **2022**, *31*, 145–158. [CrossRef]
24. Sousa, B.C.; Gleason, M.A.; Haddad, B.; Champagne, V.K.; Nardi, A.T.; Cote, D.L. Nanomechanical characterization for cold spray: From feedstock to consolidated material properties. *Metals* **2020**, *10*, 1195. [CrossRef]
25. Haden, C.V.; Collins, P.C.; Harlow, D.G. Yield Strength Prediction of Titanium Alloys. *JOM* **2015**, *67*, 1357–1361. [CrossRef]
26. Shunmugavel, M.; Polishetty, A.; Littlefair, G. Microstructure and Mechanical Properties of Wrought and Additive Manufactured Ti-6Al-4V Cylindrical Bars. *Procedia Technol.* **2015**, *20*, 231–236. [CrossRef]
27. Sousa, B.C.; Walde, C.E.; Champagne, V.K.; Cote, D.L. Initial Observation of Grain Orientation Dependent Nanoindentation Hardness of Al 6061 Gas-atomized Powder. *Int. J. Metall. Met. Phys.* **2020**, *5*, 50. [CrossRef]
28. Kostryzhev, A.G. Strengthening mechanisms in metallic materials. *Metals* **2021**, *11*, 1134. [CrossRef]

Disclaimer/Publisher's Note: The statements, opinions and data contained in all publications are solely those of the individual author(s) and contributor(s) and not of MDPI and/or the editor(s). MDPI and/or the editor(s) disclaim responsibility for any injury to people or property resulting from any ideas, methods, instructions or products referred to in the content.

materials

MDPI

Article

Characterisation of Adobe and Mud–Straw for the Restoration and Rehabilitation of Persian Historical Adobe Buildings

Bina Hejazi *, Corinna Luz, Friedrich Grüner, Jürgen Frick and Harald Garrecht

Materials Testing Institute (MPA), University of Stuttgart, 70569 Stuttgart, Germany;
corinna.luz@mpa.uni-stuttgart.de (C.L.); friedrich.gruener@mpa.uni-stuttgart.de (F.G.);
juergen.frick@mpa.uni-stuttgart.de (J.F.); harald.garrecht@mpa.uni-stuttgart.de (H.G.)
* Correspondence: bina.hejazi@mpa.uni-stuttgart.de

Abstract: In the restoration or rehabilitation of traditional buildings, compatible materials with known characteristics must be used. However, the existing literature lacks comprehensive studies on the characterisation of Persian mud–straw plaster, focusing primarily on Persian adobe. Moreover, previous research on Persian adobe has primarily employed XRF and XRD tests, neglecting ion chromatography, moisture sorption isotherm determination, and thermogravimetric analysis with differential scanning calorimetry. Consequently, there is a shortage of information regarding the elemental composition, mineralogical characteristics, moisture sorption behaviour, and thermal properties of Persian mud–straw plaster, as well as Persian adobe bricks. This paper aims to address this research gap by examining historical and new adobe bricks and mud–straw plaster used in Iran, utilising a comprehensive array of analytical techniques. The results from XRF analysis reveal relatively similar chemical compositions across all samples, while XRD analysis indicates predominantly similar mineral phases. Ion chromatography results demonstrate higher conductivity and chloride concentrations in the mud–straw samples than the adobe samples, with higher values for new samples than historical ones. Freshly used straw, clay, or soil may have higher chloride concentrations caused by the arid climate and soil salinisation in the area. Additionally, moisture sorption isotherm determination results show that adobe and mud–straw plaster with a higher salt load of chlorides have significantly higher moisture absorption. The increased straw quantity in the samples increases the moisture content. Furthermore, thermogravimetric analysis and differential scanning calorimetry indicate that, at low heating, adobe and mud–straw plaster lose water due to dehydration, and at high heating, they lose carbon dioxide due to decarboxylation. The comprehensive characterisation of Persian adobe and mud–straw plaster in this study fills a significant gap in the literature and offers invaluable insights for informing restoration and rehabilitation processes, ensuring the compatibility of the materials used.

Keywords: adobe; mud–straw plaster; ion chromatography; X-ray fluorescence; X-ray diffraction; sorption isotherm; thermogravimetric analysis; differential scanning calorimetry

Citation: Hejazi, B.; Luz, C.; Grüner, F.; Frick, J.; Garrecht, H. Characterisation of Adobe and Mud–Straw for the Restoration and Rehabilitation of Persian Historical Adobe Buildings. *Materials* **2024**, *17*, 1764. https://doi.org/10.3390/ma17081764

Academic Editors: Pavol Liptai, Jaroslav Briančin and Maroš Halama

Received: 28 February 2024
Revised: 27 March 2024
Accepted: 9 April 2024
Published: 11 April 2024

Copyright: © 2024 by the authors. Licensee MDPI, Basel, Switzerland. This article is an open access article distributed under the terms and conditions of the Creative Commons Attribution (CC BY) license (https://creativecommons.org/licenses/by/4.0/).

1. Introduction

The construction materials used as the components and main elements of buildings have historically required extensive knowledge, study, and research. Advances in the construction of new materials necessitate a fundamental understanding of materials from the past. Adobe stands as one of the oldest construction materials in the world, renowned for its specific characteristics, such as distinguished hygrothermal properties, affordability, low energy consumption for production, environmental compatibility, and high resistance to heat and fire [1].

Traditional plaster, commonly used in adobe buildings, is mud–straw made from clayed earth and wheat straw mixed with water. Apart from its primary role in protecting the substrate, mud–straw plaster significantly influences the hygrothermal performance of the building, enhancing indoor air quality and comfort [2].

For traditional buildings, which often require continuous maintenance and periodic repair, restoration, or rehabilitation, any intervention plan involving materials must prioritise compatibility with the original materials. Different forms of compatibility, including chemical, physical, mechanical, rheological, and thermal compatibilities, must be carefully considered [3].

Statistics indicate the global prevalence of adobe and earth-based housing, emphasising the importance of preserving this longstanding tradition. In Iran alone, a significant portion of housing units are constructed using clay and mud materials, underlining the need for attention to earthen architecture on both historical and contemporary fronts [4].

Today, there is a growing trend towards strengthening and constructing earthen/adobe buildings, with the use of soil in architecture finding increasing applications not only in developing countries but also in developed regions such as the United States and various European countries [5].

In light of this, comprehensive knowledge of earthen materials and an understanding of their advantages and disadvantages are imperative [6].

Despite the historical significance and widespread use of adobe and mud–straw plaster, there remains a notable gap in the literature regarding the comprehensive characterisation of these materials that can be used in the restoration and rehabilitation of adobe buildings, particularly in the context of Persian architecture. Previous studies on Persian earthen materials have focused primarily on adobe bricks, with limited exploration of mud–straw plaster. Furthermore, the existing research has predominantly utilised XRF and XRD analyses, neglecting other critical aspects, such as ion chromatography, moisture sorption behaviour, and TGA-DSC analyses.

This paper aims to bridge this gap by conducting a thorough investigation into the characterisation of Persian adobe and mud–straw plaster for restoration and rehabilitation purposes. By exploring elemental composition, mineralogical characteristics, moisture sorption behaviour, and response to thermal analysis through techniques like TGA-DSC of both historical and new adobe bricks and mud–straw plaster used in Iran, this study contributes to a deeper understanding of these traditional building materials and their suitability for restoration and rehabilitation projects.

The knowledge gained from XRF and XRD analyses provides insights into the elemental and mineralogical composition of adobe and mud–straw plaster, which are crucial for understanding their structural integrity and durability over time. Additionally, ion chromatography can offer valuable information about the presence of soluble salts and ions within these materials, which can affect their long-term stability and compatibility with restoration interventions.

Understanding the moisture sorption behaviour of adobe and mud–straw plaster is essential for assessing their response to environmental conditions and potential risks of decay, such as efflorescence or salt crystallisation. This information can inform appropriate conservation strategies to mitigate moisture-related damage and ensure the long-term preservation of historical structures.

Furthermore, the knowledge gained from TGA-DSC analyses can significantly support restoration and rehabilitation interventions for adobe and mud–straw plaster in several ways. Firstly, TGA-DSC tests provide insights into the thermal decomposition behaviour of these materials, which is crucial for understanding how they respond to heat and fire exposure. This information can inform decisions related to fire safety measures and the selection of appropriate fire-retardant treatments during restoration efforts. Additionally, TGA-DSC analyses can help identify any organic and inorganic additives present in the materials, which may influence their structural integrity, durability, and susceptibility to degradation over time. Understanding the composition and thermal characteristics of adobe and mud–straw plaster through TGA-DSC testing allows for informed decision-making regarding material selection, conservation treatments, and preservation strategies, ultimately contributing to the long-term sustainability and resilience of historical adobe structures.

Therefore, the comprehensive characterisation of adobe and mud–straw plaster through techniques such as XRF, XRD, ion chromatography, and the assessment of moisture sorption and TGA-DSC analyses enhances the understanding of these traditional building materials and facilitates informed decision-making in restoration and rehabilitation projects aimed at preserving Persian architectural heritage.

Figure 1 illustrates the traditional method of producing adobe bricks, hereafter referred to as adobe in this paper, in the Bam Citadel in Iran in 2007 during its restoration after the earthquake of 2003. To produce adobe, sand is mixed with clay. In some cases, fibrous materials are added to the mixture to prevent cracking during drying, increase tensile strength, enhance thermal insulation, or reduce weight. Fibrous materials include straw or the droppings of herbivores, such as goats, cows, horses, and camels. In Iran, adobe is generally made of clay soil containing 30% to 40% clay and 60% to 70% earth mixed with water, comprising approximately 25% of the weight of the dry soil. Typically, wheat straw is added to the mixture of clay and earth, constituting approximately 0.5% of the weight of dry soil [7] (Table 1). As shown in Figure 1, adobe can be made simply by forming the mixture and allowing it to dry in the sun. Water is added to clay and earth, left for one or two days to allow the earth mixture to soak in water to activate the clay, and mixed by a shovel or by feet, and then by hand. The mixture is then forged into moulds. After one or two days, the adobe units are placed vertically for faster drying. The dry adobe units are then collected together to be used.

(a) (b) (c) (d) (e) (f)

Figure 1. Adobe production: (**a**) adding water to soil and leaving for one or two days, (**b**) mixing adobe components using a shovel, (**c**) mixing adobe components by hand, (**d**) forging in the mould, (**e**) trimming and placing the adobe units vertically for drying faster, (**f**) dried adobe units ready to be used in construction, Bam citadel, 2007.

One of the most commonly used traditional plasters for covering adobe walls or roofs is mud–straw plaster, hereafter referred to as mud–straw in this paper. It is used as a thermal and moisture insulation layer. Adobe walls and roofs are protected by being plastered with a 2.5 cm to 3 cm thick layer of mud–straw. Mud–straw plaster is composed of clay, sand, water, and straw, typically wheat straw or grass. This plaster adheres to the adobe unit because both are made of the same materials. The straw content in mud–straw typically varies between 5% and 10% by weight. The water content for the mixture is about 25% of the weight of the clay soil (Table 1). Straw is cut into 5 cm lengths and added to the plaster. Figure 2 shows the phases of producing mud–straw plaster. Clay soil, straw, and water are mixed and left for two to four days. Then, the mud–straw is mixed again using a shovel or feet and placed on the surface of the adobe wall or roof to protect it.

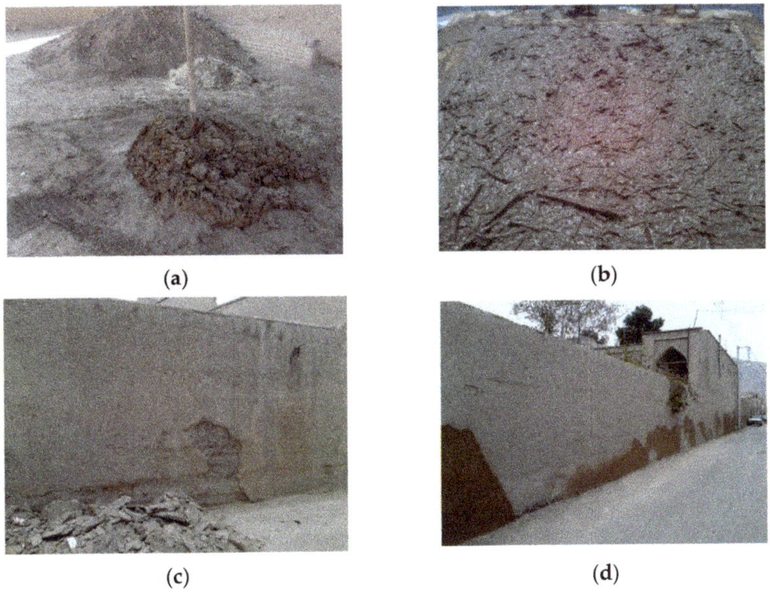

Figure 2. Mud–straw plaster production and usage: (**a**) mixing clay soil, straw, and water, (**b**) leaving mud–straw mix for one or two days, (**c**) removing the old mud–straw plaster, (**d**) plastering the adobe wall with new mud–straw plaster, Soukias adobe house (seventeenth century), Art University of Isfahan, Isfahan, 2013.

Table 1. Quantities of clay soil, adobe brick, and mud–straw plaster components [7].

Material	Clay Soil Components		Adobe Brick and Mud–Straw Plaster Components		
	Clay Content (wt%)	Earth Content (wt%)	Clay Soil Content (wt%)	Straw Content (wt%)	Water to Clay Soil Ratio (%)
Adobe brick	30–40	60–70	99.5	0.5	25
Mud–straw plaster	30–40	60–70	90–95	5–10	25

This paper investigates the characterisation of Persian adobe bricks and mud–straw plaster for rehabilitating historical adobe buildings. To achieve this, specific historical adobe buildings with original materials from the same location were selected, and their samples were analysed. The samples consisted of two original historical adobe bricks, one new adobe brick used for restoration and rehabilitation, one sample of mud–straw plaster produced two decades ago, and another sample of mud–straw plaster produced in the laboratory using materials sourced from the same region.

After providing an introduction to the significance of these traditional building materials and their widespread use, the paper proceeds to a comprehensive literature review. Following this, the methodology section details the equipment used, sample preparation procedures, and analytical methods employed, including ion chromatography, XRD analysis, XRF analysis, determination of sorption isotherms, and thermogravimetric analysis with differential scanning calorimetry. The subsequent sections present the results obtained from these analyses, discussing the ion chromatography results, XRF and XRD tests, sorption isotherm results, and TGA/DSC results in detail. The discussion section further interprets these results, highlighting key findings and their implications. Finally, the paper concludes with a summary of the findings and their relevance to the rehabilitation of Persian adobe and mud–straw materials.

2. Literature Review

In selecting the previous research projects referenced in this section, careful consideration is given to their relevance to the research gap, scope, and objectives of this study. The first part of the literature review focuses on research worldwide concerning adobe and clay-based plasters, aiming to provide a comprehensive overview of studies relevant to traditional building materials. The second part specifically examines research conducted on Persian adobe, seeking to identify gaps in knowledge regarding this particular material. Each cited work is chosen based on its contribution to the understanding of elemental composition, mineralogical characteristics, moisture sorption behaviours, and thermal properties of adobe and clay-based plaster. By examining previous research endeavours, the aim is to establish a foundation for the study, identifying areas where gaps in knowledge persist and where the investigation can make a meaningful contribution.

Studying the characteristics of adobe and clay-based plasters has been the focus of a number of researchers worldwide. For adobe bricks, Coffman et al. [8] conducted a mineralogical study of adobe bricks from many historical adobe buildings in different parts of the world. They determined mineral composition, including clay type and quantity and soil granulation for adobe samples. They studied the effect of two chemical consolidants, an alkoxysilane and an isocyanate, on adobe samples and found that chemical consolidation depends very much on clay mineralogy and grain-size distribution.

Baglioni et al. [9] conducted experimental tests on rammed earth and adobe used in the Drâa Valley, Morocco, in order to determine the mineralogical and mechanical properties of the materials. They found that the earth used for adobe is richer in clay compared to that used for rammed earth. Both types of earth showed high cohesion and low expandability, making them quite suitable for construction. They contained quartz, feldspars, and a small amount of expandable clay minerals like smectite, which increase plasticity and cohesion but can also lead to shrinkage problems. Old adobe was more clayey than new adobe and, therefore, more suitable for construction.

Fratini et al. [10] studied the mineralogical, physical, and mechanical properties of adobe samples from seven different buildings in the historical centre of Lamezia Terme, Italy. They showed that adobe samples contained quartz, feldspars, clay minerals, and accessory minerals (micas and iron hydroxides). In most of the samples, the clay minerals were constituted by minerals, such as illite and kaolinite, that do not have swelling behaviour. Only in a few samples, clay minerals with swelling characteristics, such as chlorite-vermiculite and smectite, existed. Only one sample had a significant amount of calcite.

Costa et al. [11] studied the influence of the mineralogical composition on the properties of adobe bricks from various buildings in Aveiro, Portugal. Adobe samples were characterised by their mineralogical composition, soil granulation, water uptake, durability, and mechanical strength. Kaolinite, illite, and smectite were the main clay mineralogical compositions. They concluded that the phyllosilicate content, in particular kaolinite abundance, had a positive effect on the absorption/drying and mechanical properties of adobe bricks.

El Fgaier et al. [12] performed a hygroscopic analysis of sorption and desorption isotherms and studied the effect of sorption capacity on the thermo-mechanical properties of three types of unfired clay bricks industrially produced in northern France. The studied samples had higher sorption capacity than other construction materials, such as fire clay bricks. At 95% relative humidity, the moisture content of unfired clay bricks reached 3.5%. This phenomenon would make earthen materials able to balance the indoor climate by releasing or adsorbing moisture according to changes in the relative humidity of the ambient air. However, by increasing the relative humidity, the thermal capacity increased due to the penetration of water vapour and by acting as thermal bridges.

De Castrillo et al. [13] compared pre-historical and nineteenth- and twentieth-century traditional and contemporary adobe bricks in Cyprus. Tests were performed for physical and mineralogical characterisation of raw materials used in the production of adobe bricks. According to the chemical analysis and XRF test, the dominant chemical elements in adobe samples were calcium, silicon, iron, and aluminium. Calcium had the highest concentration, which could be related to the addition of lime during the mix design in order to improve the cohesion of the adobe brick. Silicon dioxide concentration was the second dominant element, attributed to the existence of quartz in the raw material for making adobe. The XRD test was consistent with the XRF results, indicating calcite, quartz, and albite as the dominant minerals in all adobe samples. Some adobe samples had high amounts of gypsum attributed to the gypsiferous soils of the region. In general, irrespective of the period, calcite, quartz, and albite were the dominant minerals in all adobe samples. Some certain differences in the chemical composition of the adobe samples, such as high amounts of gypsum in some samples, were attributed to the type of soil of the region.

Ashour et al. [14] determined the equilibrium moisture content (EMC) of mud–straw plaster. The earth used had four different compositions of cohesive soil and sand. Three types of fibre materials, wheat straw, barley straw, and wood shavings, were added to the soil. The results show that the EMC increases with an increase in relative humidity and decreases with an increase in temperature. Additionally, the effect of relative humidity on the EMC is more pronounced than that of temperature. For a relative humidity of 43%, the EMC for different samples was approximately 1.1% to 3.6% at 10 °C and about 0.8% to 2.9% at 40 °C. By increasing the humidity to 95%, the EMC increased to approximately 2.3% to 5.5% at 10 °C and 1.9% to 4.8% at 40 °C.

Costa et al. [15] characterised adobe samples taken from the central coastal region of Portugal. Adobe samples were divided into four groups according to their mineralogical composition: lime-stabilised adobe with a high percentage of calcite, solid adobe with a high amount of phyllosilicates, medium adobe with similar quartz and calcite contents, and iron-rich adobe with higher values of iron oxides and hydroxides than the others. In general, adobe samples consisted mainly of quartz, calcite, and phyllosilicates. Quartz, calcite, phyllosilicates, and K-felspars were the main minerals of the silt–clay fraction of the adobe samples. Iron oxides and hydroxides dominated the accessory minerals. The phyllosilicates identified were clay minerals like kaolinite, illite, smectite, chlorite, vermiculite, and illite–smectite. The mineralogical composition of the adobe samples was mainly composed of silica, calcium, and aluminium.

Sanchez-Calvillo et al. [16] characterised adobe samples from damaged buildings in Jojutla de Juarez, Mexico. Mineralogical and granulometry analyses were performed. Calcite was present in most of the adobe samples. Kaolinite with high proportions was observed in a number of adobe samples. The cause of the use of kaolinite in producing adobe bricks was attributed to its low swelling and shrinkage properties.

Laborel-Préneron et al. [17] studied the hygrothermal characteristics of earthen materials containing a large volume, about 40% of the volume corresponding to about 6% of weight, of the plant aggregates of barley straw, hemp shiv, or corn cob. The results show a low thermal conductivity with this large volume of plant aggregates. The water vapour permeability did not improve. The sorption–desorption isotherms indicated that the sorption capacity improved only slightly due to the low plant matter mass.

Gomes [18] investigated the hygrothermal characteristics of earthen materials. The sorption and desorption curves were obtained at three oven-drying temperatures. The maximum water content ranged from 3% to 5%.

Mellaikhafi et al. [19] performed physical–chemical, mineralogical, and geotechnical tests on five types of soils used for making rammed earth and adobe in an oasis of south-eastern Morocco. The main minerals in the soils were quartz, calcite, ferroan, clinochlore, and muscovite. Two clay minerals, kaolinite and illite, were absent. Two minerals, smectite and vermiculite, with swelling properties, were also absent.

For earth-based plaster, Ranesi et al. [20] studied the relative humidity-related properties of five plastering mortars with clay, air lime, and natural hydraulic lime bases and three finishing pastes with gypsum and gypsum–air lime bases, in addition to cement-based plaster for comparison. The earth plaster was the most appropriate plaster for relative humidity passive regulation and showed high hygroscopicity. Thereafter, the combination of gypsum and air lime was more suitable than the pure gypsum paste, which showed a very low moisture capacity. The adsorption/desorption of the natural hydraulic lime mortar was moderately good. Hydrated air lime plaster had the lowest adsorption and desorption. The cement plaster was the least suitable, with the lowest water vapour permeability and slow adsorption and desorption.

Lima et al. [21] examined the effect of clay mineralogy on earth plaster properties. Three mortars were produced using different clayish earths to assess their influence. The results show that clay mineralogy significantly impacted plaster properties, such as vapour adsorption, drying shrinkage, mechanical strength, dry abrasion, and thermal conductivity. Illitic clayish earth exhibited balanced properties, making it suitable for earth-based plasters.

Savadogo et al. [22] investigated the physico-mechanical and durability properties of earthen plaster stabilised with fermented rice husk. Various mixtures were tested, revealing improvements in properties with the addition of fermented rice husk. However, excessive rice-husk content negatively affected water absorption and erosion resistance, suggesting an optimal ratio for enhanced performance.

Santos et al. [23] evaluated the efficiency of earth plaster on different masonry types. A commercial unstabilised earth mortar was used to plaster experimental masonry walls, demonstrating durability across various substrates. Despite initial variations, earth plaster exhibited long-term stability, proving its technical efficiency across historical and contemporary masonry structures.

Vares et al. [24] assessed the hygrothermal performance of clay–sand plaster. Different covering materials were tested to evaluate the moisture buffering and vapour-permeability properties. The results indicate variations in moisture uptake and diffusion, emphasising the importance of material selection for indoor climate control.

Santos and Faria [25] characterised various earthen plasters using laboratory and in-situ tests. Different mortar formulations were evaluated for mechanical properties and durability. The results highlight the influence of additives on plaster performance, with some formulations exhibiting improved adhesive strength and mechanical properties.

Santos et al. [26] conducted a comparative analysis of the mineralogical, mechanical, and hygroscopic properties of earthen, gypsum, and cement-based plasters. Five different plastering mortars were examined, including unstabilised and stabilised earth-based plasters, as well as gypsum and cement-based pre-mixed plasters. While earthen mortars exhibited lower mechanical strength compared to gypsum and cement-based mortars, they demonstrated the highest hygroscopicity, functioning as passive moisture buffers.

Ojo et al. [27] explored the characteristics of unfired earthen building materials using muscovite-rich soils and alkali activators. They discovered that alkali activation significantly improved the physical and mechanical properties of these materials, suggesting a potential for sustainable low-cost housing solutions.

Bass et al. [28] examined ancient earthen plaster from three Native American sites in the American Southwest. Their analysis studied the microstructure, mineral composition, and deterioration mechanisms. Their research reveals how locally sourced plaster was tailored to suit specific site conditions and functional requirements, highlighting the adaptability of ancient plaster technologies.

Muşkara and Bozbaş [29] characterised earthen building materials in vernacular houses in North-West Turkey, focusing on the Saraylı, Örcün, and Selimiye villages. They conducted archaeometric investigations on samples from earthen bricks, mud plasters, and mortars, analysing their mineralogical and chemical properties. The study aimed to understand building technologies and raw material properties for developing restoration strategies aligned with sustainable architecture principles.

Saleh [30] investigated natural adobe plaster in the Jordan Valley, using advanced techniques to analyse plaster morphology. The results reveal significant variance in plaster recipes across different regions, suggesting local soil differences or additive variations. Additionally, it was found that regardless of local recipes, three layers were typically used for exterior wall protection, each with different ratios of earth materials.

Saleh [31] conducted an experimental campaign to characterise wall plaster in Al-Ulla, Saudi Arabia. Various methods, including visual inspections, microscopic analysis, advanced imaging techniques, and mineralogical analysis, were employed. Four types of plaster were identified through the experiments, which facilitatedthe accurate diagnosis of conservation strategies for Qarh's monuments at Al-Ulla.

Silveira et al. [32] conducted studies on rehabilitating a cultural heritage district with traditional adobe constructions in the Aveiro district, Portugal. Their research efforts focused on characterising adobe's composition, mechanical behaviour, and structural performance to guide rehabilitation practices effectively. Based on their findings, they suggested specific methods for improving adobe composition to enhance durability and seismic resistance. Additionally, they provided recommendations for structural reinforcement techniques tailored to the unique characteristics of traditional adobe constructions.

Sánchez et al. [33] investigated various physical and mechanical properties of adobe for the rehabilitation of adobe buildings. They conducted a comprehensive review of experimental studies spanning the last 15 years from different countries to examine the mechanical properties of adobe masonry. By comparing the results of these studies, correlations between different physical and mechanical properties were established and compiled in their document, providing valuable insights for adobe building rehabilitation efforts. However, their review revealed significant variations in regulations to accommodate local conditions, highlighting the need for adaptable guidelines in the field.

Papayianni and Pachta [34] studied the consolidation and upgrading of historical earth block masonry constructions. They presented a methodology for analysing the building materials and techniques of historical earth block houses in northern Greece, alongside the design and testing of compatible repair materials for their rehabilitation. Laboratory tests were conducted on historical materials to determine their microstructural, physico-mechanical, and chemical properties. The results informed the development of compatible repair materials based on earth, including soil-based grouts with enhanced properties through specific additives and admixtures.

González-Sánchez et al. [35] investigated stabilised earthen mixtures to preserve traditional adobe buildings, focusing on enhancing durability for rehabilitation. Their study evaluated an earthen mixture with a vegetal origin gel from rice starch, aiming to improve mechanical strength and waterproofing. Divided into experimental and application phases, the research first developed eco-friendly mixtures and then applied them in adobe dwellings in Santa Ana Chapitiro, Michoacan, Mexico, through participatory design. Initial observations suggest promising early stage performance.

Faria et al. [36] conducted an experimental characterisation of an earth plastering mortar for the rehabilitation of adobe buildings. An extended experimental campaign was developed to assess multiple properties of a ready-mixed earth plastering mortar and to increase scientific knowledge of the influence of test procedures on those properties. The plaster satisfied the requirements of the existing German standard and seemed adequate for application as rehabilitation plaster on historical adobe buildings.

Gomes et al. [37] investigated the compatibility of earth-based repair mortars with rammed earth substrates. The study analysed the performance of eight repair mortars formulated with earth collected from rammed earth buildings in South Portugal or a commercial type of earth and tested with four types of binders. Mortars were applied on two standard defects on rammed earth blocks, representing common issues on exterior surfaces. The study evaluated mortar performance, substrate compatibility, and visual effectiveness of the intervention. The results show variations in mortar behaviour based on the type of rammed earth support, with unstabilised earth mortars demonstrating the best performance compared to their stabilised counterparts.

Jia et al. [38] analysed historical earthen plaster to enhance its characteristics for restoration. They investigated the effects of different material compositions on shrinkage and cracking properties through laboratory tests on plaster specimens. The results show that adding sand and vegetal fibres improved plaster properties by increasing the shrinkage limit and inhibiting volume shrinkage. However, wheat straw led to surface cracking, while calcined ginger nut paste displayed potential for plaster restoration with minimal shrinkage and surface cracking. The study aimed to understand the scientific nature of earthen plaster and its behaviour during desiccation.

Mattone et al. [39] conducted experimental tests on earth–gypsum plasters for the conservation and rehabilitation of earthen constructions. The study aimed to develop effective materials to protect earthen architectural heritage from weathering, considering the significance of earthen architecture worldwide. By mixing earth and gypsum with natural or synthetic additives, the research sought to design plasters capable of withstanding atmospheric agents prevalent in various locations where these constructions are situated.

A few investigations have been performed on the characteristics of Persian adobe. Hosseini et al. [40] conducted mineralogical and physical tests on two historical and two new adobe samples from Belqeis Castle in Esfarayen, North-East Iran. The results show that a part of the minerals included calcite and quartz, indicating a lack of clay minerals. In addition, the majority of minerals were silicates (muscovite, biotite, feldspar, and enstatite).

Zakavi [41] studied the soil from six mines near the Choga Zambil in Susa, South-East Iran, for making new adobe bricks. Calcite and quartz were the dominant minerals in the soil samples. The soil samples were poor in terms of high-quality clay minerals such as kaolinite and montmorillonite. The amount of chlorine and sulphate ions in four soil samples was high. Soil samples containing sodium and potassium chloride showed a higher potential for swelling of the clay.

Dormohamadi and Rahimnia [42] studied dynamic compaction on the mechanical behaviour of adobe bricks made with six different clayey–silty soil types from six different mines in the town of Ardakan, near Yazd, Central Iran. The experiments included the determination of physical, mineralogical, and chemical characterisation, as well as mechanical properties tests. The results show that the main component of all six soil samples was a mixture of silicates and calcium aluminates (feldspar) resulting from the combination of silica with lime, alumina plus magnesium oxide, and iron oxide. In all of them, the amount of silica was approximately one-third of the weight of the soil. This is due to the high amount of aeolian sand in this desert region of Iran. The amount of alumina was less than the amount of silica in all samples. The four main phases of mineralogical composition, i.e., quartz, calcite, feldspars, and clay minerals, as well as traces of dolomite were present in all soil types. The soils were rich in silt and silica due to the high amount of aeolian sand, again a characteristic of the soil of the region. The amount of swelling clay minerals, like smectite, in the soil was low.

Eskandari [43] studied the physical, mineralogical, and mechanical properties of Persian historical and new adobe bricks. Six groups of adobe bricks, consisting of three groups of historical adobe bricks from the town of Maybod, near Yazd, and from the town of Jarquyeh, near Isfahan, and three groups of new adobe bricks from Maybod and Yazd, Central Iran, were tested. The results show that quartz was a significant amount of the constituent minerals. These results are also consistent with the geological characteristics of the soil formation in the desert areas of the studied adobe bricks and the presence of abundant wind sand in those areas. Calcite was the second most abundant mineral in the soil of each study group, which is due to the characteristics of arid and semi-arid regions. Albite was the third most abundant mineral among the studied adobe bricks groups, which provides relatively good resistance to adobe bricks.

Table 2 summarises the important findings based on characterising the adobe and clay-based plaster samples.

Table 2. Summary of key findings from literature review on adobe and clay-based plasters.

Researcher	Focus	Key Findings
Coffman et al. [8]	Mineralogical study of adobe bricks from historical buildings worldwide	Determined mineral composition and grain-size distribution of adobe bricks; found consolidation effectiveness dependent on clay mineralogy and grain size.
Baglioni et al. [9]	Properties of rammed earth and adobe in Morocco	Found higher clay content in adobe compared to rammed earth; identified mineral composition, including quartz, feldspars, and expandable clay minerals; concluded old adobe more suitable for construction due to higher clay content.
Fratini et al. [10]	Properties of adobe in Italy	Identified mineral composition including quartz, feldspars, and clay minerals; observed variations in clay mineral types; noted significance of calcite in some samples.
Costa et al. [11]	Influence of mineral composition on adobe properties in Portugal	Found kaolinite, illite, and smectite as main clay minerals; noted positive effect of phyllosilicate content on absorption/drying and mechanical properties of adobe bricks.
El Fgaier et al. [12]	Hygroscopic properties of unfired clay bricks in France	Showed higher sorption capacity compared to fired clay bricks; observed moisture content increase with relative humidity; highlighted potential for indoor climate regulation.
De Castrillo et al. [13]	Comparison of adobe bricks across historical periods in Cyprus	Identified dominant minerals, including calcite, quartz, and albite; noted variations in chemical composition attributed to regional soil differences.
Ashour et al. [14]	Equilibrium moisture content of mud–straw plaster in unspecified location	Observed increase in equilibrium moisture content with humidity and decrease with temperature; found relative humidity has more pronounced effect than temperature.
Costa et al. [15]	Mineralogical characterisation of adobe in central Portugal	Identified four groups of adobe based on mineral composition; quartz, calcite, and phyllosilicates predominant; variations in iron-oxide content noted.
Sanchez-Calvillo et al. [16]	Characterisation of adobe from Jojutla de Juarez, Mexico	Identified calcite and kaolinite as major minerals; noted low swelling and shrinkage properties of kaolinite.
Laborel-Préneron et al. [17]	Hygrothermal properties of earthen materials with plant aggregates	Showed low thermal conductivity with a large volume of plant aggregates; minimal improvement in water vapour permeability observed.
Gomes [18]	Hygrothermal characteristics of earthen materials in unspecified location	Determined sorption and desorption curves; observed maximum water content of 3–5%.

Table 2. *Cont.*

Researcher	Focus	Key Findings
Mellaikhafi et al. [19]	Physical, chemical, and geotechnical tests of soils for rammed earth and adobe in Morocco	Identified quartz, calcite, and ferroan clinochlore as main minerals; absent kaolinite and illite; observed absence of swelling minerals, such as smectite and vermiculite.
Ranesi et al. [20]	Relative humidity properties of plastering mortars in unspecified location	Identified earth plaster as most suitable for humidity regulation; gypsum–air lime combination found more effective than pure gypsum.
Lima et al. [21]	Effect of clay mineralogy on earth plaster properties in unspecified location	Found significant impact of clay mineralogy on plaster properties, including vapour adsorption, drying shrinkage, and mechanical strength.
Savadogo et al. [22]	Physico-mechanical properties of earthen plaster stabilised with fermented rice husk	Observed improvements in properties with addition of rice husk; noted negative impact on water absorption and erosion resistance with excessive rice-husk content.
Santos et al. [23]	Efficiency of earth plaster on various masonry types	Demonstrated durability of earth plaster across different substrates; exhibited long-term stability.
Vares et al. [24]	Hygrothermal performance of clay–sand plaster in unspecified location	Evaluated moisture buffering and vapour-permeability properties; emphasised importance of material selection for indoor climate control.
Santos and Faria [25]	Characterisation of earthen plasters in unspecified location	Evaluated mechanical properties and durability of various mortar formulations; noted influence of additives on plaster performance.
Santos et al. [26]	Comparative analysis of mineralogical and hygroscopic properties of plasters in unspecified location	Examined mechanical and hygroscopic properties of different plasters; noted lower mechanical strength but higher hygroscopicity in earthen plasters.
Ojo et al. [27]	Characteristics of unfired earthen materials in unspecified location	Investigated characteristics and properties of earthen materials; found alkali activation improved physical and mechanical properties.
Bass et al. [28]	Microstructure and mineral composition of ancient earthen plaster in American Southwest	Studied microstructure, minerals, and deterioration mechanisms; highlighted adaptability of plaster to site conditions.
Muşkara and Bozbaş [29]	Archaeometric investigations of earthen building materials in Turkey	Analysed mineralogical and chemical properties of earthen materials from vernacular houses; aimed to understand building technologies and raw material properties for restoration strategies.
Saleh [30]	Analysis of natural adobe plaster in Jordan Valley	Analysed plaster morphology; found significant variance in plaster recipes across regions.
Saleh [31]	Experimental characterisation of wall plaster in Al-Ulla, Saudi Arabia	Characterised wall plaster using various methods; identified four types of plaster.

Table 2. *Cont.*

Researcher	Focus	Key Findings
Silveira et al. [32]	Characterisation and rehabilitation of adobe constructions in Portugal	Characterised adobe composition, mechanical behaviour, and structural performance; suggested methods for enhancing durability and seismic resistance.
Sánchez et al. [33]	Review of adobe masonry properties for rehabilitation in unspecified location	Compiled data on adobe mechanical properties from various studies; highlighted variations in regulations for local conditions.
Papayianni and Pachta [34]	Analysis and repair of historical earth block masonry in Greece	Developed methodology for analysis and repair of historical earth block houses; tested compatible repair materials based on earth.
González-Sánchez et al. [35]	Investigation of stabilised earthen mixtures for adobe preservation in Mexico	Evaluated eco-friendly earthen mixture for enhanced durability; applied mixtures in adobe dwellings with promising results.
Faria et al. [36]	Experimental characterisation of earth plastering mortar for adobe rehabilitation	Conducted tests on earth plastering mortar; found it suitable for rehabilitation of historical adobe buildings.
Gomes et al. [37]	Compatibility of earth-based repair mortars with rammed earth substrates	Tested repair mortars on rammed earth blocks; observed variations in mortar behaviour based on support type.
Jia et al. [38]	Analysis of historical earthen plaster for restoration	Investigated plaster properties; found certain additives improved properties, while others led to cracking.
Mattone et al. [39]	Testing of earth–gypsum plasters for conservation of earthen constructions	Developed plasters to protect earthen heritage; tested mixtures with natural and synthetic additives.
Hosseini et al. [40]	Mineralogical and physical tests on historical and new adobe in Iran	Determined mineral composition; found predominance of quartz and calcite.
Zakavi [41]	Mineralogical analysis of soil for new adobe in Iran	Identified dominant minerals; noted soil's poor clay content.
Dormohamadi and Rahimnia [42]	Physical, mineralogical, and mechanical tests of adobe in Iran	Identified main minerals; observed low swelling clay content in soil.
Eskandari [43]	Properties of historical and new adobe bricks in Iran	Identified mineral composition; found quartz as significant constituent.

The review of the literature indicates that there has been no study on the characterisation of Persian mud–straw, and the studies focused only on Persian adobe. Even in the case of Persian adobe, only XRF and XRD tests have been conducted. There has been no ion chromatography, moisture sorption isotherm determination, thermogravimetric analysis, or differential scanning calorimetry reported for Persian adobe bricks. Therefore, in the existing literature, there is no information about the elemental composition, mineralogical characteristics, moisture sorption behaviour, and thermal properties of Persian mud–straw plaster. In addition, the nonexistence of information about the moisture sorption behaviour and thermal properties of Persian adobe bricks provides an important gap in knowledge about Persian adobe. This paper examines the above-mentioned characteristics of both historical and new adobe bricks and mud–straw plaster produced and used in Iran in order to bridge the existing gap.

3. Samples, Method, and Equipment

3.1. Samples

Adobe bricks and mud–straw samples for testing were taken from different adobe buildings in the town of Maybod, 55 km west of the city of Yazd, Central Iran (Figure 3). The locations and periods of the samples, along with their respective straw content and corresponding geology region, are provided in Table 3. The selection of locations was based on the availability of historical buildings constructed with original adobe and mud–straw in the Maybod region. These locations were chosen to encompass a range of construction practices and historical periods representative of Persian architecture. The samples were carefully chosen to ensure the representation of both recent restorations and original structures, thereby providing a comprehensive overview of the materials used in historical building construction in the area. The number of samples was limited to ensure a manageable scope for the study while still capturing the diversity of materials and construction practices found in Persian architecture. Additionally, the dimensions of each sample are included in the table to provide further context regarding their physical characteristics and construction methods.

The town of Maybod has an arid and desert climate. The adobe bricks tested are coded as follows: A4, with dimensions of $21 \times 21 \times 5.4$ cm^3, made in 2017 in Narin Castle and used for recent restorations; C1, with dimensions of $22.5 \times 23 \times 6$ cm^3, made in 1870 in the Kazem Abad village as an original and historical adobe brick; and D2, with dimensions of $20 \times 24.5 \times 5.5$ cm^3, made in 1320 in Narin Castle as an original and historical adobe brick. The mud–straw plasters tested are coded as follows: W3, with dimensions of $20 \times 20.5 \times 2.7$ cm^3, made in 2019 in the laboratory for the present study from materials used for the restoration of the Deh-Naw Mosque, and X1, with approximate dimensions of $24 \times 18 \times 1.5$ cm^3, made in 2009 in the Deh-Naw Mosque used for previous restorations (Figure 4, Table 3). Due to limited access and the availability of samples, particularly historical ones, only one sample of each type of adobe brick and mud–straw plaster was tested. The measurement of straw content in the adobe and mud–straw samples revealed that the adobe samples A4, C1, and D2 contained 0.58 wt%, 0 wt%, and 0.02 wt% straw, respectively, while the mud–straw samples W3 and X1 contained 11 wt% and 8 wt% straw, respectively. Other researchers, as described in Section 1, have studied a number of adobe samples from other parts of Iran, including Yazd, Maybod, Ardakan, Esfarayen, Jarquyeh, and Susa (Figure 3a), whose results will be compared with results of this study in following sections.

Table 3. Types of adobe and mud–straw studied in the present study.

Material	Code	Location	Period	Comment	Length (cm)	Width (cm)	Thickness (cm)	Corresponding Geology Region	Straw Content (wt%)
Adobe	A4	Narin Castle (Maybod)	2017	Used for recent restorations	21	21	5.4	Cultivated land (Q^{cu} unit)	0.58
	C1	Kazem Abad village (Maybod)	Probably 1870	Original	22.5	23	6	Cultivated land (Q^{cu} unit)	0
	D2	Narin Castle (Maybod)	Probably 1320	Original	20	24.5	5.5	Cultivated land (Q^{cu} unit)	0.02
Mud–straw	W3	Deh-Naw Mosque (Maybod)	2019	Made in the laboratory for the present study from materials used for recent restorations	20	20.5	2.7	Cultivated land (Q^{cu} unit)	11
	X1	Deh-Naw Mosque (Maybod)	2009	Used for previous restorations	24	18	1.5	Cultivated land (Q^{cu} unit)	8

Figure 3. Locations of studied adobe and mud–straw samples: (**a**) in this study (Narin Castle and Kazem Abad village in Maybod) with sample codes, (**b**) by other researchers (Yazd, Ardakan, Esfarayen, Jarquyeh, and Susa) and in this study (Maybod) with sample codes, (**c**) Narin Castle, (**d**) Deh-Naw Mosque, (**e**) Kazem Abad village.

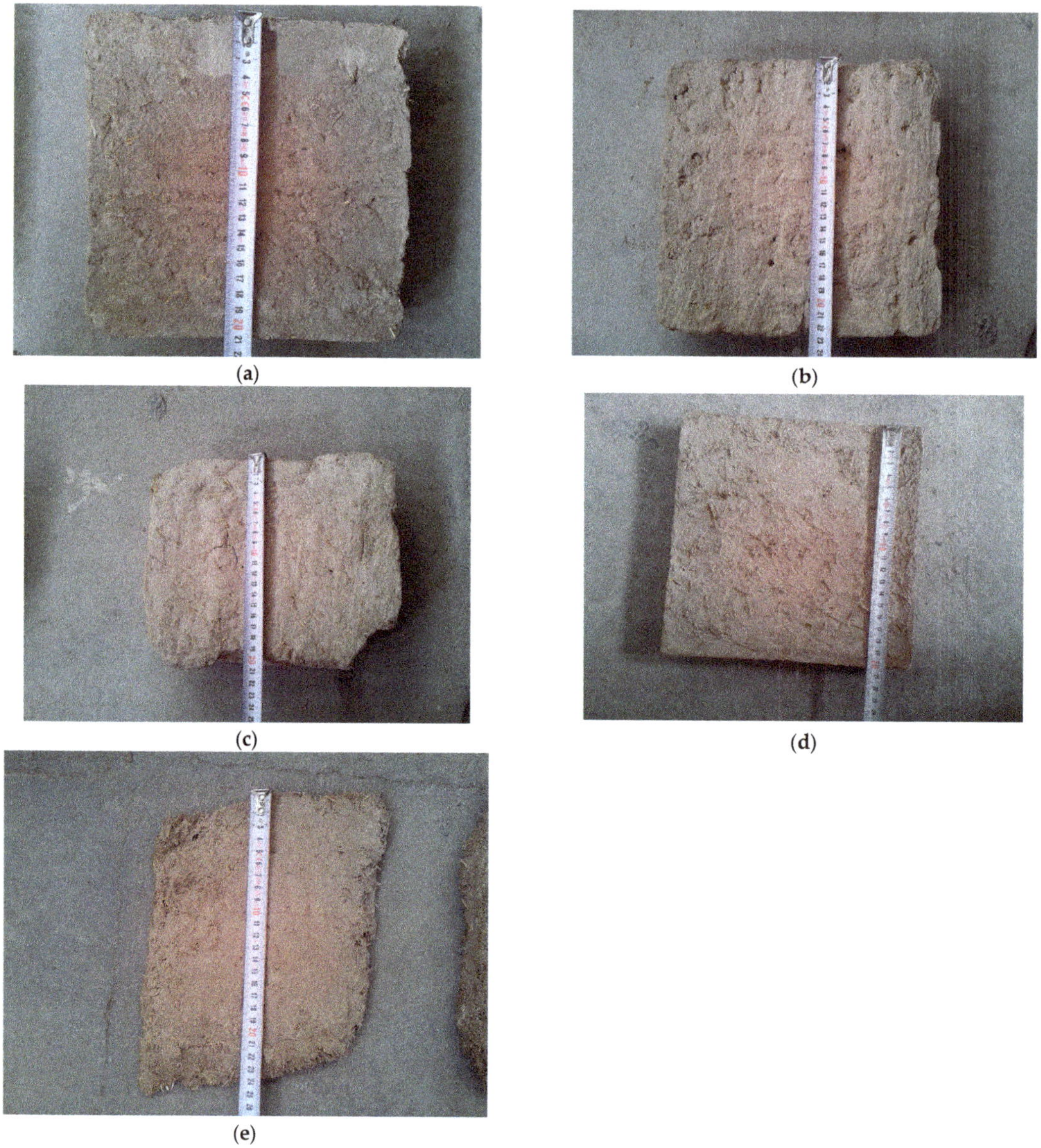

Figure 4. Tested adobe and mud–straw plaster samples: (**a**) adobe brick A4 (Narin Castle, Maybod, 2017), (**b**) adobe brick C1 (Kazem Abad village, Maybod, 1870), (**c**) adobe brick D2 (Narin Castle, Maybod, 1320), (**d**) mud–straw plaster W3 (Deh-Naw Mosque, Maybod, 2019) made in the laboratory, (**e**) mud–straw plaster X1 (Deh-Naw Mosque, Maybod, 2009).

3.2. *Geology of the Region*

Since the construction materials used in traditional buildings are often supplied from the nearest suitable location, the soil and rock materials around the site should be carefully studied. According to Ghorbani [44], the region from which the materials are taken is located in the structural zone of Central Iran. The area around the towns of Maybod and Ardakan is surrounded by the heights of the region. The heights on the eastern side are composed of Eocene formations. Figure 5 shows the satellite map of the region created from a Landsat ETM+ [45] satellite image created by the authors.

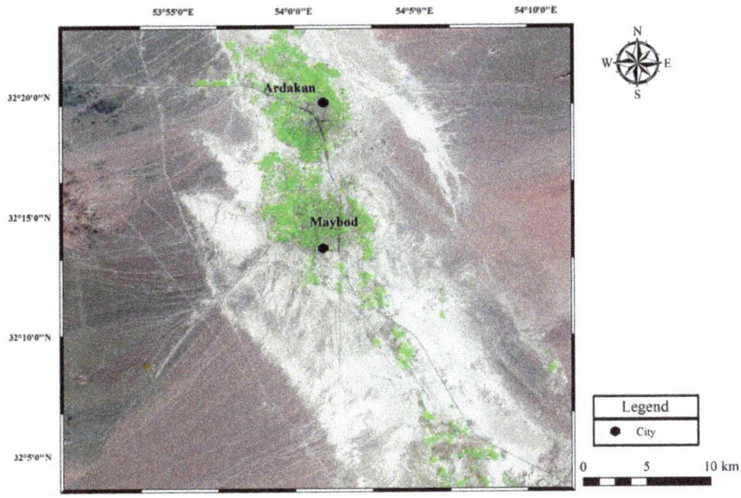

Figure 5. Satellite map of the region [45].

The heights on the west side are composed of tuff, igneous units, such as Eocene andesitic and dacitic lavas and trachyte and rhyolite lavas, as well as Palaeozoic era (Cambrian period) and black limestone. Erosion by wind and seasonal floods has caused erosion of the weaker Neogene and Eocene units, and as a result, the sediments created in the plains of the region have been deposited and caused land terraces. The towns of Maybod and Ardakan are located on the sediments of the present era, which include clay flat (Q^c unit) and cultivated lands (Q^{cu} unit). Figure 6 shows the geological map of the region prepared from satellite data and combining the existing geological maps [46,47].

It should be noted that adobe bricks are made of fine-grained soil, but different soil types have various properties that affect the quality of adobe bricks. So, the origin and geological process for the deposition and composition of soil are important. In the following, the general geological characteristics of the geological units of the region shown in Figure 6, created by the authors, are described.

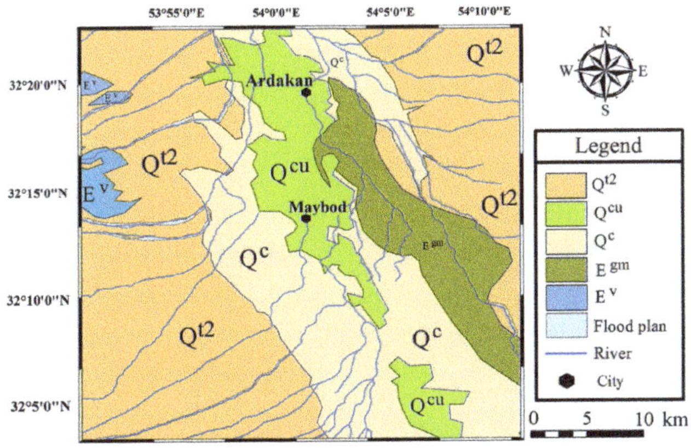

Figure 6. Geological map of the region.

3.2.1. Gypsiferous Marl (E^{gm})

This unit from the Eocene epoch is located on the unit of sandstone, marl, shale, tuff, and gypsum or unit $E^{s,m}$. The material of this unit is clay shale (mudstone), which is brick-coloured, with layers ranging from red to brown clay, and thin gypsum interlayers can be observed between them [48]. In this unit, conglomerate, red-to-brown sandstone layers can be seen, and the protrusions of these layers show the trend of the layers in this region. The main volume of this unit in the range of the map in Figure 6 is made of shale, and conglomerate layers are seen as interlayers in it, which indicates the successive changes in energy and depth of the basin at the time of the sedimentation of this unit. For most of the deposition time of this unit, a calm environment and moderate depth have been dominant. The red colour of this unit indicates the oxidation conditions in the formation environment of this unit. Also, hot and dry weather conditions must have been established in the region. In addition, the depth of water and the energy of the environment had been such that it has made it possible for oxygen to be in contact with the sediments that were forming. The exact age of this unit is attributed to the Middle Eocene (Lutetian) according to the identified fossils.

3.2.2. Lavas and Tuffs (E^v)

This unit mainly consists of grey and, sometimes, green andesitic, dacite, and rhyodacite tuffs and lavas. The texture of these rocks in the lava section is porphyry, and in the tuff sections, it is clastic and vitroclastic. The plagioclase minerals of these rocks are decomposed into epidote, chlorite, sericite, and clay minerals, and its amphiboles are decomposed into epidote, tremolite, and chlorite. Quartz mineral phenocrysts are also found in these rocks. The chemical composition of two rock samples of this unit contains the oxides presented in Table 4 [49].

Table 4. Chemical composition of lava and tuff unit samples (percentage of weight, wt%) [49].

	SiO_2	Al_2O_3	Fe_2O_3	MgO	CaO	Na_2O	K_2O	MnO	TiO_2	P_2O_5
Sample 1	65.36	14.77	2.06	2.84	3.17	3.88	4.20	0.01	0.63	0.17
Sample 2	60.75	15.26	8.45	4.68	3.27	2.71	2.29	0.04	0.69	0.23

3.2.3. Young Terraces and Gravel Fans (Q^{t2})

This unit is a new alluvial sediment that forms most of the loose flat plains. It consists of rock fragments that are older than the Quaternary formation and eroded fragments of the Q^{t1} unit (Pliocene epoch). Its matrix consists of clay, silt, and sand and is free of cement. This unit is composed of a larger grain size near the heights and is added to the middle of the plains by moving away from the heights of the larger parts and towards its clay, silt, and sand. The slope of the layering of this unit is horizontal. The age of this unit is Holocene or the present day.

3.2.4. Clay Flat (Q^c)

This clay soil unit is equivalent to the Q^{t2} unit that forms the young garrisons and alluvial fans. Due to the distance of Q^c from the heights, its coarse-grained particles are reduced, and its main volume is fine-grained, such as clay and silt particles. This unit forms flat and smooth lands, and agricultural lands are often close to this unit.

3.2.5. Cultivated Land (Q^{cu})

This agricultural land unit is actually part of the Q^{t2} unit that can be cultivated. Therefore, areas of the Q^{t2} unit that, due to the distance from the heights, their coarse grains are reduced, and the volume of sand, silt, and especially clay particles are increased are the so-called Q^{cu}. This unit is found adjacent to the Q^c unit in the region.

3.3. *Method*

Before providing a detailed explanation of each method employed in this study, it is essential to provide a general overview of the rationale behind the selection of these methods and their significance in addressing the research objectives. The characterisation of adobe and mud–straw samples necessitates a comprehensive approach that encompasses various analytical techniques to gain insights into their elemental composition, mineralogical characteristics, moisture sorption behaviours, and thermal properties. The chosen methods offer distinct advantages in explaining the structural, compositional, and thermal attributes of these traditional building materials. Additionally, reference will be made to previous studies that have analysed similar samples to provide context for the comparisons made throughout the paper.

3.3.1. Rationale for Method Selection and Standards

The methods employed in this study were carefully chosen based on their ability to provide detailed insights into different aspects of adobe and mud–straw samples. Ion chromatography, performed according to DIN EN ISO 10304-1 [50] and DIN EN ISO 10304-4 [51], facilitates the quantification of salts present in the samples, which is crucial for understanding their deterioration mechanisms and informing conservation strategies. X-ray diffraction (XRD) analysis, according to DIN EN 13925-2 [52], offers valuable information about the mineralogical composition and crystalline phases present in the samples, aiding in the identification of key mineral constituents. X-ray fluorescence (XRF) analysis, according to DIN 51001 [53], complements XRD by providing data on the elemental composition of the samples, enhancing our understanding of their chemical characteristics. The determination of hygroscopic sorption properties through gravimetric analysis, according to DIN EN ISO 12571 [54], offers insights into the moisture sorption behaviour of the materials, which is essential for assessing their durability and performance in different environmental conditions. Thermogravimetric analysis (TGA), according to ISO 11358-1 [55], and differential scanning calorimetry (DSC), according to ISO 11358-1 [56], enable the investigation of thermal decomposition processes and phase transitions in the samples, describing their thermal stability and behaviour at elevated temperatures.

The selection of characterisation methods in this study was guided by a thorough review of the available literature, which offers a comprehensive overview of different techniques and their applications across various materials. The references [57,58] provide insights into the advantages and shortcomings of different characterisation techniques, aiding in the rationale behind their selection for specific analyses. While a multitude of methods exist for material characterisation, the chosen techniques were considered suitable for their ability to provide relevant information about the adobe bricks and mud–straw plasters under investigation.

3.3.2. Sequence of Steps in Methods

Each method employed in this research follows a sequence of steps according to the specific analytical technique utilised. Ion chromatography, a method crucial for evaluating the presence and concentration of damaging salts in historical building materials, involves several sequential steps. Initially, samples containing straw are dried, crushed, and homogenised. Distilled water is then added for the elution process, followed by filtration. The soluble salts are determined by analysing the amounts of soluble ions in the solution. X-ray diffraction (XRD) involves multiple sequential steps for accurate mineralogical analysis. Initially, straw is removed from the samples to prevent interference during high-temperature analysis. The samples are then ground to approximately 5 μm, prepared on specific sample holders, and placed in the spectrometer. X-ray fluorescence (XRF) analysis requires meticulous sample preparation to ensure accurate chemical composition analysis. Sample preparation involves grinding the sample to a fine powder, mixing it with a binding/grinding aid, and pressing the mixture into a homogeneous sample pellet or using a suitable flux and heating it to make a fused tablet. The determination of hygroscopic sorption properties involves specific steps to obtain isotherm sorption curves. Samples are dried to a constant mass and then placed in the test device at a constant temperature, with relative humidity increasing or decreasing at steps. Moisture content is determined at each relative humidity, and the isotherm sorption curve is drawn accordingly. Thermogravimetric analysis (TGA) and differential scanning calorimetry (DSC) require precise sample preparation and testing procedures. Before conducting TGA and DSC, straw is removed from the samples to prevent potential interference or combustion effects at elevated temperatures. Samples are then prepared in two groups: completely dry and saturated. The samples undergo controlled temperature changes, and mass loss or heat flow is measured and recorded.

3.3.3. Equipment

The devices for ion chromatography were Metrohm (881 Compact IC Pro for cations with 863 Compact Autosampler, Herisau, Switzerland) and Dionex (Dionex ICS-1500 for anions with Dionex AS-DV autosampler, Sunnyvale, CA, USA). The chemical composition of the soil samples was determined by XRF analysis, using a Bruker AXS S4 Pioneer spectrometer with Rh-radiation (Bruker AXS GmbH, Karlsruhe, Germany). Melting tablets were prepared, and the loss of ignition (1000 °C) was determined. Major elements and some trace elements were analysed. The mineralogical composition of the adobe samples was determined by X-ray diffraction using a Bruker AXS D8 Advance X-ray diffractometer (Bruker AXS GmbH, Karlsruhe, Germany) with a Cu anode, operating at 40 kV—30 mA, a step size = 0.01°, and a scanning range 2θ between 5° and 70° (Bruker AXS GmbH, Karlsruhe, Germany). The equipment used for the determination of the hygroscopic sorption properties was POROTEC GraviSorp 120 (POROTEC GmbH, Hofheim am Taunus, Germany).

3.4. Ion Chromatography

Knowledge of the quantitative concentrations and enrichment of damaging salts in all types of inorganic historical building materials is very important. Ion chromatography is a highly sensitive and fast method to evaluate a set of interesting and common salts in historical building materials. It allows for the separation of ions and polar molecules based on their interactions with the resin (stationary phase) and the eluent (mobile phase). The system employs a stationary phase, such as an ionic resin, packed in the column, and a mobile phase, typically an eluent made of sodium hydrogen carbonate with additional neutralising strippers for chemical suppression to remove all background eluent ions. This process decreases the conductivity of the eluent, facilitating the injection of the sample into the mobile phase for analysis. In ion chromatography, conductivity values are derived from the quantitative concentrations of ions present in the samples. Through the analysis of ion concentrations, conductivity levels can be determined using established methods or standards, allowing for comparisons between different samples. This method provides valuable insights into the conductivity characteristics of historical building materials, aiding in the assessment of their properties and potential salt enrichment. Ion chromatography is a powerful tool for separating and determining low concentrations of ions and is particularly useful in environmental and water quality studies [59].

The samples containing straw for ion chromatographic analysis are first dried to a constant weight and then crushed and homogenised. Distilled water is added for the elution process and later filtered. The soluble salts are determined by analysing the amounts of soluble ions in the solution. Different columns and eluents are used for cations (Na^+, $NH_4{}^+$, K^+, Mg^{2+}, and Ca^{2+}) and anions (F^-, Cl^-, Br^-, $NO_3{}^-$, $NO_2{}^-$, $PO_4{}^{2-}$, $SO_4{}^{2-}$, and $C_2O_4{}^{2-}$). Certified reference solutions with specific amounts of interesting ions were used for calibration in this study, and the dried samples were weighed first. Then, distilled water was added in a ratio of approximately 1:10, and the samples were weighed again. An ultrasonic bath was provided for 15 min. The samples were allowed to settle for more than 12 h. Then, the solutions were filtered and analysed.

3.5. XRD Analysis

XRD is a vital technique for characterising materials, utilising X-ray wavelengths ranging from 0.01 nm to 10 nm. It works in conjunction with XRF, which provides elemental data. XRD identifies and quantifies minerals and their species, revealing crystalline phases and offering comprehensive insight into chemical composition and crystal structure. Each crystalline structure generates a unique X-ray pattern, similar to a fingerprint for identification. XRD is adept at distinguishing between compounds, such as different oxidation states or polymorphs [59]. Sample preparation for X-ray powder diffraction involves removing straw to prevent interference during high temperatures, grinding the samples to approximately 5 µm, and placing them in specific sample holders. This technique provides valuable information on crystallographic structure and chemical composition, aiding in the identification of crystalline phases, even in compounds, and potentially harmful mineralogical phases, such as salts.

3.6. XRF Analysis

XRF offers insights into the chemical composition of samples, focusing on elemental analysis rather than specific phases. Utilising an XRF spectrometer, fluorescent radiation emitted by various atoms in the sample is measured to identify and quantify material elements. This technique provides detailed information on the elemental composition, including the presence and quantity of elements like Fe and O, presented as percentages or parts per million (ppm) in a graphical output [59]. Sample preparation involves grinding the sample to a fine powder, ideally less than 75 µm, and forming a homogeneous sample pellet through pressing or fusion with flux at high temperatures.

In this study, straw was first removed from the sample to ensure accurate XRD mineralogical analysis, preventing potential organic material interference. Then, the sample was ground and dried in a drying chamber at 105 °C. The dried sample was weighted precisely (at least 1.5 g) in a constantly heated porcelain crucible and calcined at 1000 °C in a muffle furnace. Next, 1.2 g of the sample, which had been calcined at 1000 °C, was weighed into a platinum crucible. For this purpose, 6 g of Spectromelt A12 (flux, Merck, Darmstadt, Germany) was weighed out accurately. Both substances were mixed with a spatula. The mixture was melted at 1150 °C for 20 min. The melt was swirled about every 5 min in order to remove air bubbles and for homogenisation. The melt that was as free of air bubbles as possible was transferred into a platinum mould. The mould was heated for at least 5 min before the transfer. The melt tablets were analysed with a Bruker AXS S4 Pioneer spectrometer with Rh-radiation (Bruker AXS GmbH, Karlsruhe, Germany) regarding their chemical element composition, i.e., their major and trace elements.

3.7. Determination of Sorption Isotherm (GraviSorp)

Hygroscopicity refers to a material's ability to absorb moisture from its surroundings, while sorption capacity indicates its capability to absorb or release water vapour until reaching equilibrium. The relationship between moisture content and relative humidity at a constant temperature is represented by the moisture sorption isotherm or the isotherm sorption curve [60].

To obtain the isotherm sorption curve, the sample is dried to a constant mass and then placed in the test device at a constant temperature, with relative humidity increasing in steps. The moisture content is determined at each step after the sample achieves equilibrium, allowing for the construction of the isotherm sorption curve.

Similarly, to draw the isotherm desorption curve, the same procedure is followed but with a decrease in relative humidity. The starting point is a relative humidity of at least 95% or, alternatively, the last point of the sorption curve. At a constant temperature, the sample is placed in the test device, with the relative humidity decreasing in steps. By determining the moisture content at each relative humidity, the isotherm desorption curve can be drawn.

In the GraviSorp 120 from Porotec used for this research, 10 samples can be examined simultaneously under constant temperature and with predefined humidity levels

In this study, two sets of tests were conducted. The first set involved examining the sorption isotherm characteristics of adobe and mud–straw samples over a 30-day period. The second set of tests was exclusively conducted on mud–straw samples, both with and without the addition of straw, for a duration of 60 days. The second set aimed to investigate the influence of straw on the sorption isotherm performance of mud–straw samples.

All tests were conducted at a constant temperature of 22 °C, following a similar procedure. For instance, in the 30-day test, the relative humidity was incrementally increased from 10% to 95% in steps of 10% each, with each humidity level applied for 24 h during the first 15 days. Subsequently, during the latter 15 days, the relative humidity was gradually decreased from 95% to 10%. At each specific relative humidity level, the moisture content of the sample was measured. The point at which the sample reaches equilibrium with its environment, exchanging an equal amount of absorbed and desorbed water molecules, defines the moisture content known as the equilibrium moisture content (EMC). It is calculated from Equation (1), in which W_m and W_d are the moist and dry weights of the sample, respectively.

$$\mathrm{EMC} = \frac{(W_m - W_d) \times 100}{W_d} \quad (1)$$

3.8. Thermogravimetric Analysis (TGA) and Differential Scanning Calorimetry (DSC)

Thermogravimetric analysis (TGA) is a thermal analysis method used to measure the mass changes of a material with respect to temperature changes. In the TGA test, the sample is heated, and its mass changes are continuously recorded to measure the mass loss relative to the initial mass of the samples. Thermogravimetric analysis provides information about the physical characteristics of materials, such as thermal decomposition and solid–gas reactions, including oxidation, reduction, and dehydration. Differential scanning calorimetry (DSC) is another thermal technique used to investigate characteristics such as heat capacity and chemical reactions, like the oxidation behaviour of materials. This is achieved by measuring the heat flow into or out of a sample as it undergoes physical and chemical changes during controlled temperature changes [61].

In this study, prior to conducting the TGA and DSC analyses, straw was removed from the samples to prevent potential interference or combustion effects at elevated temperatures. Subsequently, two groups of samples were prepared from each of the three types of adobe and two types of mud–straw samples: one group of completely dry samples and another group of saturated samples. The dry samples were exposed to silica gel in a sealed container for four months until they reached complete dryness, as indicated by no further change in weight during periodic weighing. Conversely, the saturated samples were placed in a humidity chamber for four months until they reached saturation, confirmed by stable weights during periodic weighing. Following the preparation of these samples, an STA 409 PC Luxx Simultaneous thermal analyser (Netzsch-Gerätebau GmbH, Selb, Germany) was utilised for simultaneous TG-DSC measurements. During the test, the temperature was incrementally increased from 25 °C to 980 °C according to a controlled temperature program, and the mass loss of the samples was measured and recorded.

4. Results and Discussion

This investigation focused on traditional and contemporary Persian adobe bricks as well as mud–straw plasters, subjecting them to rigorous analysis using a range of techniques aligned with established standards. The subsequent sections delineate the findings obtained through ion chromatography, XRF, XRD, sorption isotherm analysis, and TGA-DSC measurements.

4.1. Ion Chromatography Results

The ion chromatography test results for adobe and mud–straw samples are presented in Tables 5 and 6. The conductivity of the eluates on the adobe samples ranges from 358 μS/cm for the adobe sample C1, made in 1870, to 467 μS/cm for the adobe sample A4, made in 2017. It is 393 μS/cm for adobe sample D2 made in 1320.

The conductivity values for mud–straw samples are significantly higher than those of the adobe samples, with readings of 635 μS/cm for the mud–straw sample X1, produced in 2009, and 1622 μS/cm for the mud–straw sample W3, manufactured in 2019. This notable difference in conductivity could be attributed to the higher straw content in the mud–straw samples (11 wt% for W3 and 8 wt% for X1), contrasting with the adobe samples, which have considerably lower straw content (0.58 wt% for A4, 0 wt% for C1, and 0.02 wt% for D2).

The chloride content in the newly produced mud–straw W3 (2019) is measured at 0.81 wt%. In comparison, the adobe samples A4 (2017) and D2 (1320), originating from the same location, exhibit chloride contents of 0.13 and 0.14 wt%, respectively. Meanwhile, the adobe sample C1 (1870) and mud–straw sample X1 (2009), both sourced from the same area although in different locations, show chloride contents of 0.07 and 0.06 wt%, respectively. The chloride content of the mud–straw W3 is 13.5 times the mud–straw X1, 6.2 times the adobe sample A4, 11.6 times the adobe sample C1, and 5.8 times the adobe sample D2.

It is observed that the conductivity of the new adobe sample A4 is higher than that of the historical adobe samples C1 and D2. Additionally, both mud–straw samples W3 and X1 that are newly produced have high conductivity values. The very high conductivity of sample W3 is caused by a high chloride content of 0.81 wt%. The reason for this can be attributed to the high admixture of straw (11 wt% for sample W3 compared to 8 wt% for sample X1) and the usage of fresh materials in this sample (produced in 2019, contrasting with sample X1 produced in 2009).

According to Wild [62], wheat straw can contain about 0.23 wt% of water-soluble chloride in the plant mass, while rice straw can contain even about 0.58 wt%. On the other hand, the fresh loam or soil used may also have higher chloride concentrations due to the characteristics of the arid climate of the region. Leaching effects due to precipitation after ripening or after harvesting can lead to the leaching of chlorides in wheat straw.

In a similar manner, natural weathering over the years causes the leaching of chloride from the mud-straw plaster and adobe, as observed in sample X1, which has been weathered since 2009.

Table 5. Ion chromatography test results for adobe and mud–straw samples: concentration of anions.

Sample	Cl^- (wt%)	NO_2^- (wt%)	NO_3^- (wt%)	SO_4^{2-} (wt%)	Conductivity (μS/cm)
Adobe A4	0.13	<0.01	0.01	0.14	467
Adobe C1	0.07	<0.01	0.01	0.12	358
Adobe D2	0.14	<0.01	0.02	0.06	393
Mud–straw W3	0.81	<0.01	<0.01	0.15	1622
Mud–straw X	0.06	0.01	0.03	0.35	635

Table 6. Ion chromatography test results for adobe and mud–straw samples: concentration of cations.

Sample	Na^+ (wt%)	NH_4^+ (wt%)	K^+ (wt%)	Mg^{2+} (wt%)	Ca^{2+} (wt%)
Adobe A4	0.12	<0.01	0.02	0.01	0.04
Adobe C1	0.09	<0.01	0.01	0.01	0.03
Adobe D2	0.11	<0.01	0.01	0.01	0.02
Mud–straw W3	0.54	<0.01	0.13	0.05	0.10
Mud–straw X1	0.07	0.01	0.03	0.01	0.14

Although straw has high soluble potassium content and only low concentrations of soluble sodium in the plant mass, the eluates are sodium-accentuated. This suggests that the possible source for the high chloride concentration is the common salt existing in the clay and soil.

4.2. XRF and XRD Tests

In order to compare the XRF and XRD test results of the samples in this study with those of other researchers, the results of a series of adobe samples studied by other researchers were collected. The XRD test was carried out for materials under three different moisture conditions: normal (ambient) moisture, completely dry, and saturated. The rationale behind conducting the XRD test at various moisture contents is to identify different minerals associated with distinct moisture levels. This approach allows for the detection of a wider range of minerals that might be present in the materials under different moisture conditions. To make the materials dry, they were exposed to silica gel in a sealed container for four months, during which time they underwent periodic weighing, and they were deemed dry when their weight remained unchanged. To make them saturated, the samples were exposed to humidity in a humidity chamber for four months. To confirm saturation, the samples underwent periodic weighing until their weight showed no further change.

In Table 7, samples from other researchers are coded as follows: M1 (for sample M1, six specimens, i.e., M1(1) to M1(6), were tested, as indicated in Table 8), adobe made in 2016 and used for recent restorations in the town of Ardakan, 60 km from Yazd and 5 km from the town of Maybod [42]; M2, adobe made in 1900 in Jarquyeh, 270 km east of Yazd [43]; M3, adobe made in 2019 and used for recent restoration in Yazd [43]; M4 and M5, adobes made in 1300 in Belqeis Castle in Esfarayen, North-East Iran [40]; M6 and M7, adobes made in 2016 and used for recent restorations in Belqeis Castle [40]; and M8 and M9, adobes made in 2016 and used for recent restorations in the Choga Zambil in Susa, South-East Iran [41]. Figure 3a shows the locations of adobes selected by other researchers.

Table 7. Types of adobe studied by other researchers.

Material	Code	Location	Period	Comment	Reference
Adobe	M1	Ardakan (60 km east of Yazd)	2016	Used for recent restorations	[42]
	M2	Jarquyeh (270 km east of Yazd)	Probably 1900	Original	[43]
	M3	Yazd	2019	Used for recent restorations	[43]
	M4, M5	Belqeis Castle (Esfarayen, North-East Iran)	Probably 1300	Original	[40]
	M6, M7	Belqeis Castle (Esfarayen, North-East Iran)	2016	Used for recent restorations	[40]
	M8, M9	Choga Zambil (Susa, South-East Iran)	2016	Used for recent restorations	[41]

4.2.1. XRF Results

The XRF test results obtained for adobe and mud–straw samples in this study and adobe samples by other researchers are presented in Tables 8 and 9.

For adobe and mud–straw samples in this study taken from Maybod, 55 km from Yazd, the dominant chemical compounds are SiO_2 (about 45 wt%), CaO (around 16 wt%), and Al_2O_3 (about 10 wt%). Notably, Ba, Sr, and Zr are prevailing common chemicals found in all samples, with concentrations of approximately 679 ppm, 563 ppm, and 222 ppm, respectively.

The historical and new adobe and mud–straw samples investigated in this study exhibit a relatively similar chemical composition of elements. The locations of the different samples are only a few kilometres apart.

Table 8. Chemical composition analysed by XRF for adobe and mud–straw samples in this study and adobe samples by other researchers (percentage of weight, wt%; main element composition as oxides).

Oxides	This Study					Other Researchers						
	Location											
	Maybod					Ardakan						Susa
	Adobe			Mud–Straw		Adobe						
						M1						M9
						[42]						[41]
	A4	C1	D2	W3	X1	Sample: M1(1)	M1(2)	M1(3)	M1(4)	M1(5)	M1(6)	
SiO_2	45.18	45.16	46.11	43.89	47.64	39	41	44.9	42.2	43.7	43.3	33.1
TiO_2	0.54	0.53	0.52	0.53	0.46	0.7	0.6	0.7	0.6	0.7	0.6	0.65
Al_2O_3	10.33	8.94	9.75	9.57	10.93	10.7	9.8	10.8	8.5	12.4	10.6	7.51
Fe_2O_3	4.52	3.85	4.10	4.20	4.10	3.6	3	3.6	3.2	4.3	3	4.35
MnO	0.09	0.09	0.09	0.09	0.08	-	-	-	-	0.1	-	0.07
MgO	4.34	4.09	4.31	4.17	3.39	5.7	6.6	6.8	6.3	5.6	5.6	5.77
CaO	16.31	18.20	16.61	18.02	15.57	15.7	13.3	12.9	17.8	14.3	14.3	22.07
K_2O	2.12	1.88	1.91	2.07	2.45	2.7	2.5	2.8	2.4	2.9	2.5	1.84
Na_2O	1.28	1.33	1.43	1.46	1.31	1.5	2.6	1.4	1	1.3	2	0.53
P_2O_5	0.14	0.13	0.14	0.13	0.13	<0.01	<0.01	<0.01	<0.01	<0.01	<0.01	0.14
L.O.I	17.68	17.10	17.35	18.96	16.31	18.3	18.3	15.9	17.8	14.6	17	23.49
Sum	102.54	101.29	99.93	103.09	102.37	97.0	97.7	99.8	99.8	99.9	98.9	99.52

There is a meaningful relationship between the values in Tables 4 and 8. The percentage of elements in adobe bricks and mud–straw (Table 8) is relatively similar to the percentage of elements in two geological units lava (igneous) and tuff present in the region (Table 4). Note that Table 8 includes the percentage of loss on ignition in the adobe and mud-straw samples, revealing a notable presence of calcium oxide (CaO) within them. Figure 6 shows limestone rocks at the higher altitudes of the region. Since calcium oxide has a high solubility, the presence of a significant percentage of CaO in the adobe and mud–straw samples of the Maybod area is probably due to the dissolved CaO deposited in the soil. If the loss of ignition value is subtracted from Table 8, the CaO value of this table is changed to about 3%, and the composition of the remaining oxide elements is calculated from 100%, the results are similar to Table 4, which indicates a similar chemical composition between lava and tuff rocks of the region and the soil composition used in the adobe bricks and mud–straw of Maybod.

Specimens of the adobe sample M1 [42] taken from Ardakan, 5 km from Maybod and 60 km from Yazd, have a similar chemical composition. As mentioned before, six specimens, i.e., M1(1) to M1(6), were tested for sample M1. The dominant compounds are SiO_2 with about 42 wt%, CaO with around 15 wt%, Al_2O_3 with about 10 wt%, and MgO with 6 wt%. The amount of loss of ignition is also comparable. The similarity of the samples in Ardakan and Maybod is due to comparable soil, as these two towns are only 5 km away.

In the adobe sample M9 [41], the order of abundance of chemical elements is the same, dominated by SiO_2 with 33 wt%, CaO with 22 wt%, Al_2O_3 with 8 wt%, and MgO with 6 wt%. The loss of ignition, on the other hand, is significantly higher, and the high CaO content of the sample suggests a higher carbonate content. This new adobe sample was produced in Susa, about 900 km from Maybod.

Table 9. Trace elements analysed by XRF for adobe and mud–straw samples in this study (concentrations in ppm).

Trace Elements	Adobe Type			Mud–Straw Type	
	A4	**C1**	**D2**	**W3**	**X1**
Ba	619	749	646	855	485
Co	17	14	15	16	15
Ni	85	69	77	73	80
Cu	45	50	42	132	45
Zn	115	95	98	106	94
Cr	132	119	142	121	119
Sr	558	575	628	590	466
Y	58	60	61	58	74
Zr	218	221	231	213	228
Ga	22	22	24	22	25
Nb	79	86	86	85	106
Ce	75	73	72	70	82

4.2.2. XRD Results

As an example of the XRD patterns obtained for the adobe and mud–straw samples in the present study, the XRD patterns for the adobe sample D2 and the mud–straw sample X1 at normal (ambient) moisture are illustrated in Figure 7. Mineralogical compositions identified by XRD for the adobe and mud–straw samples at normal moisture in this study and adobe samples by other researchers are presented in Table 10.

In general, the majority of compounds found in the adobe and mud–straw samples studied in this research are similar. The common compounds in all samples include quartz (SiO_2), and different feldspars such as albite ($Na(AlSi_3O_8)$) and orthoclase ($(K_{0.94}Na_{0.06})(AlSi_3O_8)$). Furthermore, calcite ($CaCO_3$), dolomite ($CaMg(CO_3)_2$) and different clay minerals, clinochlore-1Mllb ($Mg_5Al(Si, Al)_4O_{10}$), and illite ($K(Al_4Si_2O_9(OH)_3)$) are present in all samples. Hematite (Fe_2O_3) also exists in the samples except in the adobe sample C1, made in 1870; the mud–straw sample W3, made in 2019; and the mud–straw sample X1.

In the historical adobe samples, C1 made in 1870 and D2 made in 1320, additional compounds exist. Palygorskite ($Mg_5(SiAl)_8O_{20}(OH)_{2.8}$) is observed in both the adobe samples C1 and D2. The adobe sample C1 contains ferroactinolite ($(Ca,Na,K)_2Fe_5Si_8O_{22}$), while D2 contains cordierite ($Mg_2Al_4Si_5O_{18}$) and heulandite ($(X)_3(Al_3Si_9O_{24})$*7-8H_2O).

In all studied samples, swelling clay minerals, like montmorillonite and smectite, are present in very small amounts, while the amount of non-swelling clay mineral, illite, is significant. This low presence of swelling clay minerals is important for producing good quality adobe bricks, as they tend to cause fissures during drying if they contain high amounts of swelling clay minerals. The absence of significant amounts of swelling clay minerals in the samples is a positive indicator for adobe brick production. This means that the risk of deformation and cracking during the drying process is significantly reduced. Additionally, the presence of quartz is consistent with the typical composition of desert soils in the region and aligns with the geological units of the area.

The adobe sample M2 [43], made in 1900 in Jarquyeh, 270 km east of Yazd, and the adobe sample M3 [43], made in 2019 in Yazd, show the greatest similarity in mineral content. They contain quartz (SiO_2), albite ($NaAlSi_3O_8$), calcite ($CaCO_3$), clinochlore-1Mllb ($Mg_5Al(SiAl)_4O_{10}$), and dolomite ($CaMg(CO_3)_2$). Other adobe samples from other locations have a completely different mineralogical composition; the reason is the different types of soils at different locations far from each other.

The XRD results show that quartz (SiO_2) and calcite ($CaCO_3$) are present in all the adobe and mud–straw samples shown in Table 10 except the adobe sample M5.

In addition, the XRF results in Table 8 show a comparable chemical composition to the adobe sample of type M1 from Dormohamadi and Rahimnia [42]. Adobe samples of the series M1 in Table 8 are made in Ardakan with the same soil used for adobe samples studied in this research, while the M9 adobe sample, with significantly lower concentrations of SiO_2 and Al_2O_3 but higher contents of CaO, was made in Susa, very far from Ardakan.

Upon comparing the XRD patterns of completely dry and saturated samples, it becomes evident that the three adobe samples contain the same minerals; however, the varying heights of the peaks indicate differences in the mineral quantities among the different adobe samples. The comparison between the mud–straw samples reveals the presence of gypsum in the mud–straw sample X1, while the mud–straw sample W3 lacks gypsum.

Examining all adobe and mud–straw samples reveals that they possess similar minerals but differ in their respective amounts. Interestingly, only the mud–straw sample X1 contains gypsum; the other samples do not.

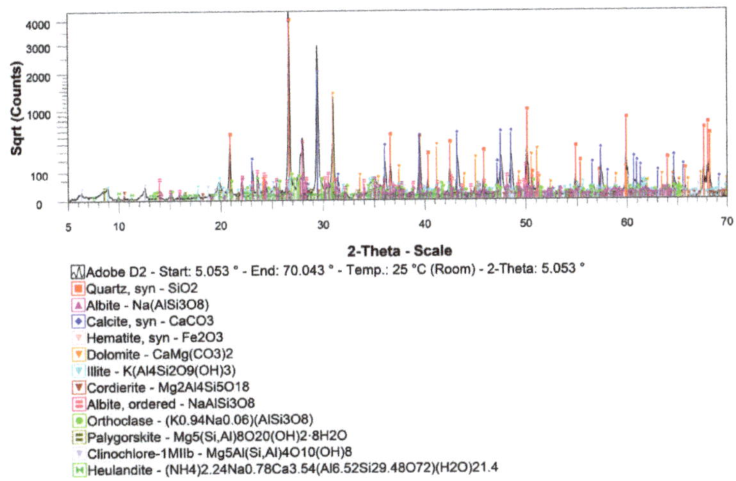

(**a**)

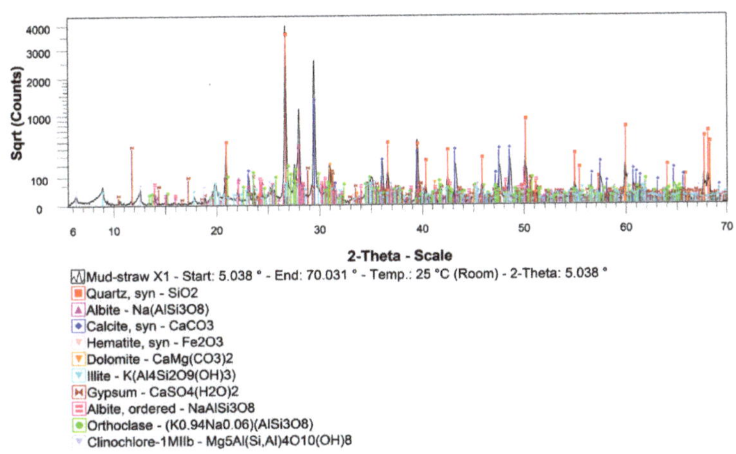

(**b**)

Figure 7. XRD pattern for materials at normal (ambient) moisture: (**a**) adobe D2, (**b**) mud–straw X1.

The presence of gypsum in the mud–straw sample X1 may be attributed to various factors, including variations in the composition of raw materials used during the construction process. It is most likely that gypsum, a naturally occurring mineral, was unintentionally added to the materials used for constructing the mud–straw sample X1. One reasonable explanation could be the accidental incorporation of gypsum from other applications within the adobe building. Gypsum is commonly used in various construction practices in adobe buildings, such as creating gypsum forms for constructing arches and vaults. These forms provide structural support and shape for adobe arches and vaults during construction. Additionally, gypsum mortar is frequently used to fill voids in mortar joints to enhance the strength and integrity of adobe vaults. It is conceivable that gypsum from these forms and mortar applications might have unintentionally been mixed with the components of the mud–straw plaster during construction. Furthermore, both the mud–straw samples X1 and W3 were used for plastering the same building. Sample X1, containing gypsum, was produced and applied to the building in 2009, while sample W3, containing no gypsum, was produced and applied to the building in 2019. Both samples were produced from the same soil using the same production technique. This similarity in production methods and materials between samples X1 and W3 further supports the notion that the presence of gypsum in sample X1 is likely accidental, possibly due to the use of gypsum for other applications within the construction process.

Table 10. Mineralogical composition identified in the XRD test for adobe and mud–straw samples in this study and by other researchers.

Compound Name	Chemical Formula	This Study					Other Researchers								
		Location													
		Maybod					Ardakan	Jarquyeh	Yazd	Esfarayen		Susa			
		Adobe			Mud–Straw		Adobe								
		A4	C1	D2	W3	X1	M1	M2	M3	M4	M5	M6	M7	M8	M9
							[42]	[43]		[40]		[41]			
Quartz	SiO_2	•	•	•	•	•	•	•	•	•		•	•	•	•
Albite	$Na(AlSi_3O_8)$	•	•	•		•					•	•	•		
Albite, ordered	$NaAlSi_3O_8$	•	•	•	•	•		•	•						
Orthoclase	$(K_{0.94}Na_{0.06})(AlSi_3O_8)$	•	•	•	•	•									
Calcite	$CaCO_3$	•	•	•	•	•	•	•	•	•	•	•	•	•	•
Clinochlore	$Mg_5Al(SiAl)_4O_{10}(OH)_8$	•	•	•	•	•		•	•						•
Dolomite	$CaMg(CO_3)_2$	•	•	•	•	•	•	•	•					•	
Illite	$K(Al_4Si_2O_9(OH)_3)$	•	•	•	•	•	•								
Hematite	Fe_2O_3	•		•		•									
Palygorskite	$Mg_5(SiAl)_8O_{20}(OH)_{2.8}$		•	•											
Ferroactinolite	$(Ca,Na,K)_2Fe_5Si_8O_{22}$		•												
Cordierite	$Mg_2Al_4Si_5O_{18}$			•											
Heulandite	$(X)_3(Al_3Si_9O_{24})$*7-8H_2O			•											
Hydrotalcite	$(Mg_{0.667}Al_{0.333})$					•									
Feldspar	$NaAlSi_3O_8$—$CaAl_2Si_2O_8$						•							•	
Kaolinite	$Si_2Al_2O_5(OH)_4$						•								
Chlorite	$(Fe,Mg,Al)_6(SiAl)_4O_{10}(OH)_8$						•								
Smectite	$(Ca,Na,H)(Al,Mg,Fe,Zn)_2$ $(SiAl)_4O_{10}(OH)_{2-x}H_2O$						•								
Muscovite	$KAl_2(AlSi_3O_{10})_3(F.OH)_2$							•	•	•	•	•	•		
Microcline	$KAlSi_3O_8$								•						
Biotite	$K(Mg,Fe)_3AlSi_3O_{10}(F,OH)_2$									•					
Enstatite	$MgFe(Si_2O_6)$										•		•		
Apatite	$Ca_5(PO_4)_3(F,Cl,OH)$										•	•	•		
Gypsum	$CaSO_4,2H_2O$					•									•

4.3. Sorption Isotherm Results

The determination of moisture sorption isotherms was conducted for two sets of samples. The first set comprised three types of adobe samples of A4, D2, and C1, along with two types of mud–straw samples of W3 and X1, for comparison. Straw was consistently present in the mud–straw samples in this set, reflecting real-world mud–straw plaster.

The second set exclusively consisted of the mud–straw samples W3 and X1 examined in two scenarios: with and without the addition of straw. This examination aimed to assess the influence of straw on the sorption isotherm characteristics of mud–straw samples.

4.3.1. Comparative Analysis of Adobe and Mud–Straw Samples

Duplicate determinations were conducted for each type of adobe and mud–straw plaster over a period of 30 days. Sorption isotherm (the lower) and desorption (the upper) curves of adobe and mud–straw samples in this study for the test duration of 30 days are shown in Figure 8. Table 11 and Figure 9 present the average moisture contents at the relative humidities of 60%, 80%, and 95%. The moisture content, called the equilibrium moisture content (EMC), at a relative humidity of 80% (EMC_{80}), is an important characteristic variable for physical simulations [1]. The moisture content at a relative humidity of 95% (EMC_{95}) is used for comparison with other researchers' findings. EMC_{60}, chosen for comparison, enriches the study by examining the materials' response to lower humidity, complementing the exploration of varied moisture environments alongside EMC_{80} and EMC_{95}.

Among the adobe samples, adobe C1, made in 1870, exhibits the lowest moisture content and adobe A4, made in 2017, shows the highest moisture content. This is in relation to the amount of straw in the adobe samples, where adobe C1 has the lowest straw content of 0 wt%, and adobe A4 has the highest straw content of 0.58%. At the relative humidity of 60%, the moisture content (EMC_{60}) for adobe samples A4, C1, and D2 are, respectively, 1.83%, 1.1%, and 1.63%. The values for the relative humidity of 80% (EMC_{80}) are 3.5%, 2.1%, and 3.1%, respectively. The values are, respectively, 7.1%, 4.4%, and 6.4% for the relative humidity of 95% (EMC_{95}). Other researchers have reported equilibrium moisture content equal to 0.5% to 7% [63], 4% to 6% [64], 3.5% [12], 5.3% [65], 3% to 5% [18], and 1.3% [20].

In the case of the mud–straw samples, mud–straw W3, produced in 2019, exhibits the highest moisture content, while mud–straw X1, produced in 2009, displays the lowest moisture content. There is once more a correlation between the moisture content and the straw amount, with mud–straw W3 containing a higher straw content of 11 wt% compared to mud–straw X1, which contains a lower straw content of 8 wt%. The moisture content at the relative humidity of 60% (EMC_{60}) for sample W3 is 2.5% and that of sample X1 is 0.59%. These values at the relative humidity of 80% (EMC_{80}) are, respectively, 4.8% and 3.1%. At the relative humidity of 95% (EMC_{95}), these values are 15.3% for sample W3 and 7.4% for sample X1. Ashour et al. [14] reported an equilibrium moisture content of less than 7% for mud–straw plaster with wheat straw, barley straw, and wood shavings.

It is observed that adobe A4, D2, and the mud–straw sample W3, which are more heavily loaded with common salt (Tables 5, 6 and 11), also have significantly higher moisture absorption. The analysis of the variation in the amount of salt mixtures with respect to relative humidity for samples was conducted using RUNSALT 1.9 software [66]. Figure 10a displays the variation in the amount of common salt content, NaCl, versus relative humidity for the adobe and mud–straw samples. Table 11 indicates that the amount of common salt in the mud–straw sample W3 (11.36×10^{-4} mol) is 9.3 times the mud–straw sample X1, 5 times the adobe sample A4, 8 times the adobe sample C1, and 5.9 times the adobe sample D2. The analysis of the variation in the amount of salt mixtures with respect to relative humidity for the mud–straw sample W3 is depicted alongside the sorption isotherms and desorption curves in Figure 10b. It can be seen that at the relative humidity of about 71%, all salt is dissolved, and from then on, the second curve shows a sudden increase in slope, which means a significant increase in moisture absorption.

Table 11. Average moisture content of adobe and mud–straw (containing straw) samples in this study for test duration of 30 days.

Material	Code	Location	Period	Straw Content (wt%)	Average Moisture Content at RH = 60% (EMC_{60}) (%)	Average Moisture Content at RH = 80% (EMC_{80}) (%)	Average Moisture Content at RH = 95% (EMC_{95}) (%)	Cl^- (wt%)	Na^+ (wt%)	NaCl at RH = 15% (mol)
Adobe	A4	Narin Castle (Maybod)	2017	0.58	1.83	3.50	7.10	0.13	0.12	2.25×10^{-4}
	C1	Kazem Abad village (Maybod)	Probably 1870	0	1.10	2.10	4.40	0.07	0.09	1.43×10^{-4}
	D2	Narin Castle (Maybod)	Probably 1320	0.02	1.63	3.10	6.40	0.14	0.11	1.94×10^{-4}
Mud–Straw	W3	Deh-Naw Mosque (Maybod)	2019	11	2.47	4.80	15.30	0.81	0.54	11.36×10^{-4}
	X1	Deh-Naw Mosque (Maybod)	2009	8	1.54	3.10	7.40	0.06	0.07	1.22×10^{-4}

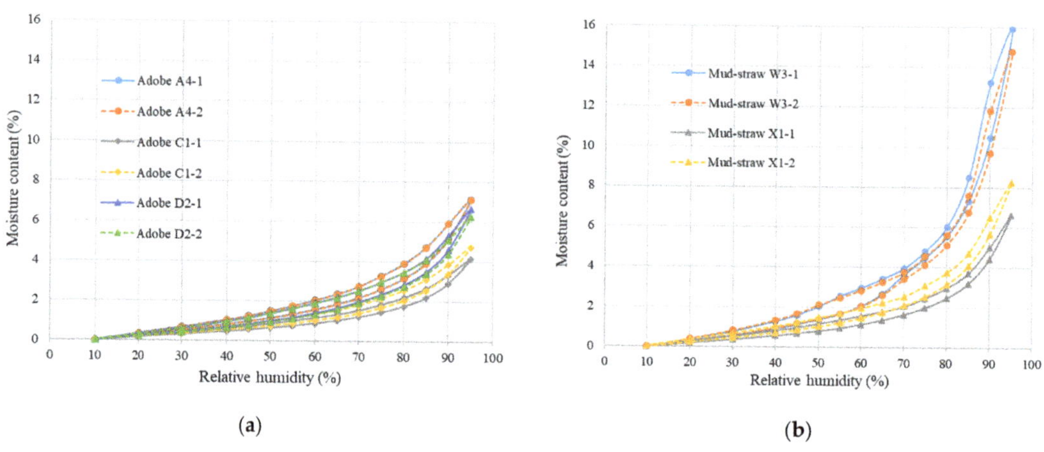

Figure 8. Sorption isotherm and desorption curves for test duration of 30 days: (**a**) adobe, (**b**) mud–straw.

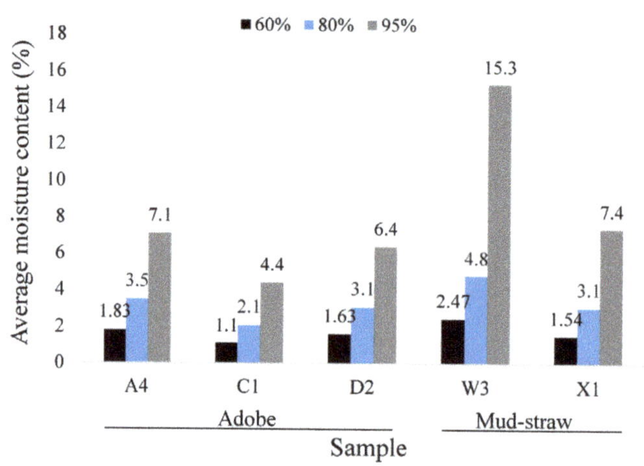

Figure 9. Average moisture content of adobe and mud–straw (containing straw) samples at relative humidities of 60%, 80%, and 95% for test duration of 30 days.

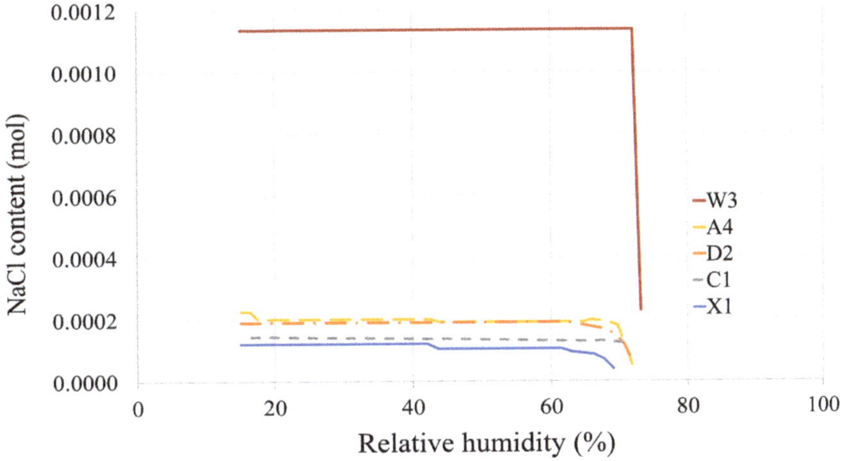

(a)

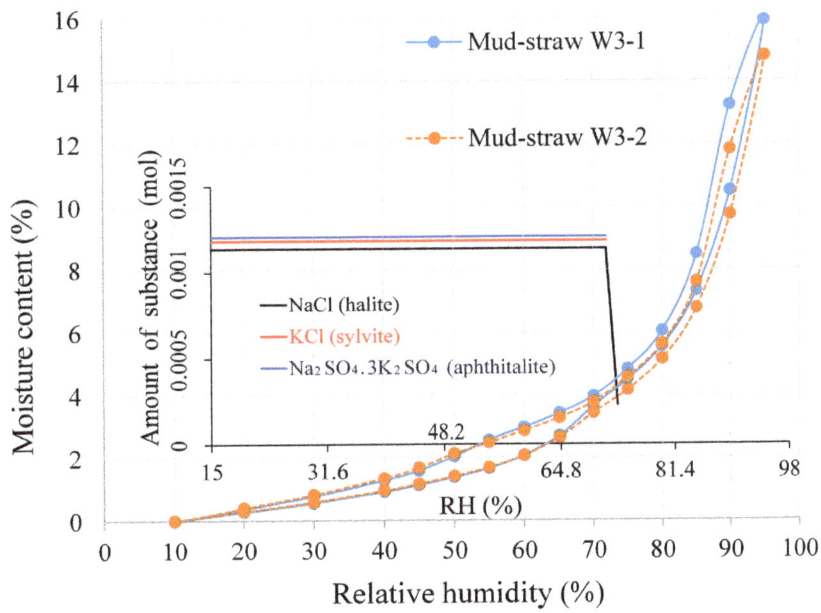

(b)

Figure 10. (**a**) Variation in amount of common salt content versus relative humidity for adobe and mud–straw samples, (**b**) variation in amount of salt mixtures versus relative humidity for mud–straw W3 (containing straw) together with sorption isotherm and desorption curves (test duration of 30 days).

4.3.2. Influence of Straw on Mud–Straw Sorption Isotherms

Two additional sets of sorption isotherm tests were conducted to study the effect of the presence of straw in the mud–straw samples. Tests were performed once on mud–straw samples with straw and once without straw. The tests were conducted for 60 days, which simulated long-term moisture exposure.

Figure 11 shows the sorption isotherm and desorption curves of the mud–straw samples with and without straw for the test duration of 60 days. The presence of straw in the mud–straw samples leads to higher moisture content at all relative humidities compared to the samples without straw.

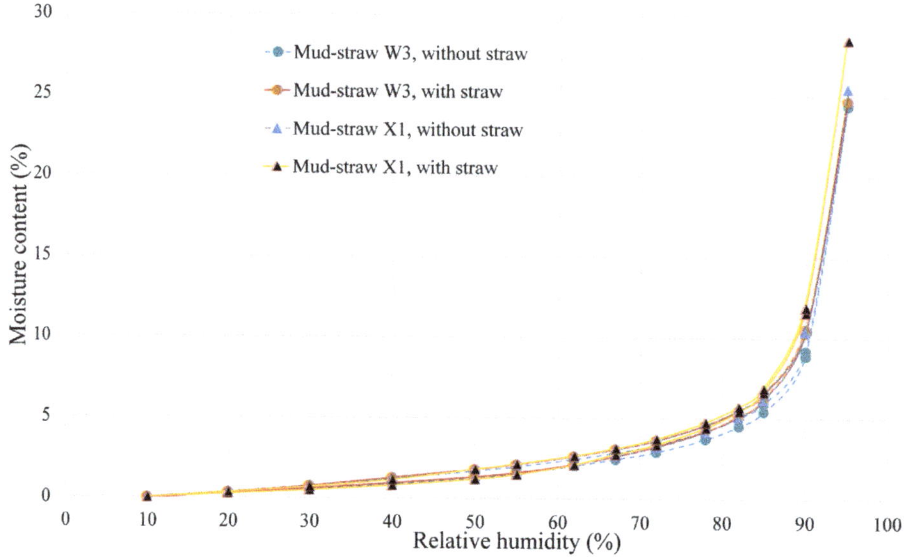

(a)

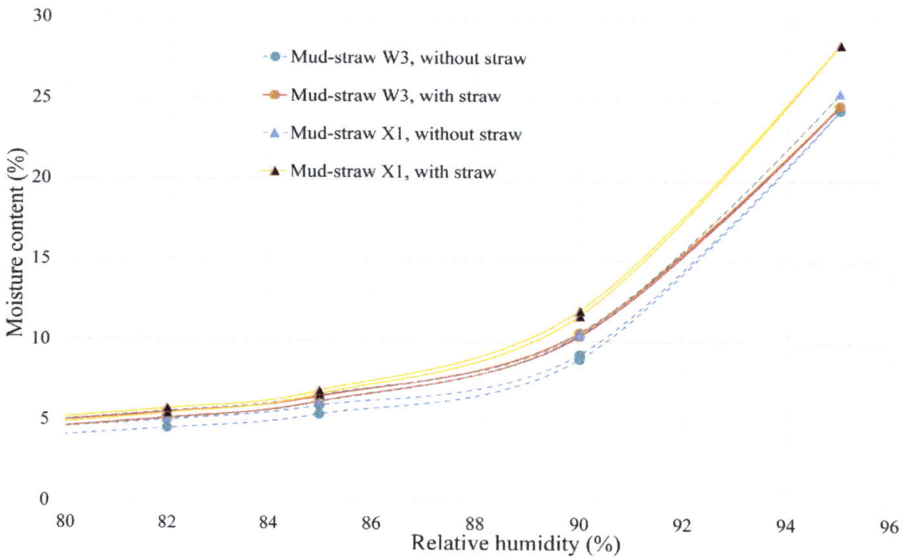

(b)

Figure 11. (**a**) Sorption isotherm and desorption curves for mud–straw samples with and without straw for a test duration of 60 days, (**b**) inset of the curve at high relative humidity.

4.4. TGA and DSC Results

The results of thermogravimetric analysis and differential scanning calorimetry for the adobe and mud–straw samples are illustrated in Figure 12 and summarised in Table 12. In the completely dry adobe samples A4, C1, and D2, water loss occurs between 30 °C and 105 °C, with mass losses of 1.14%, 1.18%, and 1.72%, respectively. For saturated adobe samples, these values increase to 2.03%, 2.33%, and 3.38% for samples A4, C1, and D2, respectively. Major carbon dioxide loss is observed between 550 °C and 800 °C, with mass losses of 12.41%, 13.91%, and 11.86% for the dry adobe samples A4, C1, and D2, respectively. For saturated samples, these values are 11.79%, 13.39%, and 11.2%.

In both the completely dry and saturated adobe samples, at temperatures between 30 °C and 105 °C, heating causes the release of water (H_2O) through the process of dehydration. This is also observable as a peak in the corresponding DSC curve between 30 °C and 105 °C. On the other hand, at temperatures between 550 °C and 800 °C, the samples undergo decarboxylation due to heating, and hence, carbon dioxide (CO_2) is released. The mass loss due to decarboxylation is observed at a peak in the corresponding DSC curve between 550 °C and 800 °C.

A similar trend holds for the completely dry and saturated mud–straw samples. At temperatures ranging from 30 °C to 105 °C, loss of water is observed. In the dry samples, water loss is 1.32% for sample W3 and 1.38% for sample X1, and in saturated samples, it is 8% for sample W3 and 14.08% for sample X1, which corresponds to the endothermic peaks of the DSC curves. In the mud–straw sample X1, an abrupt peak is observed at temperatures from 105 °C to 136 °C. This peak can be attributed to the loss of excess water from gypsum present in this sample, as shown in Figure 12c,d. The gypsum in the mud–straw sample X1 is traced in Figure 7b. Also, at temperatures ranging from 550 °C to 800 °C, loss of carbon dioxide occurs with values of 12.04% and 12.53% of the completely dry X1 and W3 adobe samples, respectively, and 10.04% of the saturated adobe sample X1 and 10.74% of the saturated adobe sample W3, corresponding to the peaks of the DSC curves between 550 °C and 800 °C (Figure 12c,d).

At low temperatures from 30 °C to 105 °C, for the mud–straw samples, which are without straw content in the TGA and DSC tests, under both completely dry and saturated conditions, the water loss of the mud–straw sample X1 is more than that of the mud–straw sample W3. This corresponds to the characteristics of their sorption isotherm and desorption curves shown in Figure 11 and the values presented in Table 12, in which the moisture content of the X1 sample without straw is higher than the W3 sample without straw. The mud–straw sample X1, with a higher moisture content, loses more water than the mud–straw sample W3, with a lower moisture content.

In both the adobe and mud–straw samples, as indicated in Table 10, the tested samples include calcite ($CaCO_3$). At high temperatures from 550 °C to 800 °C, the calcite is decomposed to calcium oxide (CaO) and carbon dioxide (CO_2), and the samples release CO_2 (Figure 12). This process leads to mass loss due to decarboxylation.

Table 8 presents the XRF analysis results, further detailed in Table 12, revealing significant variations in the CaO content within the adobe group. Among these samples, the adobe sample C1 exhibits the highest CaO content at 17.92 wt%, while the adobe sample A4 displays the lowest CaO content at 15.82 wt%. These differences suggest potential differences in the $CaCO_3$ content, with C1 potentially containing a higher proportion of $CaCO_3$ compared to A4. This correspondence aligns with the observed trend in CO_2 loss from the TGA and DSC results. Under completely dry conditions, sample C1 displays a CO_2 loss of 13.91%, whereas sample A4 exhibits 12.41%. Similarly, under saturated conditions, the CO_2 loss is 13.39% for C1 and 11.79% for A4.

Similar trends are observed within the mud–straw group, where the W3 sample demonstrates a higher CaO content at 17.35 wt% compared to the X1 sample, which exhibits a lower content of 15.14 wt%. Correspondingly, the CO_2 loss trend follows a similar pattern, with the W3 sample exhibiting higher CO_2 loss percentages compared to the X1 sample. Under completely dry conditions, the W3 sample shows a CO_2 loss of 12.53%, whereas the X1 sample exhibits 12.04%. Similarly, under saturated conditions, the CO_2 loss is 10.74% for W3 and 10.04% for X1.

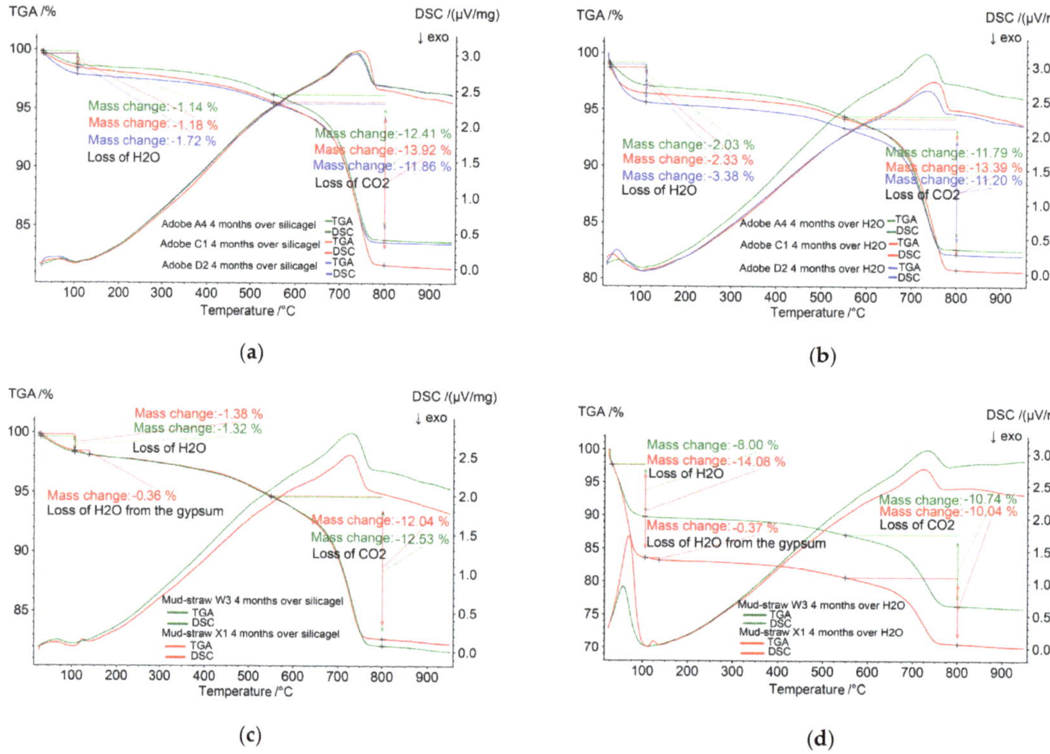

Figure 12. TGA and DSC curves: (**a**) completely dry adobe samples, (**b**) saturated adobe samples, (**c**) completely dry mud–straw samples, (**d**) saturated mud–straw samples.

Table 12. Water loss and carbon dioxide loss of completely dry and saturated adobe and mud–straw samples measured by TGA and DSC.

Material	Code	H_2O Loss (%)		CO_2 Loss (%)		CaO Analysed by XRF (wt%)
		Completely Dry	Saturated	Completely Dry	Saturated	
Adobe	A4	1.14	2.03	12.41	11.79	15.82
	C1	1.18	2.33	13.91	13.39	17.92
	D2	1.74	3.38	11.86	11.20	16.16
Mud–Straw	W3	1.32	8	12.53	10.74	17.35
	X1	1.38	14.08	12.04	10.04	15.14

5. Discussion on Results

The results obtained from XRF analysis indicate that a similar chemical composition exists in all studied adobe and mud–straw samples, both historical and new. The chemical elements with higher amounts are, respectively, Ba at about 679 ppm, then Sr (563 ppm) and Zr (222 ppm). The dominant chemical compounds are SiO_2, with about 45 wt%, followed by CaO (17 wt%) and Al_2O_3 (10 wt%). The similarity of chemical elements and compounds in the samples could be related to the same type of soil in this area.

The XRD results indicate that all samples studied have a similar mineralogical composition. The common phases in all samples are quartz, feldspars (albite and orthoclase), calcite, dolomite, and clay minerals (clinochlore and illite). Other phases are also observed in some samples. Clay minerals with high swelling potential, like montmorillonite and smectite, are absent in the studied samples, whereas the non-swelling clay mineral illite is present.

The ion chromatography results show that the mud–straw samples have higher conductivity values than the adobe samples. This can be attributed to the presence of straw in the mud–straw samples.

It is observed that the conductivity of new adobe and mud–straw samples is higher than historical ones. The very high conductivity of one of the mud–straw samples is caused by a high chloride content. The reason for this could be the high admixture of straw and the freshness of the material used in this sample. On the other hand, the fresh loam or soil used may also have higher chloride concentrations caused by the arid climate and soil salinisation. Likewise, natural weathering of the mud–straw plaster and the adobe over the years might have led to the leaching of chloride.

The results obtained from isotherm sorption determination show that adobe and mud–straw plaster with considerably higher moisture absorption are those with a higher load of chlorides. Adobe samples with higher straw contents exhibit higher moisture content.

The thermogravimetric analysis and differential scanning calorimetry results indicate that at temperatures between 100 °C and 125 °C, the loss of water of the completely dry and saturated adobe samples varies from 0.61% to 1.72% and from 2.03% to 3.38%, respectively, whereas that of the completely and saturated mud–straw samples ranges from 1.32% to 1.38% and from 8% to 14.08%, respectively. At temperatures between 650 °C and 800 °C, the range of the loss of carbon dioxide of the completely dry and saturated adobe samples is, respectively, between 11.86% and 14.4% and between 11.2% and 13.39%, while that of the completely dry and saturated mud–straw samples ranges from 12.04% to 12.53% and from 10.04% to 10.74%.

6. Implications for Restoration and Rehabilitation Practices

Preserving the architectural heritage of adobe structures requires a thorough understanding of their material characteristics and structural behaviour. By adhering to the principles outlined by the ICOMOS-ISCARSAH Committee [3], restoration and rehabilitation practices can effectively address the challenges posed by ageing and environmental factors. This section briefly explores the implications of the study findings on intervention solutions, the compatibility of material properties, and considerations for restoration and rehabilitation approaches for Persian adobe buildings.

6.1. *Influence of Sample Specificity on Intervention Solutions*

The specificity of each sample plays a crucial role in determining appropriate intervention solutions for restoration and rehabilitation projects. As recommended by the ICOMOS-ISCARSAH Committee, a comprehensive understanding of the structural behaviour and material characteristics is essential [3]. Variations in elemental composition, mineralogical composition, and moisture sorption behaviour between samples can influence the selection of conservation treatments and restoration techniques. For instance, insights gained from XRF and XRD analyses provide valuable information about the elemental and mineralogical composition of adobe and mud–straw plaster, guiding decisions on suitable consolidation methods and protective coatings to enhance structural integrity and durability over time. Furthermore, ion chromatography reveals the presence of soluble salts and ions within these materials, informing strategies to mitigate moisture-related damage and ensure long-term stability and compatibility with restoration interventions.

6.2. *Compatibility of Adobe and Mud–Straw Plaster Properties*

The compatibility of new materials, such as adobe and mud–straw plaster, with the original materials used in restoration work is paramount, as emphasised in the ICOMOS-ISCARSAH recommendations [3]. Understanding the elemental composition, mineralogical characteristics, and moisture sorption behaviour of adobe and mud–straw plaster is crucial for assessing their compatibility with existing materials and identifying potential risks of incompatibility. It is imperative that the materials used in restoration interventions complement the original construction materials to ensure structural integrity and authenticity in heritage conservation efforts.

In the case of adobe buildings, the compatibility of adobe with mud–straw plaster, which serves as a protective coating, is particularly important. Adobe is vulnerable to moisture-related damage, such as efflorescence or salt crystallisation, which can compromise its structural stability over time. Mud–straw plaster acts as a barrier against environmental elements, providing insulation and protecting the adobe from moisture ingress. Therefore, ensuring compatibility between adobe and mud–straw plaster is essential to maintain the structural integrity and longevity of adobe structures.

Furthermore, TGA-DSC analyses provide insights into the thermal decomposition behaviour of these materials, informing decisions related to fire safety measures and the selection of appropriate fire-retardant treatments during restoration efforts. Additionally, TGA-DSC analyses help identify any organic and inorganic additives present in the materials, influencing their structural integrity, durability, and susceptibility to degradation over time.

6.3. Considerations for Rehabilitation Solutions

In the domain of restoration and rehabilitation practices, an important consideration is the compatibility of materials, ensuring the integrity, authenticity, and longevity of heritage structures. This paper highlights the fundamental significance of material compatibility by providing comprehensive insights into the elemental composition, mineralogical characteristics, and thermal characteristics of adobe and mud–straw plaster used in Persian adobe buildings.

Several restoration and rehabilitation approaches can be employed to address the conservation needs of Persian adobe buildings. Drawing inspiration from established preservation practices and innovative restoration techniques, restoration practitioners can develop tailored solutions to enhance structural stability, mitigate decay risks, and preserve architectural authenticity. By incorporating sustainable building practices and modern conservation technologies, rehabilitation projects can achieve a balance between preserving heritage values and meeting contemporary standards of safety and comfort.

In all restoration and rehabilitation activities, the compatibility of materials used is essential. The selection of materials must align with the original construction methods and existing fabric of the structure to ensure harmonious integration and long-term preservation. Examples of rehabilitation solutions include consolidation treatments, protective coatings, and adaptive reuse strategies, each carefully selected to address the unique challenges posed by Persian adobe architecture.

7. Conclusions

In this study, ion chromatography, XRF analysis, XRD analysis, moisture sorption isotherm determination, thermogravimetric analysis, and differential scanning calorimetry of a number of historical and new adobe and mud–straw samples were conducted. Samples were taken from historical buildings, and new adobe bricks and mud–straw plaster were used in the same places for restoration, except for one mud–straw sample that was made in the laboratory using materials from the site. The selected historical buildings are a few kilometres apart, all located in the city of Maybod, 55 km east of the city of Yazd, Central Iran.

The comprehensive analysis conducted on the adobe and mud–straw samples has provided valuable insights into their elemental composition, mineralogical characteristics, moisture sorption behaviours, and thermal properties. The consistent chemical compositions, predominantly featuring SiO_2, CaO, and Al_2O_3, reflect the prevalent soil types in the studied region. This uniformity signifies a common geological origin, establishing a foundational understanding of construction materials within this region.

The mineralogical investigation identified key compounds, like quartz, feldspars (albite and orthoclase), calcite, dolomite, and clay minerals (clinochlore and illite), across all samples. The absence of high-swelling clay minerals, often associated with structural weaknesses during the drying process, indicates the suitability of these materials for adobe brick production.

The study of moisture sorption highlighted the impact of material composition on hygroscopic behaviour. The presence of straw notably increased moisture retention, which was especially evident in the mud–straw samples.

Moreover, the thermal analyses reveal distinctive dehydration and decomposition processes in both the adobe and mud–straw samples. The observed release patterns of water and carbon dioxide provided insights into the materials' thermal responses and chemical decomposition mechanisms.

Additionally, this study addressed a notable gap in the existing literature concerning the comprehensive characterisation of Persian adobe and mud–straw plaster. By extending the scope of analysis to include ion chromatography, moisture sorption isotherm determination, and thermogravimetric analysis with differential scanning calorimetry, this research contributed to filling this gap and advancing the knowledge base in the field of traditional building materials. Of particular significance is the lack of previous studies focusing on mud–straw, highlighting the importance of the present research.

These collective findings hold significant implications for construction practices and restoration efforts, offering a deeper understanding of the structural, compositional, and thermal attributes of adobe and mud–straw materials. The insights gleaned from this study provide valuable guidance for construction projects, facilitating informed decisions in material selection, structural design, and restoration methodologies. However, further research and experimentation in varied environmental conditions is needed to enhance this understanding and refine the application and preservation strategies of these traditional building materials in contemporary construction practices.

Author Contributions: Conceptualisation, B.H., J.F. and H.G.; methodology, B.H., J.F. and H.G.; software, B.H., C.L., F.G., J.F. and H.G.; validation, B.H. and J.F.; formal analysis, B.H., C.L., F.G., J.F. and H.G.; investigation, B.H.; resources, B.H. and J.F.; data curation, B.H.; writing—original draft preparation, B.H.; writing—review and editing, J.F. and H.G.; visualisation, B.H., C.L., F.G., J.F. and H.G.; supervision, J.F. and H.G. All authors have read and agreed to the published version of the manuscript.

Funding: This research received no external funding.

Institutional Review Board Statement: Not applicable.

Informed Consent Statement: Not applicable.

Data Availability Statement: The data that support the findings of this study are available on request from the corresponding author.

Conflicts of Interest: The authors declare no conflicts of interest.

References

1. Giada, G.; Caponetto, R.; Nocera, F. Hygrothermal properties of raw earth materials: A literature review. *Sustainability* **2019**, *11*, 5342. [CrossRef]
2. Liuzzi, S.; Stefanizzi, P. Experimental study on hygrothermal performances of indoor covering materials. *Int. J. Heat Mass Transf.* **2016**, *34*, S365–S370.
3. ICOMOS-ISCARSAH Committee. Recommendations for the Analysis, Conservation and Structural Restoration of Architectural Heritage. 2003. Available online: http://iscarsah.icomos.org/content/principles/ISCARSAH_Principles_English.pdf (accessed on 20 November 2021).
4. *Statistical Yearbook of Iran*; Statistics Center of Iran: Tehran, Iran, 2016.
5. Parisi, F.; Tarque, N.; Varum, H.; Vargas-Neumann, J. Adobe constructions in the world: A first overview. In *Structural Characterization and Seismic Retrofitting of Adobe Constructions*; Varum, A., Parisi, F., Tarque, N., Silveira, D., Eds.; Springer: Berlin/Heidelberg, Germany, 2021.

6. Adegun, O.B.; Adedeji, Y.M.D. Review of economic and environmental benefits of earthen materials for housing in Africa. *Front. Archit. Res.* **2016**, *6*, 519–528. [CrossRef]
7. Hejazi, M.; Mehdizadeh Saradj, F. *Persian Architectural Heritage: Architecture*; WIT Press: Southampton, UK; Boston, MA, USA, 2014.
8. Coffman, R.; Agnew, N.; Austin, G.; Doehne, E. Adobe Mineralogy: Characterization of Adobes from around the World. In Proceedings of the 6th International Conference on the Conservation of Earthen Architecture: Adobe 90 Preprints, Las Cruces, NM, USA, 14–19 October 1990; pp. 424–429.
9. Baglioni, E.; Fratini, F.; Rovero, L. The Materials Utilised in the Earthen Buildings Sited in the Drâa Valley (Morocco): Mineralogical and Mechanical Characteristics. In Proceedings of the 6th Seminário de Arquitectura de Terra em Portugal and 9th Seminário Ibero-Americano de Construção e Arquitectura com Terra, Coimbra, Portugal, 20–23 February 2010.
10. Fratini, F.; Pecchioni, E.; Rovero, L.; Tonietti, U. The earth in the architecture of the historical centre of Lamezia Terme (Italy): Characterization for restoration. *Appl. Clay Sci.* **2011**, *53*, 509–516. [CrossRef]
11. Costa, C.S.; Rocha, F.; Varum, H.; Velosa, A. Influence of the mineralogical composition on the properties of adobe blocks from Aveiro, Portugal. *Clay Miner.* **2013**, *48*, 749–758. [CrossRef]
12. El Fgaier, F.; Lafhaj, Z.; Chapiseau, C.; Antczak, E. Effect of sorption capacity on thermo-mechanical properties of unfired clay bricks. *J. Build. Eng.* **2016**, *6*, 86–92. [CrossRef]
13. De Castrillo, M.C.; Philokyprou, M.; Ioannou, I. Comparison of adobes from pre-history to-date. *Archaeol. Sci. Rep.* **2017**, *12*, 437–448. [CrossRef]
14. Ashour, T.; Georg, H.; Wu, W. An experimental investigation on equilibrium moisture content of earth plaster with natural reinforcement fibres for straw bale buildings. *Appl. Therm. Eng.* **2011**, *31*, 293–303. [CrossRef]
15. Costa, C.; Arduin, D.; Rocha, F.; Velosa, A. Adobe blocks in the center of Portugal: Main characteristics. *Int. J. Archit. Herit.* **2021**, *15*, 467–478. [CrossRef]
16. Sanchez-Calvillo, A.; Alonso-Guzman, E.M.; Martinez-Molina, W.; Navarrete-Seras, M.A.; Ruvalcaba-Sil, J.L.; Navarro-Ezquerra, A.; Mitrani, A. Characterization of adobe blocks: Point-load assessment as a complementary study of damaged buildings and samples. *Heritage* **2021**, *4*, 864–888. [CrossRef]
17. Laborel-Préneron, A.; Magniont, C.; Aubert, J.E. Hygrothermal properties of unfired earth bricks: Effect of barley straw, hemp shiv and corn cob addition. *Energy Build.* **2018**, *178*, 265–278. [CrossRef]
18. Gomes, R.I. Assessment of Hygrothermal Characteristics of Earthen Materials. Ph.D. Thesis, Universidade Nova de Lisboa, Lisbon, Portugal, 2019.
19. Mellaikhafi, A.; Tilioua, A.; Souli, H.; Garoum, M.; Hamdi, M.A. Characterization of different earthen construction materials in oasis of south-eastern Morocco (Errachidia Province). *Case Stud. Constr. Mater.* **2021**, *14*, e00496. [CrossRef]
20. Ranesi, A.; Faria, P.; Veiga, M.D.R. Traditional and modern plasters for built heritage: Suitability and contribution for passive relative humidity regulation. *Heritage* **2021**, *4*, 2337–2355. [CrossRef]
21. Lima, J.; Faria, P.; Santos Silva, A. Earth plasters: The influence of clay mineralogy in the plasters' properties. *Int. J. Archit. Herit.* **2020**, *14*, 948–963. [CrossRef]
22. Savadogo, N.; Traore, Y.B.; Nshimiyimana, P.; Lankoande, N.; Messan, A. Physico-mechanical and durability characterization of earthen plaster stabilized with fermented rice husk for coating adobe walls. *Cogent Eng.* **2023**, *10*, 2243740. [CrossRef]
23. Santos, T.; Faria, P.; Silva, V. Can an earth plaster be efficient when applied on different masonries? *J. Build. Eng.* **2019**, *23*, 314–323. [CrossRef]
24. Vares, O.; Ruus, A.; Raamets, J.; Tungel, E. Determination of hygrothermal performance of clay-sand plaster: Influence of covering on sorption and water vapour permeability. *Energy Proc.* **2017**, *132*, 267–272. [CrossRef]
25. Santos, T.; Faria, P. Characterization of earthen plasters Influence of formulation and experimental methods. *Rev. Grem.* **2020**, *7*, 151–168. [CrossRef]
26. Santos, T.; Gomes, M.I.; Silva, A.S.; Ferraz, E.; Faria, P. Comparison of mineralogical, mechanical and hygroscopic characteristic of earthen, gypsum and cement-based plasters. *Constr. Build. Mater.* **2020**, *254*, 119222. [CrossRef]
27. Ojo, E.B.; Mustapha, K.; Teixeira, R.S.; Savastano, H., Jr. Development of unfired earthen building materials using muscovite rich soils and alkali activators. *Case Stud. Constr. Mater.* **2019**, *11*, e00262. [CrossRef]
28. Bass, A.; Porter, D.; Spilde, M.; Guebard, M.; Shaum, K.; Ferriola, N. Characterization and comparative analysis of ancient earthen plasters from the American Southwest. *MRS Adv.* **2017**, *2*, 2145–2178. [CrossRef]
29. Muşkara, Ü.; Bozbaş, S.K. Characterization of earthen building materials in Gölcük vernacular houses. In *Conservation of Architectural Heritage*; Springer International Publishing: Cham, Switzerland, 2022; pp. 13–26.
30. Saleh, L. Material Characterization of Natural Adobe Plasters in the Jordan Valley: Mining the Past for a More Sustainable Future. Master's Thesis, Cornell University, Ithaca, NY, USA, 2015.
31. Saleh, M.M. Characterization of Qarh's wall plasters, Al-Ulla, Saudi Arabia. A case study. *Int. J. Conserv. Sci.* **2013**, *4*, 65–80.
32. Silveira, D.; Varum, H.; Costa, A. Rehabilitation of an important cultural and architectural heritage: The traditional adobe constructions in Aveiro district. *WIT Trans. Ecol. Environ.* **2007**, *102*, 705–714.
33. Sánchez, A.; Varum, H.; Martins, T.; Fernández, J. Mechanical properties of adobe masonry for the rehabilitation of buildings. *Constr. Build. Mater.* **2022**, *333*, 127330. [CrossRef]
34. Papayianni, I.; Pachta, V. Consolidation and Upgrading of Historic Earth Block Masonry Constructions. In Proceedings of the 3rd International Balkans Conference Challenges in Civil Engineering, 3-BCCCE, Tirana, Albania, 19–21 May 2016.

35. González-Sánchez, B.; Sandoval-Castro, K.; Navarro-Ezquerra, A.; Ramírez-Casas, J.; Sanchez-Calvillo, A.; Alonso-Guzmán, E.M.; Navarro-Mendoza, E.G. Development and intervention proposal with earthen refurbishments with vegetal origin gel (VOG) for the preservation of traditional adobe buildings. *Heritage* **2023**, *6*, 3025–3042. [CrossRef]
36. Faria, P.; Santos, T.; Aubert, J.E. Experimental characterization of an earth eco-efficient plastering mortar. *J. Mater. Civ. Eng.* **2016**, *28*, 04015085. [CrossRef]
37. Gomes, M.I.; Diaz Gonçalves, T.; Faria, P. The compatibility of earth-based repair mortars with rammed earth substrates. In *Historic Mortars: Advances in Research and Practical Conservation*; Hughes, J.J., Válek, J., Groot, J.W.P., Eds.; Springer: New York, NY, USA, 2019; pp. 305–318.
38. Jia, Q.; Chen, W.; Tong, Y.; Guo, Q. Laboratory study on shrinkage and cracking behavior of historic earthen plaster. *Eng. Geol.* **2023**, *318*, 107096. [CrossRef]
39. Mattone, M.; Rescic, S.; Fratini, F.; Del Fà, R.M. Experimentation of earth-gypsum plasters for the conservation of earthen constructions. *Int. J. Archit. Herit.* **2017**, *11*, 763–772. [CrossRef]
40. Hosseini, R.; Emami, S.M.A.; Zamanifard, A. The role of chemical parameters and soil texture quality in the erosion of adobe in the Belgias historical site. *Maremat Me'mari-e Iran* **2018**, *9*, 91–110.
41. Zakavi, S. Investigation of the clay resources in Chogazanbil area for preparation of the mud bricks optimized for the restoration purposes. *Archaeom. Res.* **2019**, *5*, 35–46.
42. Dormohamadi, M.; Rahimnia, R. Combined effect of compaction and clay content on the mechanical properties of adobe brick. *Case Stud. Constr. Mater.* **2020**, *13*, e00402. [CrossRef]
43. Eskandari, A. Characterisation of Persian Historical and New Adobe Bricks. Master's Thesis, University of Isfahan, Isfahan, Iran, 2021.
44. Ghorbani, M. *The Geology of Iran: Tectonic, Magmatism and Metamorphism*; Springer International Publishing: New York, NY, USA, 2021.
45. Landsat ETM+. 2011. Available online: https://landsat.gsfc.nasa.gov/satellites/landsat-7/ (accessed on 20 October 2021).
46. *Geological Map 1:100000 Aqda*; Geological Survey and Mineral Exploration Organisation of Iran: Tehran, Iran, 2007.
47. *Geological Map 1:100000 Ardakan*; Geological Survey and Mineral Exploration Organisation of Iran: Tehran, Iran, 2005.
48. Yousefi, M.; Hosseini, K.; Nabavieh, M.; Torshizian, H. *Report of 1:100000 Ardakan Geological Map*; Geological Survey and Mineral Exploration Organisation: Tehran, Iran, 2005.
49. Alaei, S.; Fodazi, M. *Report of 1: 100000 Aqda Geological Map*; Geological Survey and Mineral Exploration Organisation: Tehran, Iran, 2007.
50. *DIN EN ISO 10304-1*; Wasserbeschaffenheit—Bestimmung von gelösten Anionen mittels Flüssigkeits-Ionenchromatographie—Teil 1: Bestimmung von Bromid, Chlorid, Fluorid, Nitrat, Nitrit, Phosphat und Sulfat (Water Quality—Determination of Dissolved Anions by Liquid Chromatography of Ions—Part 1: Determination of Bromide, Chloride, Fluoride, Nitrate, Nitrite, Phosphate and Sulfate). Deutsches Institut für Normung: Berlin, Germany, 2009.
51. *DIN EN ISO 10304-4*; Wasserbeschaffenheit—Bestimmung von gelösten Anionen mittels Ionenchromatographie—Teil 4: Bestimmung von Chlorat, Chlorid und Chlorit in gering belastetem Wasser (Water Quality—Determination of Dissolved Anions by Liquid Chromatography of Ions—Part 4: Determination of Chlorate, Chloride and Chlorite in Water with Low Contamination). Deutsches Institut für Normung: Berlin, Germany, 2022.
52. *DIN EN 13925-2*; Röntgendiffraktometrie von polykristallinen und amorphen Materialien—Teil 2: Verfahrensabläufe (Non-Destructive Testing—X-ray Diffraction from Polycrystalline and Amorphous Material—Part 2: Procedures). Deutsches Institut für Normung: Berlin, Germany, 2003.
53. *DIN 51001*; Prüfung oxidischer Roh- und Werkstoffe—Allgemeine Arbeitsgrundlagen zur Röntgenfluoreszenz-Analyse (RFA) (Testing of Oxidic Raw Materials and Basic Materials—General Bases of Work for X-ray Fluorescence Method (XRF)). Deutsches Institut für Normung: Berlin, Germany, 2003.
54. *DIN EN ISO 12571*; Wärme- und feuchtetechnisches Verhalten von Baustoffen und Bauprodukten—Bestimmung der hygroskopischen Sorptionseigenschaften (Hygrothermal Performance of Building Materials and Products—Determination of Hygroscopic Sorption Properties). Deutsches Institut für Normung: Berlin, Germany, 2021.
55. *ISO 11358-1*; Plastics Thermogravimetry (TG) of Polymers—Part 1: General Principles. International Organization for Standardization: Geneva, Switzerland, 2023.
56. *ISO 11357-1*; Plastics—Differential scanning calorimetry (DSC)—Part 1: General Principles. International Organization for Standardization: Geneva, Switzerland, 2023.
57. Sultan, K. *Practical Guide to Materials Characterization: Techniques and Applications*; Wiley: New York, NY, USA, 2022.
58. Qiu, Q. A state-of-the-art review on the carbonation process in cementitious materials: Fundamentals and characterization techniques. *Constr. Build. Mater.* **2020**, *247*, 118503. [CrossRef]
59. Harris, D.C.; Lucy, C.A. *Quantitative Chemical Analysis*; Macmillan: New York, NY, USA, 2020.
60. Saidi, M.; Cherif, A.S.; Zeghmati, B.; Sediki, E. Stabilization effects on the thermal conductivity and sorption behavior of earth bricks. *Constr. Build. Mater.* **2018**, *167*, 566–577. [CrossRef]
61. Balandin, A.A.; Ivanova, O.S.; Anokhina, T.S.; Chukov, D.I. Thermogravimetric and differential scanning calorimetry analysis of polyurethane foam waste recycling. *J. Anal. Chem.* **2021**, *76*, 987–993.
62. Wild, T. Demineralisierung und Mechanisch/Thermische Entwässerung von Braunkohlen und Biobrennstoffen. Ph.D. Thesis, University of Dortmund, Dortmund, Germany, 2006.

63. Moevus, M.; Anger, R.; Fontaine, L. Hygro-Thermo-Mechanical Properties of Earthen Materials for Construction: A Literature Review. In Proceedings of the TERRA 2012, 12th SIACOT Proceedings, 11th International Conference on the Study and Conservation of Earthen Architectural Heritage, 12th Iberian-American Seminar on Earthen Architecture and Construction, Lima, Peru, 22–27 April 2012.
64. Cagnon, H.; Aubert, J.E.; Coutand, M.; Magniont, C. Hygrothermal properties of earth bricks. *Energy Build.* **2014**, *80*, 208–217. [CrossRef]
65. Medjelekh, D.; Ulmet, L.; Dubois, F. Characterization of hygrothermal transfers in the unfired earth. *Energy Procedia* **2017**, *139*, 487–492. [CrossRef]
66. Godts, S.; Steiger, M.; Orr, S.A.; Stahlbuhk, A.; Desarnaud, J.; De Clercq, H.; Cnudde, V.; De Kock, T. Modeling salt behavior with ECOS/RUNSALT: Terminology, methodology, limitations, and solutions. *Heritage* **2022**, *5*, 3648–3663. [CrossRef]

Disclaimer/Publisher's Note: The statements, opinions and data contained in all publications are solely those of the individual author(s) and contributor(s) and not of MDPI and/or the editor(s). MDPI and/or the editor(s) disclaim responsibility for any injury to people or property resulting from any ideas, methods, instructions or products referred to in the content.

MDPI AG
Grosspeteranlage 5
4052 Basel
Switzerland
Tel.: +41 61 683 77 34

Materials Editorial Office
E-mail: materials@mdpi.com
www.mdpi.com/journal/materials

Disclaimer/Publisher's Note: The statements, opinions and data contained in all publications are solely those of the individual author(s) and contributor(s) and not of MDPI and/or the editor(s). MDPI and/or the editor(s) disclaim responsibility for any injury to people or property resulting from any ideas, methods, instructions or products referred to in the content.

www.ingramcontent.com/pod-product-compliance
Lightning Source LLC
Chambersburg PA
CBHW061806210926
51850CB00006B/625

* 9 7 8 3 7 2 5 8 1 8 5 8 7 *